TOPICS IN PHOSPHORUS CHEMISTRY

Volume 7

New York · London · Sydney · Toronto

Topics in Phosphorus Chemistry

Volume 7

AUTHORS

M. Bermann
I. I. Creaser
John D. Curry
John O. Edwards
S. Y. Kalliney
D. Allan Nicholson
R. K. Osterheld
Oscar T. Quimby
Malcolm A. Shaw
Robert S. Ward

Interscience Publishers, a division of John Wiley & Sons

Preface

The chemistry of phosphorus in its fundamental and applied aspects has grown rapidly in the past decade. The progress being made in nearly every scientific discipline dealing with this key element has created a need for a continuous central forum in which progress may be reviewed rapidly on a larger, more critical scale and for a broader audience than the specialized journals permit.

It is the purpose of *Topics in Phosphorus Chemistry* to provide the general scientific reader as well as the specialist in phosphorus chemistry with a series of critical evaluations and reviews of progress in the diverse special areas of the science written by scientists actively engaged in work in the field. An attempt has been made to keep the articles timely and current to the extent that previously unpublished work can be included. Reviews are, at the same time, intended to give enough background to the specific subjects to allow the reader to cross disciplinary lines effectively. It is hoped that further developments in phosphorus chemistry will be stimulated by this approach.

No fixed pattern has been established for the *Topics in Phosphorus Chemistry*. A flexible attitude will be preserved and the course of the series will be dictated by the workers in the field. The editors of *Topics in Phosphorus Chemistry* not only welcome suggestions from our readers but eagerly solicit your advice and contributions. The series is dedicated to the phosphorus chemist in particular while providing all chemists and biochemists with a charted route into the various facets of phosphorus chemistry.

The Editors

v

Contents

TOPICS IN PHOSPHORUS CHEMISTRY

Volume 7

Addition Reactions of Tertiary Phosphorus Compounds with Electrophilic Olefins and Acetylenes

MALCOLM A. SHAW* and ROBERT S. WARD†

University Chemical Laboratory, Cambridge, England

CONTENTS

I. Introduction

The addition of nucleophiles to olefins and acetylenes which are activated by electron-withdrawing groups gives rise to products of widely different structure (1,2). This is because, in general, the initial 1:1 intermediate (e.g., **1** or **2**) is not sufficiently stable to be isolated, but is stabilized by rearrangement, cyclization, or further addition.

The reaction takes its simplest course when 'mobile' protons are available to protonate the initial intermediate. So-called 'mobile' protons may be supplied by the solvent, or in aprotic media, by prototropy.

In the absence of 'mobile' protons, stabilization of the initial intermediate frequently gives rise to products of complex structure. One

---

* Present address: Unilever Limited, Port Sunlight Laboratory.
† Present address: Department of Chemistry, University College of Swansea.

$$\ddot{Y} + R\cdot C\equiv C\cdot R \longrightarrow \begin{array}{c} Y^+ \\ \diagdown \\ C=\bar{C}\cdot R \\ \diagup \\ R \end{array} \xrightarrow{H^+} \begin{array}{c} Y^+ \\ \diagdown \\ C=CHR \\ \diagup \\ R \end{array}$$

$$(1)$$

$$\ddot{Y} + RCH=CHR \longrightarrow \begin{array}{c} Y^+ \quad H \\ | \quad | \\ R-C-\bar{C}-R \\ | \\ H \end{array} \longrightarrow \begin{array}{c} Y^+ \\ \diagdown \\ \bar{C}-CH_2R \\ \diagup \\ R \end{array}$$

$$(2)$$

$$H\ddot{Y} + R\cdot C\equiv C\cdot R \longrightarrow \begin{array}{c} HY^+ \\ \diagdown \\ C=\bar{C}\cdot R \\ \diagup \\ R \end{array} \longrightarrow \begin{array}{c} Y \\ \diagdown \\ C=CHR \\ \diagup \\ R \end{array}$$

class of nucleophiles whose addition reactions have been studied in detail are the nitrogen-containing heterocycles (3). The large number of adducts which have been isolated from the reactions of pyridine, its derivatives, and other heterocyclic nitrogen compounds, with acetylenic esters serve to emphasize the sensitive dependence of the products of nucleophilic addition upon the particular reactants and reaction conditions employed (3,4).

The nucleophilic addition reactions of tervalent phosphorus compounds have also been studied for many years. However it is only during the last decade, and with the aid of modern physical methods, that the correct structures of most of the adducts have been realized. Studies in this field have been stimulated by the increasing interest shown in organophosphorus chemistry since the application by Wittig of the olefination reactions of alkylidenephosphoranes (5).

A tervalent phosphorus center is a powerful nucleophile by virtue of its lone pair of electrons. Whether the bonds of tervalent phosphorus compounds are best described as strained $3p\sigma$ bonds, or whether they involve some $3s$ or $3d$ orbital participation, is undecided (6,7). However it is generally assumed that such participation is not appreciable and that the lone pair of electrons on the phosphorus atom comprise largely its $3s$ electrons. There is considerable evidence for both $p\pi-p\pi$ and $p\pi-d\pi$ conjugation in aryl phosphines (6–8).

Tervalent phosphorus compounds are more powerful nucleophiles than their nitrogen analogs due to the lower electronegativity and greater polarizability of tervalent phosphorus (6). The greater nucleophilicity of phosphines compared to amines is further enhanced by the more crowded conditions around the smaller nitrogen atom (7). Other reasons for the importance of their nucleophilic reactions are the strong

sp^3 bonds formed by phosphorus and the possibility of $p\pi$–$d\pi$ stabilization of the phosphonium center.

A $p\pi$–$d\pi$ bond is formed when the unhybridized 'p' electrons of a substituent are donated into the vacant $3d$ orbitals of a phosphonium center. For example, the physical properties and chemical stability of phosphine oxides and phosphonium ylids are consistent with $p\pi$–$d\pi$ bonding in such systems (9).

$$\diagdown\!\!\!\!\underset{\diagup}{P^{\pm}}\!\!-\!\!O^{-} \longleftrightarrow \diagdown\!\!\!\!\underset{\diagup}{P}\!\!=\!\!O \qquad \diagdown\!\!\!\!\underset{\diagup}{P^{\pm}}\!\!-\!\!C\!\!\underset{\diagdown}{\diagup} \longleftrightarrow \diagdown\!\!\!\!\underset{\diagup}{P}\!\!=\!\!C\!\!\underset{\diagdown}{\diagup}$$

The 'biphilicity' of tervalent phosphorus compounds (7) (i.e., the ability of a tervalent phosphorus center to donate electrons to form a σ bond, and of the phosphonium center formed to simultaneously or subsequently accept electrons to form a π or σ bond) will influence the reactions of tervalent phosphorus compounds with olefinic and acetylenic compounds in a number of important ways. The case of nucleophilic addition to an acetylene will be considered.

$$R_3P\colon + R'\cdot C\!\equiv\!C\cdot R' \longrightarrow \left[\begin{array}{c} R_3P^{\delta+} \\ \;\;\;|\;\;\;^{\delta-} \\ R'\cdot C\!\equiv\!C\cdot R' \end{array}\right]$$

(3)

$$\downarrow$$

Products $\longleftarrow$ $\underset{R'}{\overset{R_3P}{\diagdown}}\!\!C\!\!-\!\!\ddot{C}\!\!-\!\!R' \longleftrightarrow \underset{R'}{\overset{R_3P^+}{\diagdown}}\!\!C\!\!=\!\!\bar{C}\!\!-\!\!R'$

(4)

Firstly, $p\pi$–$d\pi$ bonding will lower the energy of the transition state (3) by reinforcing the incipient σ bond. Secondly, $p\pi$–$d\pi$ bonding will stabilize the initial betaine (4) as shown. Thirdly, the acceptor properties of a phosphonium center and the stability of alkylidenephosphorane or phosphoryl systems frequently determine the eventual products of nucleophilic addition. Clearly, the first and third of these considerations apply equally to nucleophilic addition to olefins. It is also worth mentioning that the comparable ease with which pyridines form adducts with acetylenes is probably due, in part, to the ability of a pyridinium group to stabilize an adjacent charged center or π system by $p\pi$–$p\pi$ resonance interaction.

Brief consideration of the tervalent derivatives of other group V elements shows that tervalent phosphorus compounds occupy a unique position. They are the most nucleophilic (10) and their 'onium' derivatives are the most amenable to $p\pi$–$d\pi$ stabilization. It would be predicted that the importance of $p\pi$–$d\pi$ bonding would decrease from phosphorus to bismuth due to the less effective overlap of the $2p$ orbitals on carbon with the increasingly diffuse 'n' d orbitals on the adjacent 'onium' center (11). The greater reactivity and basicity of arsonium ylids than the corresponding phosphonium ylids are consistent with this conclusion (12).

II. Additions to Olefins

A. Monoactivated Olefins

Tertiary phosphines readily add to the double bond of monoactivated olefins forming 1:1 zwitterions (**5**) which are unstable in aprotic solvents and initiate anionic polymerization (13). However these 1:1 intermediates can be trapped in the presence of acids, giving phosphonium salts (**6**) with an electron-attracting group in the β position (14). The diastereoisomeric dibenzoylhydrogentartrate salts of the racemic salt (**6**), prepared from racemic methylbenzylbutylphosphine and acrylonitrile, have been separated and used to prepare optically pure phosphine since the addition reaction can be reversed using sodium methoxide (15). The zwitterionic intermediates (**5**, R = alkyl) add to the carbonyl bond of aldehydes with subsequent expulsion of the phosphine moiety to yield substituted vinyl monomers (**7**) (16).

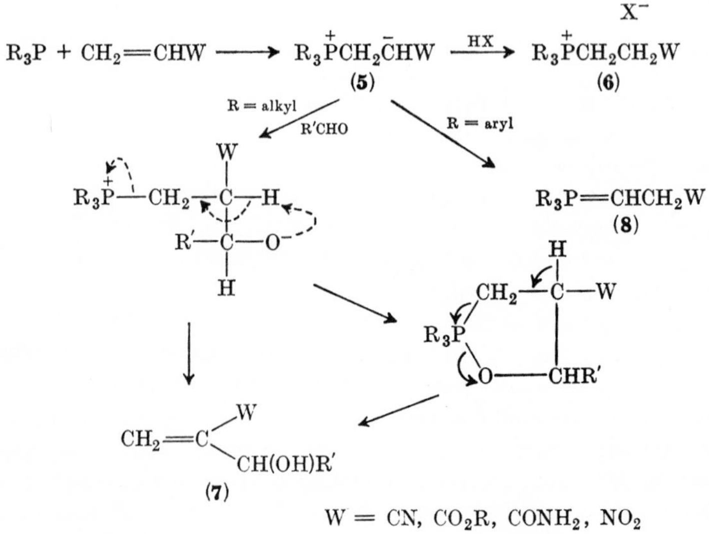

$R_3P + CH_2{=}CHW \longrightarrow R_3\overset{+}{P}CH_2\overset{-}{C}HW \xrightarrow{HX} R_3\overset{+}{P}CH_2CH_2W$ X^-
(5)
(6)
R = alkyl
R'CHO
R = aryl
$R_3P{=}CHCH_2W$
(8)
$CH_2{=}C(W)CH(OH)R'$
(7)
W = CN, CO_2R, $CONH_2$, NO_2

Two possible mechanisms were suggested for this reaction, the first involving proton transfer and elimination of the phosphine, and the second involving an intermediate pentacovalent oxaphosphorane. It seems likely that the first alternative is to be preferred.

When the intermediates (**5**, R = aryl) are prepared in protonic solvents, proton migration occurs, probably through exchange with the solvent, to give alkylidenephosphoranes (**8**) stabilized by $p\pi$–$d\pi$ bonding (17). The phosphoranes were characterized by their Wittig reaction products. Proton migration also occurs in aprotic solvents if traces of acids or bases are present.

Trippett (18) has used phosphines of the type Ph_2PCH_2R, where R is a group capable of stabilizing a carbanion, in this type of reaction and observed that the 1:1 intermediates transfer a proton giving alkylidenephosphoranes (**9**) which were used *in situ* in a Wittig olefination; aminophosphines similarly gave iminophosphoranes. The 1:1 intermediate from diphenylvinylphosphine cyclizes to the alkylidenephosphorane (**10**) (19) via addition of the generated carbanion to the activated double bond of the generated phosphonium center. Diphenylvinylphosphine forms di- and polymeric phosphonium salts in the presence of certain alkyl halides. The reactions proceed via addition of the phosphine to the generated vinylphosphonium salts (20).

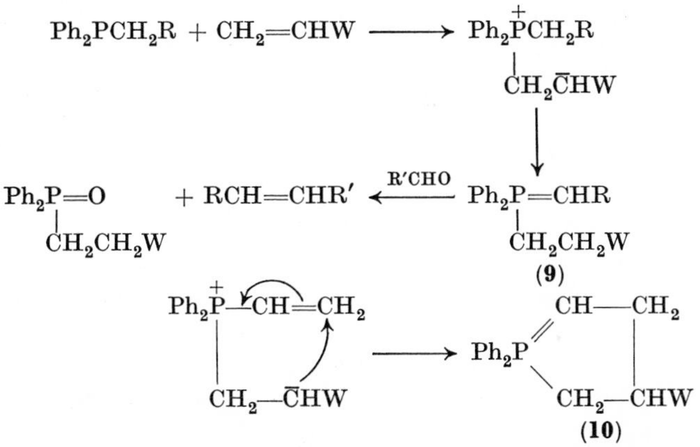

$$Ph_2PCH_2R + CH_2{=}CHW \longrightarrow \overset{+}{Ph_2PCH_2R}$$
$$\underset{\bar{C}H_2\bar{C}HW}{|}$$

$$Ph_2P{=}O \quad + RCH{=}CHR' \xleftarrow{\;R'CHO\;} Ph_2P{=}CHR$$
$$\underset{CH_2CH_2W}{|} \qquad\qquad\qquad \underset{CH_2CH_2W}{|}$$
$$(\mathbf{9})$$

$$\overset{+}{Ph_2P}{-}CH{=}CH_2 \qquad\qquad CH{-}CH_2$$
$$\underset{CH_2-\bar{C}HW}{|} \longrightarrow Ph_2P{\diagdown} \underset{CH_2-CHW}{|}$$
$$(\mathbf{10})$$

$$R = Ph, CO_2Me; \; W = CN, CO_2Me, CONH_2$$

Vullo (21) has reacted tetrakis(hydroxymethyl)phosphonium chloride with acrylic acid derivatives, in the presence of base. The reactions proceed through the formation of tris(hydroxymethyl)phosphine which adds to the electrophilic olefins, but owing to the differing basicities of the 1:1 intermediates formed, different reaction products are isolated depending on the olefin used. The 1:1 intermediate from acrylic acid is only weakly basic because the carbanion is protonated by the carboxyl group and is unable to catalyze further formation of the phosphine from the phosphonium salt. Thus the betaine (11) is isolated from the reaction mixture. However in the reactions with acrylonitrile and acrylamide the carbanions derived from addition to the olefinic bond are strongly basic species which propagate the reaction by converting hydroxymethylphosphonium intermediates to the corresponding phosphines and formaldehyde. Ultimate products are tertiary phosphines (12).

$$(HOCH_2)_3P + CH_2{=}CHCO_2H \longrightarrow (HOCH_2)_3\overset{+}{P}CH_2CHCO_2H$$

$$\downarrow$$

$$(HOCH_2)_3\overset{+}{P}CH_2CH_2CO_2^-$$
$$(11)$$

$$(HOCH_2)_3\overset{+}{P}CH_2\overset{-}{C}HW \longrightarrow (HOCH_2)_2PCH_2CH_2W$$

$$\downarrow 2\,CH_2{=}CHW$$

$$P(CH_2CH_2W)_3$$
$$W = CN,\ CONH_2 \qquad\qquad (12)$$

Tertiary phosphines and phosphites react with 3-halogenoacrylic acid derivatives to form only *trans*-vinylphosphonium salts (13) and phosphonates (14) (22), irrespective of the stereochemistry of the starting olefins. The results are consistent with an addition–elimination mechanism, the steric interaction between R_3P and CO_2R' in the intermediate (15a) controlling the stereochemical course of the reaction and favoring *trans* products. Polar interactions in the intermediate (15b) would perhaps favor *cis* products, but are apparently unimportant. The possibility that initially formed *cis* adducts are isomerized (by nucleophilic addition to the activated double bond) during the course of the reaction could not be fully excluded; however NMR examination of reaction mixtures involving the *cis* esters and phosphines at various times throughout the course of the reaction gave no suggestion that *cis* salts were first formed. In contrast, it is found that the diphenylphosphide anion (Ph_2P^-) reacts with the geometrical isomers of β-bromostyrene and α-bromostilbene to give products with largely retained configuration (23). It has been suggested (22) that the difference is due to the much smaller steric requirement of the diphenylphosphide residue in the transition states of the reactions.

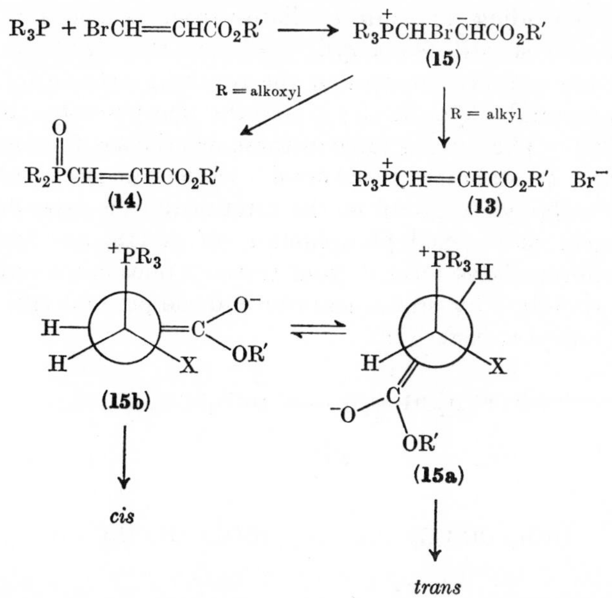

$$R_3P + BrCH\!\!=\!\!CHCO_2R' \longrightarrow R_3\overset{+}{P}CHBr\overset{-}{C}HCO_2R'$$
(15)

R = alkoxyl

R = alkyl

$$\underset{\textbf{(14)}}{R_2\overset{O}{\overset{\|}{P}}CH\!\!=\!\!CHCO_2R'} \qquad\qquad \underset{\textbf{(13)}}{R_3\overset{+}{P}CH\!\!=\!\!CHCO_2R' \quad Br^-}$$

(15b)

(15a)

cis

trans

2-Halogenoacrylic acid derivatives react with phosphites to form
cis- and *trans*-vinylphosphonates **(16)** via addition to the double bond,
followed by rearrangement and elimination of ethyl iodide (cf. the
Arbusov reaction) (24). A mechanism involving a hydride shift was
proposed (24) but Kirby and Warren (25) have suggested a more
plausible mechanism involving a proton shift, analogous to that observed in reactions with tertiary phosphines. Loss of halide ion from
the intermediate ylid then precedes elimination of ethyl iodide.

(16)

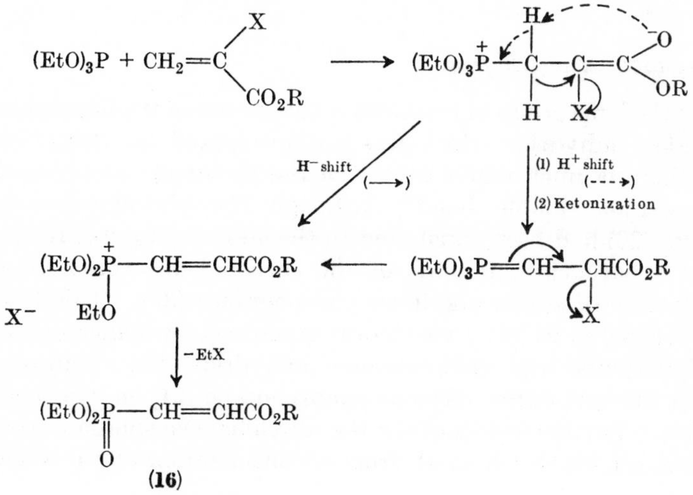

The corresponding 1:1 intermediates from reactions of tertiary phosphites and substituted acrylic acids are stabilized by a proton shift from the carboxyl group and the resulting betaine (cf. Ref. 21) undergoes internal alkyl transfer giving the phosphonates (17) as final products (26). The driving force in these reactions is the formation of the strong $p\pi$–$d\pi$ stabilized P=O bond. In the presence of added alkyl halide the carboxylate group of the intermediate is esterified by the added reagent and mixed phosphonate esters (18) are formed (27). The 1:1 intermediates formed from tertiary phosphites and acrolein cannot be stabilized by proton transfer and the product (19) is formed by internal alkyl transfer (28).

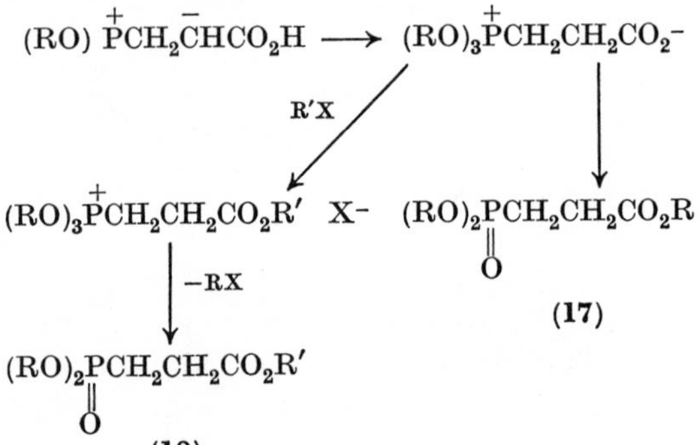

$$(RO) \overset{+}{P}CH_2\overset{-}{C}HCO_2H \longrightarrow (RO)_3\overset{+}{P}CH_2CH_2CO_2^-$$

R'X

$$(RO)_3\overset{+}{P}CH_2CH_2CO_2R' \quad X^- \qquad (RO)_2PCH_2CH_2CO_2R$$
$$\overset{\parallel}{O}$$

−RX

$$(RO)_2PCH_2CH_2CO_2R'$$
$$\overset{\parallel}{O}$$

(17)

(18)

$$(RO)_3\overset{+}{P}{-}CH_2{-}CH{=}C\overset{O^-}{\underset{H}{\diagdown}} \longrightarrow (RO)_2\overset{O}{\overset{\parallel}{P}}CH_2CH{=}C\overset{OR}{\underset{H}{\diagdown}}$$

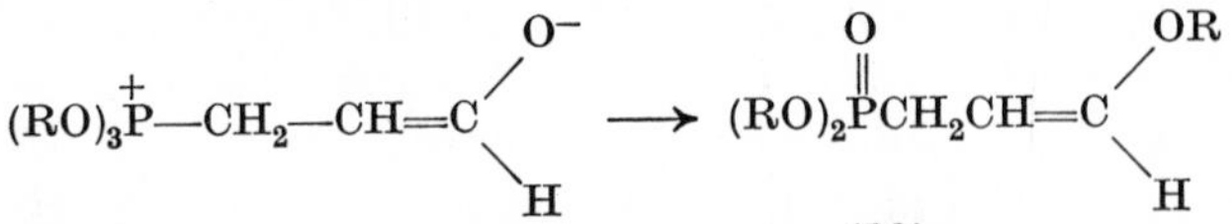

(19)

B. 1:2-Diactivated Olfefins

Typical of this group of reactions is that between triphenylphosphine and maleic anhydride which was not recognized for many years as proceeding via nucleophilic attack of the phosphine on the activated carbon–carbon double bond. Although the phosphonium enolate structure (20) had been attributed to the adduct (29), its IR spectrum showed a bathochromic shift of the carbonyl absorption frequency characteristic of stable alkylidenephosphoranes (30). This led to its reformulation as 21 (31), which was confirmed by synthesis from triphenylphosphine and chlorosuccinic anhydride, the resulting phosphonium salt giving the alkylidenephosphorane (21) on treatment with base (32). Further evidence for the alkylidenephosphorane structure was obtained by Osuch et al. from a consideration of ^{31}P NMR data and degradation of the adduct to the known phosphonium salt (**22**) (33).

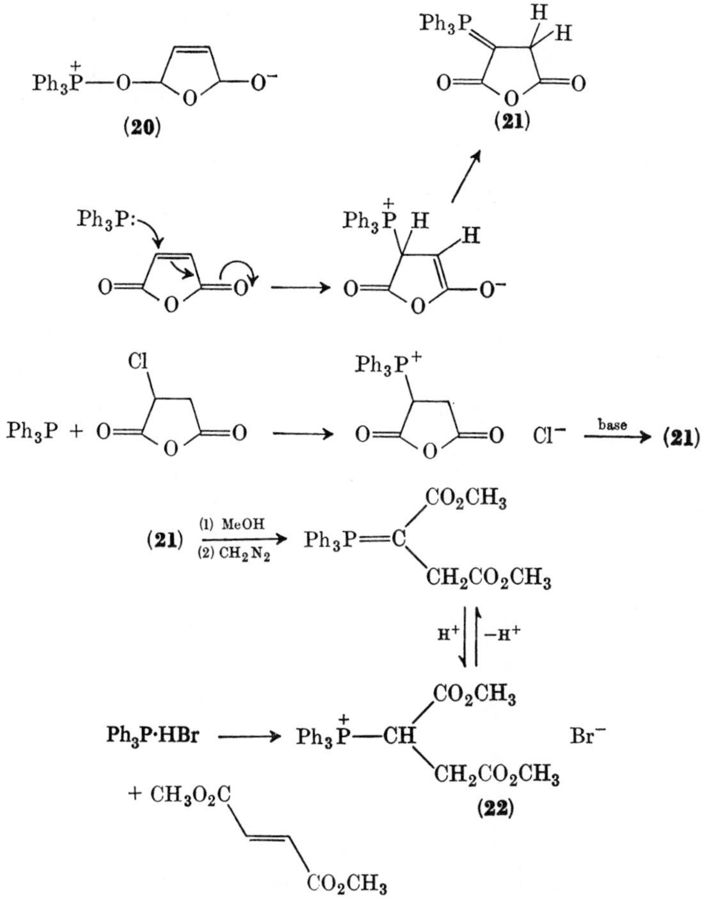

$Ph_3\overset{+}{P}$—O—⬠—O⁻

(**20**)

Ph_3P, H, H ... O=⬠=O

(**21**)

Ph_3P:

O=⬠=O ⟶ $Ph_3\overset{+}{P}$ H H ... O=⬠—O⁻

Ph_3P + O=⬠=O (Cl) ⟶ Ph_3P^+ ... O=⬠=O Cl⁻ —base→ (**21**)

$$(\textbf{21}) \xrightarrow[(2)\ CH_2N_2]{(1)\ MeOH} Ph_3P{=}C\begin{array}{l} CO_2CH_3 \\ CH_2CO_2CH_3 \end{array}$$

H⁺ ‖ –H⁺

$$Ph_3P{\cdot}HBr \longrightarrow Ph_3\overset{+}{P}{-}CH\begin{array}{l} CO_2CH_3 \\ CH_2CO_2CH_3 \end{array} \quad Br^-$$

(**22**)

+ CH_3O_2C ⌇ CO_2CH_3

Maleimide derivatives react similarly with phosphines forming alkylidenephosphoranes (**23**) which were characterized by their IR and NMR spectra and also by their Wittig reaction products; in fact the noncrystallizable derivatives of tributylphosphine were best characterized by the latter method (34). Isomaleimides (**24**) when reacted with triphenylphosphine gave the corresponding maleimide adducts (**23b,d**) although it has not been established whether prior rearrangement of the isoimide occurs, induced by triphenylphosphine, or if an initially formed triphenylphosphine–isoimide adduct rearranges to the imide isomer under the reaction conditions.

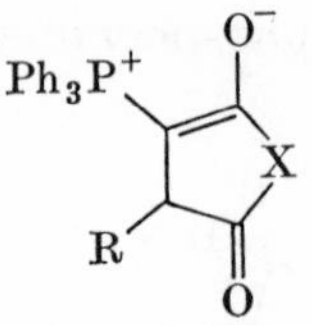

(23)

(a) R = H, X = NH
(b) R = H, X = NPh
(c) R = Me, X = NPh
(d) R = H, X = N—N

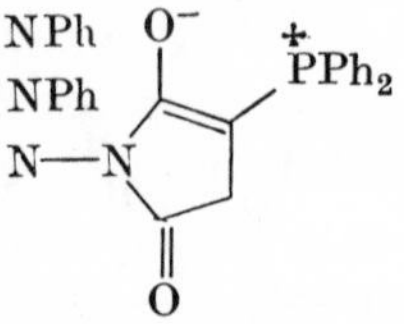

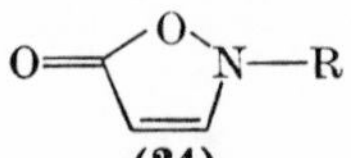

(24)

(a) R = Ph
(b) R = N=

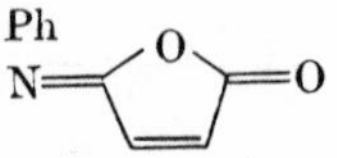

Although betaine (25) (35) and phosphonium enolate (26) (13,36) structures had been proposed for the adducts obtained from tertiary phosphines and p-benzoquinone, Ramirez et al. (37) showed that the UV spectra and chemical reactions of the adducts were more consistent with an alkylidenephosphorane structure (27). This was confirmed by an independent synthesis (38), the hydrogen iodide salt of the adduct

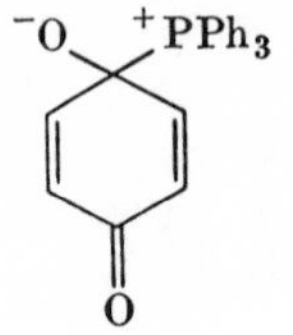

(25)

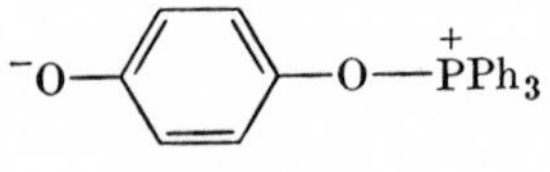

(26)

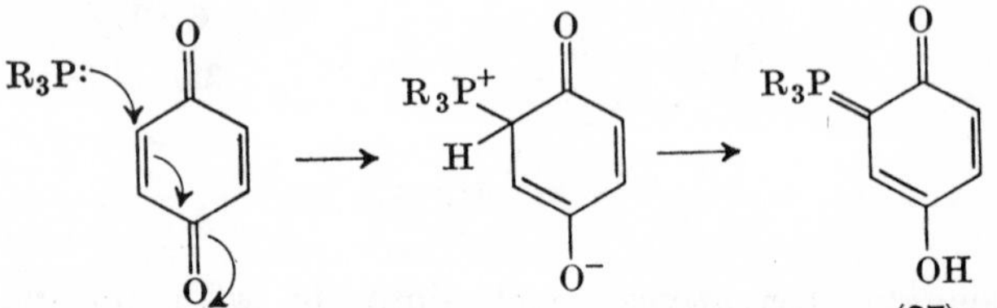

(27)

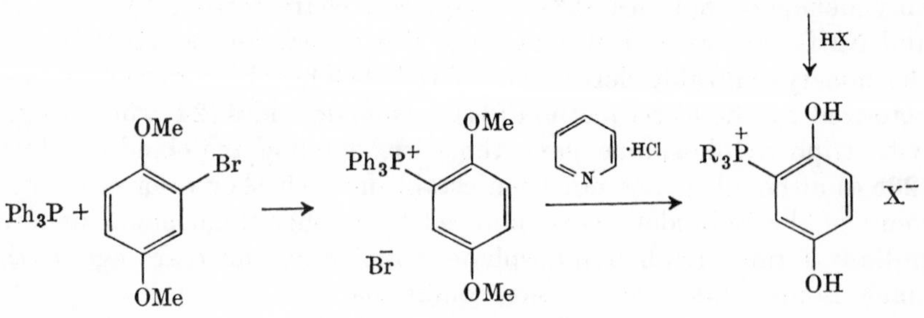

being related to the salt obtained from triphenylphosphine and the dimethyl ether or 2-bromohydroquinone. The stable alkylidenephosphorane (27) is formed by addition of the phosphine to the electrophilic double bond of the quinone and the resulting betaine is stabilized by proton migration.

Tervalent phosphites containing primary alkyl groups do not add to the carbon–carbon double bond of quinones but rather react at the oxygen atom in a biphilic manner (39). However phosphites with secondary alkyl groups do react at carbon as evidenced by the substitution of the chlorine atoms of chloranil to give the tetraphosphonate (28, R = isopropyl) (40). This is possible because the intermediate (29) can lose a secondary carbonium ion in product formation.

It has been suggested that diphenylphosphine oxide reacts as its tervalent tautomer with p-benzoquinone and adds nucleophilically to the activated double bond to give the tertiary phosphine oxide (30) after proton migration (41).

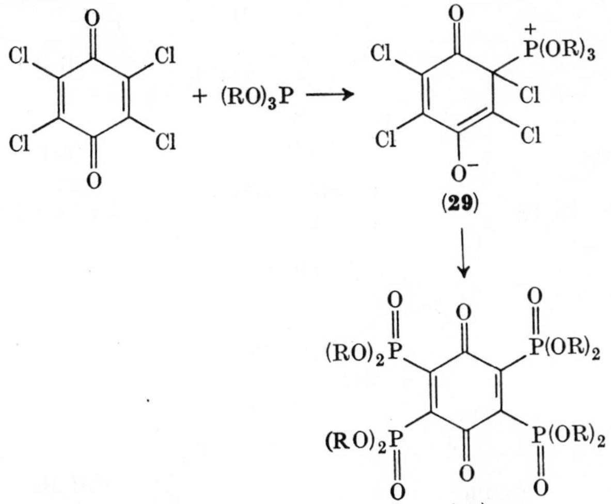

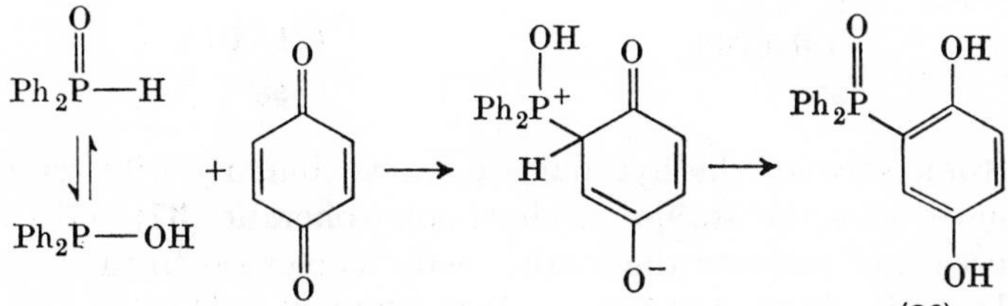

The structures of the adducts formed from tertiary phosphorus compounds and dibenzoylethylene have been the subject of much speculation. Trimethyl phosphite (42) had been assumed to yield the pentacovalent 1,2-oxaphospholene (**31**), as had diethylphenylphosphinite, while the adducts from tertiary phosphines have been formulated as *O*-phosphonium betaines (**32**) (13,44) which, it has been suggested, might equilibrate with the *C*-phosphonium betaines (**33**) via a 1,2-oxaphospholene analogous to **31** (43). However Ramirez et al. (45) have shown that in all cases the products of the reactions of tertiary phosphorus compounds with dibenzoylethylene are stable alkylidenephosphoranes (**34**) obtained from the initially formed *C*-phosphonium betaines (**33**) by proton migration. The adducts were characterized by their chemical reactions (e.g., reaction with HCl gave phosphonium salts (**35**) which reverted to the phosphoranes on treatment with base, and reaction with benzyl bromide gave the enol ethers (**36**) formed by *O*-alkylation) and IR and NMR spectra.

(32)

R = R′ = Bu, Et, Ph

(31)

R = R′ = OMe
R = OMe, R′ = Ph

(33)

R = R′ = Bu, Et, Ph

(34)

(35) **(36)**

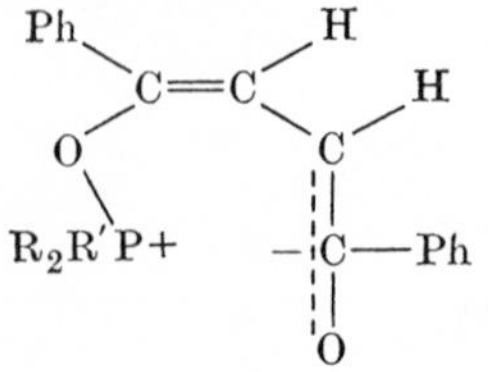

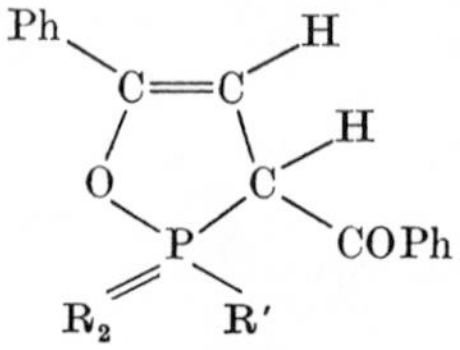

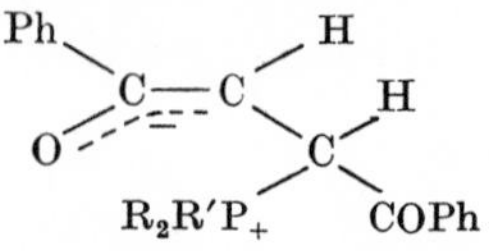

In protonic solvents diethyl maleate reacts similarly with triphenylphosphine to give the stable alkylidenephosphorane (**37**) (17b), while in the absence of solvents the methyl ester isomerizes to the fumarate via bond rotation in the 1:1 intermediate (**38**) (46). This intermediate, and the corresponding ones from cinnamic acid derivatives, can be trapped with acids to form phosphonium salts (**39**) (14).

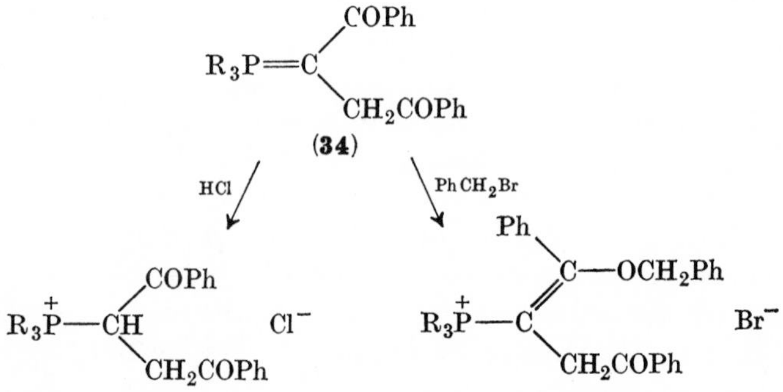

Maleic and cinnamic acids react with tertiary phosphites in a manner analogous to acrylic acid, in other words, the initial 1:1 intermediate undergoes proton and alkyl transfer to give the phosphonate (40) (47).

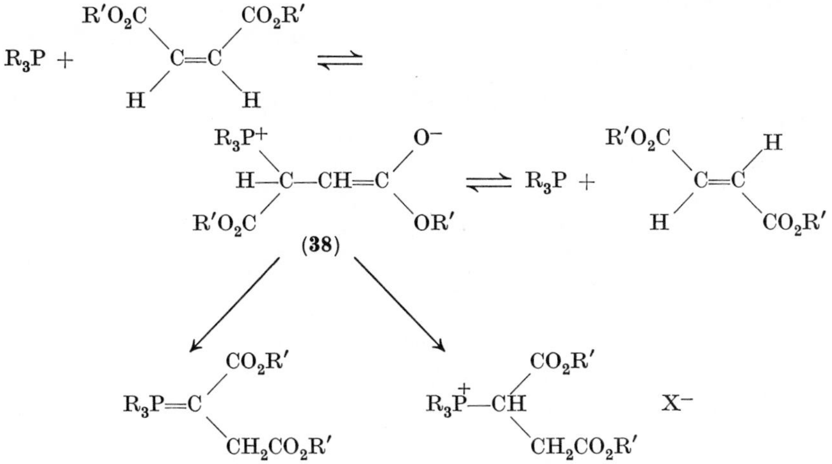

$$R_3P +$$
R'O$_2$C, CO$_2$R', C=C, H, H ⇌

R$_3$P$^+$ H—C—CH=C O$^-$ / OR' R'O$_2$C (38) ⇌ R$_3$P + R'O$_2$C, H, C=C, H, CO$_2$R'

R$_3$P=C CO$_2$R' CH$_2$CO$_2$R' (37)   R$_3$P$^+$—CH CO$_2$R' CH$_2$CO$_2$R' (39)   X$^-$

$$(RO)_3P + R''CH=CHCO_2H \longrightarrow (RO)_2\overset{O}{\overset{\|}{P}}CHR''CO_2R$$
(40)

C. 1:1-Diactivated Olefins

Horner and Klupfel (13) reacted vinylidenediacid and benzalmalonic acid derivatives with tertiary phosphines and formulated the adducts as inner salts (betaines, 41) because they could be protonated by acids to yield phosphonium salts which reverted to the betaines on treatment with base. This formulation was verified (in the case of benzalmalonitriles) by Rappoport et al. (48) who observed that the nitrile absorption of the adducts in their IR spectra was shifted to lower frequency relative to the absorption of the reactants, thus indicating substantial conjugation into the nitrile groups of the adducts. Electron-attracting substituents in the phenyl ring stabilized the betaines, whereas electron-donating substituents had the opposite effect. Similar betaines (42) from the reactions of tertiary phosphines with 3-benzylidene-2,4-pentanedione were characterized using IR and ^{31}P NMR spectra (49). In this latter case no cyclization to a pentacovalent phosphorane was observed over a wide temperature range.

The stability of these adducts is due to the strong conjugation of the negative charge with the electron-attracting groups; proton migration to give alkylidenephosphoranes (stabilized by $p\pi$–$d\pi$ bonding) does not occur because there are no electron-attracting groups on the carbon atom α to the phosphonium center which could stabilize a negative charge to a sufficient extent.

$$R_3P + CHR'{=}CR_2'' \rightarrow R_3\overset{+}{P}CHR'\overset{-}{C}R_2''$$

$$\textbf{(41)}$$

$$R' = H, Ph; R'' = CN, CO_2Me$$

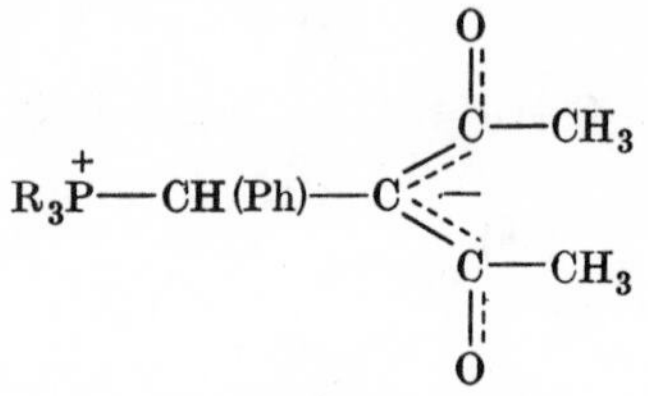

$$\textbf{(42)}$$

1:1 Intermediates from the addition of phosphite, phosphonite, and phosphinite esters to 3-benzylidene-2,4-pentanedione cyclize to penta-covalent oxaphospholanes (**43**), characterized by large positive chemical shifts in their ^{31}P NMR spectra (50). The different reaction products from phosphines and phosphite esters are attributed mainly to the higher electronegativity of oxygen compared to carbon which tends to favor the P^V state in reactions with phosphite esters (49). For example, cyclization of **42**, in which the 5-membered ring must occupy apical-equatorial sites, would require having the less electronegative carbon atom in the available apical position of the required trigonal bipyramid structure (**44**), while the adducts from phosphite esters are stable phosphoranes having oxygen atoms in the apical positions (**45**) (51). Using variable temperature NMR it was shown, however, that at elevated temperatures the P—O bond of the 1,2-oxaphosphol-4-ene ring of **43** is ruptured and an equilibrium between cyclic and open dipolar forms is observed (50).

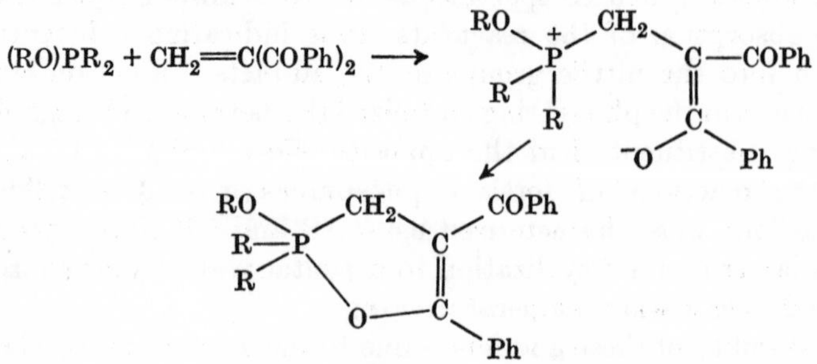

$$\textbf{(43)}$$

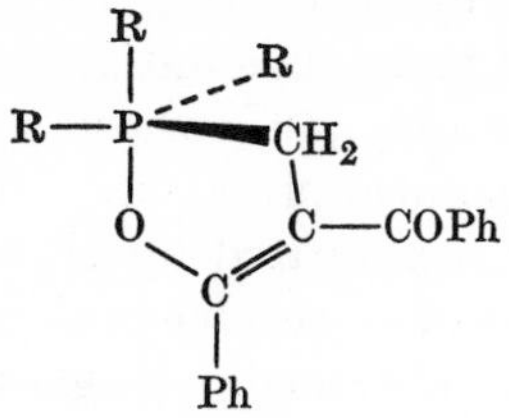

(44) 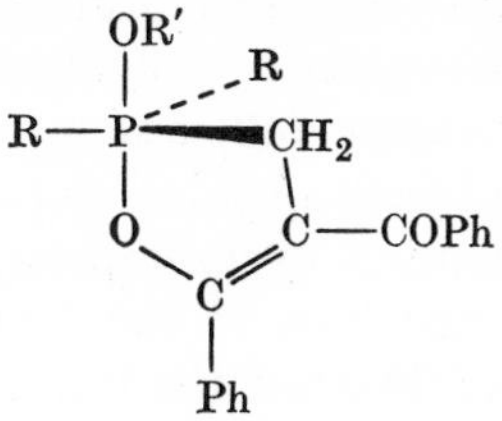   (45)

D. Other Olefinic Compounds

Triphenylphosphine reacts with tetracyanoethylene to give a 1:2 adduct formulated as the pentacovalent phospholane (46) (52). However this adduct has a negative chemical shift in its ^{31}P NMR spectrum and this is to be contrasted with the large positive chemical shifts observed for authentic P^V derivatives with five σ bonds to carbon (53), furthermore the corresponding 1:2 adducts prepared from optically active phosphines retain optical activity (54); thus considerable doubt must be cast on this formulation. Under aqueous acidic conditions tetracyanoethylene and tetracyanoquinodimethane react with phosphines to form reduced olefins and phosphine oxides. Although these reactions could proceed via hydrolysis of 1:1 betaines or phosphonium salts, a mechanism involving phosphinium radical cations is preferred since optically active phosphines yield racemic phosphine oxides (54). The zwitterion (47) was formed when triphenylphosphine added to the double bond of 2,3,5,6-tetracyano-7-oxabicyclo[2.2.1]hepta-2,5-diene (48), and readily decomposed on heating to tetracyanobenzene and triphenylphosphine oxide (55).

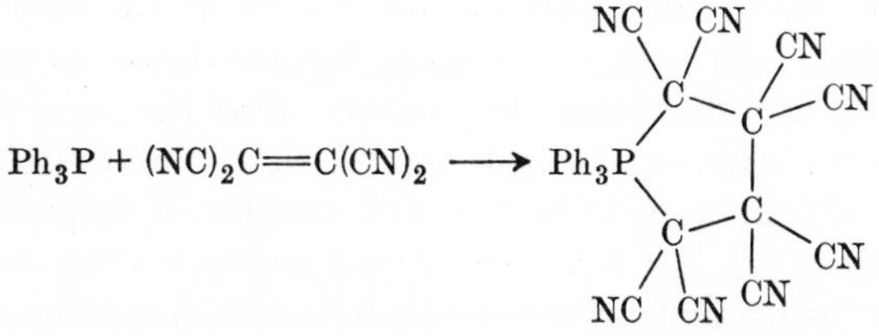

$Ph_3P + (NC)_2C{=}C(CN)_2 \longrightarrow$

(46)

$Ph_3P +$

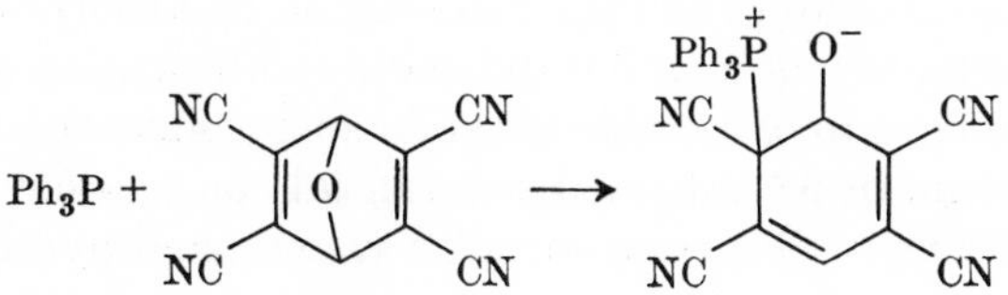

(48)   (47)

Dimethyl acetylenedicarboxylate, upon heating or standing for considerable periods, polymerizes and gives substantial quantities of a furan derivative (49) and a tetramer (50) (56), whose structure was determined from physical data and chemical properties. The presence of a substituted cyclopropene ring was indicated by IR spectroscopy while the presence of the 7-oxanorbornene residue was indicated by pyrolysis of 50 when the furan (49) was obtained. Hughes and Woods

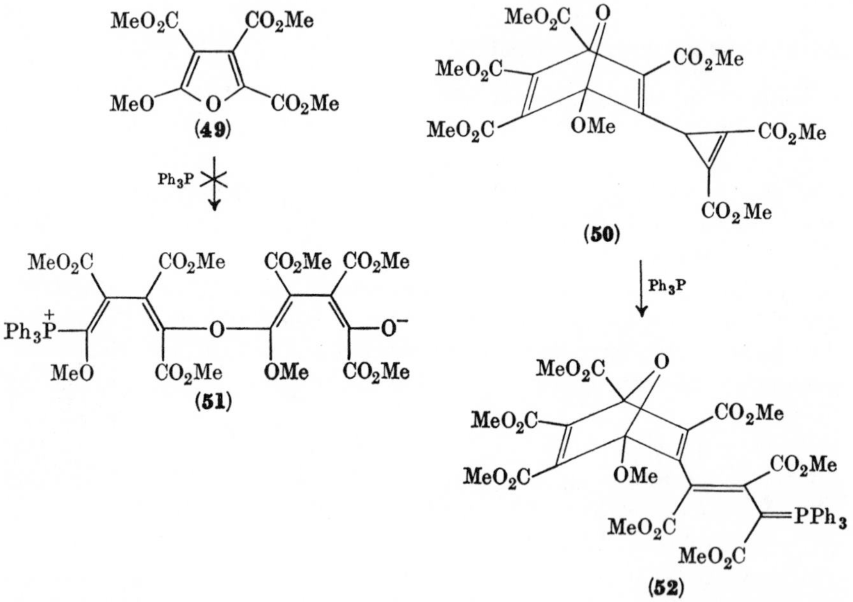

(57) reported that triphenylphosphine added to the activated double bond of the furan to give a 1:2 adduct formulated as the zwitterion (51) but Kauer and Simmons (56) showed that the pure furan did not react at all with triphenylphosphine and that the adduct obtained by Hughes and Woods was in fact a 1:1 adduct of triphenylphosphine and the tetramer 50. It was formulated as the stable alkylidenephosphorane (52) resulting from nucleophilic addition of the phosphine to the electrophilic cyclopropenyl double bond. This structure was proposed since the adduct no longer showed an IR absorption characteristic of the cyclopropene ring but did show an absorption characteristic of the carbonyl group of a stable alkylidenephosphorane. In addition, the UV spectrum of 52 indicated substantial conjugation of the alkylidenephosphorane with the unsaturated system, and pyrolysis gave the furan 49.

Triphenylphosphine also adds to the activated double bond of 2,6-diphenylpyrylium perchlorate to form the pyranylphosphonium salt (53) (58).

Phosphite esters add to the activated double bond of phenylcyclobutanedione in a manner similar to that observed with acrolein, the substituted phosphonate (54) being formed (59).

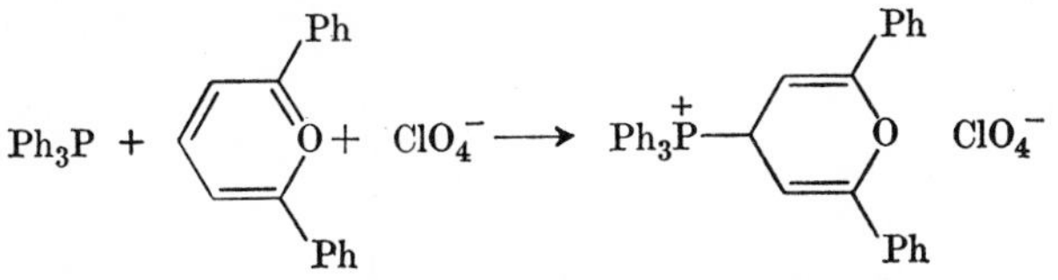

(53)

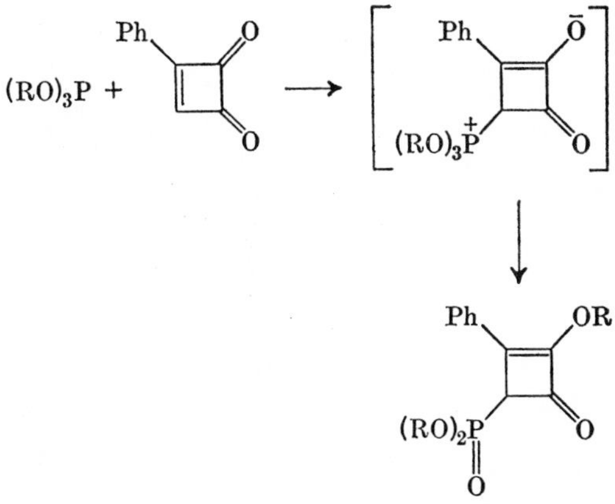

(54)

III. Additions to Acetylenes

A. 2:1 Adducts

Johnson and Tebby found that the reaction of excess triphenylphosphine with dimethyl acetylenedicarboxylate in ether gave a stable adduct (60). Analysis indicated that it was a 2:1 adduct and **55 was** suggested as a tentative structure. Subsequent investigation has produced considerable evidence to support their assignment (61). One particularly important piece of evidence in favor of structure **55** is the extremely low carbonyl stretching frequency (159.2 mm^{-1}), which is characteristic of stable alkylidenephosphoranes (30,31). Chemical evidence is also consistent with this structure. Reductive cleavage with zinc and acetic acid gave triphenylphosphine and dimethyl succinate in a 2:1 molar ratio. Hydrolysis with boiling water gave equimolar quantities of triphenylphosphine, triphenylphosphine oxide, and dimethyl fumarate. Two possible mechanisms for the latter reaction have been considered (61–63). The second (concerted) mechanism is preferred since it has been shown that the phosphoranyl-phosphonium salts involved in the alternative mechanism are stable and do not spontaneously fragment in the manner indicated (63).

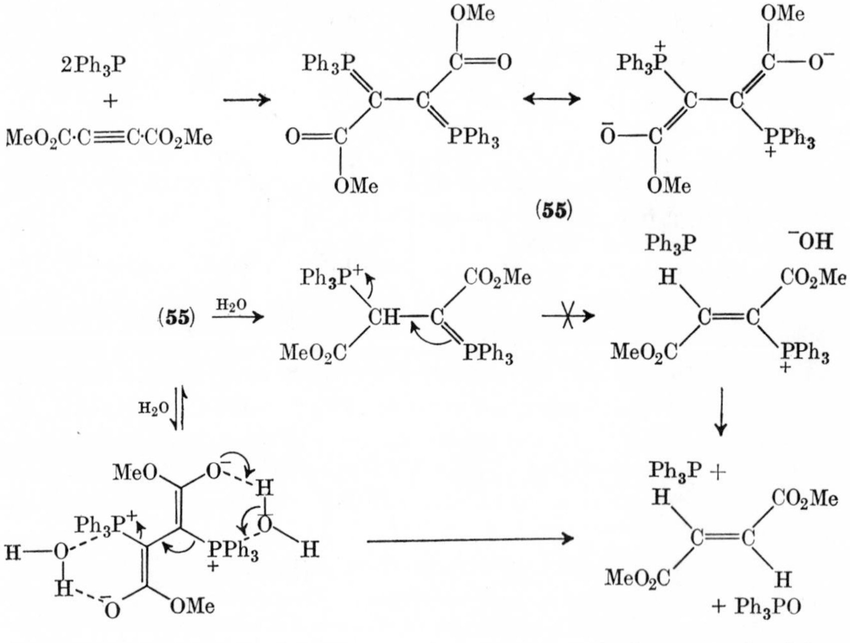

The ^{1}H NMR spectra of phosphorus compounds have received increasing attention in recent years. The temperature-dependent NMR spectrum of **55** in CDCl$_3$ shows that the alkylidene-1,2-diphosphorane exists in solution as a mixture of three slowly interconverting conformers (**55a–c**). This is due to hindered internal rotation which is a characteristic feature of α-alkoxycarbonylalkylidenephosphoranes (64), and reflects the partial double bond character of the α-carbon–carbonyl–carbon bond. In acidic solution on the other hand, the adduct is protonated and its NMR spectrum is interpretable in terms of the corresponding bisphosphonium salt.

Diphenylmethylphosphine forms an analogous 1,2-diphosphorane with dimethyl acetylenedicarboxylate (61), and similar adducts have also been prepared from acetylenic ketones (63). An earlier report by

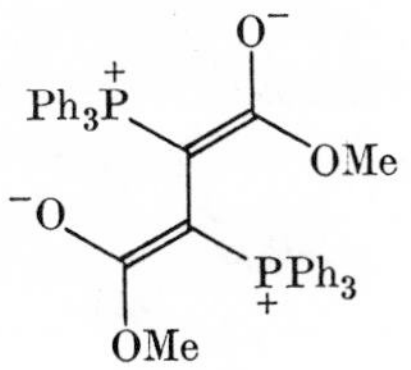

(55a) 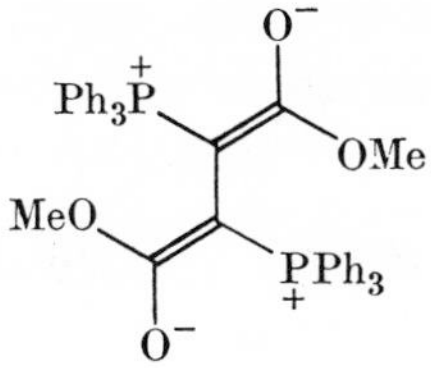   **(55b)** 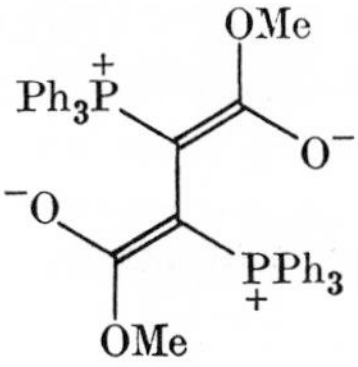   **(55c)**

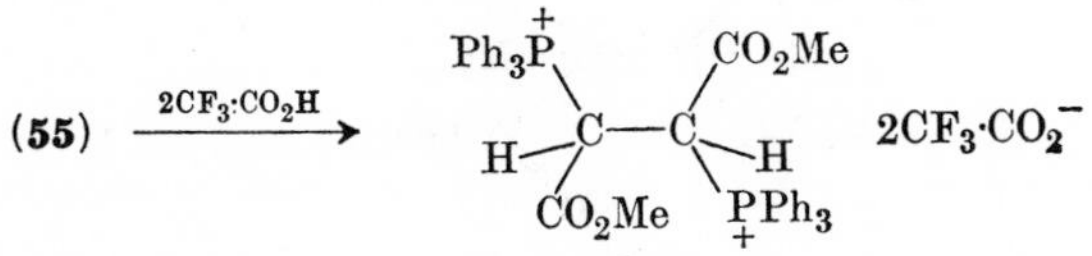

Horner and Hoffmann describes the formation of a 2:1 adduct from triphenylphosphine and dibenzoyldiacetylene (65), although their proposed structure for this adduct (56) has not been thoroughly proven.

Bisdiphenylphosphinomethane forms a cyclic 1:1 adduct with dimethyl acetylenedicarboxylate (62). IR and NMR spectra indicate that this adduct is not the expected alkylidenephosphorane (57) but its tautomer (58). The IR spectrum shows strong absorptions at 173.7 and 162.0 mm^{-1}, corresponding to saturated and conjugated carbonyl groups respectively. The NMR spectrum in CDCl$_3$ shows two distinct one-proton multiplets at δ 5.00 and 1.25 ppm. The latter signal has been assigned to the proton on the carbon atom between the two phosphorus nucleic, by comparison with the methine proton resonances of analogous phosphoranes. The effect of change of

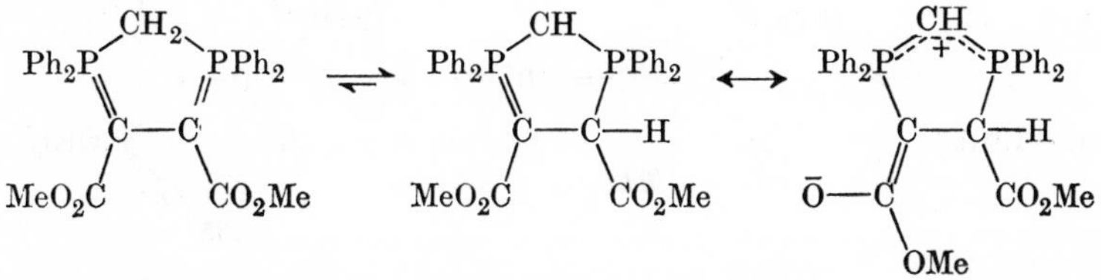

(57)   **(58)**

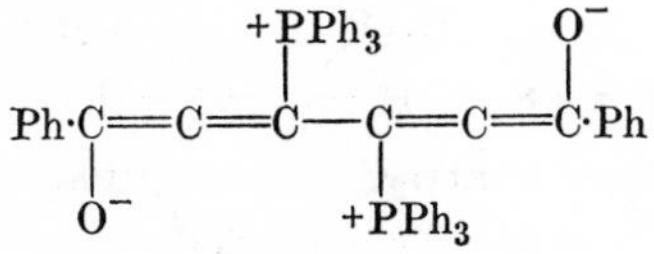

(56) 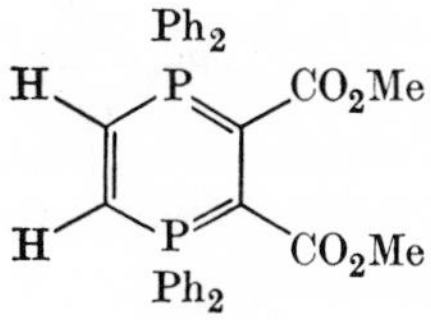   **(59)**

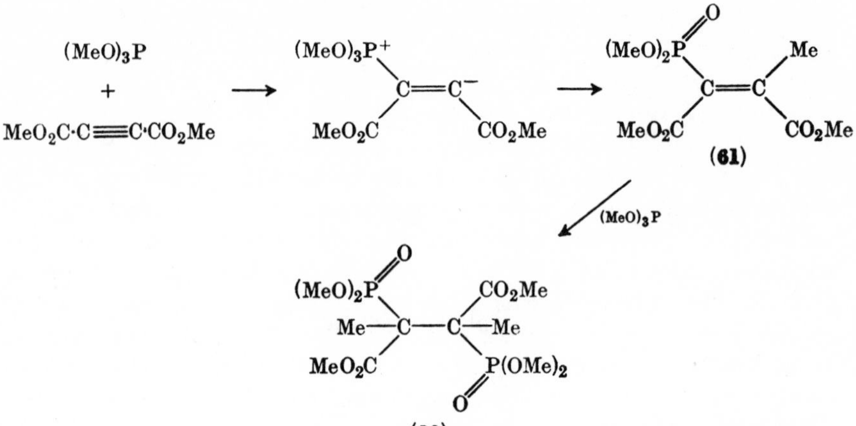

$(MeO)_3P$
+
$MeO_2C \cdot C \equiv C \cdot CO_2Me$

$\longrightarrow$

$(MeO)_3P^+$
$\underset{MeO_2C}{\diagdown} C = C \underset{CO_2Me}{\diagdown}{}^-$

$\longrightarrow$

$(MeO)_2\overset{\displaystyle O}{\overset{\|}{P}} \underset{MeO_2C}{\diagdown} C = C \overset{Me}{\diagup}{}_{CO_2Me}$

(61)

$\downarrow \; (MeO)_3P$

$(MeO)_2\overset{\displaystyle O}{\overset{\|}{P}} \; \overset{CO_2Me}{}$

$Me - \overset{|}{C} - \overset{|}{C} - Me$

$MeO_2C \qquad \underset{\displaystyle O}{\overset{|}{P}(OMe)_2}$

(60)

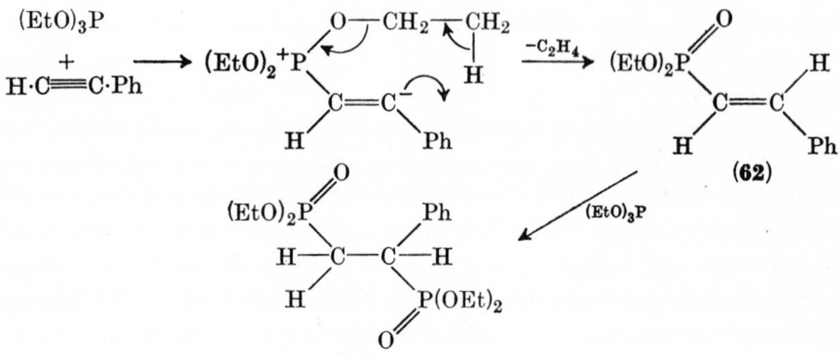

$(EtO)_3P$
+
$H \cdot C \equiv C \cdot Ph$

$\longrightarrow$

$(EtO)_2{}^+P \overset{O - CH_2 \frown CH_2}{\diagdown} \underset{H \diagdown}{} \; H$
$\overset{|}{C} = C$
$H \qquad Ph$

$\xrightarrow{-C_2H_4}$

$(EtO)_2\overset{\displaystyle O}{\overset{\|}{P}} \underset{H}{\diagdown} C = C \overset{H}{\diagup}{}_{Ph}$

(62)

$\swarrow \; (EtO)_3P$

$(EtO)_2\overset{\displaystyle O}{\overset{\|}{P}} \qquad Ph$

$H - \overset{|}{C} - \overset{|}{C} - H$

$H \qquad \underset{\displaystyle O}{\overset{|}{P}(OEt)_2}$

(63)

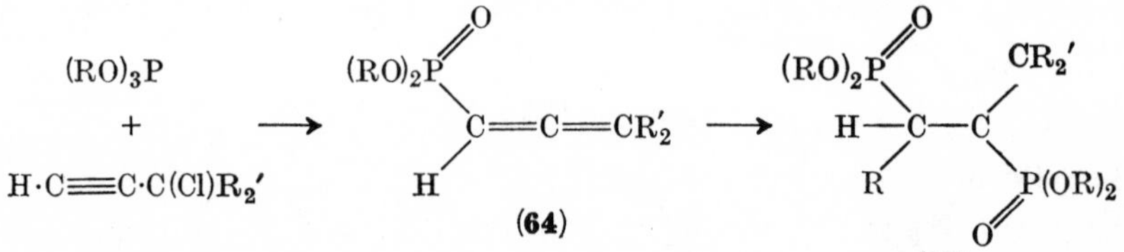

$(RO)_3P$
+
$H \cdot C \equiv C \cdot C(Cl)R_2'$

$\longrightarrow$

$(RO)_2\overset{\displaystyle O}{\overset{\|}{P}} \underset{H}{\diagdown} C = C = CR_2'$

(64)

$\longrightarrow$

$(RO)_2\overset{\displaystyle O}{\overset{\|}{P}} \qquad CR_2'$

$H - \overset{|}{C} - \overset{|}{C}$

$R \qquad \underset{\displaystyle O}{\overset{|}{P}(OR)_2}$

(65)

$Ph_2\overset{\displaystyle O}{\overset{\|}{P}} - H$
$\updownarrow$
$Ph_2P - OH$

$+ \; EtO_2C \cdot C \equiv C \cdot CO_2Et$

$\longrightarrow$

$Ph_2\overset{\displaystyle O}{\overset{\|}{P}} \qquad CO_2Et$

$H - \overset{|}{C} - \overset{|}{C} - H$

$EtO_2C \qquad \underset{\displaystyle O}{\overset{|}{P}Ph_2}$

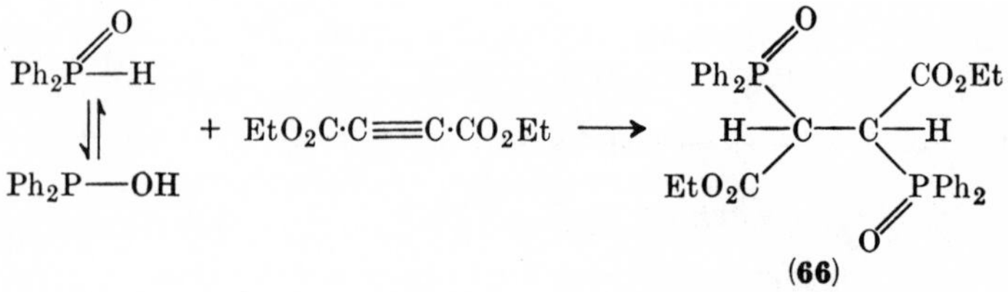

(66)

temperature on the NMR spectrum indicates the presence of two conformers resulting from restricted rotation of the conjugated ester group.

The preparation of the 1,4-diphosphorin derivative (**59**) from *cis*-1,2-bisdiphenylphosphinoethene has also been described (62). Investigation of this adduct is hindered by its instability. However the IR and mass spectra are consistent with the proposed structure.

Trimethyl phosphite also forms a 2:1 adduct with dimethyl acetylenedicarboxylate (66). The product in this case is the bisphosphonate (**60**) whose formation involves migration of two methyl groups from oxygen to carbon. It seems most probable that **60** arises by attack of a second molecule of phosphite on the initially formed vinyl phosphonate (**61**). The addition of trialkyl phosphites to less electrophilic acetylenes (e.g., phenylacetylene) requires much more vigorous conditions and does not involve alkyl migration. For example, the reaction of triethyl phosphite with phenylacetylene proceeds by way of concerted proton transfer and elimination of ethylene to give the vinyl phosphonate (**62**), which with a second molecule of phosphite gives **63** (67). Alkylacetylenes are even less reactive with the exception of tertiary acetylenic chlorides which undergo an Arbuzov reaction by S_N2' attack of the phosphite on the acetylene, to give an allenyl phosphonate (**64**), which subsequently undergoes nucleophilic addition of a second molecule of phosphite, and alkyl migration to give **65** (68). It has been suggested that diphenylphosphine oxide reacts as its tervalent tautomer (in the complete absence of base) with diethyl acetylenedicarboxylate go give the 2:1 adduct (**66**) via nucleophilic addition followed by proton transfer (41).

B. 1:1 and 2:2 Adducts

Horner and Hoffmann isolated an unstable adduct of triphenylphosphine and dimethyl acetylenedicarboxylate which gave triphenylphosphine oxide and dimethyl fumarate on hydrolysis (65). They formulated it as the straightforward 1:1 adduct (**67**). Later workers were unable to isolate a 1:1 adduct of triphenylphosphine and dimethyl acetylenedicarboxylate but have succeeded in trapping it with other reagents. For example, Johnson and Tebby obtained the colorless betaine (**68**) at −50° in the presence of carbon dioxide (60). This betaine can be protonated by anhydrous hydrogen chloride to give the phosphonium salt **69** whose IR spectrum shows it to be a mixture of *cis* and *trans* isomers, the latter predominating. Hydrolysis of **68** or **69** gives dimethyl fumarate and maleate in a 3:1 ratio.

The betaine (68) evolves carbon dioxide at 50 to give a stable insoluble orange product which was originally formulated as the cyclic
pentacovalent diphosphorane (70) but was later shown to be an
alkylidene-1,4-diphosphorane (71) (69). It is envisaged that the
formation of 71 involves combination of a reformed reactive intermediate (67) and an intact stabilized intermediate (68), followed by loss
of a further molecule of carbon dioxide.

Structures 70 and 71 are very different, and brief consideration of the
physical and chemical properties of the orange adduct allows an immediate choice between the two alternatives.   Firstly, potentiometric
titration and the isolation of a diperchlorate salt confirm the basicity
of the adduct.   Secondly, reductive cleavage with zinc and acetic acid
gave 1,2,3,4-tetramethyl butanetetracarboxylate and triphenylphosphine which confirms the presence of a four-carbon-atom chain in the
molecule.   Thirdly, the ^{31}P chemical shift of the adduct in $CDCl_3$
solution (-22.9 ppm relative to 85% H_3PO_4) falls within the accepted
range for an alkylidenephosphorane (33,45).   A compound having
structure 70 would be expected to exhibit a large positive ^{31}P chemical
shift (see Section II.D).

Although **71** was originally prepared from the carbon dioxide trapped intermediate (**68**), it may also be prepared directly from triphenylphosphine and dimethyl acetylenedicarboxylate in certain polar solvents (acetone, acetonitrile, and ethyl acetate) (70). It is possible that the 1:1 intermediate is stabilized by combination with a carbonyl or nitrile group, giving an intermediate of type **68** in which the electrophilic nature of the β carbon atom is enhanced. Nucleophilic attack by a second molecule of 1:1 intermediate would then give **71**. Some support for this hypothesis may be drawn from the work of Winterfeldt and Dillinger who have intercepted the 1:1 intermediate (**67**) by reaction with benzaldehyde (71). The products, **72** and **73**, are consistent with the intervention of **74**.

In the presence of water the reaction of triphenylphosphine with dimethyl acetylenedicarboxylate gives triphenylphosphine oxide and dimethyl fumarate, via hydration of the initial 1:1 intermediate (**67**) followed by fragmentation of the vinylphosphonium hydroxide (**75**). It has been demonstrated that other disubstituted electrophilic acetylenes can be reduced to the corresponding olefins by this procedure which affords a convenient synthesis of dideuterated olefins if deuterium oxide is used in place of water (72). Diphenylvinylphosphine reacts with dimethyl acetylenedicarboxylate and water to give a stable 1:1:1 adduct (**76**), formed by hydrolysis of the more stable tautomer (**77**) of the initial phosphine:acetylene adduct (73). A 1:2:1 adduct of the same components has also been isolated.

Monosubstituted electrophilic acetylenes also react with triphenylphosphine and water but in these cases the vinylphosphonium hydroxides formed (**78**) do not fragment, but rearrange to phosphine oxides (**79**) (74). Diphenyl-1,2-diphenylethylphosphine oxide (**79a**) has also been isolated from a number of other apparently unrelated reactions, such as the reaction of methylenetriphenylphosphorane with benzaldehyde in alcohol, and the reaction of triphenylphosphine with styrene oxide, also in alcohol (75). It has been shown that **78a** is a common intermediate (74d). Thus nucleophilic attack of triphenylphosphine on styrene oxide, or of methylenetriphenylphosphorane on benzaldehyde, first of all gives the expected Wittig reaction intermediate (**80**) but in protonic media this equilibrates with the vinylphosphonium hydroxide (**78a**) via the corresponding 2-hydroxyethylphosphonium salt (**81**). The intervention of **78a** in the reaction of triphenylphosphine with styrene oxide was demonstrated by carrying out the reaction in the presence of one equivalent of 30% $H_2^{18}O$. The product (**79a**) was found to have extensively incorporated the labeled oxygen. The phosphorus-to-carbon 1,2 shift (**78 → 79**) has been used to achieve ring expansion of a 9-phosphafluorene to a 9.10-dihydro-9-phosphaphenanthrene-9-oxide (**74c**), and of a phosphetane to a phospholane oxide (**74e**).

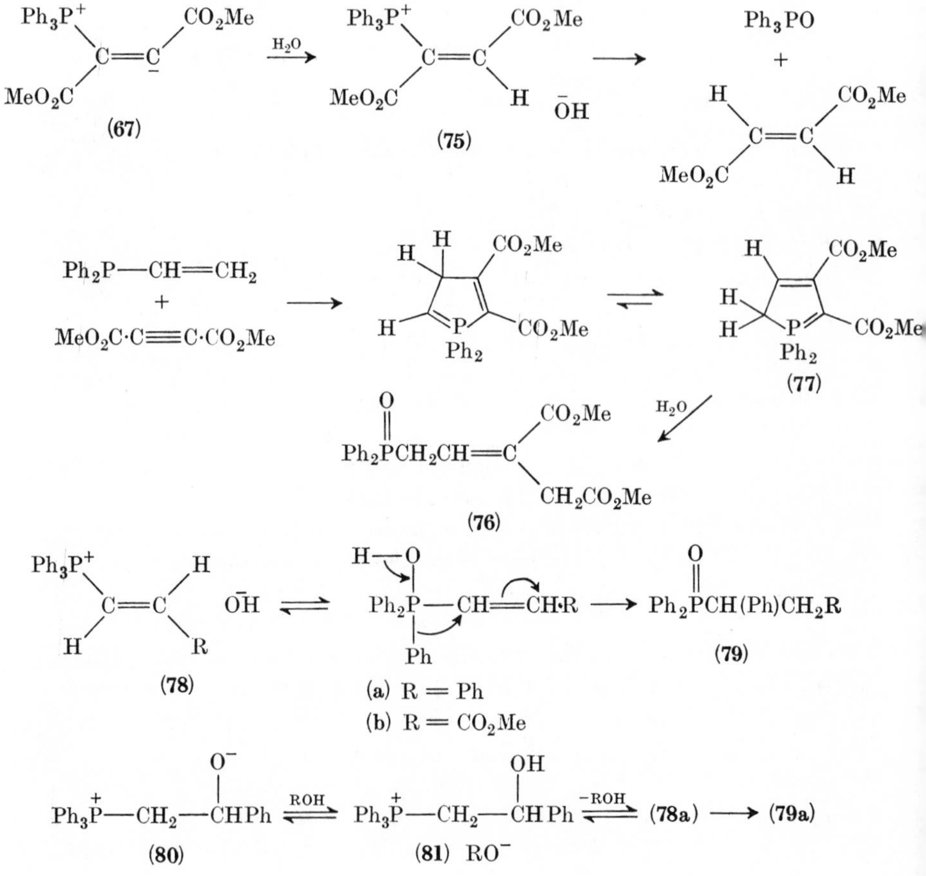

If sulfur dioxide is present in addition to water, fragmentation or aryl migration are avoided and adducts are formed which are triphenylphosphoniaethane sulfonates (**82**) (76). The interaction of sulfur dioxide and water to give the bisulfite ion prior to reaction with the vinylphosphonium hydroxide was established by use of $^2H_2^{18}O$. Furthermore, **82a** and **82b** have been synthesized by reaction of sodium bisulfite with the appropriate vinylphosphonium bromides **83** and **84**.   The betaine (**82a**) readily fragments on heating but forms a stable alkylidenephosphorane on treatment with alkali.   The lability of the α-methine proton of **82a** is further demonstrated by the slow interconversion of threo and erythro isomers in neutral solution.

$$SO_2 + 2 H_2O \rightleftharpoons HSO_3^- + H_3O^+$$

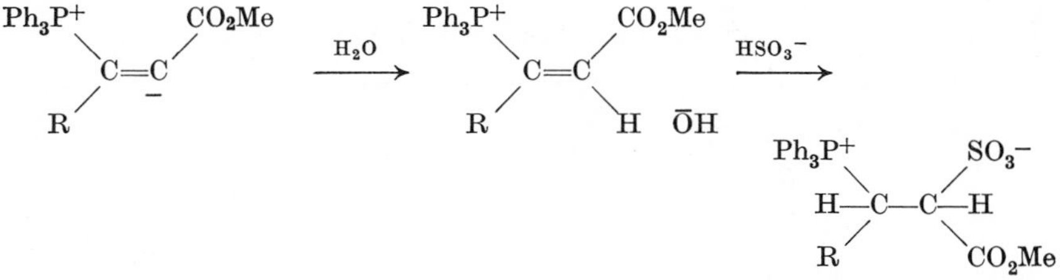

(**82a**) R = CO_2Me
(**82b**) R = H
(**82c**) R = Ph

The vinylphosphonium salts **83**, **84**, and **85** are most conveniently prepared by reacting triphenylphosphine with the appropriate acetylene in the presence of hydrogen bromide.   In this way the 1:1 intermediates are trapped by protonation (74b, 76, 77).   Although Hoffmann and Diehr claimed to have prepared **83** from triphenylphosphine hydrobromide and dimethyl acetylenedicarboxylate in acetonitrile solution (77), it has since been shown that the product which they isolated was in fact the dihydrobromide of the alkylidene-1,4-diphosphorane (**71**) (76).   The vinylphosphonium salt may however be obtained by carrying out the reaction in chloroform (76).   Hydrolysis of **83** gives dimethyl fumarate (75%), dimethyl maleate (25%), and triphenylphosphine oxide.   More recently it has been shown that the reaction of triphenylphosphine hydrobromide with propiolic acid gives **86** and not **87**, as previously reported (22).   The formation of **86** probably involves addition of a second molecule of phosphine to the cis-vinylphosphonium salt (**81**), followed by decarboxylation and protonation of the phosphoranyl–phosphonium salt so formed, since the trans isomer of **87** is resistant to attack by triphenylphosphine (22).

Under acidic conditions dialkyl- and diaryl-alkynylphosphines form salts of 1,4-diphosphacyclohexadiene (**88**).   The reaction proceeds via acid catalyzed attack of one phosphine molecule on the activated acetylenic bond of another, followed by cyclization of the 1:1 intermediate by intramolecular nucleophilic addition (78).

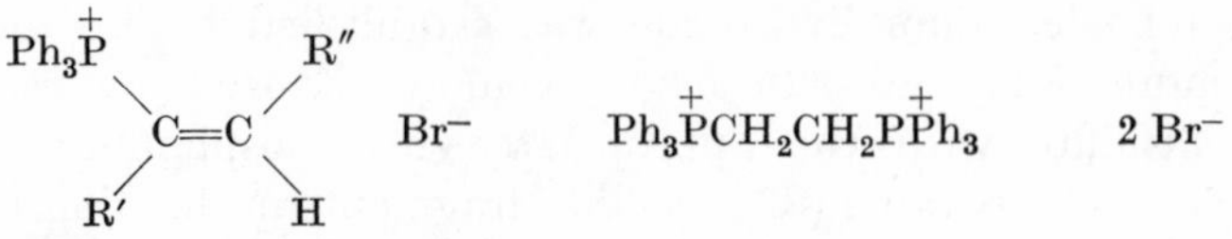

$$\begin{array}{c} \overset{+}{Ph_3P} \qquad R'' \\ \diagdown C=C \diagup \qquad Br^- \\ R' \diagup \qquad \diagdown H \end{array} \qquad \overset{+}{Ph_3PCH_2CH_2}\overset{+}{PPh_3} \qquad 2\,Br^-$$

(83) R' = R'' = CO$_2$Me
(84) R' = Ph, R'' = CO$_2$Me
(85) R' = H, R'' = Ph
(87) R' = H, R'' = CO$_2$H

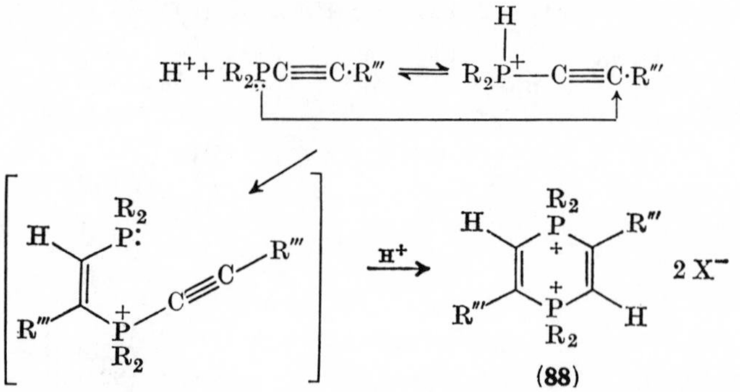

$$H^+ + R_2\ddot{P}C\equiv C\cdot R''' \rightleftharpoons R_2\overset{H}{\overset{|}{P}}{}^+\!\!-C\equiv C\cdot R'''$$

$$\left[\begin{array}{c} H \diagdown \quad \overset{R_2}{P:} \\ \quad \diagup \quad \diagdown \\ R''' \diagup \overset{+}{\underset{R_2}{P}}-C\equiv C-R''' \end{array} \right] \quad \xrightarrow{\;H^+\;} \quad \begin{array}{c} H \diagdown \quad \overset{R_2}{\overset{P}{\overset{+}{}}} \quad \diagup R''' \\ \quad \\ R''' \diagup \overset{+}{\underset{R_2}{P}} \diagdown H \end{array} \quad 2\,X^-$$

(88)

The reaction of trialkyl phosphites with acetylenic acids to give β-alkoxycarbonylvinyl phosphonates involves an oxygen-to-oxygen alkyl migration analogous to the oxygen-to-carbon migration encountered in Section III.A (79).

The formation of allenic phosphine oxides (and phosphonates) from propargyl phosphinites (and phosphites) effectively involves intramolecular addition to the triple bond (80). However as all the reported examples of this reaction involve unactivated acetylenes it is probably better classified as an $S_N i$ replacement or a 2,3 sigmatropic rearrangement.

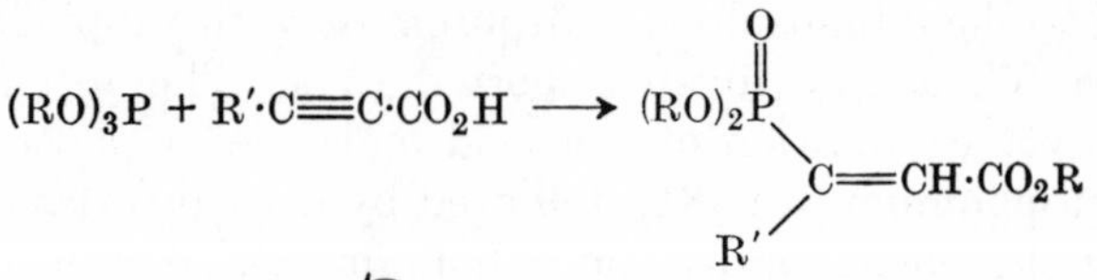

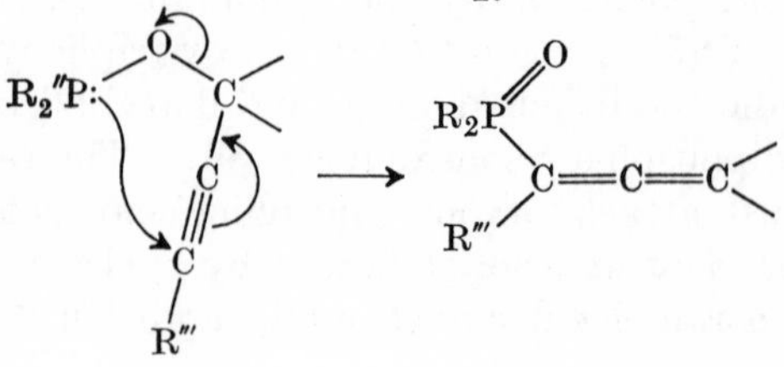

$$(RO)_3P + R'\cdot C\equiv C\cdot CO_2H \longrightarrow \begin{array}{c} O \\ \parallel \\ (RO)_2P \\ \diagdown C=CH\cdot CO_2R \\ R' \diagup \end{array}$$

$$\begin{array}{c} R_2''P: \overset{O}{\diagdown} C \diagup \\ \quad \diagdown C \diagup \\ \quad \parallel \\ \quad C \\ \quad \diagup \\ \quad R''' \end{array} \quad \longrightarrow \quad \begin{array}{c} O \\ \parallel \\ R_2P \\ \diagdown C=C=C \diagup \\ R''' \diagup \end{array}$$

C. 2:3 Adducts

The reaction of triphenylphosphine with dicyanoacetylene gives a red crystalline product for which the pentacovalent phosphole structure **89** was originally proposed (52).   However the mass spectrum of this compound shows it to be a 2:3 adduct and it has been reformulated as the alkylidene-1,6-diphosphorane (**90**) (81).   The study of **90** has revealed interesting differences in the behavior of alkylidenephosphoranes stabilized by different functional groups.   Alkylidenephosphoranes which are stabilized by ketone or ester groups (e.g., **71**) are protonated in acidic media, as shown by changes in the NMR and UV spectra.   However alkylidenephosphoranes which are stabilized by nitrile groups are much less basic and phosphoranes such as **90** which are stabilized by two or more nitrile groups remain unprotonated in acidic media (81).

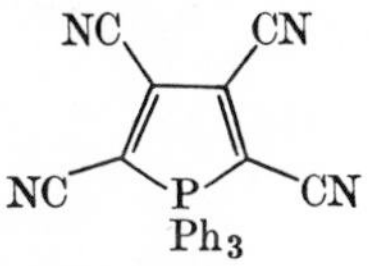

D. 1:2 Adducts

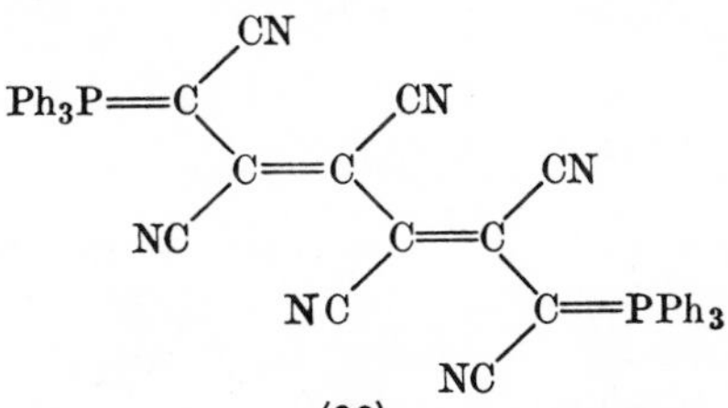

It has already been seen that triphenylphosphine and dimethyl acetylenedicarboxylate are a particularly versatile pair of reactants. In addition to the adducts previously discussed they form three 1:2 adducts (60,82,83).   At −50 a yellow unstable 1:2 adduct is obtained which has been formulated by different workers as the zwitterion **91** (60) and as the phosphole **92** (84).   On warming to room temperature the unstable 1:2 adduct is stabilized by rearrangement.   It is now known that two stable 1:2 adducts are produced by this rearrangement (82,83), although only one was originally isolated (60,84).   These are the isomeric allylidenephosphoranes **93** (originally formulated as the phosphine **94**) and **95**.

The structural elucidation of the unstable 1:2 adduct, **91** or **92**, is hindered by its ready rearrangement to **93** and **95**.   The report by Hendrickson (84) that legible NMR data, supporting the phosphole formulation, could be obtained on a very fresh sample of the unstable adduct probably deserves reinvestigation in view of the fact that only one rearrangement product (93) was then known. It is possible that the spectrum observed by Hendrickson was in fact the NMR spectrum of a mixture of the two stable adducts rather than that of the unstable adduct itself. The zero R_f of the unstable adduct on TLC would tend to favor structure 91 (70). It is interesting to note that Horner and Hoffmann proposed a similar structure for an unstable adduct which they isolated from the reaction of triethylphosphine with dimethyl acetylenedicarboxylate (65). It will be seen, however, that the formation of 93 from triphenylphosphine and dimethyl acetylenedicarboxyate (shown in the scheme below) implicates the phosphole (92) at least as a reaction intermediate. The other stable adduct (95) is formed directly from 91.

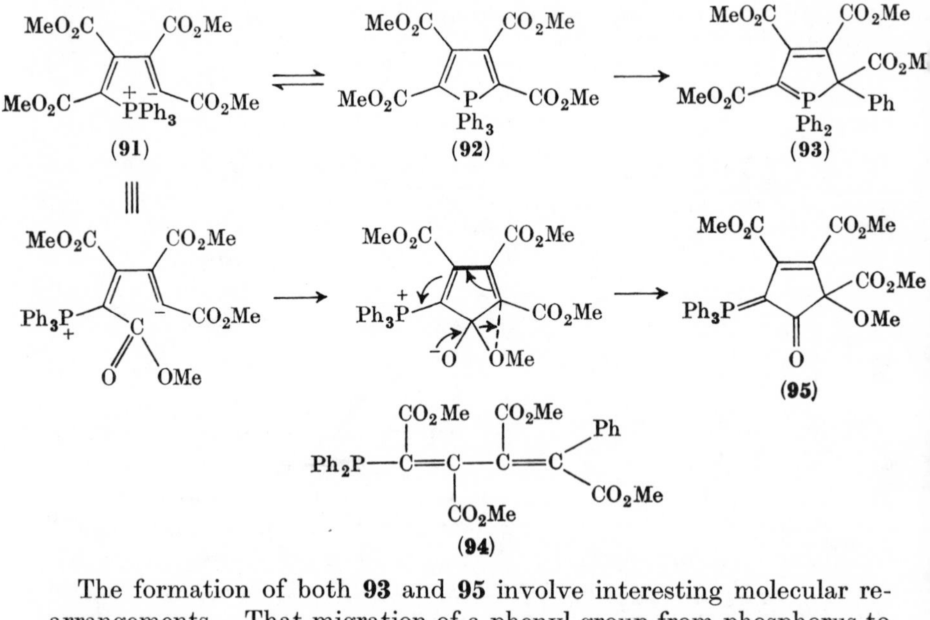

The formation of both 93 and 95 involve interesting molecular rearrangements. That migration of a phenyl group from phosphorus to carbon is involved in the formation of 93 was indicated by its oxidation to diphenylphosphinic and benzoic acids (60). It was confirmed by its mass spectrum which gave no peak at m/e 262 characteristic of compounds containing triphenylphosphine residues (85), and by its conversion into the phosphine oxide (96) on treatment with zinc and acetic acid. The formation of 96 has been shown to proceed by way of hydrolysis of the initially formed reduction product (97). A choice between the two possible products of phenyl migration, 93 and 94, is readily made from the IR spectrum of the adduct. The spectrum shows strong absorption bands at 174, 170, 167, and 153 mm^{-1} corresponding to the carbonyl stretching frequencies of the four ester groups. The extremely low frequency of two of these bands (167 and 153 mm^{-1}) is only consistent with their being due to the α and γ ester groups (respectively) of an allylidenephosphorane. A phosphine moiety does not significantly alter the carbonyl stretching frequency of a conjugated ester group.

It has been shown that in the formation of **93** the aryl group best able to support a negative charge migrates preferentially (70). Furthermore, analysis of the mass spectrum of the product obtained by reaction of a mixture of triphenyl- and tri-p-tolylphosphines with excess dimethyl acetylenedicarboxylate proves the aryl migration to be intramolecular since molecular ions corresponding to triphenyl and tri-p-tolyl adducts only are observed. In addition, oxidation of the tritolyl analog of **93** established that during the migration from phosphorus to carbon the tolyl group retains its para orientation since p-toluic and di-p-tolylphosphinic acids were obtained.

The formation of the other stable 1:2 adduct (**95**) involves cyclization of the zwitterion (**91**) followed by rearrangement of a methoxyl group. Its structure has been confirmed by X-ray analysis (83).

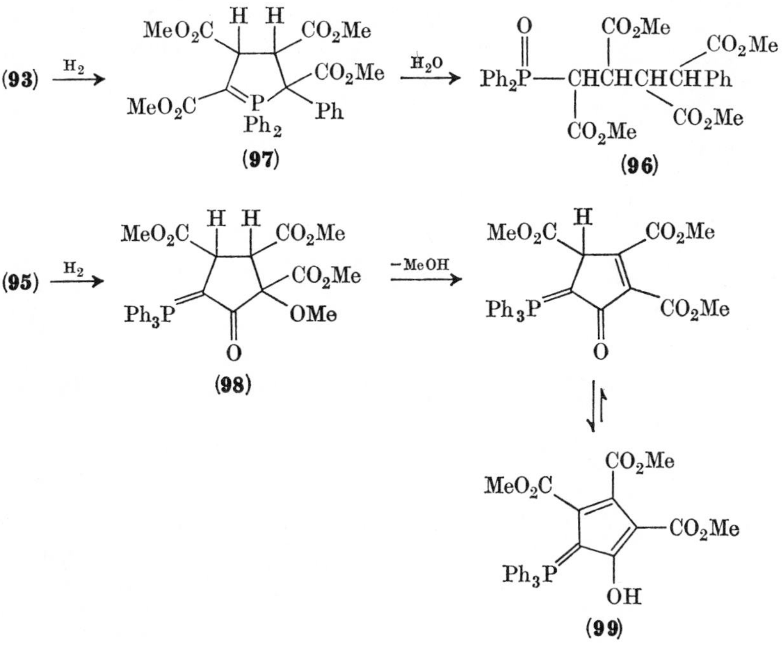

Reduction of **95** with zinc and acetic acid gives the alkylidenephosphorane (**98**), which eliminates methanol on treatment with sodium methoxide. The NMR, spectrum of the latter product, which may also be prepared by Raney nickel reduction of **95**, suggests that it probably exists as its cyclopentadienylide tautomer (**99**).

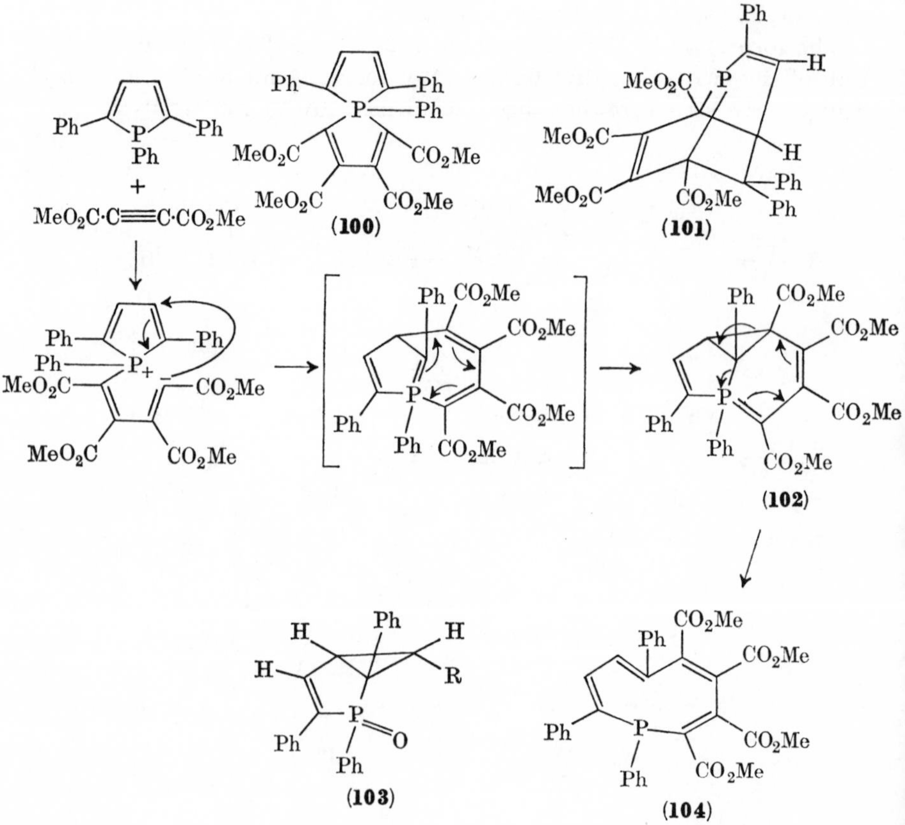

1,2,5-Triphenylphosphole forms a 1:2 adduct with dimethyl acetylenedicarboxylate which is stable indefinitely at room temperature but rearranges in boiling chloroform to a second 1:2 adduct (86). Hughes and Uaboonkul assigned structures **100** and **101** to these compounds (86). However in view of the extreme instability of the analogous 1,1,1-triphenylphosphole (**92**) it would be rather surprising if the spirobiphosphole (**100**) were to be stable. Indeed, the NMR (^{1}H and ^{31}P) and UV spectra are inconsistent with this structure, and Waite and Tebby have shown that the product is in fact an allylidenephosphorane (87). They have proposed the alternative structure, **102**, for this adduct on the basis of a comparison of its NMR and UV spectra with those of **103** (R=H, CO$_2$H, CO$_2$Me).   In boiling chloroform the phosphorane (**102**) rearranges to give *cis,cis,trans,cis*-phosph(III)onin (**104**).

The formation of a 1:2 adduct from 1-phenyl-2,2,3,3-tetramethyl-phosphetane proceeds by ring expansion of the four-membered ring to give **105** (88).

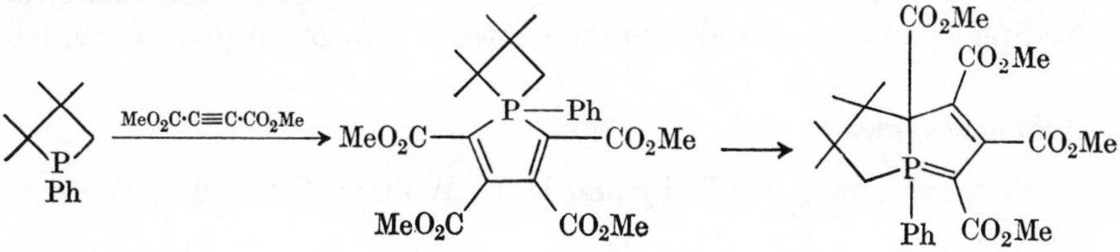

The reaction between triphenylphosphine and hexfluorobut-2-yne is very rapid even at −78° and polymerization occurs (89).

IV. Additions to Benzyne

Tertiary phosphines also add to the benzyne 'triple bond.'   Wittig et al. (90) have obtained 9-phenylphosphafluorene from such a reaction using triphenylphosphine.   When diphenylmethylphosphine is used, methylidenetriphenylphosphorane is formed via a proton shift (91). A similar proton shift occurs in the reaction with triethylphosphite resulting in elimination of ethylene and formation of the phosphonate (**106**) (67).

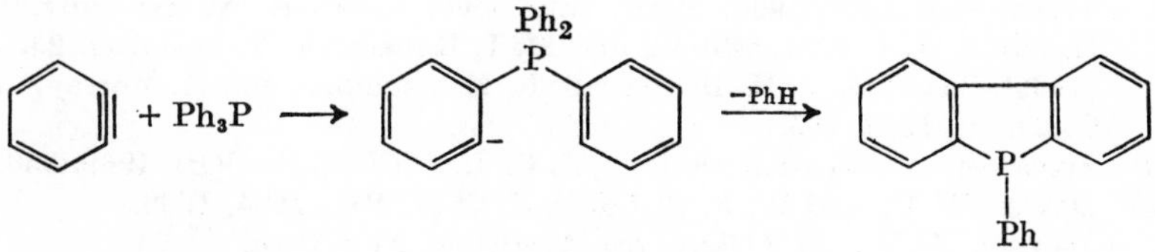

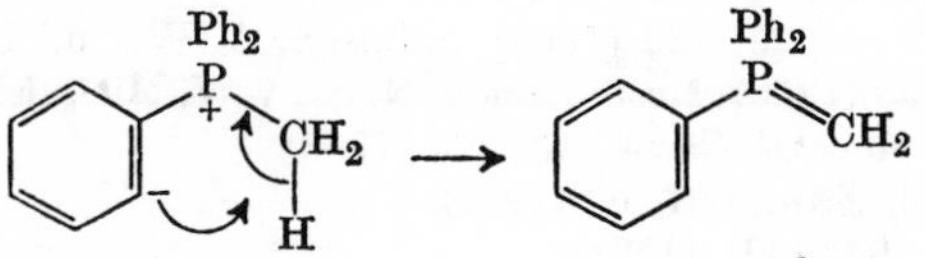

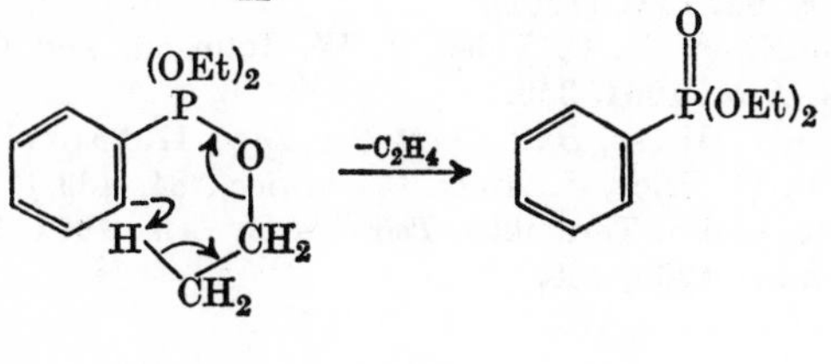

V. Additions with Primary and Secondary Phosphorus Compounds

These reactions have not been discussed in detail since they proceed simply via prototropy of 1:1 phosphine–olefin and 1:1 phosphine–acetylene intermediates. The additions of diphenylphosphine and its anions to activated acetylenes to give both vinylphosphines and 1,2-bisdiphosphinoethanes are characteristic of this group (64, 77, 89, 92).

Acknowledgments

We thank Drs. J. C. Tebby and D. H. Williams for their interest and advice.

REFERENCES

1. Bergmann, E. D., D. Ginsberg, and R. Pappo, *Org. Reactions*, **10**, 179 (1959).
2. Winterfeldt, E., *Angew. Chem. Intern. Ed.*, **6**, 423 (1967).
3. Acheson, R. M., *Adv. Heterocyclic. Chem.*, **1**, 125 (1963).
4. For recent papers, see Acheson R. M., and J. N. Bridson, *J. Chem. Soc. (C)*, **1969**, 1143; and other papers in this series.
5. For a review, see Trippett, S., *Quart. Rev.*, **17**, 406 (1963).
6. Hudson, R. F., *Structure and Mechanism in Organo-Phosphorus Chemistry*, Academic, New York, Chapters 1–3.
7. Kirby, A. J., and S. G. Warren, *The Organic Chemistry of Phosphorus*, Elsevier, Amsterdam, Chapter 1.
8. Recent papers are: Rakshys, J. W., R. W. Taft, and W. A. Sheppard, *J. Amer. Chem. Soc.*, **90**, 5236 (1968); Bissey, J. E., and H. Goldwhite, *Tetrahedron Lett.*, **1966**, 3247; Yakovleva, E. A., E. N. Tsvetkov, D. I. Lobanov, A. I. Shatenshtein, and M. I. Kabachnik, *Tetrahedron*, **25**, 1165 (1969); Shaw, G., J. K. Becconsall, R. M. Canadine, and R. Murray, *Chem. Commun.*, **1966**, 425.
9. For a recent discussion see Bart, J. C. J., *J. Chem. Soc. (B)*, **1969**, 350.
10. Davies, W. C., and W. P. G. Lewis, *J. Chem. Soc.*, **1934**, 1599.
11. Johnson, A. W., *Ylid Chemistry*, Academic, New York.
12. Johnson, A. W., *J. Org. Chem.*, **25**, 183 (1960); Johnson, A. W. and R. B. LaCount, *Tetrahedron*, **9**, 130 (1960); Nesmeyanov, N. A., V. V. Mickulshina, and O. A. Reutov, *J. Organometal. Chem.*, **13**, 263 (1968).
13. Horner, L., and K. Klupfel, *Ann.*, **591**, 69 (1959).
14. Hoffmann, H., *Chem. Ber.*, **94**, 1331 (1961).
15. Young, D. P., W. E. McEwen, D. C. Velez, J. W. Johnson, and C. A. VanderWerf, *Tetrahedron Lett.*, **1964**, 359.
16. Morita, K., Z. Suzuki, and H. Hirose, *Bull. Chem. Soc. Jap.*, **41**, 2815 (1968).
17. (a) Takashina, N., and C. C. Price, *J. Amer. Chem. Soc.*, **84**, 489 (1962); (b) Oda, R., T. Kawabata, and S. Tanimoto, *Tetrahedron Lett.*, **1964**, 1653.
18. Trippett, S., *Chem. Commun.*, **1966**, 468.

19. Savage, M. P., and S. Trippett, *J. Chem. Soc. (C)*, **1968**, 591.
20. Shutt, J. R., and S. Trippett, *J. Chem. Soc. (C)*, **1969**, 2038.
21. Vullo, W. J., *Ind. Eng. Chem. Prod. Res. Develop.*, **5**, 346 (1966).
22. Pattenden, G., and B. J. Walker, *J. Chem. Soc. (C)*, **1969**, 531.
23. Aguair, A. M., and D. Daigle, *J. Org. Chem.*, **30**, 2826 (1965); Anderson, W. A., R. Freeman, and C. A. Reilly, *J. Chem. Phys.*, **39**, 1518 (1963).
24. Coover, H. N., M. A. McColl, and J. B. Dickey, *J. Amer. Chem. Soc.*, **79**, 1963 (1957).
25. See Ref. 7, page 52.
26. Kamai, G., and V. A. Kukhtin, *Zh. Obshch. Khim.*, **27**, 2373 (1957); *Chem. Abstr.*, **52**, 7127 (1958).
27. Kamai, G., and V. A. Kukhtin, *Zh. Obshch. Khim.*, **28**, 1196 (1958); *Chem. Abstr.*, **52**, 19909 (1958).
28. Kamai, G., and V. A. Kukhtin, *Zh. Obshch. Khim.*, **27**, 2376 (1957); *Chem. Abstr.*, **52**, 7127 (1958).
29. Schonberg, A., and A. F. A. Ismail, *J. Chem. Soc.*, **1940**, 1374.
30. Ramirez, F., and S. Dershowitz, *J. Org. Chem.*, **22**, 41 (1957).
31. Aksnes, G., *Acta Chem. Scand.*, **15**, 692 (1961).
32. Hudson, R. F., and P. A. Chopard, *Helv. Chim. Acta*, **46**, 2178 (1963).
33. Osuch, C., J. E. Franz, and F. B. Zienty, *J. Org. Chem.*, **29**, 3721 (1964).
34. Hedaya, E., and S. Theodoropulos, *Tetrahedron*, **24**, 2241 (1968).
35. Davies, W. C., and W. P. Walters, *J. Chem. Soc.*, **1935**, 1786.
36. Schonberg, A., and R. Michaelis, *Chem. Ber.*, **69**, 1080 (1936).
37. Ramirez, F., and S. Dershowitz, *J. Amer. Chem. Soc.*, **78**, 5614 (1956).
38. Hoffmann, H., L. Horner, and G. Hassel, *Chem. Ber.*, **91**, 58 (1958).
39. For a discussion, see Ref. 7, pp. 77ff.
40. Reetz, T., *U.S. Pat.*, 2,935,518 (1960); *Chem. Abstr.*, **54**, 19598 (1960).
41. Campbell, I. G. M. and I. D. R. Stevens, *Chem. Commun.*, **1966**, 505.
42. Kukhtin, V. A., and K. M. Orekhova, *Zh. Obshch. Khim.*, **30**, 1539 (1960); Kamai, G. and V. A. Kukhtin, *ibid.*, **27**, 2431 (1957).
43. Harvey, R. G., and E. V. Jensen, *Tetrahedron Lett.*, **1963**, 1801.
44. Kuwajima, I., and T. Mukaiyama, *J. Org. Chem.*, **29**, 1385 (1964).
45. Ramirez, F., O. P. Madan, and C. P. Smith, *Tetrahedron*, **22**, 567 (1966).
46. Hands, A. R., *J. Chem. Soc.*, **1964**, 1181.
47. Kamai, G., and V. A. Kukhtin, *Trudy Kazah Khim. Tekhnol. Inst., im. S.M. Kirova*, **23**, 135 (1957); *Chem. Abstr.*, **52**, 9948 (1958).
48. Rappoport, Z., and S. Gertler, *J. Chem. Soc.*, **1964**, 1360.
49. Ramirez, F., J. F. Pilot, and C. P. Smith, *Tetrahedron*, **24**, 3735 (1968).
50. Ramirez, F., O. P. Madan, and S. R. Heller, *J. Amer. Chem. Soc.*, **87**, 731 (1965).
51. For further discussion on the stability of trigonal bipyramidal structures see Westheimer, F., *Acc. Chem. Res.*, **1**, 70 (1968).
52. Reddy, G. S., and C. D. Weis, *J. Org. Chem.*, **28**, 1822 (1963).
53. Hellwinkel, D., *Chem. Ber.*, **98**, 576 (1965).
54. Powell, R. L. P., and C. D. Hall, *J. Amer. Chem. Soc.*, **91**, 5403 (1969).
55. Weis, C. D., *J. Org. Chem.*, **27**, 3520 (1962).
56. Kauer, J. C., and H. E. Simmons, *J. Org. Chem.*, **33**, 2720 (1968).
57. Hughes, A. N., and M. Woods, *Tetrahedron*, **23**, 2973 (1967).
58. Krivun, S. V., *Dokl. Akad. Nauk. SSSR*, **182**, 347 (1968); *Chem. Abstr.*, **70**, 29009 (1969).

59. DeSelms, R. C., *Tetrahedron Lett.*, **1968**, 5545.
60. Johnson, A. W., and J. C. Tebby, *J. Chem. Soc.*, **1961**, 2126.
61. Shaw, M. A., J. C. Tebby, R. S. Ward, and D. H. Williams, *J. Chem. Soc. (C)*, **1967**, 2442.
62. Shaw, M. A., J. C. Tebby, R. S. Ward, and D. H. Williams, *J. Chem. Soc. (C)*, **1970**, 504.
63. Shaw, M. A., and J. C. Tebby, *J. Chem. Soc. (C)*, **1970**, 5.
64. Bestmann, H. J., G. Joachim, I. Lengyel, S. F. M. Oth, J. Mereny, and J. Weitkamp, *Tetrahedron Lett.*, **1966**, 3355; Crouse, D. M., A. T. Wehman, and E. E. Schweizer, *Chem. Commun.*, **1968**, 866; Zeliger, H. I., J. P. Snyder, and H. J. Bestmann, *Tetrahedron Lett.*, **1969**, 2199; Randall, F. J. and A. W. Johnson, *Tetrahedron Lett.*, **1968**, 2841.
65. Horner, L., and H. Hoffmann, *Angew. Chem.*, **68**, 473 (1956).
66. Griffin, C. E., and T. D. Mitchell, *J. Org. Chem.*, **30**, 2829 (1965).
67. Griffin, C. E., and T. D. Mitchell, *J. Org. Chem.*, **30**, 1935 (1965).
68. Pudovic, A. N., *Zh. Obshch. Khim.*, **20**, 92 (1950).
69. Shaw, M. A., J. C. Tebby, J. Ronayne, and D. H. Williams, *J. Chem. Soc. (C)*, **1967**, 944.
70. Tebby, J. C., personal communication.
71. Winterfeldt, E., and H. J. Dillinger, *Chem. Ber.*, **99**, 1588 (1966).
72. Richards, E. M., J. C. Tebby, R. S. Ward, and D. H. Williams, *J. Chem. Soc. (C)*, **1969**, 1542.
73. Hughes, A. N., and M. Davies, *Chem. Ind.*, **1969**, 138.
74. (a) Allen, D. W., J. C. Tebby, and D. H. Williams, *Tetrahedron Lett.*, **1965**, 2361; (b) Allen, D. W. and J. C. Tebby, *Tetrahedron*, **23**, 2795 (1967); (c) Richards, E. M. and J. C. Tebby, *Chem. Commun.*, **1967**, 957; (d) Richards, E. M. and J. C. Tebby, *Chem. Commun.*, **1969**, 494; (e) Hawes, M. and S. Trippett, *J. Chem. Soc. (C)*, **1969**, 1465.
75. Trippett, S., and B. J. Walker, *J. Chem. Soc. (C)*, **1966**, 887.
76. Shaw, M. A., J. C. Tebby, R. S. Ward, and D. H. Williams, *J. Chem. Soc. (C)*, **1968**, 2795.
77. Hoffmann, H., and H. J. Diehr, *Chem. Ber.*, **98**, 363 (1965).
78. Aguair, A. M., and K. C. Hansen, *J. Amer. Chem. Soc.*, **89**, 3067 (1967); **89**, 4235 (1967); Aguair, A. M., J. R. S. Irelan, G. W. Prejean, J. P. John, and C. J. Morrow, *J. Org. Chem.*, **34**, 2681 (1969).
79. Kirillova, K. M., V. A. Kukhtin, and T. M. Sudakova, *Dokl. Akad. Nauk. SSSR*, **149**, 316 (1963).
80. Pudovic, A. N., and O. S. Shulyndina, *Zh. Obshch. Khim.*, **38**, 2074 (1968); Mark, V., *Tetrahedron Lett.*, **1962**, 281; Pudovic, A. N. and I. M. Aladzheva, *Dokl. Akad. Nauk. SSSR*, **151**, 1110 (1963); Boiselle, A. P. and N. A. Meinhardt, *J. Org. Chem.*, **28**, 1828 (1962).
81. Shaw, M. A., J. C. Tebby, R. S. Ward, and D. H. Williams, *J. Chem. Soc.*, *(C)*, **1968**, 1609.
82. Waite, N. E., J. C. Tebby, R. S. Ward, and D. H. Williams, *J. Chem. Soc. (C)*, **1969**, 1100.
83. Waite, N. E., J. C. Tebby, R. S. Ward, M. A. Shaw, and D. H. Williams, *J. Chem. Soc. (C)*, **1971**, 1620; Kennard, O., W. D. S. Motherwell, and J. C. Coppola, *J. Chem. Soc. (C)*, **1971**, 2461.
84. Hendrickson, J. B., R. E. Spenger, and J. L. Sims, *Tetrahedron Lett.*, **1961**, 477.

85. Cooks, R. G., R. S. Ward, D. H. Williams, M. A. Shaw, and J. C. Tebby, *Tetrahedron*, **24**, 3289 (1968); Williams, D. H., R. S. Ward, and R. G. Cooks, *J. Amer. Chem. Soc.*, **90**, 966 (1968).
86. Hughes, A. N., and S. Uaboonkul, *Tetrahedron*, **24**, 3437 (1968).
87. Waite, N. E., and J. C. Tebby, *J. Chem. Soc. (C)*, **1970**, 386.
88. Harger, M. J. P., and S. Trippett, personal communication.
89. Cullen, W. R., and D. S. Dawson, *Canad. J. Chem.*, **45**, 2887 (1967).
90. Wittig, G., and E. Benz, *Chem. Ber.*, **92**, 1999 (1959).
91. Seyferth, D., and J. M. Burlitch, *J. Org. Chem.*, **28**, 2403 (1963).
92. Aguair, A. M., and T. G. Archibald, *Tetrahedron Lett.*, **1966**, 5471; **1966**, 5541.

Oligophosphonates

JOHN D. CURRY and D. ALLAN NICHOLSON

The Procter & Gamble Company, Miami Valley Laboratories, Cincinnati, Ohio

OSCAR T. QUIMBY

Retired, Cincinnati, Ohio

CONTENTS

I. Preparation

A. Introduction

This chapter is primarily a review devoted to low molecular poly-phosphonates containing *gem-* and/or *vic*-diphosphonate groups, namely, R_2O_3P—$\overset{|}{\underset{|}{C}}$—$PO_3R_2$ or R_2O_3P—$\overset{|}{\underset{|}{C}}$—$\overset{|}{\underset{|}{C}}$—$PO_3R_2$; it emphasizes preparation more than properties. In many cases these compounds have been prepared with ester reagents by either the familiar Michaelis-Arbuzov reaction or the Michaelis-Becker-Nylen reaction. The former involves reaction of a trialkyl phosphite with a dihaloalkane, yielding a tetraalkyl ester; the latter involves the reaction of a dihaloalkane or a haloalkylphosphonate with a metal dialkyl phosphite, often yielding a product with some of the alkyls displaced by metal ions. Additions of hydrogen dialkyl phosphites or of trialkyl phosphites to unsaturates have also been employed in phosphonate preparations.

Certain compounds can be prepared as the acids (or their condensates). This applies to 1-hydroxy-1,1-diphosphonates and 1-amino-1,1-diphosphonates, which can be made by a fourth method not generally recognized as a highly versatile synthetic tool. This method involves rearrangements of mixed carboxylic–phosphorous anhydrides to form condensates which will hydrolyze to one of the above acids.

Because the syntheses of several of the compounds discussed in this review have not been published, the section on phosphonation via mixed anhydrides is supported by an experimental section (located in Appendix I) reporting original work.

B. Phosphonation by Arbuzov Reaction

A recent review (53) examined several aspects of the Arbuzov reaction critically, but, understandably, gave little attention to its use in making polyphosphonates.

1. METHYLENEDIPHOSPHONATES

Since derivatives of methylenediphosphonate were first prepared by Nylen in 1930 (100), several variations on the Arbuzov reaction have been tested in this preparation (17,36,76,138,146,149,165). Of these the Roy synthesis (138) is the most satisfactory. If triisopropyl phosphite and dibromomethane react under controlled conditions, yields of 80–90% of $CH_2(PO_3R_2)_2$ may be obtained. The isopropyl bromide formed in such a reaction will not isomerize the triisopropyl phosphite in competition with the desired reaction. Temperatures above 185° are to be avoided, for propylene is eliminated and the acid results.

Pyrolysis of isopropyl esters (20) has proved quite useful in making *gem*-diphosphonic acids. This pyrolysis is acid catalyzed and is aided by electron-withdrawing groups attached to the bridge carbon. The acids may be reesterified to the desired ester with trialkyl orthoformates (32). Any phosphonate ester may be hydrolyzed with concentrated mineral acids in water, but often it is difficult to remove all of the residual water from the resulting acids. Recently the hydrolysis of tetraethyl methylenediphosphonate has been studied quantitatively by Bel'skii and Kudryavtseva (6).

2. OTHER *GEM*-DIPHOSPHONATES

Neidlein and co-workers (98) have carried out an Arbuzov reaction on N-dichloromethylenebenzene sulfonamide (ϕSO_2NCCl_2) and prepared the corresponding diphosphonates in 95–96% yields. A similar p-chlorophenyl derivative is claimed (85).

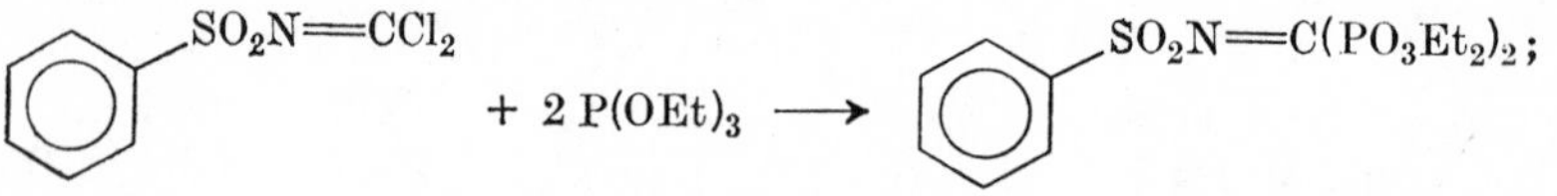

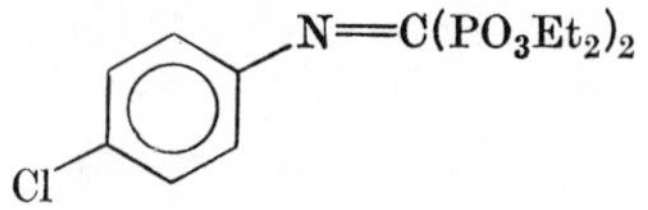

3. VICINAL DIPHOSPHONATES

When triethyl phosphite reacts with dichloroacetylene, tetraethyl acetylenediphosphonate ($:CPO_3Et_2)_2$ results (62). Using trimethyl phosphite the two-step synthesis occurred in 88% yield for the monophosphonate and 72% for the second step (145). Tetramethyl acetylenediphosphonate will add to cyclopentadiene, in a Diels-Alder manner, to yield the norbornadienyl compound.

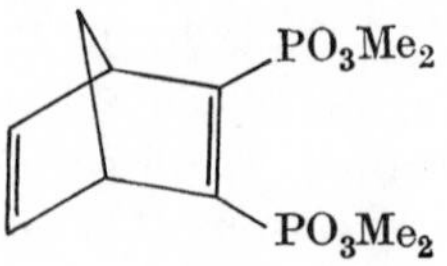

Similarly the diphosphonate adds to 1,3-cyclohexadiene; this product converts, via ethylene loss, to the phenylene compound at 150°. Acetylenediphosphonate reacts with diazomethane to yield a 2-pyrazoline (145).

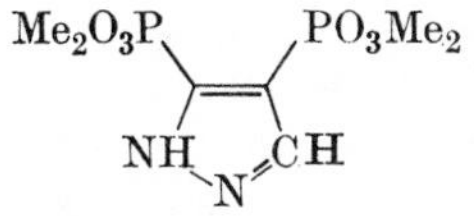

$$Me_2O_3P\diagdown\diagup PO_3Me_2$$

NH_N CH

A related study (160) involves the addition of tetraethyl *trans*-ethene-1,2-diphosphonate (cf. section on hydroxydiphosphonate esters) to dienes to yield materials of the type:

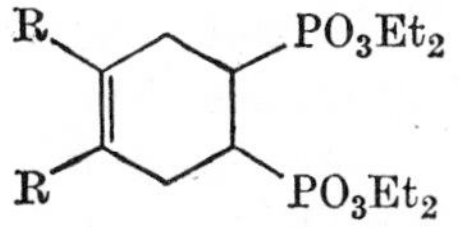

R. — PO_3Et_2

R — PO_3Et_2

The cyclohexenyl diphosphonate may be reduced to the corresponding cyclohexane derivative. Tavs (160) believes that the phosphorus groups exist in the *trans* configuration relative to the saturated ring.

Obrycki and Griffin have reacted trimethyl phosphite with isomeric bromoiodobenzenes and related materials to prepare phenylenediphosphonates using photolysis. The ortho, meta, and para isomers were obtained (101).

Phosphonates of the type $R_2O_3P(CH_2)_nPO_3R_2$, $n = 1$ or 2, may be made from an omega-chloromonophosphonate and a trialkyl phosphite (33).

An extensive series of 1,2-diphosphonoethane derivatives has been prepared by Frank (37,38). Trialkyl phosphites react with 1,2-dichloroperfluorocycloalkenes to give diphosphonates:

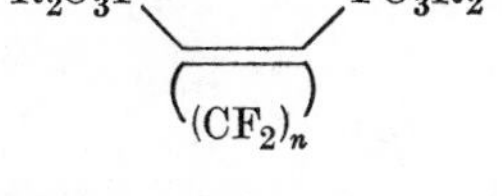

$$R_2O_3P\diagdown\diagup PO_3R_2$$

(CF_2)_n

$n = 2, 3, 4$

$R = CH_3, C_2H_5$, etc.

Unlike the dichloroacetylene case mentioned previously, no monophosphonate was isolated. One phosphonate moiety, however, is removed by phosphorus pentachloride, thus providing a good way to obtain the elusive monophosphonate (38). The diphosphonate ($n = 3$, $R = C_2H_5$) has a phosphorus chemical shift of $+1.5$ ppm consistent with the highly electron-withdrawing perfluorinated ring (173). These diphosphonates, interestingly, usually show little evidence for $C=C$ unsaturation in the infrared (38).

4. TRIPHOSPHONATES OF TYPE $RC(PO_3R_2)_3$

Burn and co-workers (15) have reported the synthesis of $C_6H_5C(PO_3Et_2)_3$ from benzoic anhydride and triethyl phosphite. For steric reasons such a structure seems rather unlikely; the product would involve less crowding around the bridge carbon if it were a phosphate-diphosphonate, for example, $C_6H_5C(PO_3Et_2)_2OPO_3Et_2$.

Similarly, numerous 1,1,1-triphosphonates have been reported in the patent literature as being derived from trichloromethyl groups and trialkyl phosphites (7,157–159). More structural evidence is needed on the products to make this convincing. Since rather forcing conditions had to be used in making the tetraisopropyl methylenediphosphonate (138), it seems unlikely that a third bulky phosphonate group could be placed on the bridge carbon. Nevertheless, an attempt was made to get the triphosphonate from the reaction of bromoform and triisopropyl phosphite (175,176); no phosphorus species was detected in ^{31}P NMR spectra on such products at chemical shifts appropriate for the triphosphonate.

C. Derivatives Made from Diphosphonates

Several new *gem*-diphosphonates have been made by halogenation or metalation at the bridge carbon of tetraalkyl methylenediphosphonates. The former were converted to hydroxy or carbonyl derivatives by hydrolysis. The latter were derivatized by alkylation.

1. BY HALOGENATION

Several halogenated methylenediphosphonates have been cited in the literature. Bunyan and Cadogan (14a) have reported the formation of $Cl_2C(PO_3Et_2)_2$ from the reaction of Cl_3CBr and $P(OEt)_3$; this is not useful as a preparative method as there are several products. Burn and co-workers (16) have suggested $ClCH(PO_3Et_2)_2$ as an intermediate (unisolated) in the reaction of triethyl phosphite with chloroform; $CH(PO_3Et_2)_3$, the potential product, was not demonstrated.

Reaction of molecular halogen with the sodium carbanion of tetraisopropyl methylenediphosphonate does not yield exclusively the monohalo derivative (131); the product is a ternary mixture of the parent ester and its mono- and dihalogenated products. Alteration of the method of combining reagents fails to alter product composition appreciably.

For materials of the type $X_2C(PO_3R_2)_2$, $X = Cl$, Br, or I and $R = CH_3$, C_2H_5, or iC_3H_7, the most convenient method (131,170) involves stirring the ester with an excess of aqueous sodium hypohalite, the latter added slowly unless the reaction mixture is cooled. All six dichloro and dibromo esters have been prepared. With the isopropyl ester no added solvent is necessary; in the other two cases an organic solvent, such as chloroform or carbon tetrachloride, should be used to prevent hydrolysis of the lower esters which are more water soluble (131). The dichloro and dibromo esters may be distilled *in vacuo*, using care to prevent pyrolysis in the case of isopropyl esters. None of the iodo derivatives was isolated in the pure state; they appear to be quite unstable.

With the isopropyl ester it is possible to obtain products with the concentration of monohalo derivative at 55–60% (131). To do this the enhanced reactivity of the remaining methylene hydrogen must be offset by adding electrolyte to repress its solubility in the aqueous phase.

If one pyrolyzes tetraisopropyl dihalomethylenediphosphonate to the acid, adds sodium hydroxide, and refluxes the solution (pH $\sim$ 11), yellow carbonyldiphosphonate, $OC(PO_3Na_2)_2$, is obtained (130). By lowering the solution pH to 4.5, colorless $(HO)_2C(PO_3HNa)_2$ is obtained; this conversion is reversible. Reduction of $OC(PO_3Na_2)_2$ to $HC(OH)(PO_3Na_2)_2$ may be carried out; the methanehydroxydiphosphonate is purified by crystallization as $HC(OH)(PO_3HNa)_2$. Reaction of phosgene with trialkyl phosphites does not yield carbonyldiphosphonate (67,130). The corresponding reaction of thiophosgene with triethyl phosphite was reported to yield $SC(PO_3Et_2)_2$, but proof of structure was not given (12). In another patent (8a) Birum describes the preparation of thiophosphate–diphosphonates from thiophosgene, trialkyl phosphites, and certain active halogen compounds by a two-step process.

2. BY METALATION

Esters of methylenediphosphonic acid may be metalated to give the carbanion $M[CH(PO_3R_2)_2]$. Metals such as potassium or sodium have been used (75,131); bases such as butyl lithium, sodium hydride, and potassium *t*-butoxide will function (57,131). The carbanion has been alkylated to give a wide range of derivatives (see following section).

The reaction to obtain the carbanion is easy to execute, especially with isopropyl esters. Temperature control below 25° should be observed to guard against undesirable side reactions. Once prepared, the carbanion is fairly stable. Yields of over 90% are not unusual (131).

Several carbanion salts have been isolated. Baldeschwieler and co-workers (5) isolated $Na[CH(PO_3Et_2)_2]$; this material is a white hygroscopic solid, which is a hexamer in carbon tetrachloride (low temperature) and benzene. The ^{1}H NMR spectra for the ester and its sodio-carbanion appear in Table I. These NMR data are consistent with a strong P-P coupling in the sodio derivative. Cotton and Schunn (24) isolated the potassium salt of this ester and reported it as more reactive than the sodium salt; molecular weight measurements in benzene again indicate a high level of association. Calcium amalgam and calcium hydride are not sufficiently strong bases to form an analogous derivative.

Brophy and Gallagher (14) have reported the proton spectrum of tetramethyl methylenediphosphonate and its sodium carbanion. These results, like those for the ethyl derivative, indicate a high level of P-P interaction in the carbanion.

Quimby and co-workers (131) have prepared the sodio-isopropyl derivative; since it is the most soluble of the three in solvents like toluene, there is less danger of interrupting a reaction of the carbanion by its precipitation. However, the carbanion of the methyl ester appears to be more reactive. The ^{1}H NMR spectrum has been studied by Siddall (150) and by Quimby and co-workers (131). Typical data for all three esters and their carbanions appear in Table I. For all three materials note the shift upfield for H_α upon carbanion formation. A magnetic nonequivalence of the methyl groups in the isopropoxy moieties has been used to explain the doubling for H_γ (131).

Horner and co-workers (59) have shown that the vicinal diphosphonate, $Et_2O_3PCH_2CH_2PO_3Et_2$, cleaves upon attempted metalation with the base potassium t-butoxide and, unlike the phosphine oxide, does not give olefin as the product with carbonyl compounds.

3. BY ALKYLATION

The carbanion of methylenediphosphonate has been derivatized by alkylation using various materials. Kosolapoff (75) was the first to demonstrate this reaction, using the potassium carbanion of tetraethyl methylenediphosphonate to prepare tetraethyl pentane-1,1-diphosphonate. Numerous other derivatives have been reported (24,34,54, 131,132,132a,150,170); most of them are included in the table of physical properties given in Appendix II. Monoalkylate is seldom the sole product, but it usually accounts for about 80% of the product. The remainder is methylenediphosphonate and dialkylate $R_2C(PO_3R_2)_2$; the former comes from base exchange between starting carbanion and monoalkylate, the latter from alkylation of the carbanion of monoalkylate.

TABLE I
Proton Magnetic Resonance Data[a]

Species	Proton	Ester Multiplicity	Ester τ	Ester J (Hz) H–H	Ester J (Hz) H–P	Carbanion Multiplicity	Carbanion τ	Carbanion J (Hz) H–H	Carbanion J (Hz) H–P
Tetramethyl methylene-diphosphonate	α	3	7.34		21.0	3[b]	9.07		6.5
	β	2	6.26		10.5	3	6.37		11.0[c]
Tetraethyl methylene-diphosphonate	α	3	7.71	0	20.3	3	9.52	0	7.4
	β	~5	5.95	6.9	8.3	~9	6.13	7.1	7.2[c]
	γ	3	8.67	6.9	0	3	8.78	7.1	0
Tetraisopropyl methylene-diphosphonate	α	3	7.68	0	20.75	3[b]	(8.87)[b]	0	(6.75)[b]
	β	14	5.25	6.25[e]	8.0[e]	≥14	5.46	f	f
	γ	4	{8.65, 8.67}	6.5	0	4	{8.60, 8.72}	6.0	0
			Δτ = 0.017 ppm				Δτ = 0.12 ppm		

$$\overset{O}{\underset{\uparrow}{}}\quad \overset{O}{\underset{\uparrow}{}}$$

$$P_2 \!-\! \underset{H_\alpha}{C} \!-\! P_1 \!-\! O \!-\! \underset{H_\beta}{C} \!-\! \underset{H_\gamma}{C}$$

[a] Numbering system refers to [structure]. All τ values are relative to TMS (10.00 ppm) as internal standard.

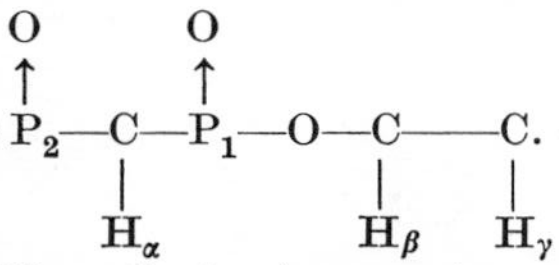

J values are accurate to ±0.25 Hz. Carbanion spectra were measured in benzene.

[b] Not easily resolved. The proximity of the H_γ absorption to the H_α absorption makes this assignment somewhat tenuous for the isopropyl derivative.

[c] Assuming $J(H_\beta - P_2) = 0$ and $J(P_1 - P_2)$ is large.

[d] Ref. 5.

[e] Decoupling used to make assignments.   [f] Decoupling reveals a broad multiplet.

An interesting variation in the predicted product of alkylation occurs with carbon tetrachloride. The initial product, $Cl_3CCH(PO_3R_2)_2$, dehydrochlorinates to give the vinyl derivative $Cl_2C{=}C(PO_3R_2)_2$ (131).

Upon reaction of the carbanion of $CH_2(PO_3Et_2)_2$ with various aldehydes, Henning and Gloyna (57) found a vinyl monophosphonate as a main product.

$$M[CH(PO_3Et_2)_2] + RCHO \longrightarrow MO\overset{\overset{O}{\|}}{P}(OEt)_2 + RCH{=}CHPO_3Et_2$$

Ethane-2-hydroxy-1,1,2-triphosphonate may be prepared from the carbanion of methylenediphosphonate by the following sequence (132,176,177):

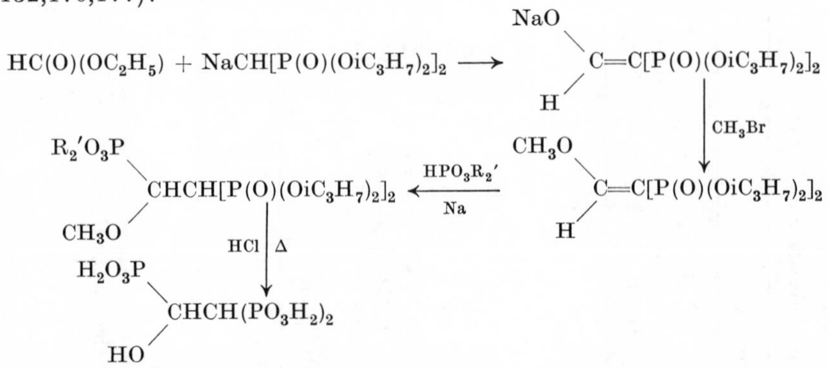

$$HC(O)(OC_2H_5) + NaCH[P(O)(OiC_3H_7)_2]_2 \longrightarrow \underset{H}{\overset{NaO}{\diagdown}}C{=}C[P(O)(OiC_3H_7)_2]_2$$

$$\Big\downarrow CH_3Br$$

$$\underset{H}{\overset{CH_3O}{\diagdown}}C{=}C[P(O)(OiC_3H_7)_2]_2 \xrightarrow[\ \ Na\ \ ]{HPO_3R_2'} \underset{\overset{CH_3O}{\diagup}}{\overset{R_2'O_3P}{\diagdown}}CHCH[P(O)(OiC_3H_7)_2]_2$$

$$\Big\downarrow {\scriptstyle HCl\ \Delta}$$

$$\underset{\overset{HO}{\diagup}}{\overset{H_2O_3P}{\diagdown}}CHCH(PO_3H_2)_2$$

No report was found in the literature dealing with attempted alkylation of ethane-1,2-diphosphonates, possibly because attempted metalation of such diphosphonates causes cleavage (59).

4. BY REDUCTION

Tetraalkyl esters of methylene and ethylenediphosphonic acids may be reduced with lithium aluminum hydride to the diphosphines, $H_2P(CH_2)_nPH_2$ (54,84,169). Hays and Logan (54) have extended the study to include alkylated methylenediphosphonates, using the acid chlorides as starting materials. The reduction of $CH_2(POCl_2)_2$ would seem to be preferred over reduction of the ester because of the possibility of metalating the latter.

5. DIPHOSPHONIC TETRACHLORIDES

Reaction of a mixture of methylenediphosphonic acid and a tetraalkyl methylenediphosphonate with phosphorus pentachloride yields the diphosphonic tetrachloride (54,79,83,136,165), there being more tendency toward chlorination at the bridge carbon if ester alone is used (cf. also Ref. 103). This procedure is also applicable to the ethylene derivative (79,83,104). Hays and Logan (54) have also reported that α-chlorination may occur during the formation of 1,1-pentylidenediphosphonic tetrachloride. The acid chlorides react with alcohols as expected (149), and with P_4S_{10} to give the thio acid chlorides, $Cl_2(S)P(CH_2)_nP(S)Cl_2$ (83).

6. DIPHOSPHONIC ANHYDRIDES

The anhydride of ethylenediphosphonic acid has been prepared by controlled hydrolysis of the tetrachloride (106); upon treatment with alcohols, the anhydride yields dialkyl esters, $HRO_3PCH_2CH_2PO_3RH$.

7. BIOCHEMICAL STUDIES

Several investigators (35,95,97) have prepared and tested the biological activity of a phosphate–diphosphonate anhydride in which a methylenediphosphonate unit replaces a pyrophosphate unit in adenosine triphosphate (ATP) to give AMDP whose acid has the structure:

$$\text{adenosine-O}\overset{\overset{\displaystyle O}{\|}}{\underset{\underset{\displaystyle H}{O}}{P}}\text{O}\overset{\overset{\displaystyle O}{\|}}{\underset{\underset{\displaystyle H}{O}}{P}}\text{CH}_2\overset{\overset{\displaystyle O}{\|}}{\underset{\underset{\displaystyle H}{O}}{P}}\text{OH}$$

Flesher, Oester, and Myers (35) found that the vasodepressor action of AMDP is greater than that of ATP. This methylenediphosphonate derivative has also been studied in other systems (94,151,152,156).

The synthesis of 5'-guanylyl-methylenediphosphonate, adenosine-5'-methylenediphosphonate, and the phosphonate derivative of arabinofuranosylcytosine have been reported (58,96,168). Related ^{31}P NMR data appear in Ref. 86.

Various phosphorus-containing materials have been studied as potential inhibitors of beef-heart succinic dehydrogenase (137), among them methylenediphosphonic and ethane-1,2-diphosphonic acids. Both phosphonates are less potent than pyrophosphate.

D. Phosphonation by Nylen Reaction

This type of reaction, namely, $-\overset{|}{\underset{|}{C}}X + MOP(OR)_2 \rightarrow -\overset{|}{\underset{|}{C}}PO_3R_2 +$

MX, with X = a halogen and M = an alkali metal, is variously called the Michaelis-Becker, the Michaelis-Becker-Nylen or the Nylen reaction. In the context of this review, the latter name is appropriate because Nylen was the first to employ a salt of a hydrogen dialkyl phosphite in the synthesis of a diphosphonate from an organic dihalide (100).

1. GEMINAL DIPHOSPHONATES

Nylen described the reaction of sodium diethyl phosphite with methylene diiodide, yielding disodium diethyl methylenediphosphonate without first preparing the intermediate iodomethylphosphonate (100); no tetraethyl ester was isolated. Using sodium diallyl phosphite and

$$2\,NaOP(OC_2H_5)_2 + CH_2I_2 \rightarrow CH_2[P(O)(OC_2H_5)ONa]_2 + 2\,C_2H_5I$$

the same dihalide, Cade (17) obtained the disodium diallyl methylenediphosphonate; substitution of the dibromide for the diiodide gave the same product. Arbuzov and Kushkova (1) could not, however, isolate the diphosphonate upon reaction of sodium diethyl phosphite with diethyl iodomethylphosphonate. Schwarzenbach and Zurc (142) were able to prepare tetraalkyl methylenediphosphonates from sodium dialkyl phosphites and dialkyl chloromethylphosphonates. Several

$$(RO)_2PONa + ClCH_2PO_3R_2 \rightarrow CH_2(PO_3R_2)_2 + NaCl$$

other workers have successfully carried out such syntheses (56,91,103, 136). In some instances the use of the potassium salt gave a higher yield (91,93).

Sodium dialkyl phosphites have been reacted with other reagents, mostly substituted dihalides, to yield *gem*-diphosphonates. When sodium diethyl phosphite is allowed to react with $Br_2C(SO_2C_2H_5)_2$, Arbuzov and Bogonostseva (4) report that tetraethyl methylenediphosphonate is formed in low yield. Phosgene reacts with excess sodium dialkyl phosphite to give a phosphate-*gem*-diphosphonate (130).

In some rather recent work Gross and Costisella (49–51) have allowed hydrogen diethyl phosphite to react with the dimethyl acetal of dimethylformamide, forming the dimethylamino derivative of methylenediphosphonate:

$$2\,HPO_3Et_2 + (CH_3O)_2CHN(CH_3)_2 \rightarrow (CH_3)_2NCH(PO_3Et_2)_2 + 2\,CH_3OH$$

The aminodiphosphonate reacts with sodium hydride and aldehydes to yield diethyl 1-dimethylaminoalkenylphosphonates:

$$RCH{=}C\begin{array}{l} \diagup PO_3Et_2 \\ \diagdown N(CH_3)_2 \end{array}$$

Since the monophosphonate is readily converted to the carboxylic acid, RCH_2COOH, this reaction sequence provides a way of converting aldehydes to acids with one more carbon. The amino diphosphonate is readily converted to a quaternary salt, which, in turn, may be converted to a stable N-ylide by treatment with aqueous potassium carbonate solution.

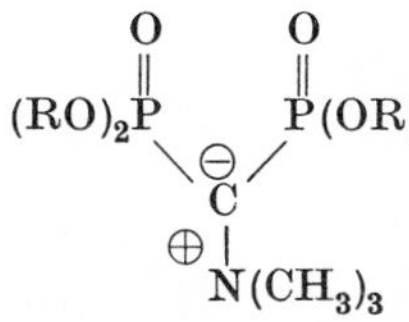

$$(RO)_2\overset{\overset{O}{\|}}{P} \quad \overset{\overset{O}{\|}}{P}(OR)_2$$
$$\underset{\underset{\underset{N(CH_3)_3}{|}}{\overset{\oplus}{}}}{\overset{\ominus}{C}}$$

2. VICINAL DIPHOSPHONATES

The Nylen reaction has been extended to the synthesis of 1,2-diphosphonates. The first report of the reaction of 1,2-dihaloethanes with sodium dialkyl phosphites to make tetraalkyl ethane-1,2-di-phosphonate was made by Boyer and Mangham in 1953 (11). Other workers have confirmed this report (33,91,103). Another patent (93) indicates that when using a salt of hydrogen diethyl phosphite, potassium is sometimes superior to sodium; sodium works well with hydrogen dibutyl phosphite (74a). Yields range from 50 to 90%.

In a rather unique reaction, two patents (155,163) describe the preparation of diphosphonates from esters of carboxylic acids and metal dialkyl phosphites. For example, sodium diethyl phosphite and β-chloroethyl myristate yield tetraethyl ethane-1,2-diphosphonate, thus:

$$ClCH_2CH_2O\overset{\overset{O}{\|}}{C}(CH_2)_{12}CH_3 + 2\ NaOP(OEt)_2 \longrightarrow$$
$$(CH_2PO_3Et_2)_2 + NaOOC(CH_2)_{12}CH_3 + NaCl$$

Substituted 1,2-dihaloethanes may or may not give a 1,2-diphos-phonate upon reaction with sodium dialkyl phosphites. Arbuzov and Lugovkin (3) could not isolate a diphosphonate from the reaction of $BrCH_2CHBrCN$ with $NaOP(OEt)_2$. Using methyl 2,3-dichloropro-pionate, Kamai and Kukhtin (71) were able to isolate the corresponding vicinal diphosphonate, but the dibromo ester gave an undistillable product.

$$2\ (EtO)_2PONa + ClCH_2\overset{\overset{Cl}{|}}{C}HCOOCH_3 \longrightarrow 2\ NaCl + Et_2O_3PCH_2\overset{\overset{COOCH_3}{|}}{C}HPO_3Et_2$$

3. TRIPHOSPHONATES OF TYPE $RC(PO_3R_2)_3$

Birum claimed the preparation of a thiophosphate-triphosphonate in accordance with the following reaction (7):

$$Cl_3CSCl + 4\,NaOP(OR)_2 \xrightarrow[80°]{benzene} (R_2O_3P)_3CSPO_3R_2 + 4\,NaCl$$

No structure proof was offered.   It is doubtful that such a crowded species could be made, especially under the relatively mild conditions described.

Attempts were made by Roy (177) to prepare such triphosphonates by both of the following reactions (R = isopropyl):

$$(RO)_2PONa + BrCH(PO_3R_2)_2 \nrightarrow HC(PO_3R_2)_3 + NaBr$$

$$R_2O_3PCl + NaCH(PO_3R_2)_2 \nrightarrow HC(PO_3R_2)_3 + NaCl$$

Neither reaction gave more than traces of a phosphorus species whose position in the ^{31}P NMR spectrum was appropriate for the triphosphonate; nor was such a species concentrated by fractional distillation under pressures of 50μ or less.

In this connection it is well to recall the alkylation of methylenediphosphonate by reaction of its carbanion with isopropyl chloroformate.

$$RO_2CCl + NaCH(PO_3R_2)_2 \rightarrow NaCl + RO_2CCH(PO_3R_2)_2$$

The pentaisopropyl ester was stable enough to allow partial purification to 90–95% by vacuum distillation (131), but the acid readily decarboxylates, leaving methylenediphosphonic acid.

E. Phosphonation by Addition of Hydrogen Dialkyl Phosphites

Di- and higher polyphosphonates result from the addition of hydrogen dialkyl phosphites to olefins, olefin oxides, allenes, acetylenes, and carbonyl compounds; the latter are discussed in the next section (hydroxydiphosphonates).   Such reactions are generally catalyzed by bases such as alkali metal alkoxides and alkali metals; in the case of activated acetylenes addition will occur simply on heating.

1. TO OLEFINS

One of the earliest reports of a preparation of a diphosphonate via this type of reaction concerned the addition of a mixture of hydrogen and sodium dibutyl phosphite to allyl bromide (139); a high-boiling material was obtained which was proposed to be either 1,2- or 1,3-diphosphonopropane.   Schwarzenbach and co-workers (143) used a mixture of hydrogen and sodium diethyl phosphite with allyl chloride; they obtained essentially a single diphosphonate whose acid dissociation constants corresponded with those of a 1,2 isomer.

$$CH_2{=}CHCH_2Cl + HPO_3Et_2 + NaOP(OEt)_2 \longrightarrow NaCl + CH_3\underset{\underset{PO_3Et_2}{|}}{CH}CH_2PO_3Et_2$$

Pudovik and Frolova (113) later showed that when the phosphite was added exclusively as sodium diethyl phosphite to allyl bromide, the major product was again the 1,2 isomer, but a minor amount of the 1,3-diphosphonate also formed. The reaction is also discussed by Petrov and co-workers (104).

The propensity of allylic halides to form both 1,2- and 1,3-diphosphonates on reaction with sodium dialkyl phosphite is further demonstrated by the work of Pudovik and Arbuzov (109) on $CH_3OCH_2CH_2CH{=}CHCH_2Cl$; on addition of $NaOP(OR)_2$ the 1,2 and 1,3 isomers were formed in approximately 85:15 ratio.

A number of workers have discussed the preparation of tetraalkyl ethane-1,2-diphosphonates from vinylphosphonate esters (81,111,115). Pudovik and co-workers (126) have described the similar addition of hydrogen dialkyl phosphites to α-cyanovinylphosphonate esters.

Kreutzkamp and Schindler (80) reported the formation of a *gem*-diphosphonate from the base-catalyzed addition of hydrogen dibutyl phosphite to dibutyl β-acetylvinylphosphonate, but no structure proof was given. From spatial considerations the addition at C=C might be expected to yield a 1,2-diphosphonate; also there is the possibility of addition at the C=O.

$$\overset{\overset{O}{\|}}{CH_3C}CH{=}CHPO_3Bu_2 + H{-}PO_3Bu_2 \longrightarrow \overset{\overset{O}{\|}}{CH_3C}CH_2CH(PO_3Bu_2)_2$$

A vinylphosphonate is presumably an intermediate in the reported synthesis of a *vic*-diphosphonate from $C_6H_5CH{=}CHBr$ and two equivalents of hydrogen dialkyl phosphite (3).

1,3-Butadiene-2,3-diphosphonate adds one equivalent of hydrogen diethyl phosphite to make the 1,2,3-triphosphonate (127).

$$CH_2{=}\underset{\underset{PO_3Et_2}{|}}{\overset{\overset{PO_3Et_2}{|}}{C}}{-}C{=}CH_2 + HPO_3Et_2 \longrightarrow Et_2O_3PCH_2{-}\underset{\underset{PO_3Et_2}{|}}{\overset{\overset{PO_3Et_2}{|}}{CH}}{-}C{=}CH_2$$

2. TO OLEFIN OXIDES

Preis and co-workers (108) reported that the reaction of propylene oxide with excess sodium diethyl phosphite yields tetraethyl propane-1,2-diphosphonate.   Presumably   the β-hydroxypropylphos-

$$CH_3-\underset{\underset{H}{|}}{\overset{\overset{O}{\diagup \diagdown}}{C}}-CH_2 + 2\,NaOP(OEt)_2 \longrightarrow CH_3-\underset{\underset{PO_3Et_2}{|}}{CH}-CH_2PO_3Et_2$$

phonate intermediate undergoes dehydration followed by addition of a
second mole of phosphite to the generated double bond (see Ref. 22a
for a similar dehydration).

3. TO ALLENES

There have been two reports of hydrogen dialkyl phosphite addition
to allenes.   Both concern reaction of phosphites with γ,γ-dimethyl-
allenylphosphonates.   Pudovik and co-workers first reported in 1964
(120), and later in 1966 (125), that unsaturated 1,2-diphosphonates
resulted.   No other isomers were described.

4. TO ACETYLENES

In   1950   Pudovik   (110)   reported   that   attempted   reaction   of
$(CH_3)_2\underset{\underset{Cl}{|}}{C}-C\equiv CH$ with sodium diethyl phosphite gave a mixture of
unidentified products boiling in the correct range for di- and tri-
phosphonates.   Two years later he studied the reaction of esters of
acetylenedicarboxylic acids with hydrogen dialkyl phosphites catalyzed
by sodium alkoxide and successfully prepared ethane-1,2-dicarboxy-
1,2-diphosphonate esters (114).   In our hands this reaction has been
induced to proceed at 130° with no added catalyst.   In 1954 Pudovik
(116) also reported the addition of hydrogen dialkyl phosphites to
$C_6H_5C\equiv CCOOR$ to yield the corresponding vic-diphosphonates.

A 1963 patent (90) described the preparation of dialkyl 2-carbo-
alkoxyvinylphosphonates from alkyl propiolates and hydrogen dialkyl
phosphites.   When excess phosphite is employed the 2:1 adduct is
obtained.   The   structure   suggested   is   the gem-diphosphonate
$ROOCCH_2CH(PO_3R_2)_2$, but no evidence is presented to substantiate
the assignment or to rule out the vicinal isomer.

Saunders and Simpson prepared diisopropyl ethynylphosphonate and
successfully added two equivalents of hydrogen diisopropyl phosphite
to form hexaisopropyl ethane-1,1,2-triphosphonate (140).   More recent
reports of this type of reaction are again from the laboratories of
Pudovik (121,122).   He and his co-workers have studied the reactions
of $CH_3C\equiv CPO_3R_2$ with hydrogen dialkyl phosphites.   Addition of one

equivalent of phosphite yields the unsaturated *vic*-diphosphonate; two equivalents of phosphite led to a triphosphonate.

The most recent acetylene/phosphite addition is reported by Cilley, Nicholson, and Campbell (22a). It involves the base-catalyzed reaction of acetylenic alcohols with hydrogen dialkyl phosphites to produce *vic*-polyphosphonates. The equation for the formation of the propane-1,2,3-triphosphonate follows. Butane-1,2,3,4-tetra-

$$HC{\equiv}CCH_2OH + 3\ HPO_3R_2 \xrightarrow[\Delta]{Na} R_2O_3PCH_2{-}CH{-}CH_2PO_3R_2$$
$$\underset{PO_3R_2}{|}$$

phosphonate was also prepared from 1,4-dihydroxy-2-butyne.

F. Phosphonation by Addition of Trialkyl Phosphites

So far trialkyl phosphites have been added only to acetylenes, making *vic*-diphosphonates in each case. In 1950 Pudovik reported (110) that $(CH_3)_2\overset{\overset{\displaystyle Cl}{|}}{C}{-}C{\equiv}CH$ reacted with tributyl phosphite to form a dialkyl allenylphosphonate initially; this intermediate then underwent further reaction with the phosphite to form the diphosphonate.

$$(RO)_3P + (CH_3)_2\overset{\overset{\displaystyle Cl}{|}}{C}{-}C{\equiv}CH \xrightarrow{-RCl} (CH_3)_2C{=}C{=}\overset{\overset{\displaystyle H}{|}}{C}PO_3R_2 \xrightarrow{(RO)_3P}$$

$$(CH_3)_2C{=}\overset{\overset{\displaystyle R_2O_3P}{|}}{C}{-}\underset{\underset{\displaystyle PO_3R_2}{|}}{\overset{\overset{\displaystyle H}{|}}{C}}{-}R$$

Notice that addition of the second mole of phosphite occurred so as to yield the product wherein one of the alkyl groups R has migrated from oxygen to carbon.

Recently Griffin and Mitchell (48) described a similar migration in the reaction of trimethyl phosphite with dimethyl acetylenedicarboxylate. In this instance, two moles of trimethyl phosphite add

$$2\ (CH_3O)_3P + CH_3OOC{-}C{\equiv}C{-}COOCH_3 \longrightarrow CH_3OOC{-}\underset{\underset{\displaystyle (CH_3)_2O_3P}{|}}{\overset{\overset{\displaystyle CH_3}{|}}{C}}{-}\underset{\underset{\displaystyle PO_3(CH_3)_2}{|}}{\overset{\overset{\displaystyle CH_3}{|}}{C}}{-}COOCH_3$$

to the triple bond forming the *vic*-diphosphonate shown in the equation.

The only other reaction of this type which has been discussed differs somewhat in that the alkyl group of the phosphite does not migrate, but suffers elimination (47). For example when triethyl phosphite is heated with phenylacetylene, ethylene is evolved and an intermediate monophosphonate can be isolated.   In the presence of excess phosphite

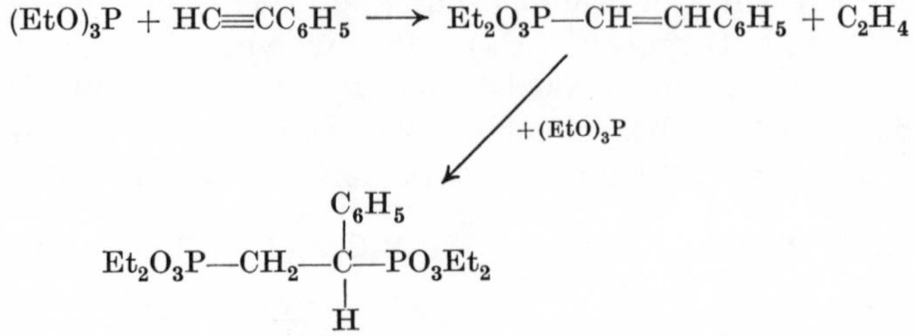

$$(EtO)_3P + HC\equiv CC_6H_5 \longrightarrow Et_2O_3P\text{—}CH\text{=}CHC_6H_5 + C_2H_4$$

$$+(EtO)_3P$$

$$Et_2O_3P\text{—}CH_2\text{—}\underset{\underset{H}{|}}{\overset{\overset{C_6H_5}{|}}{C}}\text{—}PO_3Et_2$$

more ethylene was evolved and the vicinal diphosphonate was formed.

G. Esters of Hydroxydiphosphonates (HDP)

These compounds are discussed in a separate section for three reasons: (1) their method of preparation in the literature before 1965 has involved two of the previously described methods of preparing phosphonates (combination of Arbuzov reaction and hydrogen dialkyl phosphite addition); (2) their preparation and chemical behavior provide a natural introduction for the next section on phosphonation by mixed anhydrides; (3) confusion exists in the literature regarding the preparation of these materials—for example, Hudson (61) was evidently unaware that these earlier claims (88,141) of having isolated $CH_3C(OH)(PO_3R_2)_2$ were discredited by later work (31,89,118).

1. 1-HYDROXY-1,1-DIPHOSPHONATES

The unsubstituted alkane-1-hydroxy-1,1-diphosphonic acids and their salts are quite stable compounds (see pp. 66–67 for a summary of the data available).   Such is not the case for the corresponding esters, however.   Fitch and Moedritzer (31) have described the rearrangement of these compounds to the corresponding phosphate-phosphonate.   Phosphorus NMR showed conclusively the transition

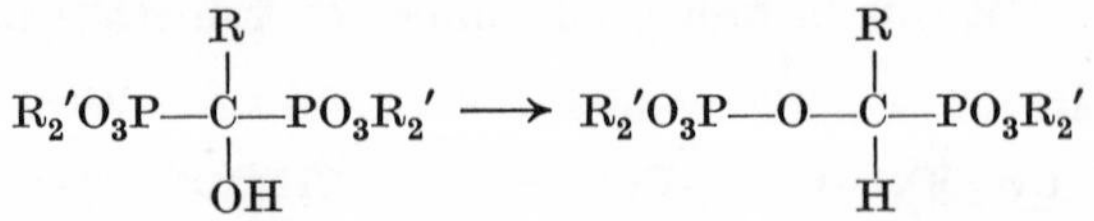

$$R_2{}'O_3P\text{—}\underset{\underset{OH}{|}}{\overset{\overset{R}{|}}{C}}\text{—}PO_3R_2{}' \longrightarrow R_2{}'O_3P\text{—}O\text{—}\underset{\underset{H}{|}}{\overset{\overset{R}{|}}{C}}\text{—}PO_3R_2{}'$$

from one type of phosphorus resonance for the diphosphonate to two equal resonances for the phosphate-phosphonate.   Pudovik and Konovalova (118) have presented chemical evidence for the same phenomenon.

Conversion of these HDPs to the phosphate-phosphonates is

catalyzed by heat and by base. Pudovik and co-workers (119) showed that sodium alkoxides and dialkyl amines promote rearrangement. When R is an alkyl, rearrangement of $RC(OH)(PO_3R_2')_2$ occurs without added base if the temperature exceeds 120° (129,174). Available evidence also suggests that the conversion occurs more readily when R is electron-withdrawing; for example, Fitch and Moedritzer (31) did not succeed in making $C_6H_5C(OH)(PO_3R_2)_2$, obtaining only the phosphate–phosphonate. By allowing the addition of hydrogen dimethyl phosphite to dimethyl benzoylphosphonate to proceed at 0° in the presence of dibutyl amine, we were able to prepare the tetramethyl ester of the HDP in high yield (174) (see Appendix I).

There are several reports in the literature dealing with the supposed preparation of HDP esters which, based on the preceding evidence (31,118), are probably not correct. A 1956 patent (141) describes HDP ester preparation from dialkyl acylphosphonates and hydrogen dialkyl phosphites catalyzed by amine bases. It is likely that the desired 1-hydroxy-1,1-diphosphonates were prepared by this method, but that they were rearranged during workup, which involved vacuum distillation at elevated temperatures (150–200°). McConnell and Coover (88) and Cade (18) employed dialkyl amines and sodium, respectively, to catalyze the addition of hydrogen dialkyl phosphites to acylphosphonates. Workup by distillation probably induced rearrangement. The former authors identified their product by assigning an IR band at 3400–3500 cm^{-1} to the presumed OH group. This assignment was criticized by Miller and co-workers (89) and it was later shown (31) that the authentic hydroxydiphosphonate has an OH absorption at a frequency of 3180 cm^{-1}.

A 1967 patent (65) describes the use of HDP esters in fire-retardant polyesters. The data presented are not sufficient to determine whether the esters described were rearranged or not.

2. ACETYLATED 1-HYDROXY-1,1-DIPHOSPHONATES

Acetylated hydroxydiphosphonates are also known, e.g., $RC(OAc)(PO_3R_2)_2$. Pudovik and co-workers (124,128) have prepared these materials via isomerization of the mixed anhydride

$$(EtO)_2P\overset{\displaystyle O}{\overset{\displaystyle \|}{-}}O-CCH_3$$

and from the reaction of acetyl chloride with tetraethyl ethane-1-hydroxy-1,1-diphosphonate in the presence of triethyl amine (117). Two earlier reports of the preparation of these materials (2,70) were shown to be incorrect (117). Reactions of acetyl chloride with excess of sodium diethyl phosphite (2) and of trialkyl phosphites with carboxylic acid anhydrides (70) were shown to yield phosphate-phosphonates rather than acylated hydroxydiphosphonates.

From molecular weight measurements on acetylphosphonate, Hall and Stephens (52) were led to suggest a reversible dimerization of acylphosphonates thus:

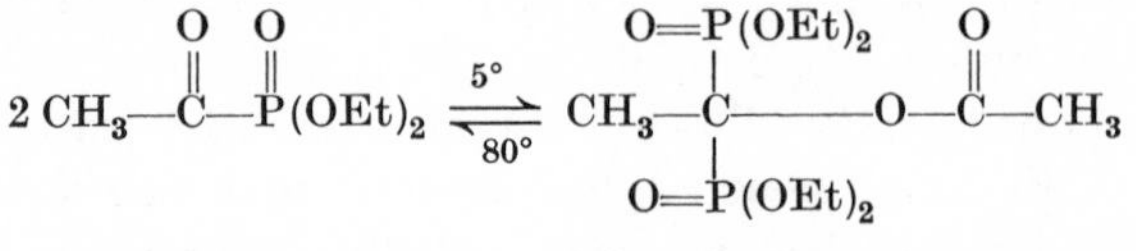

$$
2\ CH_3{-}\overset{\overset{O}{\|}}{C}{-}\overset{\overset{O}{\|}}{P}(OEt)_2 \underset{80°}{\overset{5°}{\rightleftharpoons}} CH_3{-}\underset{\underset{O=P(OEt)_2}{|}}{\overset{\overset{O=P(OEt)_2}{|}}{C}}{-}O{-}\overset{\overset{O}{\|}}{C}{-}CH_3
$$

This is unlikely because Fitch and Moedritzer (31) gave the ^{31}P NMR chemical shift for $CH_3C(O)PO_3Et_2$ as $+3$ ppm and that for $CH_3C(OH)(PO_3Et_2)_2$ as -20.8 ppm; our own experience (108a) shows that the acetylated esters, $CH_3C(OAc)(PO_3Et_2)_2$, should have a chemical shift which is about 4–6 ppm upfield from that for the hydroxy ester, i.e., at about -15 to -17 ppm. A ^{31}P NMR study on each ester over a range of temperatures would show whether the Hall-Stephens hypothesis is tenable or not.

3. PHOSPHATED 1-HYDROXY-1,1-DIPHOSPHONATES

Quimby and co-workers (130) have reported the reaction of phosgene with 3 moles of sodium diisopropyl phosphite to yield a phosphate–diphosphonate. The equation below shows a conceivable sequence,

$$
COCl_2 + 3\ NaOP(OR)_2 \longrightarrow [(R_2O_3P)_3COH] \xrightarrow{\text{base}} (R_2O_3P)_2\overset{\overset{H}{|}}{C}{-}OPO_3R_2
$$

i.e., formation of the methanehydroxytriphosphonate intermediate which rearranges under the reaction conditions. On products made at 10–20° and never warmed above room temperature the ^{31}P NMR spectra showed nearly all of the P under two peaks, the one at a chemical shift of -12 ppm (diphosphonate) having twice the area of the one at $+2$ ppm (phosphate). Distillation did not change the relative peak areas. Here again is evidence that three phosphonate groups can be crowded around a single carbon atom only with difficulty. If the hypothesized triphosphonate forms at all, it must have only fleeting existence.

4. PHOSPHITED 1-HYDROXY-1,1-DIPHOSPHONATES

In a recent patent Birum (8) has described the formation of phosphite–diphosphonates from PCl_3, an olefin oxide, and an acyl halide in a 3:8:1 molar ratio. With the epoxide as ethylene oxide and

$R = ClCH_2CH_2$ the equations become

$$3 PCl_3 + 8 C_2H_4O \longrightarrow 2 P(OR)_3 + ClP(OR)_2$$

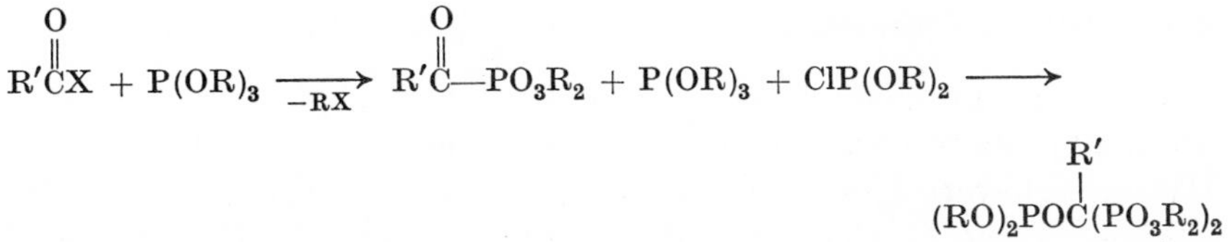

$$\overset{\overset{\textstyle O}{\|}}{R'CX} + P(OR)_3 \xrightarrow[-RX]{} \overset{\overset{\textstyle O}{\|}}{R'C}-PO_3R_2 + P(OR)_3 + ClP(OR)_2 \longrightarrow$$

$$\overset{\overset{\textstyle R'}{|}}{(RO)_2POC(PO_3R_2)_2}$$

Phosphorus NMR spectra for certain examples were consistent with the proposed structures. Rearrangement to $R'C(PO_3R_2)_3$ is unlikely on steric grounds, as already suggested in other cases.

5. 1-HYDROXY-1,2-DIPHOSPHONATES

Tavs (160) reported the formation of a *vic*-diphosphonate from reaction of phosphonoacetaldehyde with hydrogen dialkyl phosphite (catalyst = Et_3N).

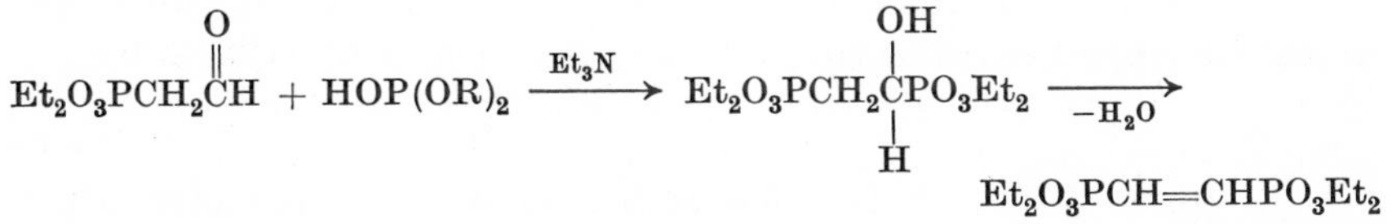

$$\overset{\overset{\textstyle O}{\|}}{Et_2O_3PCH_2CH} + HOP(OR)_2 \xrightarrow{Et_3N} Et_2O_3PCH_2\overset{\overset{\textstyle OH}{|}}{\underset{\underset{\textstyle H}{|}}{C}}PO_3Et_2 \xrightarrow{-H_2O}$$

$$Et_2O_3PCH{=}CHPO_3Et_2$$

H. Phosphonation by Mixed Anhydrides

Mixed carboxylic–phosphorous anhydrides may involve phosphorus which is either triply or quadruply bonded to other groups. Both kinds of P(III) reagents are capable of phosphonation. This and other trends stated in this section are based largely on the opinion of the authors, for little work on phosphonation by mixed anhydride chemistry has been published except in patents.

1. TRIPLY COORDINATED P

When an alkyl group of a trialkyl phosphite ester $P(OR)_3$ is replaced by an acyl (A) group to yield an ester that is partly an anhydride, for example, $(RO)_2POA$, heating causes preferential formation of a PC bond at the carbonyl carbon of the acyl, namely, $AP(O)(OR)_2$; the reaction of this intermediate with another mole of ester anhydride then yields a C-acylated 1-hydroxy-1,1-diphosphonate (123,124). The net reaction is $2(RO)_2POA \rightarrow R'C(OA)(PO_3R_2)_2$, where R' is the non-carbonyl part of the acyl group. Aside from this hybrid between ester and anhydride phosphonation, no example of *gem*- or *vic*-diphosphonate formation via mixed anhydrides containing exclusively triply co-ordinated phosphorus has come to our attention.*

2. QUADRUPLY COORDINATED P

While the number of groups coordinated to phosphorus in mixed acetic-phosphorous anhydrides was not known to von Baeyer and Hofmann, nevertheless, their 1897 paper (164), coupled with more recent findings, demonstrated that it is neither necessary to use mixed carboxylic–phosphorous anhydrides in which the phosphorus is triply coordinated, nor necessary to have any alkyl or other hydrocarbon radicals on oxygens attached to phosphorus. All that is necessary is that heat be applied to a mixed anhydride of carboxylic and phosphorous acids; the carboxyl group is then converted to a 1-hydroxy-1,1-diphosphonate group or a derivative thereof, thus allowing representation of the essential change by the equation:

$$RCOOH \rightarrow RC(OH)(PO_3H_2)_2$$

As will be shown later an additional phosphonate group can be put on the carbon adjacent to the HDP group only by sequential reactions. However, a phosphonate can be added simultaneously to a carbon farther away from the HDP group, provided that the reacting acyl and/or the resultant HDP condensate contains a suitably active unsaturated carbon-to-carbon bond (25a).

3. EARLY STUDIES OF CARBOXYLIC-PHOSPHOROUS ANHYDRIDES

When our studies began, the only HDP described in the literature was ethane-1-hydroxy-1,1-diphosphonate† (EHDP) (164). Henkel et Cie. patents on EHDP and related compounds began issuing at about this time (10).

Two closely related studies (13,161) have been published, but their relation to hydroxydiphosphonates was not recognized. Thus Brooks (13) not only made no reference to the HDP work of von Baeyer and Hofmann (164), but characterized the solid which precipitated at 50° from a liquid reaction mixture containing 5 g phosphorous acid, 40 ml

---

* The mixed anhydride triacetyl phosphite $(AcO)_3P$, made by Nerdel and Burghardt (99), would be expected to go even more readily through the same two steps as indicated for $(RO)_2POA$ by Pudovik and co-workers (123,124), yielding the fully acetylated product, $CH_3C(OAc)(PO_3Ac_2)_2$.

† Another name for this compound is 1-hydroxyethylidenediphosphonate. The writers prefer the "hydrocarbon" form of the name because the beginning of the name is then the same for di-, tri-, and tetraphosphonates of ethane.

acetic anhydride, and 10 ml acetyl chloride as "the mono-acetyl derivative of phosphorous acid."

Van Druten (161) cited von Baeyer and Hofmann but made no use of their findings, for he was concerned primarily with the yield of acetyl chloride (88%) from the reaction of phosphorus trichloride with acetic anhydride. He elucidated the structure of the solid, which remained after volatile materials were distilled under reduced pressure, only to the extent of showing that it could not be phosphorous oxide, P_4O_6.

4. REAGENTS FOR MAKING ETHANE-1-HYDROXY-1,1-DIPHOS-PHONIC ACID (EHDP)

Since EHDP has been more extensively studied than other hydroxy-diphosphonates, it will be used as the basis for describing the stages involved in HDP preparation by mixed anhydride chemistry. For making EHDP the carbon reagent either contains the acetyl (Ac) group, as in acetic acid, acetic anhydride, or acetyl chloride, or it will generate Ac upon preparation of the reaction mixture, as when ketene is bubbled through an anhydrous liquid containing phosphorous acid (molten or in solution). The phosphorus reagent has to be in the $+3$ oxidation state; if the anhydrizing power* is supplied by the carbon reagent, the P(III) reagent may be phosphorous acid, $HP(O)(OH)_2$. If acetic acid is used to supply the acetyl group, then the P(III) reagent must supply the anhydrizing links directly, for example, as P_4O_6, or indirectly, for example, as PCl_3. The anhydrizing power may be introduced partly as carbon and partly as P(III) reagent. For reasonably rapid reaction and high completeness of conversion to HDP derivatives, the combination of carbon and P(III) reagents should be chosen so as to yield a mixture having an initial AR* of 2 or more, but rarely more than 5.

* The anhydrizing power can be expressed quantitatively in terms of the anhydrizing ratio (AR), which is defined as the ratio of the number of moles of anhydride links (actual or potential) to the number of "phosphonatable" acyl groups; the latter number is equivalent to the maximum number of HDP units that can be formed. Each mole of a ketone or of an acid halide link, for example, —C(O)Cl, or >PCl, yields 1 mole of anhydride. Sample AR calculation: Consider a 1:1 molar mixture of acetic anhydride and phosphorous acid. Take 2 moles of each, so that there is enough P(III) to form 1 mole of HDP. On this basis, there are 4 moles of Ac, only one of which is phosphonatable (i.e., can acquire 2 phosphonate groups). Therefore, the AR = 2(moles anhydride)/1(mole phosphonatable Ac) = 2.

5. MIXED ANHYDRIDE INTERMEDIATES

A mixture of acetyl and P(III) reagents with AR > 0 forms anhydrides at once or does so when hydrogen halide is removed (108a); a metastable equilibrium is quickly established and changes occur slowly unless the AR or the temperature is high. Both simple anhydrides (e.g., pyrophosphorous acid, polyphosphorous acid, acetic anhydride) and mixed anhydrides (e.g., acetyl phosphorous acid, diacetyl phosphite, triacetyl phosphite) may be present. However, the chemical equilibria involved favor the simple over the mixed anhydrides, so that by ^{31}P NMR analyses on mixtures of AR $= 2$, the concentration of monoacetyl phosphorous acid is considerably lower than that of pyrophosphorous acid (108a). The phosphorus in such P(III) anhydrides is overwhelmingly coordinated to four groups, one of which is hydrogen as in diacetyl phosphite, AcOP(H)(O)OAc. NMR analyses show that while the 4-coordinate mixed anhydrides are less abundant than COC and POP anhydrides, nevertheless they are appreciably more abundant than mixed anhydrides with 3-coordinated phosphorus, for example, P(OAc)$_3$. Hence triply coordinated phosphite anhydrides are rare in such media, except at very high AR values. They become detectable by ^{31}P NMR at an AR in the 4 to 6 range and reach concentrations of 10–15 mole % of the phosphorus at an AR of 10 or more.

a. Thermal Limits. In making EHDP a common upper limit for the temperature is near 120°, set by refluxing of the excess of acetic acid, which is often used both for solvent action and for fluidizing the heterogeneous reaction mixtures. By use of a more inert solvent, such as di-n-propyl sulfone, excess of acetic acid can be eliminated; temperatures as high as 150° can then be used. This is about the upper limit, for small amounts of orthophosphate are produced at 160°; at 175° much of the phosphorus is converted to H$_3$PO$_4$ and its condensates.*

b. Active Phosphite Groups. By following EHDP formation by ^{31}P NMR during heating of mixtures of acetic anhydride and phosphorous acid, it was found probable that all of the mixed anhydrides

---

* Phosphite anhydrides alone can form phosphate at such temperatures (60,144,176). Thus molten phosphorous acid, held at 175° under a blanket of flowing nitrogen, first formed pyrophosphorous acid, but after 16–25 hr gave a viscous liquid with all of its phosphorus present as H$_3$PO$_4$ and condensed phosphoric acids. A compound with phosphorus in a lower oxidation state than $+3$ must also have been formed; presumably this was PH$_3$ (162), which would have been carried away by the nitrogen stream. Similar heating of HP(O)(OH)$_2$ at 160° gave pyrophosphorous acid but very little phosphate.

(mono-, di-, and triacetylated phosphorous acid) are active, including the least acetylated species, namely,

Consider a mixture of 1 mole Ac_2O with 2 moles $HP(O)(OH)_2$; its AR is 1.0, the lowest value which still has the potential of converting all the P(III) to EHDP. While monoacetyl phosphorous acid is the only mixed anhydride detectable by ^{31}P NMR initially, its concentration is low (10% of the total P) and, as the reaction proceeds, it soon becomes undetectable. When the mixture has been heated long enough so that about half of the phosphorus initially bonded to hydrogen is now bonded to carbon, the remaining P(III) appears to be almost wholly present as $HP(O)(OH)_2$ (detectable traces of pyrophosphorous acid are also present); ^{31}P NMR shows that with continued heating phosphonation is still going on albeit very slowly. Such observations, combined with the relative abundance of mixed anhydride species mentioned previously, strongly suggest that monoacetyl phosphorous acid is capable of initiating and participating in the reactions necessary to make $CH_3C(OH)(PO_3H_2)_2$ or derivatives of it.

Because of the mobile chemical equilibria between the various anhydrides, no comparison is possible at some arbitrarily chosen concentration for each anhydride, in the absence of the others. Nevertheless, one gets the impression that the triacetylated compound reacts faster than the diacetylated, which in turn reacts faster than the monoacetylated. In fact, increasing the anhydrizing power can substitute partially for heating; thus at an AR in the 5–10 range, a temperature in the 30–65° range may be used, but the resulting EHDP derivative is necessarily more highly condensed.

6. REORGANIZATION TO CONDENSATES OF ETHANE-1-HYDROXY-1,1-DIPHOSPHONIC ACID

Since the initial AR is usually adjusted to a value of 2 or a little higher and since formation of 1 mole of $CH_3C(OH)(PO_3H_2)_2$ consumes but one unit of the AR, the remainder must show in the product as anhydride, ester, or ether oxy links. An EHDP condensate resulting from an initial AR of 2.0 can be represented by the formula $[CH_3C(OH)(PO_3H_2)_2-H_2O]$, without specifying its structure.

For condensates involved in EHDP making, ester links are formed in greatest numbers (108a); these exist as a mixture of chain polyesters, namely,

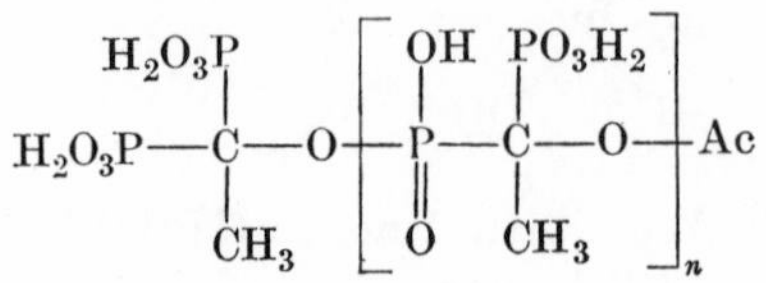

$$
\mathrm{H_2O_3P-\overset{\displaystyle H_2O_3P}{\underset{\displaystyle CH_3}{\overset{|}{C}}}-O}\left[\mathrm{\overset{\displaystyle OH}{\underset{\displaystyle O}{\overset{|}{P}}}-\overset{\displaystyle PO_3H_2}{\underset{\displaystyle CH_3}{\overset{|}{C}}}-O}\right]_n\mathrm{Ac} \qquad n = 0, 1, 2, \text{etc.}
$$

If the AR of the reaction mixture is high, some of the acidic hydrogens will be displaced to yield POC anhydrides, but these will be in equilibrium with COC and POP anhydrides, or tend to be. If the AR of the reaction mixture is about 2 and much acetic acid is present, the product may be simply the carbon-acetylated EHDP $CH_3C(OAc)(PO_3H_2)_2$, which corresponds to $n = 0$ in the chain formula given above. Prolonged heating of reaction mixtures with an initial AR near 2 at temperatures of 140–150° is likely to cause precipitation of a cyclic dimer (108a) having an ether and an anhydride link, namely,

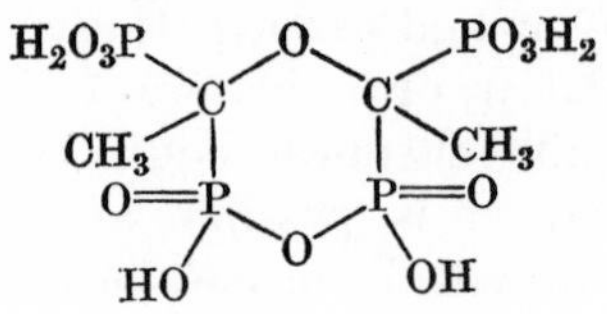

7. HYDROLYSIS OF CONDENSATES

Since the product obtained under conditions designed to convert over 90% of the H-P in phosphites to C-P in EHDP units is a condensate of EHDP, it must be subjected to hydrolysis to obtain the free acid, $CH_3C(OH)(PO_3H_2)_2$. The chain polyester is much easier to hydrolyze than the cyclic ether anhydride (108a).

8. GENERALITY OF CARBOXYL CONVERSION TO HYDROXYDIPHOSPHONATE

From the work of Blaser and Worms, as reported in patents (10,167), it appears that any monocarboxylic acid RCOOH can be converted into an HDP group. However work in these laboratories has revealed exceptions.   Table II summarizes our findings, which will be described in more detail below.

Formic acid does not yield MHDP, methanehydroxydiphosphonate (175,176), via mixed anhydride chemistry; instead the formyl radical decomposes as would be expected, yielding carbon monoxide.†

Trifluoroacetic acid presents a different kind of exception.   It yields not the expected HDP as the hydrolysis product, but rather the hydroxymonophosphonate, plus an equimolar amount of H_3PO_4;

$$CF_3—\overset{\overset{\textstyle H}{\textstyle |}}{\underset{\underset{\textstyle OH}{\textstyle |}}{C}}—PO_3H_2$$

presumably the phosphate-phosphonate, $CF_3C(H)(PO_3H_2)OPO_3H_2$, forms first and is cleaved by hydrolysis.   The CF bond is too strong to allow F^- elimination as in the Perkow reaction, for example, as in the reaction between chloral and a trialkyl phosphite which yields $Cl_2C{=}C(H)OPO_3R_2$ (102).   Several other carboxylic acids yield products in which one third to one half of the phosphorus has been converted to orthophosphate, doing this at temperatures well below the phosphite disproportionation temperature (175°) where phosphite anhydrides alone yield much orthophosphate.   Presumably these acids (e.g., sulfoacetic, trichloroacetic) are converting extensively to phosphate-phosphonates which cleave upon hydrolysis.   In spite of these exceptions, much of the work described in Appendix I supports the prior indications that carboxyl-to-HDP conversion via mixed carboxylic-phosphorous anhydrides is a rather general reaction.   Thus ^{31}P NMR analyses of crude reaction products indicated such a conversion for each of the following acids (173): acrylic,* benzoic, trans-cinnamic, crotonic, cyclopentanecarboxylic,* cyclohexanecarboxylic,* cycloheptanecarboxylic,* lauric,* nitrilotriacetic, and propiolic; only the starred compounds have been isolated in reasonably pure form (cf. Appendix I).   Both the cyclic dimer of the HDP from pivalic acid

---

† Attempts were made to overcome this tendency of $(HCO)_2O$ and HCOCl to decompose, by using CO as a possible reactant in an autoclave under pressures of 700–1500 psi.   The reaction was attempted at temperatures of 100–150°, using a variety of P(III) reagents (disodium phosphite, sodium acid phosphite, phosphorous acid alone and in combination with PCl_3).   At least one trial was made with each of two CO complexers [CuCl and $Ni(CO)_4$] as possible catalysts. No phosphonation occurred in any of the experiments.   For ways of making MHDP see Refs. 130 and 131.

TABLE II

Classes of Carboxyls in Anhydride Phosphonation

RCOOH is readily converted to $RC(OH)(PO_3H_2)_2$; little phosphate in product	RCOOH harder to convert to $RC(OH)(PO_3H_2)_2$; some phosphate in product	RCOOH formed little or no $RC(OH)(PO_3H_2)_2$; much phosphate in product	RCOOH suffered decomposition or will not react with phosphites
$CH_3(CH_2)_nCOOH$	$ClCH_2COOH$	CF_3COOH	H_2CO_3[a]
$H_2O_3PCH_2COOH$	$CH_2(COOH)_2$	CCl_3COOH[c]	HCOOH (cf. footnote, p. 000)
$N(CH_2COOH)_3$[c]	$HC\!::\!CCOOH$	HO_3SCH_2COOH	
$H_2C\!=\!CHCOOH$	$(CH_3)_3CCOOH$	$(COOH)_2$	
	$n=4,5,6$	$HOOCC(OH)(CH_2COOH)_2$	
		$(CH_2COOH)_2$[c]	
		$HOOCCH\!=\!CHCOOH$ cis and trans polymaleic anhydride[d]	

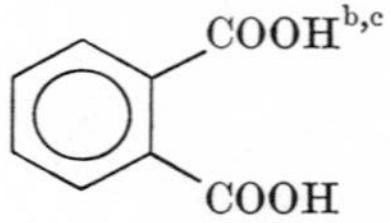

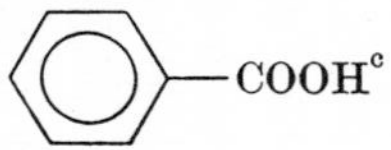

[a] Experiment tried with $COCl_2$ and HPO_3H_2, lost CO_2 and HCl.

[b] Was inert (possibly insoluble) in anhydrizing media; this may be due to formation of a stable carboxylic anhydride rather than a mixed anhydride. Isolated experiments produced some C—P compounds.

[c] Located tentatively—based on 1–5 experiments.

[d] Based on unpublished observations of J. S. Berry.

64

(175,176) and the monomeric acid have been isolated (174) in fairly pure form.

The R in RCOOH need not be a hydrocarbon radical; the data in Appendix I show that phosphonoacetic (132) and monochloroacetic acids yield $H_2O_3PCH_2C(OH)(PO_3H_2)_2$ and $ClCH_2C(OH)(PO_3H_2)_2$, respectively.

Turning to dicarboxylic acids, one encounters some common acids that do not readily form HDPs (173,175), for example, cyclohexane-1,2-dicarboxylic, phthalic, succinic, and maleic acids. Since all of them can form a relatively stable carboxylic anhydride, the hindrance may come from its limited solubility in the mixed anhydride systems or from being too stable to form mixed anhydrides readily; in addition some reaction converts much of the phosphorus to orthophosphate.

The examples given by Blaser and Worms (167) suggest no difference between mono- and dicarboxylic acids in regard to mixed anhydride phosphonation; however the data given are too limited to determine the course of the phosphonation. On the basis of exploratory experiments done here, it seems likely that carboxyls far separated in the molecule, for example, as in adipic acid, will react as in monocarboxylic acids. However compounds with the carboxyls closer together as in oxalic, malonic, and citric acids will inevitably allow side reactions to occur, therefore complicating the use of anhydride chemistry for conversion of their carboxyls to HDP units. Our attempts to phosphonate the carboxyls of malonic acid illustrate this point (cf. Appendix I). The carboxyls in nitrilotriacetic acid (NTA) may undergo this reaction (173), for the purified NTA phosphonation product had the correct C:P ratio and gave a single ^{31}P NMR peak at $\delta = -7.5$ ppm (triplet). However, the ^1H NMR spectrum, a simple doublet, is not consistent with this formulation.

9. EXTENSIONS OF ANHYDRIDE PHOSPHONATION

In the conversion of the carboxyl of acrylic acid to an HDP group, it was found that a third phosphonate group had been added (25a); this occurred without consuming anhydrizing power and put the phosphonate on the terminal carbon, giving the triphosphonic acid $H_2O_3PCH_2CH_2C(OH)(PO_3H_2)_2$. The addition of this third phosphonate group was not demonstrated in the patent information on the phosphonation of acrylic acid. Crotonic acid also adds a third phosphonate group (position not determined, but *presumably* on carbon three by the similarity of its ^{31}P NMR spectrum to that of the phosphonate from acrylic acid). Not all double bonds respond in this way to phosphite anhydrides; 1-decene failed to form any phosphonate.

Anhydride phosphonation has been extended to nitriles, which yield 1-amino-1,1-diphosphonates (ADP). The first reports come from two patents (9,82); from these and our limited experience (see Appendix I) in making $CH_3C(NH_2)(PO_3H_2)_2$ from acetonitrile, phosphorus trichloride, and acetic acid, it appears that the nitrile must first be attacked by the phosphorus trihalide. In order to obtain the ADP, however, a hydroxyl-containing acid must be added at a later stage; the patents mention phosphorous and acetic acids. Acetic acid was used in our experiments and hence the product was largely EHDP. The ADP was easy to isolate, however, because of the very low solubility of the free acid, or more accurately, inner salt $CH_3C(NH_3^+)$ $(PO_3H^-)PO_3H_2$. EADP and eleven other ADPs were mentioned by Cummins (25), but no details of preparation were given; EADP has also been made by Callis and co-workers (19). The implication of the literature is that phosphonation of nitriles is also a rather general reaction.

II. Properties of Polyphosphonates

One interesting property of these compounds is their tendency to form complexes with metal ions. This has led to some investigation of the strength of their acids. In addition, some spectroscopic information has been accumulated, chiefly, nuclear magnetic resonance, infrared, and X-ray diffraction powder patterns.

A. Stability to Heat and Hydrolysis

Most of the polyphosphonates mentioned in this review have been studied so little that information on stability comes primarily from the preparational work.

Except for a few cases (e.g., acyl phosphonates, 1-hydroxy-1-phosphonates), the C—P bond in monophosphonate esters, acids, and salts is relatively stable toward heat and most chemical reagents. While the same is often true of *gem*- and *vic*-diphosphonate groupings, the higher hydrocarbon substitution brings, as expected, more signs of instability. Thus methylenediphosphonic acid (mp 203–206°) and ethane-1,2-diphosphonic acid (mp 220–223°) can be melted without cleavage of C—P bonds, but ethane-1-hydroxy-1,1-diphosphonic acid cannot; all three acids tend to lose water and condense at temperatures between 100° and their melting points.

In spite of the above comment, in the class of diphosphonic acids having the formula $H(CH_2)_nC(OH)(PO_3H_2)_2$ where $n = 0, 1, 2$, etc., both the free acid and its salts are relatively stable. To illustrate this let us examine the information available for $CH_3C(OH)(PO_3H_2)_2$ and its salts. In thermogravimetric tests (171) the monohydrate of the acid loses its water of hydration as the temperature approaches 100° and loses another mole of water between 100 and 175°, forming a condensate; only above 175° does the odor of phosphine become evident, being quite noticeable when the temperature reaches 215°. The anhydrous Na_3H salt did not begin condensing (losing water of constitution) until the temperature exceeded 160°; extensive decomposition occurs between 285 and 350° and the darkened product may ignite spontaneously in air.

In acidic water solutions, none of the polyphosphonic acids described show any tendency toward hydrolysis at the boiling point, except for *gem*-diphosphonic acids with halogen on the carbon alpha or beta to the phosphonate group, and possibly for $(CH_3)_3CC(OH)(PO_3H_2)_2$. Prolonged boiling caused no change in the ^{31}P NMR spectrum of $RC(OH)(PO_3H_2)_2$ in which R is either H (130), a radical given in the first column of Table II, $HC\equiv C$, C_6H_5, or cyclohexyl. Where R is the bulky $(CH_3)_3C$, the HDP may be slowly cleaved; prolonged boiling of a very crude sample, containing much $HP(O)(OH)_2$ and some H_3PO_4, eliminated all of the monomeric acid, increased the $HP(O)(OH)_2$, and had no effect on the H_3PO_4.

Because of their sensitivity to base in ester form, some of the 1-hydroxy-1,1-diphosphonate anions were compared with other *gem*-diphosphonate anions in boiling sodium hydroxide solutions. Thus 8 hr of boiling of 15% solutions of Na salts in 10% aqueous NaOH caused no change (175,176) detectable by ^{31}P NMR for any of the following five anions: $^-O_2CCH_2CH(PO_3^{2-})_2$, $(^{2-}O_3P)_2CHCH_2CH(PO_3^{2-})_2$, $HC(OH)(PO_3^{2-})_2$, $CH_3C(OH)(PO_3^{2-})_2$, and $^{2-}O_3PCH_2C(OH)(PO_3^{2-})_2$. These data suggest that, when R is hydrogen or a hydrocarbyl radical, anions of the formula $RC(OH)(PO_3^{2-})_2$ are considerably more stable toward base than the corresponding esters (see earlier section on HDP esters); ethane-1-hydroxy-1,1,2-triphosphonic acid also belongs to this group.

B. Acid Dissociation Constants

So far these are available mostly for the *gem*-diphosphonic acids (Table III), but a few values are available for vicinal polyphosphonic

TABLE III

Dissociation Constants of *Gem*-Diphosphonic Acids at 25°

$$\underset{\underset{Y}{|}}{\overset{\overset{X}{|}}{H_2O_3P-C-PO_3H_2}}$$

X	Y	Ionic strength	pK_1	pK_2	pK_3	pK_4	pK_5	Ref.
H	H	0.0	2.2	2.87	7.45	10.69		63
H	H	0.1	1.7	2.75	7.33	10.42		63
H	H	0.5		2.49	6.87	10.54		21
H	H	0.0		3.05	7.35	10.96		45
H	H	0.1		2.78	7.00	10.57		45
CH_3	H	0.0		3.14	7.49	11.97		45
CH_3	H	0.1		2.88	7.22	11.67		45
CH_3	H	0.5		2.66	7.18	11.54		21
$n\text{-}C_7H_{15}$	H	0.5			7.45	11.9		21
CH_3	CH_3	0.0		3.16	8.04	12.10		45
CH_3	CH_3	0.1		2.98	7.78	11.76		45
CH_3	CH_3	0.5		2.94	7.75	12.4		21
H	OH	0.0		2.74	7.05	10.56		45
H	OH	0.1		2.60	6.78	10.26		45
CH_3	OH	0.0		3.03	7.31	11.52		45
CH_3	OH	0.1		2.80	7.00	11.16		45
CH_3	OH	0.1	1.7	2.47	7.28	10.29	11.13[a]	68
CH_3	OH	0.5		2.54	6.97	11.41		21
CH_3	OH	unspec.	1.4	2.81	7.03	11.3		72
C_2H_5	OH	unspec.	1.5	3.09	7.10	12.0		72
$n\text{-}C_3H_7$	OH	unspec.	1.6	3.17	7.40	12		72
C_6H_5	NH_2	0.1		5.29	8.17	10.29[b]		30
Cl	Cl	0.0			6.11	9.78		45
Cl	Cl	0.1			5.89	9.50		45
XY equals O=		0.0			5.81	8.42		45
		0.1			5.50	8.16		45

[a] Authors (68) ascribe this value to the dissociation of H from COH.

[b] Authors (30) ascribe this value to the betaine proton.

acids (Table IV). Constants based on extrapolation to zero ionic strength are given for several compounds (45,63); several determinations have been made only at a single ionic strength (21,22a,25,30,38, 68,72,137), sometimes unspecified (25,72). Only one study covered a range of temperatures (63). Approximate pK_a values have been reported (25) for several compounds containing the 1-amino-1,1-diphosphonic acid group; values obtained by Dyatlova and co-workers

TABLE IV

Acid Dissociation Constants at 25° for Vicinal Polyphosphonic Acids

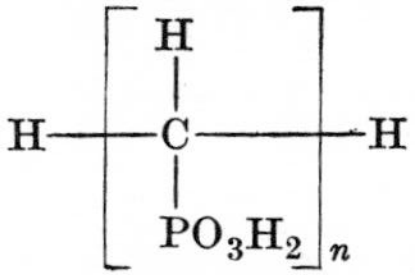

$$\left[\begin{array}{c} H \\ | \\ H\text{—}C\text{—}H \\ | \\ PO_3H_2 \end{array} \right]_n \quad\quad \text{vs Diphosphonic Acids } H_2O_3P\text{—}(\text{—}CH_2)_m\text{—}PO_3H_2$$

m or n	Ionic strength	pK_1	pK_2	pK_3	pK_4	pK_5	pK_6	pK_7	pK_8	Ref.
$n = 1$	0.005 to 0.014	2.38	7.74							24a
$n = 2$	0.0	1.5	3.18	7.62	9.28					63
$n = 2$	1.0	1.5	2.74	7.42	8.96					63
$n = 3$	1.0		1.64	2.92	6.22	8.68	12.86			22a
$n = 4$	1.0			2.39	3.49	6.38	8.25	12.53		22a
$m = 1$	0.0	2.2	2.87	7.45	10.69					63
$m = 1$	1.0	1.7	2.67	7.28	10.32					63
$m = 2$	Same as $n = 2$, see above.									
$m = 3$	0.0	1.6	3.06	7.65	8.63					63
$m = 3$	1.0	1.7	2.64	7.41	8.33					63
$m = 4$	0.0	1.7	3.19	7.78	8.58					63
$m = 4$	1.0	1.7	2.70	7.47	8.30					63

(30) for $C_6H_5C(NH_2)(PO_3H_2)_2$ are given in Table III. Frank gave approximate pK_3 and pK_4 values for a highly fluorinated vicinal diphosphonic acid (38).

Kabachnik and co-workers (68) report a pK_5 value for the dissociation of the proton from the COH group of $CH_3C(OH)(PO_3H_2)_2$, but no evidence for the dissociation of this proton was offered by other investigators (21,45,68).

Looking at successive dissociations of a particular acid in Tables III and IV, one notes a large increase of 3 to 5 pK units, corresponding to a large decrease in acid strength, which occurs when the first proton dissociates from a $[-PO_3H]^{1-}$ group. A second such change occurs when the first proton dissociates from a $[C(PO_3)_2H]^{3-}$ or a $[(CPO_3)_2H]^{3-}$ group; this is probably associated with H-bonding between two phosphonate groups (45,142) and also with the large energy required to remove an H^+ from HA^{3-} to form A^{4-}. The H bonding is strongest for gem-diphosphonates and weakens as the phosphonate groups get farther apart (see last half of Table IV) (63).

From the representative pK values in Table III one can see the usual trends with ionic strength, chain length of an alkyl group R, and electron-withdrawing power of substituents on the bridge carbon. Two studies have attempted to correlate pK values with measures of electronegativity, such as Taft's σ^* (45,87).

C. Metal Derivatives of Polyphosphonates

Work on the interaction of metal ions with polyphosphonates involves considerable variation in the ionic character of the phosphonate groups themselves: wholly ester, part ester and part acid and/or salt, or wholly acid or salt. Such interactions were probably first observed by von Baeyer and Hofmann (164).

1. METAL POLYPHOSPHONATE ESTER COMPOUNDS

Walmsley and Tyree (165) made an extensive study of first-row transition-metal complexes with the tetraisopropyl ester of methylene-diphosphonic acid. This ester was shown to be a weak-field ligand by virtue of the magnetic properties of the complexes with transition metals. Adducts were prepared from chlorides (also some per-chlorates) of Co(II), Fe(III), Ni(II), Cu(II), Zn(II), and Sb(V). Calcium chloride will also form a crystalline adduct of the composition CaL_2Cl_2, where L is tetraisopropyl methylenediphosphonate (177); it is soluble in acetone. Stewart and Siddall (154) have reported adducts of this ester with rare earth salts.

2. EXTRACTION OF METAL IONS

Healy and Kennedy (55,73) have studied the partitioning of uranyl nitrate between a number of phosphonate esters and dilute nitric acid. The distribution coefficient (ratio of uranyl ion in the organic phase to that in the aqueous phase) is higher for $Bu_2O_3PCH_2CH_2PO_3Bu_2$ than for $CH_2(PO_3Bu_2)_2$; both diphosphonates were less effective than $CH_3(CH_2)_3PO_3Bu_2$.

Half-ionic methylenediphosphonates $[CH_2(PO_3RH)_2$ where R is alkyl] have been studied as metal extractants; Gorican and Grdenic (41) have found the octyl derivatives generally more efficient than n-butyl esters. Gorican and co-workers, using the dioctyl derivative, have studied the separation of germanium from arsenic (46) and of zirconium from niobium (44); they have also investigated the extraction of titanium (43) and of niobium and tantalum (28,29). Salts of such dialkyl methylenediphosphonates with Th(IV), U(IV), Ce(IV), Fe(III), and Co(II) have been isolated (42); they have little or no solubility in water or organic solvents, but dissolve in solutions of the reagent with which they form complexes.

Other examples of the use of methylenediphosphonate (or other polyphosphonates) as extractants for metal ions are cited (37,64,105, 146–148).

3. METAL ION ASSOCIATION CONSTANTS

Complex formation between *gem*-diphosphonates and metal cations has been claimed in patents for a decade (10), but only very recently have formation constants begun to appear in the literature. Their determination has been mostly by acid base titration in the presence and absence of a metal cation (21,22,30,68,69), with a few by nephelometric methods (19,25). Some determinations have also been made on alkylenediphosphonates, $^{2-}O_3P(CH_2)_nPO_3{}^{2-}$ with $n = 2$ or more (21,69).

Some typical values for *gem*-diphosphonate anions are collected in Table V; they are given as the negative logarithm of the association constant β for the complex having the formula indicated. Values for a few other polyphosphonates and for several other metals are to be found in the papers cited.

TABLE V

Association Constants for Metal Ions with *Gem*-diphosphonates

$$\text{Where } {}^{2-}O_3P-\underset{Y}{\overset{X}{\underset{|}{\overset{|}{C}}}}-PO_3{}^{2-} \text{ is the Ligand (L)}$$

X	Y	Ionic strength	Temp. (°C)	Cation M	Log of association constant β			Ref.
					MHL	ML	ML$_2$	
H	H	0.5	25	Li$^+$	0.82	2.48		21
H	H	0.5	25	Na$^+$	0.39	1.13		21
H	H	0.5	25	Mg^{2+}	2.92	5.78		22
H	H	0.1	25	Mg^{2+}	4.02	6.38		68
H	H	0.5	25	Ca^{2+}	2.46	4.70		22
H	H	0.1	25	Ca^{2+}		6.02		63
H	H	0.1	25	Ca^{2+}	3.88	6.03		68
H	H	0.1	25	Fe^{2+}	6.6	12.6		68
H	H	0.1	25	Fe^{3+}	19.9	26.6		68
CH$_3$	H	0.5	25	Li$^+$	0.99	3.12		21
CH$_3$	H	0.5	25	Na$^+$	0.50	1.51		21
CH$_3$	H	0.5	25	Mg^{2+}	2.99	6.26		22
CH$_3$	H	0.5	25	Ca^{2+}	2.74	5.21		22
CH$_3$	CH$_3$	0.5	25	Li$^+$	1.38	3.83		21
CH$_3$	CH$_3$	0.5	25	Na$^+$	0.57	2.08		21
CH$_3$	CH$_3$	0.5	25	Mg^{2+}	3.33	6.83		22
CH$_3$	CH$_3$	0.5	25	Ca^{2+}	3.14	6.33		22

CH$_3$	OH	0.5	25	Li$^+$	1.08	3.35		21
CH$_3$	OH	0.5	25	Na$^+$	0.54	2.07		21
CH$_3$	OH	0.5	25	Mg^{2+}	3.32	6.39		22
CH$_3$	OH	0.1	25	Mg^{2+}		6.55		68
CH$_3$	OH	0.5	25	Ca^{2+}	3.58	5.74		22
CH$_3$	OH	0.1	25	Ca^{2+}		6.04		68
CH$_3$	OH	0.1	25	Ca^{2+}		7.09		19
CH$_3$	OH	0.1	50	Ca^{2+}		5.53	6.20	19
CH$_3$	OH	unspec.	25	Ca^{2+}		4.4		25
CH$_3$	OH	0.1	25	Fe^{2+}	5.31	9.05		68
CH$_3$	OH	0.1	25	Fe^{3+}		16.21		68
H	NH$_2$	unspec.	25	Ca^{2+}		5.7		25
CH$_3$	NH$_2$	unspec.	25	Ca^{2+}		6.0		25
CH$_3$	NH$_2$	0.1	25	Ca^{2+}		6.71		19
CH$_3$	NH$_2$	0.1	50	Ca^{2+}		6.08	4.26	19
C$_6$H$_5$	NH$_2$	unspec.	25	Ca^{2+}		5.6		25
C$_6$H$_5$	NH$_2$	0.1	25	Mg^{2+}	5.46	7.39		30
C$_6$H$_5$	NH$_2$	0.1	25	Ca^{2+}	4.82	6.56		30
C$_6$H$_5$	NH$_2$	0.1	25	Fe^{2+}	7.37	10.40		30
C$_6$H$_5$	NH$_2$	0.1	25	Fe^{3+}	15.08	20.15		30

D. Spectroscopic Data

Nuclear magnetic resonance data for ^{31}P in many of the compounds cited are given in the compilation of Mark and co-workers (86); when available, ^{31}P NMR data are given for the compounds listed in Appendix II. NMR data, melting points, etc., on polyphosphonic acids and salts are also likely to be found in Appendix I.

Investigators too numerous to cite have used IR (occasionally Raman) spectra to detect the presence of functional groups. However the fairly extensive study by Frank (37,38) on perfluorinated carbon rings with vicinal diphosphonate groups should be cited. An extensive study of methylenediphosphonate (salts, acid salts, free acid, and the tetraethyl ester) by IR and a few Raman spectra allowed Steger and Rehak (153) to assign frequencies for CH_2, CP_2, and PO_3 vibrations, "the stretching of the latter being slightly lower than for diphosphates." A recent compilation of IR spectra for phosphorus compounds by Corbridge (23a) includes data for P-C linkages, including a few gem-diphosphonates.

Ultraviolet spectral measurements on the carbanion of tetraethyl methylenediphosphonate indicate resonance interaction involving the various canonical forms contributing to the resonance hybrid (24).

X-ray diffraction (powder) patterns are given in a few cases, for example, for alkylenediphosphonic acids $H_2O_3P(CH_2)_nPO_3H_2$ by Moedritzer and Irani (91), for $CH_2(PO_3HNa)_2$ by Corbridge and Tromans (23), for salts of $CH_3C(OH)(PO_3H_2)_2$ by Kasparek (72), and for adducts of tetraisopropyl methylenediphosphonate with rare earth cations by Stewart and Siddall (154).

Appendix I

A. Nuclear Magnetic Resonance Spectra

These were obtained on one of the following instruments: (1) Varian DP-60 (field strength 14,100 gauss) with probes for ^1H, ^{19}F, and ^{31}P; (2) Varian A-60, giving calibrated spectra for ^1H; and (3) Varian HA-100 (field strength 23,500 gauss). The standards were as follows: (1) tetramethylsilane (TMS) $= 10.0$ ppm for ^1H (called τ when the TMS was dissolved in the sample, but τ' when the TMS was in a glass capillary; (2) trifluoroacetic acid in a capillary $= 0.0$ ppm for ^{19}F; and (3) 85% H_3PO_4 in a capillary $= 0.0$ ppm for ^{31}P.

B. Phosphonation of Acyls by Mixed Anhydrides

1. CYCLOHEXYL METHANEHYDROXYDIPHOSPHONATE

a. Dimeric Ether Anhydride. In a $3:4:2$ molar ratio cyclohexane carboxylic acid, phosphorous acid, and phosphorus trichloride supply the correct AR and ratio of acyls to P(III) reagents to make one half mole of the cyclic dimer (ring sequence COCPOP) for each mole of RCOOH taken, where R is cyclohexyl.

In one case the quantities used were: 128.2 g $C_6H_{11}COOH$, 105.9 g phosphorous acid (0.76% H_2O), and 97.5 g phosphorus trichloride in 200 ml dry di-n-propyl sulfone. Heating this mixture for 16 hr at 85° first removed the HCl, then induced reorganization of the mixed anhydrides and later initiated precipitation of a white solid (the condensate). After an additional digestion of the mixture at 115° for 24 hr, the white solid was removed by filtration. Analysis of a portion in basic solution by ^{31}P NMR showed that the phosphonate was all in the desired form (2 peaks of equal intensity at -12 and -9 ppm), but 40% of the P was present as HPO_3H_2. Since some condensate was discarded along with the liquid phase, crude product recovery was only 35% of the possible yield.

The condensate was crystallized twice from water; the first crop (40.9 g or 16% of theory) proved to be relatively pure; the mother liquor from the second crystallization contained too little phosphorous acid to be detected by NMR. The twice-recrystallized sample (yield 10.8 g), analyzed by NMR in excess of aqueous base (Na$_6$ salt), showed only the two phosphonate peaks of equal intensity at -14 and -10 ppm.

Analysis. Calculated for $C_{14}H_{28}O_{12}P_4$: C, 32.8; H, 5.5; P, 24.2%; equiv wt 85. Found: C, 33.1, H, 6.1; P, 24.3; equiv wt 83.

b. Preparation of $C_6H_{11}C(OH)(PO_3H_2)_2$ *and Its* Na_3H *Salt.* The hydrolysis was accomplished by refluxing with concentrated HCl solution for 18 hr; characterization of the hydrolysate by ^{31}P NMR showed essentially one P species at $\delta = -19.5$ ppm plus a detectable trace of HPO_3H_2. After conversion to the Na$_3$H salt by treatment with aqueous NaOH (pH 10.3), the monomer was recrystallized from water solution, using methanol and acetone as nonsolvents. In water solution the purified product showed a single P species by ^{31}P NMR (a doublet at $\delta = -19.5$ ppm, $J_{PCCH} = 9$ Hz).

Analysis. Calculated for $C_7H_{13}O_7Na_3P_2$: C, 24.7; H, 3.9; Na, 20.3; P, 18.2. Found: C, 24.8; H, 3.3; Na, 20.2; P, 18.0%.

2. OTHER CYCLOALKYL METHANEHYDROXYDIPHOSPHONATES

Cyclopentanecarboxylic and cycloheptanecarboxylic acids were tested sufficiently to indicate that they behave like cyclohexane carboxylic acid in these mixed anhydride phosphonations.

3. t-BUTYL METHANEHYDROXYDIPHOSPHONATE

This crowded molecule was difficult to make either as the monomer, $(CH_3)_3CC(OH)(PO_3H_2)_2$, or as the cyclic ether anhydride (cf. 108a). Unexpectedly, a fairly pure sample of the latter was obtained, presumably because it precipitated from the anhydrizing medium.

a. Dimeric Ether Anhydride. Pivaloyl chloride and phosphorous acid were combined in 1:1 molar ratio (AR = 2) and heated cautiously at first (copious HCl evolution at 55–60°), then for 6 hr at 110–115°. The mixture was then thinned by addition of dibutyl ether (0.5 g/g reactants) and heated for 5 hr at 110–118°. Addition of 1.85 moles water/mole of pivaloyl chloride was found to initiate crystallization in the hot mixture; upon cooling more crystals formed. Filtering, washing with ethyl ether, and air drying, gave a 9% yield; by ^{31}P NMR this crude material was highly contaminated with HPO_3H_2 (22% of the P) and there was more dangling phosphonate (doublet at −18.3 ppm, $J = 38$ Hz, 43% of P) than ring phosphonate (doublet at −6.0 ppm, $J = 38$ Hz, 36% of P) (see 108a); evidently closure of POP anhydride links was incomplete or hydrolysis had opened some of them.

Two more crystallizations (first from methanol-ether, then from water-acetone) gave a 3% yield of a purer sample of the cyclic ether anhydride (mp rose from 170° for the crude product to 202°).

Analysis. Calculated for $C_{10}H_{24}O_{12}P_4$: C, 26.1; H, 5.3; P, 26.9%. Found: C, 27.0; H, 6.9; P, 27.7%. In D_2O solution the ^1H NMR spectrum shows a large singlet in the CH_3C region at $\tau' = 8.65$ ppm. The ^{31}P NMR spectrum of this sample showed two doublets of equal intensity at $\delta = -18$ and $\delta = -6$ ppm (J about 34 Hz).

b. Hydrolysis of Ether Anhydride. The preparational work, all involving complex mixtures, had shown that the HDP moiety is not completely stable toward water at reflux temperatures in acid media; prolonged boiling had converted all phosphonate phosphorus to HPO_3H_2. Hence hydrolysis of 3.8 g of the purified dimer in 15 ml water was followed by ^{31}P NMR to determine how much refluxing should be done in order to obtain the maximum yield of the monomer.

From the NMR analyses it was evident that the cyclic dimer required about 20 hr of boiling for complete decomposition; after 23 hr about 41% of its phosphorus had been converted to the monomeric acid, $(CH_3)_3CC(OH)(PO_3H_2)_2$, and the other 59% to phosphorous acid. It was not established whether the monomeric acid itself was decomposing, but, if so, it was at a rate not greater than 20% in 24 hr.

Some intermediate gave a phosphorus peak at $+3$ ppm; its concentration rose from 0 to 9% of the total P at 4 hr, then fell to 0 again at 23 hr, a fact which shows that it can not be coming from degradation of the monomeric acid, which attained its maximum concentration at 23 hr. Whatever its structure, this intermediate probably comes from degradation of the cyclic dimer; writing R for t-butyl, it might be due to $RC(OH)_2PO_3H_2$ or to the phosphonate starred in the cyclic intermediate shown below.

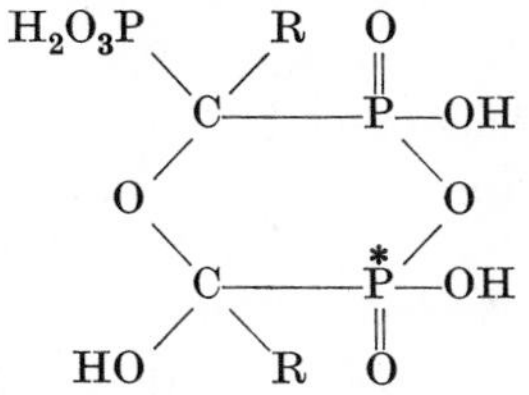

4. CHLORORETHYL METHANEHYDROXYDIPHOSPHONATE

To avoid loss of chlorine from the $ClCH_2$ group and to minimize orthophosphate formation, the reaction variables should be arranged as follows:

1. Temperature not in excess of $100°$, unless the time is short, and
2. Anhydrizing ratio well above 2, regardless of the time-temperature choice.

One sample of $ClCH_2C(OH)(PO_3H_2)_2$ was prepared by reaction of 0.10 moles phosphorous oxide (P_4O_6) with a large excess of chloroacetic acid (1.2 moles) at $65°$ for 18 hr. The resulting HDP condensates, $2\ ClCH_2C(OH)(PO_3H_2)_2 - 2\ H_2O$, were hydrolyzed to the free acid by adding 0.80 moles water and maintaining the mixture at $65°$ for 21 hr; mild hydrolysis conditions are required to avoid loss of chlorine.

To the clear solution of the crude hydroxydiphosphonic acid at $60°$ was added 0.18 moles ammonium acetate to induce precipitation of the monoammonium trihydrogen salt after the method of Pflaumer and

Filcik (107). After 2 hr of digestion at 60°, the precipitate was filtered off, washed first with acetic acid, then with ethyl ether, and air dried (yield 66% based on phosphorus).

The product assayed as pure $ClCH_2C(OH)(PO_3)_2H_3(NH_4)$ both by titration with base and by ^{31}P NMR (triplet at $\delta = -15.3$ ppm, $J = $ ca. 11 Hz). It had a melting point of 159–160°.

Analysis. Calculated for $C_2H_{10}O_7NP_2Cl$: C, 9.3; H, 3.5; N, 5.4; P, 24.1; Cl, 13.8%. Found: C, 9.5; H, 3.9; N, 5.5; P, 24.6; Cl, 13.8%.

5. ETHANE-1-HYDROXY-1,1,2-TRIPHOSPHONATE

A condensate corresponding to $H_2O_3PCH_2C)OH)(PO_3H_2)_2$—2 H_2O is precipitated almost quantitatively when an equimolar mixture of phosphonoacetic acid, phosphorous acid, and phosphorus trichloride (AR = 3) in di-n-propyl sulfone (4 g/g reaction mixture) reacts at a temperature in the range 100–115°. Starting with 1 mole of phosphonoacetic acid, the PCl_3 was added from a dropping funnel to the other components at 85°; copious HCl evolution and thickening of the mixture occurred, but after 11 min the mixture broke into two layers which were kept in intimate contact by vigorous stirring. The temperature kept rising as the PCl_3 addition continued, reaching 116° by the time the mole of PCl_3 had been added. Much crystallization of a white solid (condensate) took place during the latter part of this period. The slurry was digested for 6 hr at 116°, then cooled to room temperature; other runs show that this digestion can be cut to 2 hr or less.

The crude solid from filtration, ether washing, and drying under nitrogen weighed 307 g; subtracting its sulfone content (63 g) left 244 g of condensate, a yield of 97% if the product had the composition indicated above. On a water solution of the crude mixture (partially hydrolyzed to the free acid) a ^{31}P NMR spectrum revealed little P-containing impurity; HPO_3H_2 accounted for 5% of the total P.

The crystalline condensate is readily hydrolyzed to the free acid; for example, in 33% aqueous solution it is 50% hydrolyzed in 30 min at the boiling point; 2 hr at boiling was generally used to ensure complete hydrolysis. The acid $H_2O_3PCH_2C(OH)(PO_3H_2)_2$ was converted to the Na_5H salt (pH 10) and purified by repeated crystallization from water by adding methanol as nonsolvent. One such sample had a water content of 32.9% by weight loss at 139° in a vacuum oven (33.3% by Karl Fischer method), which corresponds to an 11 H_2O hydrate of the Na_5H salt (33.4% H_2O).

Analysis. Calculated for $C_2H_4O_{10}P_3Na_5$: C, 6.1; H, 1.0; P, 23.5; Na, 29.0%; mol wt 396. Found: C, 6.1; H, 0.9; P, 23.2; Na, 30.6%; mol wt by $Na_2SO_4 \cdot 10 \, H_2O$ transition lowering 414.

Structural analysis by ^{1}H and ^{31}P NMR is consistent with an AB_2X_2 system, where A stands for the monophosphonate P, B_2 for the two diphosphonate Ps, and X_2 for the methylene protons. Since the two kinds of phosphonate P differ in chemical shift more for the acid than for the salts, the free acid was used to obtain the following data (interpretation by T. J. Flautt):

$$\delta_A = -23.0 \text{ ppm} \qquad J_{AB} = 32 \text{ Hz}$$
$$\delta_B = -17.9 \text{ ppm} \qquad J_{AX} = 19 \text{ Hz}$$
$$\tau' = 7.3 \text{ ppm} \qquad J_{BX} = 14 \text{ Hz}$$

Area of -17.9 ppm peak is double that of the -23.0 ppm peak.

6. ETHANE-1-AMINO-1,1-DIPHOSPHONATE

According to directions in the patent of Lerch and Kottler (82) phosphorus trihalides are not effective in converting nitriles to amino-diphosphonates (ADP) unless a compound containing hydroxyls is present (water, alcohols, phenols). Blaser and co-workers (9) suggested an organic acid such as acetic or an inorganic acid such as phosphorous as the source of the OH. ' This brings the over-all reaction into the domain of mixed anhydride chemistry.

Acetonitrile, phosphorus trichloride, and acetic acid were combined in $1:1:3$ molar ratio to give a heterogeneous mixture which was heated at 50° for 6.5 hr, when the mixture became too viscous to stir. The top layer, which contained no P species by an NMR test, was removed and discarded. A large excess of water (5 moles/mole CH_3CN) was added and the solution refluxed for 5 hr to hydrolyze the EADP and EHDP condensates. As the solution cooled to room temperature EADP crystals formed as the anhydrous acid (inner salt); the crystals were recovered by filtration, washed with water, water–acetone mixture, then acetone, and finally air dried. The yield was low (6%), but the crystals were free of impurity as judged by NMR. The product was recrystallized by solution in aqueous NaOH and reacidifying with HCl; these crystals were washed three times with water and air dried.

Analysis. Calculated for $C_2H_9O_6NP_2$: C, 11.7; H, 4.4; N, 6.8; P, 30.2%. Found: C, 12.3; H, 4.7; N, 6.8; P, 30.2%. Equiv wt by titration (1st to 2nd or 2nd to 3rd endpoint) 207 (theory 205). NMR data on 1 g EADP crystals in 3 ml 50% NaOH (Na_4 salt):

$$^{31}\text{P}: \quad \delta = -22.7 \text{ ppm (quartet,} \quad J = 14 \text{ Hz)}$$
$$^1\text{H}: \quad \tau' = 8.21 \text{ ppm (triplet,} \quad J = 14 \text{ Hz)}$$

7. TRIFLUOROMETHYL METHANEHYDROXYMONOPHOSPHONATE

When trifluoroacetic acid (or its acyl derivatives) are subjected to mixed anhydride phosphonation, a hydroxymonophosphonic acid, $CF_3CH(OH)PO_3H_2$, and H_3PO_4 are isolated in equimolar amounts. The expected hydroxydiphosphonate may be formed initially, but, if so, the CF_3 group is so electron-withdrawing that HDP condensates rearrange to the phosphonate and yield the above acids upon hydrolysis. Writing these materials as the acids instead of condensates, we may symbolize the changes thus:

$$CF_3C(OH)(PO_3H_2)_2 \longrightarrow CF_3CH(OPO_3H_2)PO_3H_2 \xrightarrow{H_2O}$$
$$CF_3CH(OH)PO_3H_2 + H_3PO_4$$

Would this reaction, like that of Perkow (102), proceed at a lower than normal temperature? And would the production of phosphate be thus reduced? The answer to both questions is no, based on observations such as the following: (1) trifluoroacetyl phosphites in a mixture with AR = 2 remained essentially unchanged for a period of 10 months at room temperature and for a few hours at 50°; (2) raising the anhydrizing ratio to 20 (more PCl_3) did not bring about detectable C—P bond formation in 2 hr at 38–46° because reasonably rapid reaction temperatures of 80–130° are needed, preferably 110–130°. The need of high temperature coupled with the high volatility of CF_3COCl (bp −13°) and $(CF_3CO)_2O$ (bp 40°), means that the heating of these mixtures must be done in an autoclave or in solution in a high-boiling solvent in a vessel provided with an efficient reflux condenser.

A mixture of trifluoroacetic anhydride and phosphorous acid in 1:1 molar ratio (AR = 2) was heated for 24 hr at 125° in a glass-lined rocking autoclave under a nitrogen pressure of 500 psi. By ^{31}P NMR analysis on the hydrolyzed product, 40% of the P was present as H_3PO_4, 44% as the monophosphonic acid $CF_3CH(OH)PO_3H_2$ (multiplet at −10 ppm, $J = 8$ Hz), and the other 16% as HPO_3H_2. This analysis is typical of similar runs at atmospheric pressure in di-n-propyl sulfone, the moles of phosphate always being nearly equal to the moles of monophosphonate regardless of varying AR and source of anhydrizing power [P_4O_6, PCl_3, or $(CF_3CO)_2O$].

Purification of the monophosphonate was not carried beyond the 93% level. One sample, recrystallized as the acid from trifluoroacetic acid, gave a crystalline product in which 90–95% of the P was present as monophosphonate by ^{31}P NMR ($\delta = -10.2$ ppm); chemical analysis showed 6.5% of the P present as orthophosphoric acid. The molecular weight of the Na_2 salt (by $Na_2SO_4 \cdot 10\ H_2O$ transition lowering and corrected for 6.5 mole % Na_2HPO_4) was 215 (theory 224). The NMR spectra (^1H, ^{19}F, and ^{31}P) are in complete agreement with the acid formula $CF_3CH(OH)PO_3H_2$. As can be seen from T. J. Flautt's interpretation of the NMR spectra in the table below, no halide ion has been eliminated as it was in the Perkow reaction (102). Thus the proton spectrum was cleanly resolved into a quartet (H coupled to 3 equivalent F atoms) which is doubled by the coupling of the H to a single P atom. The fluorine spectrum is a triplet (doubled doublet), which becomes a doublet when the phosphorus is decoupled. The phosphorus spectrum also has the appearance of two overlapping quartets, but was not well resolved.

NMR Spectra of the Monophosphonic Acid

^1H
Quartet a,	$\tau' =$	5.48 ppm,	$J_{HF} =$	8.8 Hz
Quartet b,	$\tau' =$	5.70 ppm,	$J_{HF} =$	8.7 Hz
Doublet ab,	$\tau' =$	5.59 ppm,	$J_{HP} =$	13.6 Hz

^{19}F
| Doublet a, | $\delta = -6.65$ ppm, | $J_{FH} =$ | 8.7 Hz |
| Doublet b, | $\delta = -6.95$ ppm, | $J_{FP} =$ | 8.4 Hz |

^{31}P Multiplet unresolved at $\delta = -10.2$ ppm

C. Empirical Observations—Scope of the Mixed Anhydride Reaction

The general scope of the mixed anhydride reaction may be somewhat more fully defined by examining several relatively unsuccessful experiments. Distribution of species was determined by ^{31}P NMR. In no case were completely acceptable analyses obtained.

Sodium salts were prepared by adding base to an aqueous solution of the aniline salt until the pH is about 10. Aniline may be removed by extraction with ether.

1. BENZOIC ACID

Benzoic acid, phosphorus trichloride, and water (3:2:3 molar ratio) were heated together for several hours at temperatures up to 130°. The water-soluble fraction was hydrolyzed (16 hr, refluxing 20% HCl) and partially purified by washing the aniline salts. Formation and recrystallization of the sodium salt gave small amounts of a material whose ^{31}P chemical shift (-16 ppm) and C/P ratio (7:2) were consistent with the formulation of this material as a hydroxydiphosphonate.

2. NITRILOTRIACETIC ACID

Nitrilotriacetic acid, phosphorous acid, and phosphorus trichloride (1:4:2 molar ratio) were cautiously allowed to react in di-n-propyl sulfone. The maximum temperature was 100° (22 hr). After hydrolysis of the reaction mixture, and removal of the sulfone (chloroform extraction), the product was crystallized as the aniline salt. The ^{31}P NMR chemical shift (triplet at -7.5 ppm) and C/P ratio (1:1.06), for the sodium salt, were consistent with its formulation as a HDP (see p. 65).

3. PROPIOLIC ACID

Propiolic acid, phosphorous acid, and phosphorus trichloride (1:3:1 molar ratio) were heated in di-n-propyl sulfone at 83° (70 hr). The dense product was separated, hydrolyzed, converted to the aniline salt, and recrystallized. The resultant product showed numerous diphosphonate-like species ($\delta = -22$ to -9 ppm). The sodium salt(s) had a P/C ratio of 1.4:1.

4. CINNAMIC ACID

Cinnamic acid was allowed to react with phosphorous acid and phosphorus trichloride (3:4:2 molar ratio) at 65° (48 hr) and 100–105° (16 hr). Hydrolysis of the resultant dense mass gave a crude material containing several different phosphorus species (NMR). After sulfone removal, crystallization of the aniline salt was followed by conversion to the sodium salt. The latter gave a single phosphorus-containing species ($\delta = -17$ ppm). Proton NMR *suggested* $C_6H_5CH=CH-CH(PO_3)_2HNa_3$ as the structure of the product.

5. MALONIC ACID

By keeping the temperature at 100° or a little below and by using an AR near 2, one can accomplish the following by mixed anhydride reactions:

1. Attain a maximum of 65–75% of the phosphorus in the product as phosphonates (counting the several ^{31}P NMR peaks from $\delta = -24$ to -5 ppm).

2. Minimize decarboxylation, which becomes appreciable above 100° and would lead to formation of EHDP in the product.

3. Hold the orthophosphate formed to less than 10% of the phosphorus.

4. Hold the unconverted phosphorous acid to a minimum (<10% of the phosphorus).

Another phosphorus species ($\delta = -37$ ppm), possibly a phosphinate, is sometimes present to the extent of 5–10% of the phosphorus.

All products obtained were complex mixtures, as shown both by the numerous peaks in the phosphorus NMR spectrum and by the failure of crystallization methods, on the acid or various salts, to yield a pure product. Even conversion to methyl esters by heating with excess of trimethyl orthoformate did not lead to well-purified di- or higher polyphosphonates.

The orthophosphate content did not increase (^{31}P NMR) during esterification with the trimethyl orthoformate (175,176), as was the case for similar esterification of $H_2O_3PCH_2C(OH)(PO_3H_2)_2$ (172). Vacuum distillation effectively removed the phosphate impurity as $OP(OCH_3)_3$ and the phosphite impurity as the monophosphonate, $(CH_3O)_2CHP(O)(OCH_3)_2$,* and some of the higher phosphonates, but soon thereafter rose to a temperature (about 200°) where the liquid in the distillation flask began rearranging so as to make more orthophosphate (cf. earlier section on esters of 1-hydroxy-1,1-diphosphonates).

D. Tetramethyl Phenylmethanehydroxydiphosphonate

Hydrogen dimethyl phosphite (5.16 g, 0.047 mole) was placed in a reaction flask to which 100 ml of diethyl ether had been added. Dibutyl amine (0.5 g, 0.0026 mole) was added and the solution was cooled to 0°. Dimethyl benzoylphosphonate (10 g, 0.047 mole) was added slowly with rapid stirring. The reaction was moderately exothermic; external cooling was required to maintain the temperature at 0°. A white solid began to form almost immediately. After all the dimethyl benzoylphosphonate had been added, the reaction mixture was allowed to warm to room temperature. Filtration yielded 14.6 g (96%) of the title compound [mp 130–133°, δ (^{31}P NMR) $= -18.0$ ppm].

Analysis. Calculated for $C_{11}H_{18}O_7P_2$: C, 40.8; H, 5.6; P, 19.1%; mol wt 324. Found: C, 40.9; H, 5.6; P, 19.6%; mol wt 340.

The tetramethyl ester of phenylmethanehydroxydiphosphonic acid was hydrolyzed by refluxing for 3 hr with an excess of concentrated HCl. The monosodium salt was crystallized by the method of Pflaumer and Filcik (107). Analysis and ^{31}P NMR ($\delta = -15.5$ ppm) indicated a pure compound.

* This monophosphonate was formed from the $HP(O)(OH)_2$ impurity in the crude products from malonic acid, and was isolated by vacuum distillation (175,176). In the NMR spectra its single phosphorus peak fell at -14.5 ppm and it had three kinds of protons in 1:6:6 ratio, namely, a doublet at $\tau' = 5.39$ ppm ($J_{HCP} = 5.5$ Hz), a doublet at $\tau' = 6.41$ ppm ($J_{H,COP} = 10.5$ Hz), and a singlet at $\tau' = 6.68$ ppm (H,COC). This compound was also prepared (172) from pure $HP(O)(OH)_2$ and excess $HC(OCH_3)_3$.

APPENDIX II

Properties of Polyphosphonates from the Literature[a]

$\pi = PO_3G_2$

Compound[b]	G_n or $G_n(G_m')$	Phys. description	bp/mm or mp (°C)	n_D^{25}	δ (ppm)	J (Hz)	Refs.
					31P NMR data[c]		
C₁ acids, gem-DP							
$CH_2\pi_2$	H₄	S^d	203–206		−17.6 ± 0.9	22	40,86,91,92, 131,165,177
$CH_2\pi_2$	H₄						45
$CH_2\pi_2$	(HEt^d)₂	L^d		1.470^e	−17.8 (0.5M)		42
$CH_2\pi_2$	(NaAllyl)₂	S	209–211				17
$CH_2\pi_2$	Me₄	L	90/0.05	1.452	−23.0*	21	131
$CH_2\pi_2$	Et₄	L	94/0.1	1.441	−19.0*	20	5,131,149
$CH_2\pi_2$	iPr₄	L	100/0.1	1.432	−17.5*	21	131,165,176, 177
$CH_2\pi_2$	(Allyl)₄	L	140/0.6	1.468, 20	−15.2		17
$CH_2\pi_2$	Na₂H₂	S			−13.6		91
$CH_2\pi_2$	Na₄	S					91
$CH_2\pi_2$	(Me₄N)₂H₂	S			−15.7 (0.5M)		45
$CH_2\pi_2$	(Me₄N)₃H	S			−15.7 (0.5M)		45
$CH_2\pi_2$	(Me₄N)₄	S			−15.3 (0.5M)		45
$NaCH\pi_2$	Me₄	S			−45.5		14,131
$NaCH\pi_2$	Et₄	S	138–140		−41.5		5,14,131
$NaCH\pi_2$	iPr₄	S			−40.5		131
$KCH\pi_2$	Et₄	S	122–125		−45.0		24
$CH_2(POCl_2)_2$		S	104–105		−24.2		79,83,136,165
$HC(OH)\pi_2$	H₄	S			−16.1		45

HC(OH)π_2	(HNa)$_2$	S			-15.0	15	130
HC(Oπ)π_2	iPr$_6$	L			$-12.0, +2.0$*		130
HC(Cl)π_2	iPr$_4$	L			-11.5*	17	131
HC(Br)π_2	H$_4$	S			$-12.$		45
HC(Br)π_2	iPr$_4$	L		1.459	-12.0*	17	131,170
HC(I)π_2	iPr$_4$	L			-13.5*	17	131
Cl$_2$Cπ_2	H$_4$	S	249–251		-7.9	singlet	45,130,170
Cl$_2$Cπ_2	Et$_4$	L	120/0.05	1.462	-8.5*	singlet	14a,131
Cl$_2$Cπ_2	iPr$_4$	S	50		-6.5	singlet	131
Cl$_2$C(POCl$_2$)$_2$		S	75–76				103
Br$_2$Cπ_2	Et$_4$	L	120/0.08	1.491^f	-8.5*	singlet	131
Br$_2$Cπ_2	iPr$_4$	L	90/$<$0.04	1.475^f	-6.5*	singlet	131
I$_2$Cπ_2	iPr$_4$	L, U			-10.5*	singlet	131
(HO)$_2$Cπ_2	(HNa)$_2$	S			-14.5	singlet	130
O=Cπ_2	Na$_4$	S			0.0	singlet	130
N$_2$Cπ_2	Et$_4$	L	133/0.4	1.455, 20	(CN$_2$ at 2114 cm^{-1} in infrared)		135
C$_2$ acids, *gem*-DP							
CH$_3$CHπ_2	H$_4$	S	179–181		-22.8 ± 0.2	17, 22	45,131,170, 176,177
CH$_3$CHπ_2	iPr$_4$	L			-22.0*		131,170
CH$_3$CHπ_2	HNa$_3$	S			-22.6		176,177
CH$_3$CHπ_2	(Me$_4$N)$_4$	S			-20.5		45
CH$_3$C(OH)π_2	H$_4$	S	69–70		-19.8	16	45,86
CH$_3$C(OH)π_2	Me$_4$	S			-22.0		174
CH$_3$C(OH)π_2	Et$_4$	L, S	38–39		-20.8*		31,118
CH$_3$C(OH)π_2	(Me$_4$N)$_4$	S			-18.6		45
CH$_3$C(OAc)π_2	Et$_4$	L	148/1.5	1.442, 20			117,124,128

85

| Compound[b] | G_n or $G_n(G_m')$ | Phys. description | bp/mm or mp (°C) | n_D^{25} | 31P NNR data[c] | | Refs. |
					δ (ppm)	J (Hz)	
$Cl_2C=C\pi_2$	H_4	S			-7.5	singlet	131
$Cl_2C=C\pi_2$	iPr_4	L			-6.0*	singlet	131
$\pi_2CHCOOG$	$(iPr)_5$	L			-11.0*		131
$\pi_2CHCH_2PO_3G_2'$	H_6	S			-24.0, -19.0	(DP)	131
$\pi_2CHCH_2PO_3G_2'$	$(iPr)_4Me_2$	L			-30.5, -24.0	(DP)	131
$\pi_2CHCH_2PO_3G_2'$	iPr_6	L	142/0.0006	1.443, 20			140
$\pi_2CHCH(OH)\pi$	H_6	S			-17.9, 1 peak		132
$\pi_2CHCH(OH)\pi$	$(HNa)_3$	S			-16.7, -15.6		176
$\pi_2C(OH)CH_2\pi$	H_6	S			-23.0, -17.9	(DP)	132
$\pi_2C(OH)CH_2\pi$	$(HNa)_3$	S			-21.1, -18.6	(DP)	132
C_2 acids, vic-DP							
$(\pi CH_2)_2$	H_4	S	220-223		-27.4		91
$(\pi CH_2)_2$	$(HEt)_2$	S	48		-26.8		91,166
$(\pi CH_2)_2$	Me_4	L	158/2	1.443, 20			111,115
$(\pi CH_2)_2$	Et_4	L	160/2	1.436	-26.8*		14,36,39,40, 76,91,92,104
$(\pi CH_2)_2$	iPr_4	L	145/0.4	1.432, 20			111
$(\pi CH_2)_2$	Bu_4	L	207-210/7	1.440, 20			112
$(\pi CH_2)_2$	Ph_4	S	155-156				66,78
$(\pi CH_2)_2$	$(HNa)_2$	S			-23.3		86,91
$(\pi CH_2)_2$	Na_4	S			-22.4		86,91
$(\pi CH_2)_2$	$(Me_4N)_4$	S			-23.2		86
$(Cl_2OPCH_2)_2$		S	167-170				77,79,83,104
$(GOP(O)ClCH_2)_2$	Et_2	S	51-52				74
$trans$-$\pi CH=CH\pi$	Et_4	S	146/0.3	1.449, 20			160

Compound		State	bp/mm	n_D			Refs
$\pi C{\equiv}C\pi$	Me$_4$	L	13–15	1.448			145
$\pi C{\equiv}C\pi$	Et$_4$	L	184/2.5	1.448, 20			62
C$_3$ acids, gem-DP							
$\pi_2C(CH_3)_2$	H$_4$	S	229		-27.2 ± 0.3	16.5	45,131,170, 176,177
$\pi_2C(CH_3)_2$	iPr$_4$	L	69/0.005		$-25.3 \pm 0.3^*$	16	54,131,170
$\pi_2C(CH_3)_2$	HNa$_3$	S			-27.5	16	131,176,177
$\pi_2C(CH_3)_2$	(Me$_4$N)$_4$	S			-25.1		45
$[Cl_2P(O)]_2C(CH_3)_2$		S	>100		-45.1		54
$\pi_2CHN(CH_3)_2$ Na$^+$	Et$_4$	L	115/0.03		-18.0		49,50,51
$\pi_2\overset{\vert}{C}{-}N(CH_3)_2$	Et$_4$	S			-33.0		51
π_2CHCH_2COOG'	H$_5$	S			-20.0		131,170
π_2CHCH_2COOG'	Et$_4$Me	L	156/0.4	1.444			90
π_2CHCH_2COOG'	(iPr)$_4$Et	L			-20.0^*		131,170
π_2CHCH_2COOG'	HNa$_4$	S			-19.5		131
$CH_3{-}C\pi_2$	Me$_4$Et$_2$	L	195/0.6	1.461, 20			121
$\underset{CH_2PO_3G_2'}{CH_3{-}C\pi_2}$	(iPr)$_4$Et$_2$	L	194/0.5	1.455, 20			121
$\underset{CH_2PO_3G_2'}{\phantom{CH_3{-}C\pi_2}}$							
$(\pi_2CH)_2CH_2$	H$_8$	S	227–229		-20.5		131
$(\pi_2CH)_2CH_2$	iPr$_8$	L			-22.0^*		131

APPENDIX II contd.

Compound[b]	G_n or $G'_n(G_m')$	Phys. description	bp/mm or mp (°C)	n_D^{25}	³¹P NMR data[c] δ (ppm)	J (Hz)	Refs.
C₃ acids, vic-PP only							
$\pi CH_2CH(\pi)CH_3$	H_4	S	123	1.441			108,143
$\pi CH_2CH(\pi)CH_3$	Et_4	L	133/0.9	1.447, 20			103,108,143
$\pi CH_2CH(\pi)COOG'$	Et_4Me	L	164/4	1.451, 20			71
$\pi CH_2CH(\pi)CN$	Et_4	L	174/2	1.459, 20			126
$\pi CH=C—CH_3$	Et_2Me_2	L	145/0.8				121
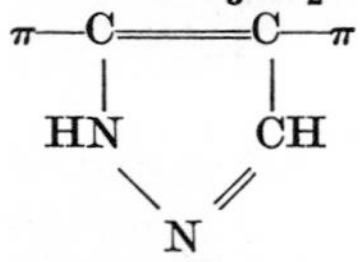 ![structure with PO₃Me₂, C=C—π, HN, CH, N] Me_4	Me_4	S	131.5				145
$\pi CH_2CH(\pi)CH_2\pi$	H_6	S			−27.5		22a
$\pi CH_2CH(\pi)CH_2\pi$	Et_6	L	170/0.1		−28.5		22a
$\pi_2CH_2CHCH_3$	Me_4Et_2	L	195/0.6	1.461, 20			121
$\pi_2CH_2CHCH_3$ PO_3G_2'	Pr_4Et_2	L	194/0.6	1.455, 20			121
PO_3G_2'							

C4 acids, gem-DP						
[(CH$_3$)$_3$NCHπ_2]I	Et$_4$	S	116–118		−11.0	51
(CH$_3$)$_3$N̄Cπ_2 (+ −)	Et$_4$	L	38–40	1.421	−28.0	51
πC=Cπ / F$_2$C—CF$_2$	Et$_4$	(weak C=C at 1595 cm^{-1})				37
C4 acids, *vic*-PP only						
$\pi\cdot$CHCOOG′ / $\pi\cdot$CHCOOG′	Me$_4$Et$_2$		203/3	1.464, 20		114
$\pi\cdot$CHCOOG′ / $\pi\cdot$CHCOOG′	Et$_6$		214/5	1.470, 20		114
π CH—CH$_2$PO$_3$G$_2$′ / π C=CH$_2$	Et$_4$Me$_2$	L	210/1	1.470, 20		127
π CH—CH$_2$PO$_3$G$_2$′ / π C=CH$_2$	Et$_6$	L	217/1	1.462, 20		127
π CH—CH$_2\pi$ / π C=CH$_2$	H$_8$	S			−27.2	22a
π CH—CH$_2\pi$ / π CH—CH$_2\pi$	Et$_8$	L			−30.3	22a
C5 acids, gem-DP						
(n-Bu)CHπ_2	H$_4$	S	163–165		−23 ± 0.5	54,75,131
(n-Bu)CHπ_2	Et$_4$	L	149/0.3	1.443, 20		75
(n-Bu)CHπ_2	iPr$_4$	L	95/0.012		−21.8 ± 0.2	54,131

APPENDIX II contd.

Compound[b]	G_n or $G_n(G_m')$	Phys. description	bp/mm or mp (°C)	n_D^{25}	^{31}P NMR data[c] δ (ppm)	J (Hz)	Refs.
POCl₂ | (n-Bu)CH | POCl₂		L	120/0.028		−34.6	21	54
(n-Bu)C(Cl)π₂ | POCl₂	H₄	S			−15.9	singlet	54
(n-Bu)—C—Cl | POCl₂		?			−32.9	20	54
C₅ acids, *vic*-DP πCH=CCH(CH₃)₂ | π	Et₄	L	139/0.8	1.462, 20			120
π CH₂C=C(CH₃)₂ | PO₃G₂'	Et₂Me₂	L	147/1	1.467, 20			125
π CH₂C=C(CH₃)₂ | PO₃G₂'	Pr₄	L	165/1.5	1.458, 20			125
πC=Cπ (CF₂)₃	H₄	S	194–199				38

90

$\pi \text{C}=\text{C}\pi$ $(\text{CF}_2)_3$	Me$_4$	L	134/0.4	1.416, 20		37,38
$\text{Cl}_2\text{PC}=\text{CPCl}_2$ $(\text{CF}_2)_3$ (O, O)	Et$_4$	L	112/0.1	1.417, 20	+1.5	37,38,173
$\text{Cl}_2\text{PC}=\text{CPCl}_2$ $(\text{CF}_2)_3$ (O, O)		S	78.5–81.5			38
C$_6$ acids, vic-PP only π CH(CH$_2$)$_3$OCH$_3$	Et$_4$	L	217/11	1.448, 20		109
—CH$_2$, π C(CH$_3$)COOG'	Me$_6$	S	164.5–165.5			48
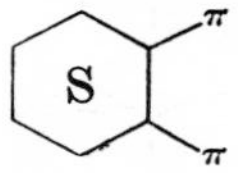 π C(CH$_3$)COOG' ring (S)	H$_4$	S	216–219			160
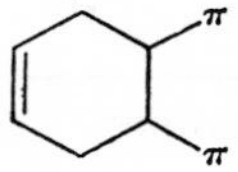 π / π cyclohexene ring	Et$_4$	L	157/0.03	1.465, 20		160
$\pi \text{C}=\text{C}\pi$ $(\text{CF}_2)_4$	Et$_4$	L	135/0.5	1.414		37

91

APPENDIX II contd.

Compound[b]	G_n or $G_n(G_m')$	Phys. description	bp/mm or mp (°C)	n_D^{25}	31P NMR data[c] δ(ppm)	J(Hz)	Refs.
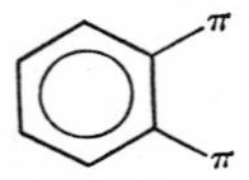	Me$_4$	S	82–84				101,145
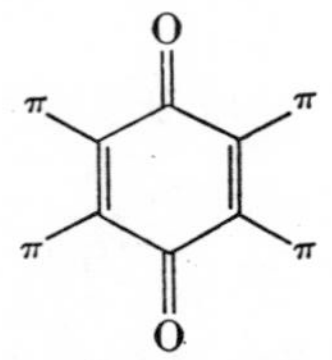	iPr$_8$	S	154.5–155.0				134
C$_7$ acids, gem-DP C$_6$H$_5$CH(OH)π_2 C$_7$ acids, vic-DP	Me$_4$	S	132–133		−18.0		174
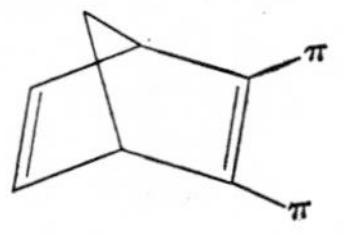	Me$_4$	L	137/0.01	1.495			145
πC=C(CH$_3$)$_2$							110
πCHCH$_2$CH$_3$	Me$_4$	L	182/6	1.478, 20			110

92

$\pi C{=}C(CH_3)_2$	Et_4	L	185/6	1.460, 20		110
$\pi CHCH_2CH_3$, C_{7+} acids, *gem*-DP						
$C_6H_5CH_2CH_2CH\pi_2$	H_4	S	210–212		−21.0	131
$C_6H_5CH_2CH_2CH\pi_2$	iPr_4	L			−20.5	131,150,170
$(n\text{-}C_4H_9)_2C\pi_2$	H_4	S	239–241		−26.0	131
$(n\text{-}C_4H_9)_2C\pi_2$	iPr_4	L			−25*	54,131
$(n\text{-}C_8H_{17})CH\pi_2$	iPr_4	L			−22.2*	54
$(n\text{-}C_9H_{19})CH\pi_2$	Et_4	L	140/0.004		−23.4*	54
$(n\text{-}C_{10}H_{21})CH\pi_2$	iPr_4	L			−22.2*	54
$(n\text{-}C_{14}H_{29})CH\pi_2$	H_4	S	156–162			131
C_{7+} acids, *vic*-DP						
$\underset{\pi}{C_6H_5CHCH_2\pi}$	H_4	S	212–214			3
$\underset{\pi}{C_6H_5CHCH_2\pi}$	Et_4	L	181/1	1.492, 20	Str. OK by ^{1}H NMR	3,47
$\underset{\pi}{C_6H_5CHCHCH_3}$	Et_4	L	196/1		Str. OK by ^{1}H NMR	47
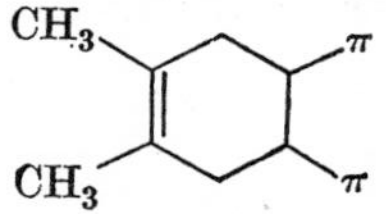 cyclohexene structure	Et_4	L	146/0.01	1.469, 20		160
$\underset{\pi}{\overset{\pi}{C_6H_5CHCHCOOG'}}$	H_5	S	113–115			116

93

APPENDIX II contd.

Compound[b]	G_n or $G_n(G_m')$	Phys. description	bp/mm or mp (°C)	n_D^{25}	[31P NMR data][c] δ (ppm)	J (Hz)	Refs.
$\overset{\pi}{C_6H_5}\overset{\pi}{CH}CHCOOG'$	Me$_5$	L	212/4	1.507, 20			116
$\overset{\pi}{C_6H_5}\overset{\pi}{CH}CHCOOG'$	Et$_4$Me	L	208/2	1.494, 20			116
$\overset{\pi}{C_6H_5}\overset{\pi}{CH}CHCOOG'$	Et$_5$	L	202/0.9	1.492, 20			116
$C_6H_5CHCHC_6H_5$	Bu$_4$	L	154/0.2		Str. OK by ^{1}H NMR		47
$(\overset{\pi}{C_6H_5}C(O)CH\pi)_2$	Me$_4$	S	134–135				26,27

[a] See Experimental Section for properties of compounds described there.

[b] If the column for G_n or $G_n(G_m')$ is blank, properties given refer to compound given in first column.

[c] Starred ^{31}P NMR data taken on undiluted liquid; others taken on solutions, mostly at concentrations of 10–50%. ^{1}H NMR data will be found for one or more of the compounds cited in each of the following references: 5,14,24,31,47–51,54,131, 135,145,149,150.

[d] Abbreviations are those used in *Chemical Abstracts*, except as follows: iPr = isopropyl, DP = diphosphonate, PP = polyphosphonate, π = PO$_3$G$_2$ where G may be H, Me$_4$N, alkali metal or R (alkyl or aryl), S = solid, L = liquid, U = unstable.

[e] Since indices of refraction from different workers may differ appreciably in the 3rd decimal place, the 4th decimal place is not given. The number following the comma gives the temperature when it is not 25°.

[f] Misquoted in original publication.

94

REFERENCES

1. Arbuzov, A. E., and N. P. Kushkova, *J. Gen. Chem. (USSR)*, **6**, 283 (1936); *Chem. Abstr.*, **30**, 4813 (1936).
2. Arbuzov, A. E., and M. M. Azanovskaya, *Dokl. Akad. Nauk SSSR*, **58**, 1961 (1947); *Chem. Abstr.*, **46**, 8606 (1952).
3. Arbuzov, B. A., and B. P. Lugovkin, *Zh. Obshch. Khim.*, **21**, 99 (1951); *Chem. Abstr.*, **45**, 7002e (1951).
4. Arbuzov, B. A., and N. P. Bogonostseva, *Zh. Obshch. Khim.*, **27**, 2356 (1957); *Chem. Abstr.*, **52**, 7129 (1958).
5. Baldeschwieler, J. D., F. A. Cotton, B. D. N. Rao, and R. A. Schunn, *J. Amer. Chem. Soc.*, **84**, 4454 (1962).
6. Bel'skii, V. E., and L. A. Kudryavtseva, *Izv. Akad. Nauk SSSR, Ser. Khim.*, 2160 (1968); *Chem. Abstr.*, **70**, 10734 (1969).
7. Birum, G., U.S. Pat. 2,857,415 (1958).
8. Birum, G. H., U.S. Pat. 3,029,271 (1962).
8a. Birum, G. H., U.S. Pat. 3,183,256 (1965).
9. Blaser, B., H. G. Germscheid, and K. H. Worms, U.S. Pat. 3,303,139 (1967).
10. Blaser, B., and K. H. Worms, Ger. Pat. 1,082,235 (1960).
11. Boyer, W. P., and J. R. Mangham, U.S. Pat. 2,634,288 (1953).
12. Brokke, M. E., and D. G. Stoffey, U.S. Pat. 3,247,052 (1966).
13. Brooks, B. T., *J. Amer. Chem. Soc.*, **34**, 492 (1912).
14. Brophy, J. J., and M. J. Gallagher, *Aust. J. Chem.*, **20**, 503 (1967).
14a. Bunyan, P. J., and J. I. G. Cadogan, *J. Chem. Soc.*, **1962**, 2953.
15. Burn, A. J., J. I. G. Cadogan, and P. J. Bunyan, *J. Chem. Soc.*, **1963**, 1527.
16. Burn, A. J., J. I. G. Cadogan, and P. J. Bunyan, *J. Chem. Soc.*, **1964**, 4369.
17. Cade, J. A., *J. Chem. Soc.*, **1959**, 2266.
18. Cade, J. A., *J. Chem. Soc.*, **1959**, 2272.
19. Callis, C. F., A. F. Kerst, and J. W. Lyons in *Coordination Chemistry*, Stanley Kirschner, Ed., Plenum Press, New York, 1969, pp. 223–240.
20. Canavan, A. E., B. F. Dowden, and C. Eaborn, *J. Chem. Soc.*, **1962**, 331.
21. Carroll, R. L., and R. R. Irani, *Inorg. Chem.*, **6**, 1994 (1967).
22. Carroll, R. L., and R. R. Irani, *J. Inorg. Nucl. Chem.*, **30**, 2971 (1968).
22a. Cilley, W. A., D. A. Nicholson, and D. Campbell, *J. Amer. Chem. Soc.*, **92**, 1685 (1970).
23. Corbridge, D. E. C., and F. R. Tromans, *Anal. Chem.*, **30**, 1101 (1958).
23a. Corbridge, D. E. C., in *Topics in Phosphorus Chemistry*, Vol. 6, M. Grayson and E. J. Griffith, Eds., Interscience, New York, 1969, pp. 237–366.
24. Cotton, F. A., and R. A. Schunn, *J. Amer. Chem. Soc.*, **85**, 2394 (1963).
24a. Crofts, P. C., and G. M. Kosolapoff, *J. Amer. Chem. Soc.*, **75**, 3379 (1953); see also **75**, 5738 (1953).
25. Cummins, R. W., *Detergent Age*, March, 1968, pp. 22–27.
25a. Curry, J. D., and W. A. Cilley, paper in preparation on propane-1-hydroxy-1,1,3-triphosphonic acid.
26. Dershowitz, S., and S. Proskauer, *J. Org. Chem.*, **26**, 3595 (1961).
27. Dershowitz, S., U.S. Pat. 3,104,257 (1963).

28. Djordjevic, C., and H. Gorican, *J. Inorg. Nucl. Chem.*, **28**, 1451 (1966).
29. Djordjevic, C., H. Gorican, and S. L. Tan, *J. Inorg. Nucl. Chem.*, **29**, 1505 (1967).
30. Dyatlova, N. M., V. V. Medyntsev, T. M. Balashova, T. Ya. Medved, and M. I. Kabachnik, *Zh. Obshch. Khim.*, **39**, 329 (1969); *Chem. Abstr.*, **70**, 118663 (1969).
31. Fitch, S. J., and K. Moedritzer, *J. Amer. Chem. Soc.*, **84**, 1876 (1962).
32. Fitch, S. J., *J. Amer. Chem. Soc.*, **86**, 61 (1961).
33. Fitch, S. J., C. Coeur, and S. K. Liu, U.S. Pat. 3,256,370 (1966).
34. Fitch, S. J., C. Coeur, and R. Irani, U.S. Pat. 3,299,123 (1967).
35. Flesher, J. W., Y. T. Oester, and T. C. Myers, *Nature*, **185**, 772 (1960).
36. Ford-Moore, A. H., and J. H. Williams, *J. Chem. Soc.*, **1947**, 1465.
37. Frank, A. W., *J. Org. Chem.*, **30**, 3663 (1965).
38. Frank, A. W., *J. Org. Chem.*, **31**, 1521 (1966).
39. Garner, A. Y., E. C. Chapin, and P. M. Scanlon, *J. Org. Chem.*, **24**, 532 (1959).
40. Ginsburg, V. A., and A. Ya. Yakubovich, *Zh. Obshch. Khim.*, **28**, 728 (1958); *Chem. Abstr.*, **52**, 17091 (1958).
41. Gorican, H., and D. Grdenic, *Proc. Chem. Soc.*, **1960**, 288.
42. Gorican, H., and D. Grdenic, *J. Chem. Soc.*, **1964**, 513.
43. Gorican, H., and D. Grdenic, *Anal. Chem.*, **36**, 330 (1964).
44. Gorican, H., and C. Djordjevic, *Mikrochim. Acta*, **1966**, 767.
45. Grabenstetter, R. J., O. T. Quimby, and T. J. Flautt, *J. Phys. Chem.*, **71**, 4194 (1967).
46. Grdenic, D., and V. Jagodic, *J. Inorg. Nucl. Chem.*, **26**, 167 (1964).
47. Griffin, C. E., and T. D. Mitchell, *J. Org. Chem.*, **30**, 1935 (1965).
48. Griffin, C. E., and T. D. Mitchell, *J. Org. Chem.*, **30**, 2829 (1965).
49. Gross, H., and B. Costisella, *Angew. Chem.*, **80**, 364 (1968).
50. Gross, H., and B. Costisella, *Angew. Chem. Int. Ed. Engl.*, **7**, 391 (1968).
51. Gross, H., and B. Costisella, *Angew. Chem. Int. Ed. Engl.*, **7**, 463 (1968).
52. Hall, L. A. R., and C. W. Stephens, *J. Amer. Chem. Soc.*, **78**, 2565 (1956).
53. Harvey, R. G., and E. R. DeSombre in *Topics in Phosphorus Chemistry*, Vol. 1, M. Grayson and E. J. Griffith, Eds., Interscience, New York, 1964, pp. 57–111.
54. Hays, H. R., and T. J. Logan, *J. Org. Chem.*, **31**, 3391 (1966).
55. Healy, T. V., and J. Kennedy, *J. Inorg. Nucl. Chem.*, **10**, 128 (1959).
56. Henning, H. G., and G. Petzold, *Z. Chem.*, **5**, 419 (1965).
57. Henning, H. G., and D. Gloyna, *Z. Chem.*, **6**, 28 (1966); *Chem. Abstr.*, **64**, 17628 (1966).
58. Hershey, J. W. B., and R. E. Monro, *J. Mol. Biol.*, **18**, 68 (1966).
59. Horner, L., H. Hoffmann, W. Klink, H. Ertel, and V. G. Toscano, *Chem. Ber.*, **95**, 581 (1962); *Chem. Abstr.*, **57**, 2248 (1962).
60. Hossenlopp, F., M. McPartlin, and J. P. Ebel, *Bull. Soc. Chim. Fr.*, **1965**, 2221.
61. Hudson, R. F., *Structure and Mechanism in Organo-Phosphorus Chemistry*, Academic, New York, 1965, Ch. 5, pp. 140–141.
62. Ionin, B. I., and A. A. Petrov, *Zh. Obshch. Khim.*, **35**, 1917 (1965); *Chem. Abstr.*, **64**, 6683 (1966).
63. Irani, R. R., and K. Moedritzer, *J. Phys. Chem.*, **66**, 1349 (1962).
64. Ishimori, T., K. Kimura, E. Nakamura, J. Akatsu, and T. Kobune,

Nippon Genshiryoku Gakkaishi, **5**, 633 (1963); *Chem. Abstr.*, **61**, 10102 (1964).

65. Jacques, J. K., Canad. Pat. 770,678 (1967).
66. Kabachnik, M. I., *Bull. Acad. Sci. USSR, Classe Sci. Chim.*, **1947**, 631; *Chem. Abstr.*, **42**, 5845 (1948).
67. Kabachnik, M. I., and P. A. Rossiiskaya, *Izv. Akad. Nauk SSSR, Otd. Khim. Nauk*, **48**, 1398 (1958); *Chem. Abstr.*, **53**, 6988 (1959).
68. Kabachnik, M. I., R. P. Lastovskii, T. Ya. Medved, V. V. Medyntsev, I. D. Kolpakova, and N. M. Dyatlova, *Dokl. Akad. Nauk SSSR*, **1967**, 177, 582; *Chem. Abstr.*, **69**, 5682 (1968).
69. Kabachnik, M. I., N. M. Dyatlova, T. Ya. Medved, and M. V. Rudomino, *Zh. Vses. Khim. Obshchest.*, **13**, 518 (1968); *Chem. Abstr.*, **70**, 63707 (1969).
70. Kamai, G., and V. A. Kukhtin, *Khim. i Primenenie Fosfororgan. Soedinenii, Akad. Nauk SSSR, Trudy 1-oi, Konferents.*, **1955**, 91; *Chem. Abstr.*, **52**, 241 (1958).
71. Kamai, G., and V. A. Kukhtin, *Trudy. Kazan. Khim. Tekhnol. Inst. im. S. M. Kirova*, **1956**, 141; *Chem. Abstr.*, **51**, 11985 (1957).
72. Kasparek, F., *Monats. Chem.*, **99**, 2016 (1968); *Chem. Abstr.*, **70**, 4222s (1969).
73. Kennedy, J., *Chem., Ind.* (*London*), **1958**, 950.
74. Knunyants, I. L., A. P. Puzerauskas, O. V. Kil'disheva, and E. Ya. Perova, *Izv. Akad. Nauk SSSR, Ser. Khim.*, **1966**, 1115; *Chem. Abstr.*, **65**, 10613 (1966).
74a. Kosolapoff, G. M., *Organophosphorus Compounds*, Wiley, New York, 1950, p. 124.
75. Kosolapoff, G. M., *J. Amer. Chem. Soc.*, **75**, 1500 (1953).
76. Kosolapoff, G. M., *J. Chem. Soc.*, **1955**, 3092.
77. Kosolapoff, G. M., and R. F. Struck, *J. Chem. Soc.*, **1961**, 2423.
78. Kosolapoff, G. M., *J. Chem. Soc.*, **1965**, 6638.
79. Kosolapoff, G. M., and A. D. Brown, Jr., *J. Chem. Soc.*, **1966**, 757.
80. Kreutzkamp, N., and H. Schindler, *Chem. Ber.*, **92**, 1695 (1959).
81. Ladd, E. C., and M. P. Harvey, U.S. Pat. 2,651,656 (1953).
82. Lerch, I., and A. Kottler, Ger. Pat. 1,002,355 (1957); *Chem. Abstr.*, **53**, 21814 (1959).
83. Maier, L., *Helv. Chim. Acta*, **48**, 133 (1965).
84. Maier, L., *Helv. Chim. Acta*, **49**, 842 (1966); *Chem. Abstr.*, **64**, 14210 (1966).
85. Malz, H., E. Kühle, and O. Bayer, U.S. Pat. 3,053,876 (1962).
86. Mark, V., C. Dungan, M. Crutchfield, and J. R. Van Wazer in *Topics in Phosphorus Chemistry*, Vol. 5, M. Grayson and E. J. Griffith, Eds., Interscience, New York, 1967.
87. Martin, D. J., and C. E. Griffin, *J. Organometal. Chem.*, **1**, 292 (1964).
88. McConnell, R. L., and H. W. Coover, *J. Amer. Chem. Soc.*, **78**, 4450 (1956).
89. Miller, C. D., R. C. Miller, and W. Rogers, Jr., *J. Amer. Chem. Soc.*, **80**, 1562 (1958).
90. Miller, L. A., U.S. Pat. 3,093,672 (1963).
91. Moedritzer, K., and R. R. Irani, *J. Inorg. Nucl. Chem.*, **22**, 297 (1961).
92. Moedritzer, K., L. Maier, and L. C. D. Groenweghe, *J. Chem. Eng. Data*, **7**, 307 (1962).
93. Moedritzer, K., U.S. Pat. 3,242,236 (1966).

98   J. D. CURRY, D. A. NICHOLSON, AND O. T. QUIMBY

94. Moos, C., N. R. Alpert, and T. C. Myers, *Arch. Biochem. Biophys.*, **88**, 183 (1960).
95. Myers, T. C., K. Nakamura, and J. W. Flesher, *J. Amer. Chem. Soc.*, **85**, 3292 (1963).
96. Myers, T. C., K. Nakamura, and A. B. Danielzadeh, *J. Org. Chem.*, **30**, 1517 (1965).
97. Myers, T. C., U.S. Pat. 3,238,191 (1966).
98. Neidlein, R., W. Haussmann, and E. Heukelbach, *Chem. Ber.*, **99**, 1252 (1966).
99. Nerdel, F., and W. Burghardt, *Naturwissenschaften*, **47**, 178 (1960).
100. Nylen, P., *Dissertation*, Uppsala, 1930, pp. 77–84.
101. Obrycki, R., and C. E. Griffin, *Tetrahedron Lett.*, **41**, 5049 (1966).
102. Perkow, W., *Chem. Ber.*, **87**, 755 (1954); see also M. S. Kharasch and I. S. Bengelsdorf, *J. Org. Chem.*, **20**, 1356 (1955).
103. Petrov, K. A., F. L. Maklyaev, and N. K. Bliznyuk, *Zh. Obshch. Khim.*, **30**, 1602 (1960); *Chem. Abstr.*, **55**, 1414 (1961).
104. Petrov, K. A., F. L. Maklyaeva, and N. K. Bliznyuk, *Zh. Obshch. Khim.*, **30**, 1608 (1960); *Chem. Abstr.*, **55**, 1414 (1961).
105. Petrov, K. A., V. B. Shevchenko, V. G. Timoshev, F. L. Maklyaev, A. V. El'kina, Z. I. Nagnibeda, and A. A. Volkova, *Zh. Neorg. Khim.*, **5**, 498 (1960); *Chem. Abstr.*, **55**, 3178 (1961).
106. Petrov, K. A., R. A. Baksova, L. V. Khorkhoyanu, and I. F. Rebus, *Probl. Organ. Sinteza, Akad. Nauk SSSR, Otd. Obshch. i Tekhn. Khim.*, **1965**, 310; *Chem. Abstr.*, **64**, 9757 (1966).
107. Pflaumer, P. F., and J. P. Filcik, Belg. Pat. 712,159 (1968).
108. Preis, S., T. C. Myers, and E. V. Jensen, *J. Amer. Chem. Soc.*, **77**, 6225 (1955).
108a. Prentice, J. B., and O. T. Quimby, paper in preparation on condensates of ethane-1-hydroxy-1,1-diphosphonic acid.
109. Pudovik, A. N., and B. A. Arbuzov, *Izv. Akad. Nauk SSSR, Otd. Khim. Nauk*, **1949**, 522; *Chem. Abstr.*, **44**, 1893 (1950).
110. Pudovik, A. N., *Zh. Obshch. Khim.*, **20**, 92 (1950); *Chem. Abstr.*, **44**, 5800 (1950).
111. Pudovik, A. N., *Dokl. Akad. Nauk SSSR*, **80**, 65 (1951); *Chem. Abstr.*, **50**, 4143 (1956).
112. Pudovik, A. N., and M. G. Imaev, *Izv. Akad. Nauk SSSR, Otd. Khim. Nauk*, **1952**, 916; *Chem. Abstr.*, **47**, 10463 (1953).
113. Pudovik, A. N., and M. M. Frolova, *Zh. Obshch. Khim.*, **22**, 2052 (1952); *Chem. Abstr.*, **47**, 9910 (1953).
114. Pudovik, A. N., *Bull. Akad. Sci. USSR, Div. Chem. Sci.*, (Eng. Transl.) **1952**, 821; same as *Chem. Abstr.*, **47**, 10467 (1953).
115. Pudovik, A. N., and G. M. Denisova, *Zh. Obshch. Khim.*, **23**, 263 (1953); *Chem. Abstr.*, **48**, 2572 (1954).
116. Pudovik, A. N., and D. Kh. Yarmukhametova, *Izv. Akad. Nauk SSSR, Otd. Khim. Nauk*, **1954**, 636; *Chem. Abstr.*, **49**, 8789 (1955).
117. Pudovik, A. N., and I. V. Konovalova, *Zh. Obshch. Khim.*, **33**, 98 (1963); *Chem. Abstr.*, **59**, 656 (1963).
118. Pudovik, A. N., and I. V. Konovalova, *Dokl. Akad. Nauk SSSR*, **143**, 875 (1962); *Chem. Abstr.*, **57**, 3480 (1962).

119. Pudovik, A. N., I. V. Konovalova, and L. V. Dedova, *Dokl. Akad. Nauk SSSR*, **153**, 616 (1963); *Chem. Abstr.*, **60**, 8060 (1964).
120. Pudovik, A. N., N. G. Khusainova, and I. M. Aladzheva, *Zh. Obshch. Khim.*, **34**, 2470 (1964); *Chem. Abstr.*, **61**, 9522 (1964).
121. Pudovik, A. N., N. G. Khusainova, and A. B. Ageeva, *Zh. Obshch. Khim.*, **34**, 3938 (1964); *Chem. Abstr.*, **62**, 7792 (1965).
122. Pudovik, A. N., N. G. Khusainova, and R. G. Galeeva, *Zh. Obshch. Khim.*, **36**, 69 (1966); *Chem. Abstr.*, **64**, 14210 (1966).
123. Pudovik, A. N., T. Kh. Gazizov, and A. P. Pashinkin, *Zh. Obshch. Khim.*, **36**, 563 (1966); *Chem. Abstr.*, **65**, 736 (1966).
124. Pudovik, A. N., T. Kh. Gazizov, and A. P. Pashinkin, *Zh. Obshch. Khim.*, **36**, 951 (1966); *Chem. Abstr.*, **65**, 8951 (1966).
125. Pudovik, A. N., and N. G. Khusainova, *Zh. Obshch. Khim.*, **36**, 1236 (1966); *Chem. Abstr.*, **65**, 16994 (1966).
126. Pudovik, A. N., G. E. Yastrebova, and V. I. Nikitina, *Zh. Obshch. Khim.*, **36**, 1232 (1966); *Chem. Abstr.*, **65**, 15418 (1966).
127. Pudovik, A. N., E. A. Ishmaeva, R. S. Akhmerova, and I. M. Aladzheva, *Zh. Obshch. Khim.*, **36**, 161 (1966); *Chem. Abstr.*, **64**, 14208 (1966).
128. Pudovik, A. N., T. Kh. Gazizov, *Zh. Obshch. Khim.*, **38**, 140 (1968); *Chem. Abstr.*, **69**, 96836 (1968).
129. Pudovik, A. N., I. V. Guryanova, L. V. Banderova, and G. V. Romanov, *Zh. Obshch. Khim.*, **38**, 143 (1968); *Chem. Abstr.*, **69**, 96839 (1968).
130. Quimby, O. T., J. B. Prentice, and D. A. Nicholson, *J. Org. Chem.*, **32**, 4111 (1967).
131. Quimby, O. T., J. D. Curry, D. A. Nicholson, J. B. Prentice, and C. H. Roy, *J. Organometal. Chem.*, **13**, 199 (1968).
132. Quimby, O. T., U.S. Pat. 3,400,148 (1968).
132a. Quimby, O. T., U.S. Pat. 3,400,176 (1968).
133. Quimby, O. T., U.S. Pat. 3,451,937 (1969).
134. Reetz, T., U.S. Pat. 2,935,518 (1960); *Chem. Abstr.*, **54**, 19598 (1960).
135. Regitz, M., W. Anschütz, and A. Liedhegener, *Chem. Ber.*, **101**, 3734 (1968); *Chem. Abstr.*, **70**, 19987 (1969).
136. Richard, J. J., K. E. Burke, J. W. O'Laughlin, and C. V. Banks, *J. Amer. Chem. Soc.*, **83**, 1722 (1961).
137. Rosen, S. W., and I. M. Klotz, *Arch. Biochem. Biophys.*, **67**, 161 (1957).
138. Roy, C. H., U.S. Pat. 3,251,907 (1966).
139. Rueggeberg, W. H. C., J. Chemak, and I. M. Rose, *J. Amer. Chem. Soc.*, **72**, 5336 (1950).
140. Saunders, B. C. and P. Simpson, *J. Chem. Soc.*, 3351 (1963).
141. Schmidt, P., Swiss Pat. 321,397 (1956).
142. Schwarzenbach, G., and J. Zurc, *Monatsh. Chem.*, **81**, 202 (1950).
143. Schwarzenbach, G., P. Ruckstuhl, and J. Zurc, *Helv. Chim. Acta.*, **34**, 455 (1951).
144. Schwarzmann, E., and J. R. Van Wazer, *J. Inorg. Nucl. Chem.*, **14**, 296 (1960).
145. Seyferth, D., and J. D. H. Paetsch, *J. Org. Chem.*, **34**, 1483 (1969).
146. Siddall, T. H., III, *J. Inorg. Nucl. Chem.*, **25**, 883 (1963).
147. Siddall, T. H., III, *J. Inorg. Nucl. Chem.*, **26**, 1991 (1964).

148. Siddall, T. H., III, and C. A. Prohaska, *Nature*, **202**, 1088 (1964).
149. Siddall, T. H., III, and C. A. Prohaska, *Inorg. Chem.*, **4**, 783 (1965).
150. Siddall, T. H., III, *J. Phys. Chem.*, **70**, 2249 (1966).
151. Simon, L. N., and T. C. Myers, *Biochem. Biophys. Acta*, **51**, 178 (1961).
152. Simon, L., T. Myers, and M. Mednieks, *Biochem. Biophys. Acta*, **103**, 189 (1965).
153. Steger, E., and J. Rehak, *Z. Anorg. Allg. Chem.*, **336**, 156 (1965).
154. Stewart, W. E., and T. H. Siddall, III, *J. Inorg. Nucl. Chem.*, **30**, 1513 (1968).
155. Stiles, A. R., U.S. Pat. 2,957,904 (1960).
156. Swanson, J. R., and R. G. Yount, *Biochem. Z.*, **345**, 395 (1966).
157. Szabo, K., and J. G. Brady, Fr. Pat. 1,343,998 (1963).
158. Szabo, K., U.S. Pat. 3,169,973 (1965).
159. Szabo, K., and J. G. Brady, U.S. Pat. 3,246,005 (1966).
160. Tavs, P., *Chem. Ber.*, **100**, 1571 (1967).
161. Van Druten, A., *Rec. Trav. Chim.*, **48**, 312 (1929).
162. Van Wazer, J. R., *Phosphorus and Its Compounds*, Vol. 1, Interscience, New York, 1958, pp. 479–480.
163. Van Winkle, J. L., and R. C. Morris, U.S. Pat. 2,681,920 (1954).
164. von Baeyer, H., and K. A. Hofmann, *Ber.*, **30**, 1973 (1897).
165. Walmsley, J. A., and S. Y. Tyree, *Inorg. Chem.*, **2**, 312 (1963).
166. Ciba, Ltd., Fr. Pat. 1,503,429 (1967); *Chem. Abstr.*, **69**, 96874 (1968).
167. Henkel et Cie., Belg. Pats. 619,619; 619,620; 619,621 (1962).
168. Merck and Co., Neth. Appl. 6,507,496 (1965); *Chem. Abstr.*, **64**, 17702 (1966).
169. Monsanto Co., Brit. Pat. 1,130,487 (1968); *Chem. Abstr.*, **70**, 37915 (1969).
170. Procter & Gamble Co. Brit. Pat. 1,026,366 (1966).

Unpublished work of the following authors is discussed in the text:
171. Brawn, D.
172. Cilley, W. A.
173. Curry, J. D.
174. Nicholson, D. A.
175. Prentice, J. B.
176. Quimby, O. T.
177. Roy, C. H.

SUPPLEMENTARY BIBLIOGRAPHY

Because of a delay in publication, this chapter only surveys the oligophosphonate literature through 1968. The following is a topical listing of more recent references. A few additional references from 1966 to 1968 are also included.

Synthesis and Physical Properties of Polyphosphonic Acids and/or the Corresponding Esters

Nicholson, D. A., and H. Vaughn, *J. Org. Chem.*, **36**, 3843 (1971).
Nicholson, D. A., and H. Vaughn, *J. Org. Chem.*, **36**, 1835 (1971).
Nicholson, D. A., W. A. Cilley, and O. T. Quimby, *J. Org. Chem.*, **35**, 3149 (1970).
Cilley, W. A., D. A. Nicholson, and D. Campbell, *J. Amer. Chem. Soc.*, **92**, 1685 (1970).
Blaser, B., K. H. Worms, H. G. Germscheid, and K. Wollmann, *Z. Anorg. Allg. Chem.*, **381**, 247 (1971).
Lapitskii, G. A., S. M. Makin, E. K. Lyapina, and A. S. Chebotarev, *Vysokomol. Soedin.*, *Ser. B*, **11**, 266 (1969).
Gross, H., and B. Costisella, *J. Prakt. Chem.*, **311**, 925 (1969).
Sturtz, G., *Bull. Soc. Chim. Fr.*, **4**, 1345 (1967).
Tavs, P., *Chem. Ber.*, **100**, 1571 (1967).
Whyte, D. D., P. F. Pflaumer, and T. S. Roberts, Fr. Pat. 1,531,913 (1968).
Dyer, J. K., U.S. Pat. 3,366,675 (1968).
Irani, R. R., and R. E. Mesmer, Fr. Pat. 1,546,145 (1968).
Pflaumer, P. F. and J. P. Filcik, Fr. Pat. 1,558,729 (1969).
Procter & Gamble Co., Neth. Appl. 6,604,176 (1966).
Procter & Gamble Co., Neth. Appl. 6,604,178 (1966).
Procter & Gamble Co., Neth. Appl. 6,604,219 (1966).
Procter & Gamble Co., Neth. Appl. 6,604,221 (1966).
Procter & Gamble Co., Neth. Appl. 6,606,548 (1966).
Procter & Gamble Co., Neth. Appl. 6,610,762 (1967).
Brun, G., and C. Blanchard, *C. R. Acad. Sci.*, *Ser. C*, **272**, 2154 (1971).
Wollmann, K., W. Ploeger, and K. W. Worms, *Ger. Offen.* 1,958,124 (1971).
Wollmann, K., W. Ploeger, and K. W. Worms, *Ger. Offen.* 1,957,300 (1971).
Nicholson, D. A., and D. Campbell, U.S. Pat. 3,579,570 (1971).
Whelan, D. J., and J. C. Johannessen, *Aust. J. Chem.*, **24**, 887 (1971).
Libby, R. A., *Inorg. Chem.*, **10**, 386 (1971).
Wiers, B. H., *J. Phys. Chem.*, **75**, 682 (1971).

Metal Complexes

Butter, E., W. Seifert, and H. Holzapfel, *Z. Chem.*, **7**, 23 (1967).
Grabenstetter, R. J. and W. A. Cilley, *J. Phys. Chem.*, **75**, 676 (1971).
Djordjevic, C., *Proc. 8th In. Conf. Coord. Chem.*, Vienna, 407 (1964).
Zur Nedden, P., *Fresenius' Z. Anal. Chem.*, **247**, 236 (1969).
Pribl, R., and V. Vesely, *Talanta*, **14**, 591 (1967).
Wiers, B. H., *J. Phys. Chem.*, **75**, 682 (1971).
Liggett, S. J., and R. A. Libby, *Talanta*, **17**, 1135 (1970).
Wiers, B. H., *Inorg. Chem.*, **10**, 2581 (1971).
Stewart, W. E., and T. H. Siddall, III. *J. Inorg. Nucl. Chem.*, **33**, 2965 (1971).

Biological Applications

Atkinson, M. R., and A. W. Murray, *Biochem. J.*, **194**, 10C (1967).
Fleisch, H., R. G. G. Russell, and M. D. Francis, *Science*, **165**, 1262 (1969).
Francis, M. D., R. G. G. Russell, and H. Fleisch, *Science*, **165**, 1264 (1969).
Nolen, G. A., and E. V. Buehler, *Toxicol. Appl. Pharmacol.*, **18**, 548 (1971).
Smith, R., R. G. G. Russell, and M. Bishop, *Lancet*, #7706, 945 (May 8, 1971).
Weiss, I. W., L. Fisher, and J. M. Phang, *Ann. Intern. Med.*, **74**, 933 (1971).

Russell, R. G. G., R. C. Mühlbauer, S. Bisaz, D. A. Williams, and H. Fleisch, *Calcif. Tissue Res.*, **6**, 183 (1970).

Francis, M. D., and L. Flora, *Isr. J. Med. Sci.*, **7**, 502 (1971).

Fleisch, H. A., R. G. G. Russell, S. Bisaz, R. C. Mühlbauer, and D. A. Williams, *Eur. J. Clin. Invest.*, **1**, 12 (1970).

Gasser, A. B., H. Fleisch, and L. J. Richelle, *Calcif. Tissue Res.*, **4** (Supplement), 96 (1970).

King, W. R., M. D. Francis, and W. R. Michael, *Clin. Orthop*, #78, 251 (July–August 1971).

Cram, R. L., R. Barmada, W. B. Geho, and R. D. Ray, *New Engl. J. Med.*, **285**, 1012 (1971).

Mühlbauer, R. C., R. G. G. Russell, D. A. Williams, and H. Fleisch, *Eur. J. Clin. Invest.*, **1**, 336 (1971).

Fleisch, H., S. Bisaz, A. D. Care, R. C. Mühlbauer, and R. G. G. Russell, *Calcitonin: Proc. 2nd Int. Symp.*, 409 (1969).

Fleisch, H., R. G. G. Russell, S. Bisaz, P. A. Casey, and R. C. Mühlbauer, *Calcif. Tissue Res.*, **2** (Supplement), 10 (1968).

Fleisch, H., R. G. G. Russell, B. Simpson, and R. C. Mühlbauer, *Nature*, **223**, 211 (1969).

Bassett, C. A. L., A. Donath, F. Macagno, R. Preisig, H. Fleisch, and M. D. Francis, *Lancet*, #7625, 845 (October 18, 1969).

Francis, M. D., *Calcif. Tissue Res.*, **3**, 151 (1969).

Jowsey, J., B. L. Riggs, P. J. Kelly, D. L. Hoffman, and P. Bordier, *J. Lab. Clin. Med.*, **78**, 574 (1971).

Cabanela, M. E., and J. Jowsey, *Mayo Clin. Proc.*, **46**, 492 (1971).

Russell, R. G. G., S. Bisaz, and H. Fleisch, *Arch. Intern. Med.*, **124**, 571 (1969).

Morgan, D. B., J.-P-. Bonjour, A. B. Gasser, K. O'Brien, and H. A. Fleisch, *Isr. J. Med. Sci.*, **7**, 384 (1971).

Russell, R. G. G., S. Bisaz, and H. Fleisch, *Proc. 4th Int. Congr. Nephrol.*, Stockholm 1969, **2**, 182 (1970).

Russell, R. G. G., and H. Fleisch, *Proc. Roy. Soc. Med.*, **63**, 876 (1970).

Other Applications

Partridge, L. K., and A. C. Tansley, Brit. Pat. 1,201,334 (1970).

Griebstein, W. J., R. J. Grabenstetter, and J. S. Widder, U.S. Pat. 3,549,677 (1970).

Procter & Gamble Co., Brit. Pat. 1,110,987 (1968).

Smith, R. A., and J. T. Dixon, Brit. Pat.1,143,123 (1969).

Procter & Gamble Co., Brit. Pat. 1,201,984 (1970).

Albright & Wilson, Fr. Addn. 88,915 (1967).

Albright & Wilson, Fr. Pat. 1,455,978 (1966).

Albright & Wilson, Fr. Pat. 1,455,979 (1966).

Monsanto Co., Fr. Pat. 1,458,566 (1966).

Procter & Gamble Co., Fr. Pat. 1,525,038 (1968).

Unilever, N. V., Neth. Appl. 6,703,756 (1967).

Francis, M. D., U.S. Pat. 3,584,116 (1971).

Francis, M. D., U.S. Pat. 3,584,124 (1971).

Kerst, A. F., U.S. Pat. 3,579,444 (1971).

Anbar, M., and G. A. St. John, *J. Dent. Res.*, **50**, 778 (1971).

Francis, M. D., U.S. Pat. 3,584,125 (1971).

Nonenzymic Hydrolysis at Phosphate Tetrahedra

R. K. OSTERHELD,

Department of Chemistry, University of Montana, Missoula, Montana

CONTENTS

In this article the subject of hydrolysis of phosphates is considered to include the reaction of water with any species containing the PO_4 tetrahedron to replace an oxygen of the tetrahedron by an oxygen atom from water. This process is often referred to as reversion when it involves the P—O—P linkage of condensed phosphates [Eq. (1)] or as

$$\left(\begin{array}{c} O\;\;O \\ OPOPO \\ O_-\;O_- \end{array}\right) + H_2O \longrightarrow 2 \left(\begin{array}{c} O \\ HOPO \\ O_- \end{array}\right) \tag{1}$$

saponification when it involves the ester linkage of a phosphate ester [Eq. (2)].

$$\underset{\underset{\text{O}_-}{\overset{\text{O}}{\parallel}}}{\text{ROPOH}} + \text{H}_2\text{O} \longrightarrow \text{ROH} + \underset{\underset{\text{O}_-}{\overset{\text{O}}{\parallel}}}{\text{HOPOH}} \qquad (2)$$

Reaction (2) can occur by cleavage of the CO or the PO bond. In this article we are only concerned with PO cleavage.

The condensed phosphates contain anions which can be considered to be built of PO_4 tetrahedra sharing oxygen atoms. Structural studies show that three major classes of phosphate anion structures exist (109,432,439): polyphosphates, metaphosphates, and ultraphosphates.

Polyphosphates have linear (unbranched) chain structures, as in the diphosphate (1) and triphosphate (2), also known as pyrophosphate and

$$\begin{bmatrix} \text{O} \ \ \text{O} \\ \text{OPOPO} \\ \text{O} \ \ \text{O} \end{bmatrix}^{-4} \qquad \begin{bmatrix} \text{O} \ \ \text{O} \ \ \text{O} \\ \text{OPOPOPO} \\ \text{O} \ \ \text{O} \ \ \text{O} \end{bmatrix}^{-5}$$

$$\qquad (1) \qquad\qquad\qquad (2)$$

tripolyphosphate, respectively. In the general formula for the polyphosphates, $P_nO_{3n+1}^{-(n+2)}$, n can have values in the hundreds or thousands. The shorter chains are classed as oligophosphates, with individual examples through $n = 8$ characterized (168). Longer chains are encountered as members of mixtures of chain lengths. High-molecular-weight polyphosphate samples, although mixtures, have been well characterized. They form many crystalline compounds and their configurations in solution have been studied by the methods of polymer chemistry. In this review the term polyphosphate will be reserved for unbranched chain anions. "Condensed phosphates" will be our term for POP bridged anions of whatever form.

The metaphosphates have ring structures with the general anion formula, $(PO_3^-)_n$. Because of the characteristic covalence of four for phosphorus, only ring structures can have the stoichiometry of this general formula. Trimetaphosphate (3) and tetrametaphosphate

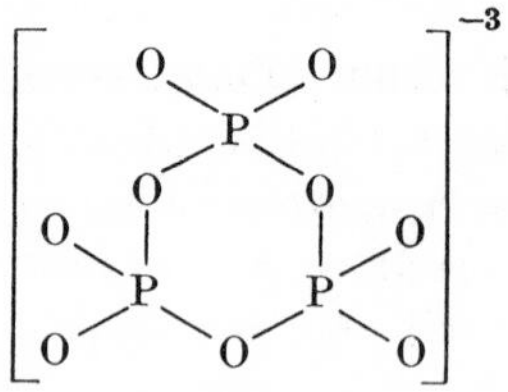

(3)

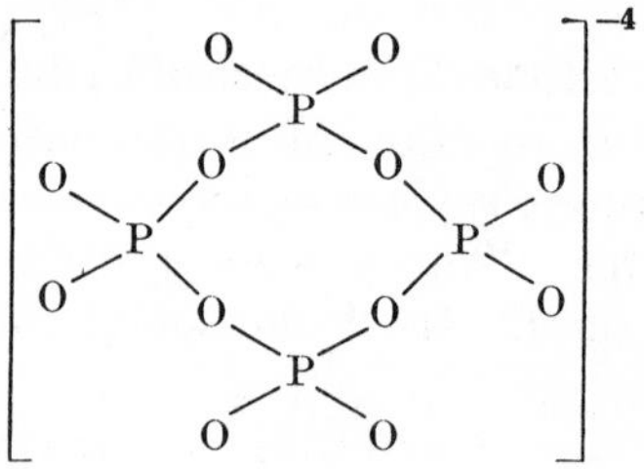

(4)

(4) are well known. The pentameta- and hexametaphosphates (408) and the octametaphosphate have been prepared (365). Because the long-chain polyphosphates approach metaphosphate stoichiometry, they once were incorrectly referred to as "hexametaphosphate" or "polymetaphosphate." Monometa- and dimetaphosphate appear in the early literature as supposedly isolatable species. The four-membered dimetaphosphate ring would be excessively strained and the mono-metaphosphate anion is coordinately unsaturated. No evidence exists for these as enduring species, but monometaphosphate is a presumed intermediate in some phosphate hydrolysis mechanisms.

Branched structures are the characteristic of the ultraphosphates, which contain phosphorus atoms involved in three P—O—P linkages (434). Ultraphosphate species need not be large (160) and can contain rings (155) and cross-linked chains (464).

The P—O—P linkages of all the condensed phosphates are unstable with respect to hydrolysis in aqueous solution. The utility of the anions in solution is determined by the kinetics of their hydrolysis.

Reviews have appeared in the last decade on the hydrolysis of condensed phosphate anions (393,396,397), of phosphate esters (60,111), and of five-membered cyclic esters (455).

I. Thermodynamics

Discussion of the thermodynamics of phosphate hydrolysis reactions is limited by the paucity of thermochemical data for phosphates and by a lower standard of accuracy for these data, compared to those for carbon compounds. Most of the thermochemical information for phosphates is not derived from combustion calorimetry but from reaction calorimetry, a technique handicapped by the occurrence of side reactions (180). This lack of attention to the thermodynamic aspects of phosphate chemistry has its consequences in the topic of hydrolysis. Statements concerning the stability of phosphates are commonly based on kinetic observations. Thus condensed phosphates are often said to be stable in neutral solution at ordinary temperatures, when in fact they are thermodynamically unstable but are slow to react.

Studies of the thermodynamics of inorganic diphosphate hydrolysis are summarized in Table I, together with calculated changes in the thermodynamic properties. The two experimental free energy of hydrolysis values are based on measurements of the value of the

　　　　　　　　　　R. K. OSTERHELD

TABLE I

Thermodynamics of Diphosphate Hydrolysis

A. Experimental results

pH	Temp. (°C)	Ionic strength	ΔH°_{hyd} (kcal/mole)	ΔS°_{hyd} (eu)	ΔG°_{hyd} (kcal/mole)	Ref.
71 % H_2SO_4			-4.42[a]			159
7.5	30	0.25			-6.6 ± 1	378
7.5	25	0.1	-7.3 ± 0.2[a]			157
7.3	25	0.6–1	-5.8 ± 0.13[a]			157
7.2	29		-9.0[a]			325
7	25	0.07	-3.7 ± 0.2			469
4–5	25	0.07	-4.5 ± 0.5			469
4.5[b]	25		-4.7	-5.0	-3.2	361

B. Calculated[c] from the values of Table III

Reaction	ΔH°_{hyd} (kcal/mole)	ΔS°_{hyd} (eu)	ΔG°_{hyd} (kcal/mole)
$H_4P_2O_7 + H_2O = 2\,H_3PO_4$	-5.3	-5	-3.8
$H_2P_2O_7^{2-} + H_2O = 2\,H_2PO_4^{-}$	-6.7	-12	-3.1
$P_2O_7^{4-} + H_2O = 2\,HPO_4^{2-}$	-6.6	-5	-5.3

[a] Questionable because of enthalpy effects that were ignored.
[b] The reported results apply at pH 4.5, although the study was not at this pH.
[c] Calculated values apply at 25°. The three reactions can be considered to occur at pH values of 1, 4.5, and 9, respectively.

diphosphate–monophosphate equilibrium constant. The thorough study by Schmulbach, Van Wazer, and Irani (361) of the reaction:

$$H_2P_2O_7^{2-} + H_2O = 2\,H_2PO_4^{-} \qquad (3)$$

is of special interest as it involved the particular protonated species shown. Concentration equilibrium constants were obtained by chromatographic analysis of samples that had reached equilibrium at 80, 100, and 110°. Approach to the equilibrium from both sides verified that a true equilibrium was being studied. True thermodynamic equilibrium constants were calculated through the use of activity co-efficients derived from the Gibbs-Duhem equation and the extended Debye-Hückel theory. The enthalpy of hydrolysis, calculated from the temperature dependence of the true equilibrium constants, was used to extrapolate to K_{true} at 25°. From this, ΔG°_{298} for the hydrolysis was obtained. Possible sources of error would include the estimation of the activity coefficients and the relatively lengthy extrapolation of results obtained over a narrow temperature range. The small negative $\Delta S°$ found is about that for the binding of an additional water molecule to the products. The equilibrium constant values involved appear in Table II, together with values calculated from the $\Delta G°_{hyd}$ values of Table I.B.

TABLE II

Equilibrium Constants for Diphosphate Hydrolysis

A. Experimental, for Eq. (3) (361)

Temp. (°C)	K_{conc}	K_{true}
25		2.2×10^2
80	2.18×10^2	3.7×10^1
100	1.52×10^2	2.2×10^1
110	1.30×10^2	1.8×10^1

B. Calculated for 25° from $\Delta G°_{hyd}$ values of Table I

Reaction	K
$H_4P_2O_7 + H_2O = 2\ H_3PO_4$	6×10^2
$H_2P_2O_7^{2-} + H_2O = 2\ H_2PO_4^{-}$	2×10^2
$P_2O_7^{4-} + H_2O = HPO_4^{2-}$	8×10^3

A K_{conc} of 2.8×10^2 has been reported for a concentrated system having a mole ratio of NaH_2PO_4–$Na_2HPO_4 = 2.37$ (361). In dilute solution this would correspond to a pH of about 7. Huhti and Gartagnis reported the formation of diphosphoric acid in concentrated monophosphoric acid (197). From their data the concentration equilibrium constant at 100° was estimated to be about 3×10^2 by Van Wazer (433) for the reaction:

$$H_4P_2O_7 + H_2O = 2\ H_3PO_4 \qquad (4)$$

From the data of Keisch, et al., (227) we have estimated a value of 2×10^2 for the same reaction at 100°. The similarity among all the equilibrium constant values cited is at first surprising in view of the wide range of conditions employed. The effects of changes in the degree of protonation of diphosphate and monophosphate must be compensatory.

The equilibrium constant for Eq. (3) is

$$K_I = \frac{a^2_{H_2PO_4^-}}{a_{H_2P_2O_7^{2-}} \cdot a_{H_2O}} \qquad (5)$$

and that for Eq. (4) is

$$K_U = \frac{a_{H_3PO_4}^2}{a_{H_4P_2O_7}a_{H_2O}} \tag{6}$$

where the subscripts I and U refer to the ionized and un-ionized species, respectively. From the acid ionization equilibria we have the following:

$$H_3PO_4 = H^+ + H_2PO_4^- \qquad K_{H_3PO_4} = \frac{a_{H^+}a_{H_2PO_4^-}}{a_{H_3PO_4}}$$

$$H_4P_2O_7 = H^+ + H_3P_2O_7^- \qquad K_{H_4P_2O_7} = \frac{a_{H^+}a_{H_3P_2O_7^-}}{a_{H_4P_2O_7}}$$

$$H_3P_2O_7^- = H^+ + H_2P_2O_7^{-2} \qquad K_{H_2P_2O_7^{-2}} = \frac{a_{H^+}a_{H_2P_2O_7^{-2}}}{a_{H_3P_2O_7^-}}$$

Using substitutions derived from the acid ionization expressions, one can convert Eq. (5) to the following form:

$$K_I = \frac{K_{H_3PO_4}^2}{K_{H_4P_2O_7}K_{H_3P_2O_7^-}} \cdot \frac{a_{H_3PO_4}^2}{a_{H_4P_2O_7}a_{H_2O}} \tag{7}$$

The ratio of the equilibrium constant values for Eqs. (3) and (4) then is

$$\frac{K_I}{K_U} = \frac{K_{H_3PO_4}^2}{K_{H_4P_2O_7}K_{H_3P_2O_7^-}} = \frac{(7.6 \times 10^{-3})^2}{(2.8)(2.3 \times 10^{-3})} = 9.0 \times 10^{-3} \tag{8}$$

The ionization constant values used in Eq. (8) are from the selection by Irani and Tulli (210).

The constancy of K_{hyd} values with changes of degree of ionization should hold for most condensed phosphates. Because the first acid ionization constant is about 10^{-2} and the second about 10^{-8} for any phosphate tetrahedron (210,443), one can show that there should be little difference in the hydrolysis equilibrium constant for the reaction involving the un-ionized species, that involving species with one H^+ ionized per phosphate tetrahedron, or that involving species with two H^+ ionized per tetrahedron—provided both reactant and product are so ionized—regardless of the degree of polymerization of the tetrahedra, short of the ultraphosphates. For hydrolysis of ultraphosphate linkages an analysis such as the above shows

$$\frac{K_I}{K_U} = [\text{Eq. (8) ratio of ionization constants}]a_{H^+}$$

The strong acid proton ($K_a \sim 10^{-2}$) created by hydrolysis of the ultraphosphate site causes the equilibrium constant ratio to depend on the hydrogen ion activity.

The enthalpy of diphosphate hydrolysis is negative and, as can be seen from Table II.A, the equilibrium constant for this hydrolysis is larger at ordinary temperatures than at the temperatures usually employed in hydrolysis studies. Although the higher experimental temperatures are needed for kinetic reasons, diphosphate must be unstable with respect to hydrolysis throughout the range of acidities, basicities, and temperatures associated with the aqueous medium. Assuming that the Schmulbach (361) values for ΔH°_{hyd} and ΔS°_{hyd} are independent of temperature, the equilibrium constant for Eq. (3) would be one at 667°. Even if the assumption is not good, one can be confident that an impossibly high temperature is needed to make the condensation of monophosphate to diphosphate the favored process in an aqueous medium. In fact, it is reasonable to assume that hydrolysis is spontaneous for all the condensed phosphates over the entire aqueous range.

The general preparative method for condensed phosphates is the thermal decomposition of a solid hydrogen phosphate, for example:

$$2\,Na_2HPO_4(c) = Na_4P_2O_7(c) + H_2O(g) \qquad (9)$$

Condensed phosphoric acids can similarly be prepared by the thermal decomposition of H_3PO_4 (406). Enough thermodynamic data are available (Table III) for a consideration of some of the reactions preparing the condensed phosphates to be fruitful.

Table IV gives values for the changes in enthalpy, entropy and free energy accompanying a number of condensed-phosphate preparative reactions. Each reaction has been written for the formation of one mole of water to simplify comparisons. The constancy of ΔS° for these reactions on this basis is interesting. The major entropy change of course, is in the liberation of the mole of gaseous water. In a method for estimating entropies, constituent water in a solid hydrogen phosphate has been assigned an entropy of 7.45 cal/(deg)(mole) (6). Using this and the value of S° for one mole of gaseous water, 45.0 cal/(deg)(mole), ΔS° for the water alone in these reactions is 37.6 eu, in close agreement with the ΔS° values of Table IV.

For comparison with the reactions of Table IV, one can reverse the reactions of Table I.B and reverse the sign on each of the thermodynamic values. In general, the formation of pyrophosphate at 25° in the solid-state reactions of Table IV is even more unfavorable than in the solution reaction of Table I.B. However, because ΔS° is so much more favorable for the solid-state decompositions, moderate temperature

TABLE III

Selected Thermodynamic Data (25°)

Compound	Condition	ΔH_f° (kcal/mole)		S° (eu)		ΔG_f° (kcal/mole)
NaH$_2$PO$_4$	(c)	−369.0	(211)[a]	30.47	(6)	
Na$_2$HPO$_4$	(c)	−419.4	(211)	35.97	(6)	
		−417.4	(303)			
Na$_2$H$_2$P$_2$O$_7$	(c)	−664.7	(213)	52.63	(6)	
Na$_4$P$_2$O$_7$	(c)	−763.7	(213)	64.60	(6)	
		−760.8	(303)			
		−766[b]	(287)			
Na$_5$P$_3$O$_{10}$—I	(c)	−1056.5	(212)	91.25	(6)	
Na$_5$P$_3$O$_{10}$—II	(c)	−1059.1	(212)	87.37	(6)	
Na$_5$P$_3$O$_{10}$·6 H$_2$O	(c)	−1485.7	(212)	146.1	(6)	
(NaPO$_3$)$_3$	(c)	−878.6	(213)	68.47	(6)	
		−877	(287)			
(NaPO$_3$)$_4$	(c)	−1169.4	(213)			
CaHPO$_4$	(c)	−435	(303)	25	(303)	
Ca$_2$P$_2$O$_7$	(c)	−797[b]	(287)	45	(303)	
H$_3$PO$_4$	(lm)	−307.92	(446)	37.8	(446)	−273.10 (446)
H$_2$PO$_4^-$	(lm)	−309.82	(446)	21.6	(446)	−270.17 (447)
HPO$_4^{2-}$	(lm)	−308.83	(446)	−8.0	(446)	−260.34 (446)
H$_4$P$_2$O$_7$	(lm)	−542.2	(446)	64	(446)	−485.7 (446)
H$_2$P$_2$O$_7^{2-}$	(lm)	−544.6	(446)	39	(446)	−480.5 (446)
P$_2$O$_7^{4-}$	(lm)	542.8	(446)	−28	(446)	−458.7 (446)

[a] References in parentheses.   Preferred value given first.
[b] At 35°.

increases cause these to become spontaneous, as shown in the last column of Table IV.   The temperature of spontaneity is not only lower than for the solution reactions, but it is more easily accessible with solids. Furthermore, movement of the volatile product, water, out of the preparative system shifts the equilibrium towards completion in the decompositions of the solids.   The gaseous product is the key to the success of these preparative methods for two reasons: (1) providing a very favorable entropy change and (2) allowing a shift in the equilibrium towards completion.   A further benefit, of course, is the inability of the condensed phosphates of different complexity to form solid solutions.   For this reason, a single, well-defined anion species results from the solid decomposition reactions, in contrast to the mixture of anionic species (32) that is produced by solution aggregation reactions in accord with reorganization theory (432).   The calculated temperatures at which $K_p = 1$ agree reasonably well with vapor

TABLE IV

Thermodynamic Changes in Preparative Reactions at 25° for Condensed Phosphates[a]

Reaction	ΔH° (kcal/mole)	ΔS° (eu)	ΔG° (kcal/mole)	K_p (atm)	Temperature at which $K_p = 1$ atm[b]
$2\ NaH_2PO_4(c)$ $= Na_2H_2P_2O_7(c) + H_2O(g)$	$+15.5$	$+36.7$	$+4.6$	4×10^{-4}	$150°C$
$Na_2H_2P_2O_7(c)$ $= \frac{2}{3}(NaPO_3)_3(c) + H_2O(g)$	21.2	38.2	9.8	6×10^{-8}	283
$2\ Na_2HPO_4(c)$ $= Na_4P_2O_7(c) + H_2O(g)$	17.3	37.7	6.1	3×10^{-5}	186
$Na_2HPO_4(c) + \frac{1}{2}NaH_2PO_4(c)$ $= \frac{1}{3}Na_5P_3O_{10}(c) + H_2O(g)$	I: 17.8^c II: 16.5	39.5 37.6	6.0 5.3	4×10^{-5} 1×10^{-4}	178 166
$2\ CaHPO_4(c)$ $= Ca_2P_2O_7(c) + H_2O(g)$	15	40	3	10^{-2}	100

[a] Calculated from the first value given for each compound in Table III.
[b] Assuming ΔH° and ΔS° are independent of temperature.
[c] Roman numerals refer to forms I and II of $Na_5P_3O_{10}$.

111

pressure data reported by Kiehl and Wallace (241) for the first two systems of Table IV.

For each of the reactions, as written in Table IV, the equilibrium constant, K_p, is simply p_{H_2O}. The intermediate appearance of $Na_2H_2P_2O_7$ in the thermal decomposition of NaH_2PO_4 is logical, given the high dissociation pressure of NaH_2PO_4 relative to that of $Na_2H_2P_2O_7$. Table IV clarifies the question of the form of $Na_5P_3O_{10}$ to be expected in the preparation of that compound. In general, form I is prepared at temperatures in the range 500–600°, and form II at temperatures near 350°. Morey fixed the transformation temperature at $417 \pm 8°$ (304). Table IV shows that the reaction of the orthophosphate mixture to produce form II is thermodynamically favored at 25°, as $\Delta G°$ for the reaction is less positive than for production of form I. That water vapor has been observed to facilitate the conversion of one form to the other can be due to its presence providing a route through the hydrogen monophosphates (or diphosphates) as a possible faster alternate to the direct crystal transformation. In the equilibrium with Na_2HPO_4 and NaH_2PO_4 at temperatures below about 400°, form I attempts to maintain a partial pressure of water lower than that demanded by the equilibrium involving form II (compare K_p values). The consumption of form I in removing water vapor from the system will result in elimination of form I from the system, while form II is produced by the thermal decomposition of the hydrogen phosphates created by the form I reaction. Above about 400°, the situation is reversed. The occasional formation of form I at temperatures well below 400° (393) must be an example of the well-known formation of a metastable phase, perhaps because of kinetic control through nucleation phenomena. For the conversion of form I to form II, the data of Table III give $\Delta H° = -2.6$ kcal/mole and $\Delta S° = -3.88$ cal/(deg)(mole). Using these values to estimate the transition temperature between these forms, we obtain a value of 397°, in very satisfactory agreement with Morey's experimental value.

In the oven dehydration at about 100° of hydrates of the condensed phosphates, hydrolysis of the anion is a common observation. Under these dehydration conditions the partial pressure of water within the crystal can be expected to reach atmospheric pressure. For all the condensed phosphates of Table IV, except possibly $Ca_2P_2O_7$, this would reverse the preparative reaction shown; for all of them K_p is less than one at 100°. Even at 25°, the equilibrium dissociation pressures for some hydrates can be expected to exceed the K_{298} value shown, making the condensed phosphate hydrate thermodynamically unstable with respect to the hydrogen phosphate in storage at ordinary temperatures.

The formation of P—O—P linkages at more ordinary temperatures can be accomplished by a number of phosphorylating reagents, including intermediates in the oxidation of phenyl or benzyl hydrogen phosphorohydrazidate by iodine or N-bromosuccinimide (53), in the oxidation of bromine of naphthaquinol monophosphate or of a naphthaquinol ester of adenosine-5 phosphate (99,100), by phosphoramidic acid and its salts and esters (90,91,102,103). In these reactions monometaphosphate is assumed to be an intermediate. The absence of, or limitation of, water in the systems leaves oxygen of another phosphate species as the best nucleophile available to the monometaphosphate. The P—O—P linkage has also been produced by the reaction of strong anhydrides, such as acetic anhydride or acetyl chloride (133,237) or p-toluenesulfonyl chloride (154) with monoesters of phosphoric acid. For these reactions, the proposed mechanism involves intermediate formation of an acyl phosphate C—O—P linkage, and the reaction of this with a second mole of phosphate ester to form a P—O—P link (237). The monometaphosphate mechanism appears to be an alternate. Acetic anhydride causes the cyclization of triphosphate to produce trimetaphosphate (400). Trimeta- and tetrametaphosphate form from less highly condensed phosphates in sulfuric acid oleums of $>20\%$ SO_3 and in chlorosulfuric acid (130). Other chemical dehydrating agents, such as trichloroacetonitrile and dicyclohexylcarbodiimide will cause condensation of monophosphoric acid well into the ultraphosphate region (160). The formation of P—O—P bridges in these reactions in essentially anhydrous media does not contradict the thermodynamic discussion presented earlier of P—O—P cleavage in aqueous medium. The reported formation of condensed phosphates in relatively dilute aqueous solution (358) could not be reproduced (361).

There has been a long-standing polemic over the appropriateness and utility of the biochemist's characterization of the oxygen bridge to phosphorus as a "high-energy bond" (271). It has been said that energy is stored in the energy-rich compound or bond in the same sense that energy is stored in a battery (214). This points directly at the one aspect of this controversy on which we wish to comment. We do not think of the energy stored in a battery as stored in a particular compound or bond, but instead as stored in the combination of chemical species present, which make possible a redox reaction accompanied by a negative free-energy change. One does not think of the energy of a Daniell cell as stored in the zinc metal or in the cupric ions. The free energy of polyphosphate hydrolysis is a property of the reaction, not of a reactant. Certain bonds profitably can be identified on structural bases as contributing in an important way to the free-energy change (48,333), but the ΔG is dependent on the *change* in these bonds, that is, on the reaction as a whole. This emphasis is crucial. The topic of high-energy phosphate linkages is the subject of a recent series of articles in *Chemistry in Britain* (21,22,199,333,354,463). Ironically, P—O—P linkages do not have the high free energies of hydrolysis once estimated (-9 to -12 kcal/mole). In the controversy over the thermodynamics of hydrolysis in biologically important systems there is danger the kinetics will be overlooked. The importance of the P—O—P linkage biologically is to a large extent a consequence of the slowness of its nonenzymic hydrolysis at *in vivo* pH and temperature (361).

II. Kinetics and Mechanisms

This section will treat this topic in a general way, to give a perspective on phosphate hydrolyses, and to bring out correlations that are not so easily noted in the sections dealing with individual phosphate categories. A fuller treatment of individual species will be found in the sections that follow, along with references to original sources.

For carboxylic ester hydrolysis Ingold lists eight possible mechanisms (only six of which have been observed) based on differentiation of acidic or basic catalysis, acyl–oxygen or alkyl–oxygen cleavage, and uni- or bimolecular involvement in the transition state (207). The molecularity of a reaction is based on the rate-determining step alone. It is defined as the number of molecules necessarily undergoing co-valency change (207), which is distinct from, though not necessarily different in value from, the reaction order. Ingold's symbols for these mechanisms appear in Table V. A detailed description of each

TABLE V

Carboxylic Ester Hydrolysis Mechanisms

| Catalysis | Acyl cleavage | | Alkyl cleavage | |
	Unimol.	Bimol.	Unimol.	Bimol.
Acidic	$A_{Ac}1 = A1^a$	$A_{Ac}2 = A2^c$	$A_{Al}1^a$	$A_{Al}2^{b,d}$
Basic	$B_{Ac}1^{a,d}$	$B_{Ac}2^c$	$B_{Al}1^a$	$B_{Al}2^b$

[a] S_N1 mechanism.
[b] S_N2 mechanism.
[c] Tetrahedral mechanism.
[d] Not observed.

mechanism has been given in convenient reference form by March (282). Very naturally, discussions of condensed phosphate and phosphate ester hydrolysis mechanisms have been based on an understanding of organic mechanisms. Until recently phosphate hydrolysis mechanisms have been characterized as S_N1 or S_N2. It is interesting that the S_N2 mechanism, which involves concerted entry and departure of reagent and hydrolysis product at a tetrahedral atom, is not the expected mechanism at an acyl carbon. The phosphorus atom in compounds of interest to us corresponds to the acyl carbon of a carboxylic ester, and it is now evident that many, perhaps most or even all, bimolecular P—O hydrolysis reactions are not S_N2, but instead, involve a more stable pentacoordinate transition state [Eq. (10)].

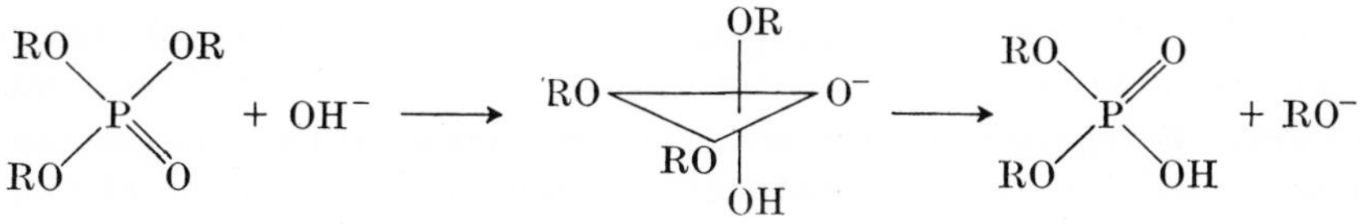

$$(10)$$

The intermediate increase, by one, in the phosphorus coordination number in this addition–elimination mechanism parallels the increase of one in the coordination number of acyl carbon in the tetrahedral mechanism [Eq. (11)].

Two significant differences must be considered in adjusting from hydrolysis mechanisms at acyl carbon to those at the phosphate tetrahedron. First, the phosphorus atom has valence d orbitals, which the carbon atom has not. These orbitals allow increased covalency for phosphorus, as in PF_5 and PF_6^-. Phosphorus generally shows a coordination number of four towards oxygen, but the pentaoxyphosphoranes, $(RO)_5P$, are well characterized (347). Second, the phosphate tetrahedron is derived from a triprotic acid, compared to the monoprotic carboxylic group. For phosphates, then, a greater variety of protonation levels arise than for carboxylic esters. In spite of this greater variety of reactive species to be considered, it now appears possible that just as the acyl hydrolyses of carboxylic esters occur by only two mechanism types (S_N1 and tetrahedral), all hydrolyses at phosphate tetrahedra occur by two types of mechanisms, S_N1 and pentacoordinate. It is also possible that the S_N2 mechanism has a place.

Based on the available evidence, all unimolecular hydrolyses at phosphate tetrahedra are S_N1 "monometaphosphate" mechanisms [Eqs. 67–70]. This mechanism involves unimolecular elimination of PO_3^- as the rate-limiting step. First advanced by Westheimer, Bunton, Barnard, Vernon, and colleagues for the hydrolysis of phosphate monoester monoanions, this mechanism is discussed in detail in Section IV.C.4. The mechanism does not require the formation of a completely free monometaphosphate ion, but does require that the P—O cleavage step be rate-limiting.

The key evidence for the occurrence of bimolecular hydrolysis at phosphorus through a pentacoordinate transition state is the observation that the strained five-membered ring of ethylene phosphate enormously increases not only the rate of ring-opening but also the rate of oxygen exchange with the solvent at the phosphoryl oxygens and the rate of methoxyl loss from methyl ethylene phosphate (Section IV.C.5). For an S_N1 or S_N2 mechainsm, the ring strain should influence only the rate of ring-opening. In a pentacoordinate transition state the ring strain is relieved, and through pseudorotation the transition state serves for the oxygen exchange and methoxyl elimination reactions, as well as for the ring-opening. Sections IV.C.2 and IV.C.5 describe the pentacoordinate mechanism and pseudorotation more fully. The transition state is presumed to have a trigonal bipyramidal configuration, using sp^3d hybrid orbitals of the phosphorus.

The pentacoordinate addition–elimination mechanism has been demonstrated for some bimolecular hydrolyses either through a rate-limiting factor that can be attributed to pseudorotation (233,250), or through a requirement for pseudorotation to explain the products. No method has been developed to demonstrate the absence of this mechanism, however, or the involvement of an S_N2 mechanism in its place. Solvent oxygen exchange of phosphoryl oxygen is not an unambiguous diagnostic, because whether such exchange occurs depends upon the relative stabilities of forms in the pseudorotational equilibria. Whether retention or inversion of configuration occurs in the mechanism is determined by the number of pseudorotations that have occurred (at the same or different pivot atoms), retention occurring for an odd number and inversion for an even number, including zero. Of course, only a matter of degree in the timing separates the S_N2 mechanism, with its concerted addition–elimination, from the pentacoordinate mechanism, with an intermediate of some lifetime.

The elucidation of reaction mechanisms at silicon is considerably further advanced (374) than at phosphorus. One can hope that a uniform pattern of mechanism ultimately is developed for reaction at silicon, phosphorus, and sulfur.

Reaction order and molecularity must not be confused. All hydrolyses at a phosphate tetrahedron are first order in the phosphate reactant. For a given species the rate expression, determined necessarily at a given pH and usually in dilute aqueous solution, is either genuinely first order or is pseudo first order, and the two cannot be distinguished. The activation entropy has been an important method for determining molecularity. Small values of $\Delta S^{\ddagger}$, about zero, are expected for unimolecular reactions. Larger negative values, generally -15 to -30 eu, are expected for bimolecular reactions (273,274,359).

Changes in hydrolysis rate with pH have generally been attributed to changing degrees of protonation of the phosphate tetrahedra involved rather than to a kinetically equivalent formulation in terms of acid catalysis [Eq. (14)], except in very acid or alkaline solution. For acid catalysis in very acid solution, authors differ in preference for a rate expression in terms of acid catalysis or in terms of the kinetically equivalent conjugate acid species. Both formulations appear in Eq. (12), where k_{H^+} and k_{CA} are, respectively, the second-order rate constant for acid-catalyzed hydrolysis of the phosphate

$$\text{Rate} = k_{H^+}[H^+][S] = k_{CA}K_{CA}[HS^+] \qquad (12)$$

substrate, S, and the first-order constant for hydrolysis of its conjugate acid. K_{CA} is the acid ionization constant for the conjugate acid. We have generally used the conjugate acid formulation in this review. The conjugate acid species certainly exist.

The major mechanistic advances in understanding of phosphate hydrolyses have been made in studies of phosphate ester hydrolyses. Not only can the number of possible species in solution be reduced by blocking of possible protonation sites, but the influence of changing substituents can be determined. Table VI shows the molecularities assigned hydrolyses at the phosphate tetrahedra of esters. The monometaphosphate mechanism has been assigned to the hydrolysis of the monoanions of monoacyl, monoaryl, and monoalkyl esters, of monoacyl dianions, and of monoaryl dianions for which the leaving aryloxide ion is particularly stable. In the absence of an exceptionally good leaving group, proton transfer to the leaving group is necessary and the mechanism is then limited to hydrolysis of species in which the $-OP(OH)O_2^-$ group is present. The pentacoordinate mechanism has been demonstrated for the hydroxide-catalyzed reaction of

TABLE VI

Molecularity of Ester P—O Cleavage Reactions[a]

Ester type	Species[b]	Acyl	Aryl	Alkyl
Triesters	+1		(2)	
	O		2	d
	O + OH⁻	d	2	2
Diesters	+1		2	d
	O		(2)	d
	−1	d	2	d
	−1 + OH⁻		2c	d
Cyclic[e]	+1			
	O			2
	−1			
	−1 + OH⁻			(1)
Monoesters	+1	d	2	d
	O	d		c
	−1	1	1	1
	−2	1	1	1c
	−2 + OH⁻	d		

[a] Parentheses enclose probable values.

[b] The symbols +1, O, −1, and −2 refer to the conjugate acid, neutral molecule, the monoanion, and the dianion, respectively. One of these symbols " +OH⁻" refers to the OH⁻-catalyzed reaction.

[c] Molecularity given although the point of cleavage is not established or is mixed.

[d] Cleavage predominantly at C—O bond.

[e] Five-membered cyclic esters.

triesters and for the hydrolysis of neutral ester species containing five-membered rings.   As noted earlier, it might very well apply to all the bimolecular reactions of Table VI.

Linear free-energy relationships have been evaluated for a number of classes of phosphate ester hydrolyses.   Table VII shows values of the slope of plots of the log of the rate constant against the pK_a of the parent of the leaving group, and shows the relative importance of the leaving substituent in determining rate.   The dependence of phosphate ester hydrolysis rates on pH is complex.   It is discussed in the later sections on esters.

The hydrolysis of condensed phosphates has not been as systematically studied as that of phosphate esters.   For example, comparatively few inorganic studies have determined kinetics for the individual species present.   Table VIII shows the molecularity of some of the condensed-phosphate hydrolysis reactions.   More than anything else,

TABLE VII

Linear Free-Energy Relationships for Phosphate
Ester Hydrolyses

Ester type	Species[a]	Slope $(\Delta \log k/\Delta \ pK_a)$	Ref.
Triesters	0	-0.99	233
	-1	-0.4	233
Diesters	-1	-0.97	248
	$-1 + OH^-$	-0.56	58
Monoesters	$+1$	~ -0.5	71
	-1	-0.32	72
		-0.27	245
		~ -0.3	71
	-2	-1.23	245
		~ -1.7	71

[a] See footnote (b), Table VI.

TABLE VIII

Molecularity of Condensed Phosphate P—O Cleavage Reactions[a]

Phosphate	Conjugate acid	Diprotonated anion	Unprotonated anion
PO_4^{3-} (oxygen exchange)	(2)	1	
$P_2O_7^{4-}$	2	1	(2)
$P_3O_{10}^{5-}$		(1)	(2)
Oligomers (end-group clipping)		(2)	[b]
Long-chain (end-group clipping)		(1)	
$P_3O_9^{3-}$		1[c]	2
Tertiary tetrahedra			2
Oligomers (meta abstraction)		[b]	(2)
Long-chain (meta abstraction)		(1)	

[a] Probable values in parentheses. The principal species about pH 4 and 11, respectively, are assumed to be diprotonated and nonprotonated.

[b] Very negative $\Delta S^{\ddagger}$ value.

[c] Partly protonated, pH about 1.

the table emphasizes current uncertainties. Among inorganic P—O hydrolyses, the monometaphosphate mechanism has been assigned to oxygen exchange between $H_2PO_4^-$ and water and to the hydrolysis of $H_2P_2O_7^{2-}$, $HP_2O_7^{3-}$, $H_2P_3O_{10}^{3-}$, $HP_3O_{10}^{4-}$, symmetrical and unsymmetrical diphosphate diester dianions, the monoprotonated diphosphate monoester anion, and monoprotonated triphosphate monoester anions. The rate constant versus pH profile for oxygen exchange between monophosphate and water is very similar to that for mono-methyl phosphate hydrolysis, both curves showing pronounced rate maxima for reaction with the monoanion, for example. Presumably the monometaphosphate mechanism is generally applicable to the uni-molecular hydrolyses at phosphate tetrahedra. Unimolecular kinetics are only observed then when this route is fast: (1) with the $-OP(OH)O_2^-$ group present or (2) with a very good leaving group.

The mechanism assigned in the literature to biomolecular P—O hydrolyses has generally been S_N2, clear or implied. It is more likely that these processes are by a pentacoordinate mechanism, but diag-nostic methods are unavailable in inorganic systems. Except at tertiary tetrahedra, the hydrolysis of completely inorganic condensed phosphates increases in rate with a decrease in pH and with an increase in chain length. A limiting rate is approached as the chain lengthens. The rate of hydrolysis at tertiary phosphate tetrahedra (with three P—O—P bridges) is much more rapid than for less complex structures and is independent of pH. The degradation (not hydrolysis) of oligophosphates by metaphosphate abstraction increases in rate with an increase in pH, which is a cation effect, at least in part. In contrast to the linear phosphates, hydrolysis of the cyclic metaphosphates decreases in rate with increased size of the ring.

In spite of a wide range of activation energies among phosphate hydrolyses, many of these reactions have comparable rates as a result of compensatory differences in A or $\Delta S^{\ddagger}$ Because of the activation energy differences, however, rate comparisons at one temperature can be misleading for other temperatures (25,38,111). The compensatory differences in E_a and A must arise to a large degree from changes in the electrophilicity of the phosphorus atom. For a more electrophilic phosphorus atom, the value of E_a for nucleophilic attack is decreased, but $\Delta S^{\ddagger}$ is also made more negative (A smaller) because the increased polarity of the transition state induces a greater degree of solvent ordering in the transition state.

Units are consistent throughout the following sections to facilitate comparisons the reader might wish to consider. It is not practical to tabulate rate-constant values because of the tremendous number of them that have been reported and that would have to be recorded to cover the possible combinations of species, pH, and temperature. Rate comparisons can most easily be made by comparing log k values, recognizing that $\log k = \log A - E_a/4.58T$.

It is difficult to separate ionic strength effects from those due to catalysis or inhibition by the ions used to establish the ionic strength. An increased concentration of tetraalkylammonium salts slightly decreases the rate of hydrolysis for all polyphosphate–pH combinations studied, undoubtedly approximating the true ionic strength effect (296,299,441). In many cases metal salts slightly decrease the hydrolysis rates at low pH values (389,441). Although studies have not been carried out within the range of applicability of the Bronsted-Bjerrum rule, this inhibition with increased ionic strength is of the direction expected if it were due to an inhibition of the formation of a transition state complex from oppositely charged ions. Inhibition through formation of a more slowly hydrolyzing metal complex is not reasonable because tetraalkylammonium ions are not believed to complex with polyphosphates (440), and at the low pH values (<4) for which metal ions commonly show inhibition competition from protons will limit the available interaction sites on the phosphate. Increased ionic strength increases the ionization constant values for polyphosphoric acids (208); the increases are great enough to cause concentration shifts to less protonated species of the order required to account for the rate decreases.

Many metallic salts increase the hydrolysis rate significantly above pH 4 (67,147,389,439). The catalysis is attributable to the cations; anions have little effect on hydrolysis kinetics (441,461). Superficially, the catalytic role of cations is easily explained. Since even alkali metal ions form complexes (258,301,437,438,448,450,451,468), it can be expected that all metal ions will do so. Although the metal ion–diphosphate interaction is largely electrostatic, ^{31}P NMR shows complex formation significantly alters the electronic environment of the phosphorus atom (118). Metal-ion catalysis might relate closely to hydrogen-ion catalysis, increasing the electropositive character of the phosphorus atom or improving the leaving ability of a too basic leaving group. Added cations might accelerate alkaline hydrolysis by decreasing the negative potential near the phosphate anion to facilitate approach by OH^-.

The catalytic effect of metal ions has been correlated with the formation constants for their polyphosphate complexes (393). The relationship certainly derives from the requirement that the cation be associated with the anion to affect it and, presumably, from a dependence of the effect on the nature and strength of the metal ion–phosphate interaction. However the latter does not relate straightforwardly to the formation constant for the complex. As the metal ion–phosphate interaction develops, both metal ion and phosphate must give up some degree of interaction with the solvent. It has been estimated that about two water molecules are lost per Ca^{2+} in forming $CaP_2O_7^{2-}$ or $CaP_3O_{10}^{3-}$ (209). The change for the phosphate ion will be more or less independent of the metal ion identity. Differences in the over-all $\Delta G°$ (or ratios of the equilibrium constants) for complex ion formation can be considered to depend primarily on the balance between the unfavorable $\Delta G°$ for partial deaquation of the metal ion and the favorable $\Delta G°$ for metal ion–phosphate interaction. There is no assurance that the over-all $\Delta G°$, or the equilibrium constant value, will reflect simply the strength of the metal ion–phosphate interaction.

TABLE IX

Structural Data for Solid Diphosphates

Compound	∢P—O—P	Average bond lengths (Å)		Ref.
		P—O$_{bridge}$	P—O$_{terminal}$	
$Mn_2P_2O_7$	180°	1.57	1.53	87, 427
β-$Zn_2P_2O_7$	180°	1.57	1.56	86
β-$Mg_2P_2O_7$	180°	1.56	1.54	85
β-$Cu_2P_2O_7$	180°	1.54	1.51	352
α-$Zn_2P_2O_7$[a]	145°	1.57	1.52	352
α-$Mg_2P_2O_7$	144°	1.59	1.52	87
α-$Co_2P_2O_7$	143°	1.57	1.52	352
β-$Ca_2P_2O_7$[b]	138°	1.61	1.53	454
α-$Ni_2P_2O_7$	137°	1.60	1.54	428
$Na_4P_2O_7 \cdot 10\ H_2O$	134°	1.61	1.51	116
$Cd_2P_2O_7$	132°	1.52	1.62	59
β-$Ca_2P_2O_7$[b]	131°	1.63	1.51	454
α-$Cu_2P_2O_7$	131°	1.58	1.53	351
α-$Sr_2P_2O_7$	131°	1.60	1.50	173
α-$Ca_2P_2O_7$	130°	1.60	1.51	88

[a] Three-anion structure, averaged.
[b] Two-anion structure, listed separately.

The alkaline hydrolysis of di- and triphosphate is inhibited by Mg^{2+} but is catalyzed by Ca^{2+} (97,161,315), yet the formation constant for the complex of either of these anions with Mg^{2+} is larger than with Ca^{2+}. Furthermore, the ^{31}P NMR chemical shift data for these complexes suggest that Mg^{2+} is more effective than Ca^{2+} in reducing the electron density around the phosphorus atoms of diphosphate, but that the situation is reversed for triphosphate (118). These observations cannot be reconciled on the basis of previously considered metal-ion catalytic effects.

Table IX summarizes structural data for simple solid diphosphates. Although one cannot expect the structure of the anion in Table IX to be preserved in the corresponding metal chelate in solution, these data are all that is available and are suggestive of possible structural variations. The range of lengths for the bridge P—O bond in Table IX, 1.52 to 1.61 Å, is appreciable from an energetic viewpoint and might be considered a basis for a metal ion effect, particularly in the mono-metaphosphate mechanism. The 130 to 180° range of reported POP bond angles is remarkable. Similar modifications in the phosphate complexes of these metal ions would represent significant changes in the transition state.

Of greater pertinence to solution configuration are the recent determinations of the structures of several octamethylpyrophosphoramide chelates (198,215). The P—O (bridge) bond length was unaffected in solid chelates with Mg^{2+}, Co^{2+}, or Cu^{2+}. In the tris chelates with the same ions, the P—O—P angles were 134, 135, and 135°, respectively. In a bis complex of Cu^{2+} it was 132°. The chelate rings were nearly planar in the tris complexes and slightly puckered in the bis complex. It is likely that unchelated diphosphate in solution would exhibit a more obtuse P—O—P angle, as this would favor $dp\ \pi$ delocalization.

A sound interpretation of the role of metal cations in these hydrolyses will await systematic studies of the dependence of hydrolysis rates on cation concentration and pH, on temperature, and on blocking of ligand positions by organic substituents. Points of cleavage and mechanisms deduced in the absence of metal ions need not apply to the complexes. Table XVIII and Figs. 4 and 7 illustrate the importance of some of these points.

III. Hydrolysis of the P—O—P Bridge

A. Diphosphate

The hydrolysis of diphosphate is a thermodynamically spontaneous process in aqueous solution over the ordinary aqueous temperature range (cf. Section I). The rate of hydrolysis is very slow, however, near room temperature at pH ≥ 7. At elevated temperatures the rate increases markedly, whereas the extent of hydrolysis at equilibrium decreases slightly.

Conditions under which the kinetics have been studied for inorganic diphosphate hydrolysis in the absence of catalysts other than hydrogen ion are shown in Table X. The results of these studies are summarized in Figure 1, which can be used to estimate first-order rate constants and half-lives at conditions of interest. The lines are dashed for the higher pH values because the data from different investigators are

TABLE X

Inorganic Diphosphate Hydrolysis Kinetics Studies

pH	Temperature (°C)	Ref.
1, 7, 13	50, 100	3
4.5, 13	70, 100	31
(1–5M H$^+$)	25–54	67
0.8–3.3	50–70	89
2–11	65.5	114
1	40–50	147
5–9	66, 88	161
1–13	30–125	165, 440, 441
	~25	226
(Concd. H$_3$PO$_4$)	25, 60	227
0.3–0.5	30–90	238
(2M NaOH)	75–100	239
2–5.5	45	240
0.9–11	65.5	278
1.1–9.8	95	296
0.07–0.3	20, 40	314
~0–1.4	30, 60	316
0.4–7	60	326
(4 × 10^{-4} − 12.6 m. HClO$_4$)	39.5	327
2–7	100	337
3–5	65.5	373
	60–100	453

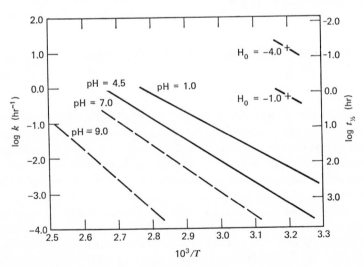

Fig. 1. First-order rate constants for diphosphate hydrolysis. Dashed lines are less well established.

more scattered, probably because of a greater sensitivity of the kinetics to cations present. The short lines for H_0 values -1.0 and -4.0 are based on studies at $39.5°$ in $HClO_4$ solutions (327). The hydrolysis of diphosphate has long been known to be first order when studied at constant pH in dilute aqueous solution (1,159,238).

The influence of hydrogen ion activity on the hydrolysis kinetics is shown in Figure 2. The dashed curves are based on relatively few

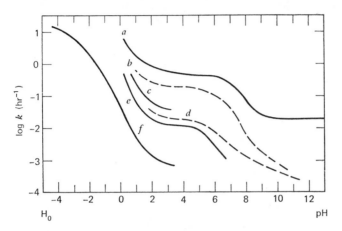

Fig. 2. First-order rate constants for diphosphate hydrolysis. Curve (a) is for γ-phenylpropyl diphosphate at $95°$ (296). Remaining curves are for inorganic diphosphate at (b) $90°$ (441); (c) $69°$ (89); (d) $65.5°$ (278); (e) $60°$ (326); and (f) $40°$ (327).

experimental points and are less well defined than the solid lines. The broad plateau in each curve at about pH 4 corresponds to a pH region over which $H_2P_2O_7^{2-}$ is the most abundant diphosphate species. This plateau is strong evidence that the change in rate with pH is the result of a change in the relative abundance of the various protonated species. Curve a, for γ-phenylpropyl diphosphate, provides further evidence. This species cannot undergo the fourth ionization at high pH that occurs for inorganic diphosphate, and its rate levels off at pH ≥ 10, where the others continue to drop.

Except for $P_2O_7^{4-}$ in very basic solutions, no diphosphate species other than $H_2P_2O_7^{2-}$ has an extended pH range of abundance. One might expect the abundance of $H_4P_2O_7$ to become roughly constant at high acidities, pH < 0, but evidence exists for further protonation of neutral acid species. It has been suggested that the compound $H_3PO_4 \cdot HClO_4$ is $P(OH)_4^+ \ ClO_4^-$ (7). Evidence for the species $H_4PO_4^+$ has been found in 100% H_3PO_4 (162), in dilute H_3PO_4 (228), in concentrated H_2SO_4 (156), and in solutions of BF_3 in H_3PO_4 (163). Ultraviolet spectra of solutions of $[Co(NH_3)_4(OH_2)PO_4H_3]^{3+}$ in concentrated $HClO_4$ indicate further protonation to $[Co(NH_3)_4(OH_2)PO_4H_4]^{4+}$, with this last stage not complete even at $11.7M$ $HClO_4$ (267). This fourth protonation is kinetically important in the hydrolysis of the Co—O—P linkage (see Section IV.A). NMR studies of solutions of phosphate in strong acids show the presence of $H_4PO_4^+$ and of $H_5P_2O_7^+$ and $H_6P_2O_7^{2+}$ (130). In sulfuric acid oleums of $>20\%$ SO_3, there is evidence for the condensation of diphosphate to tetrametaphosphate through a tetraphosphate intermediate. Formation of the conjugate

TABLE XI

First Order Hydrolysis Rate Constants[a] For Individual Species hr^{-1}

$H_4P_2O_7^+$	$H_4P_2O_7$	$H_3P_2O_7^-$	$H_2P_2O_7^{2-}$	$HP_2O_7^{3-}$	$P_2O_7^{4-}$	Temp. (°C)	Ref.
≥ 115	0.037	0.024	0.0112	0.00036		60	326
		0.0120	0.00956			60	89
	0.23	0.044	0.017	0.00315	0.00028	65.5	278
	0.0498	0.0144	0.00864	0.00468		100	337
	0.96	0.62	0.56	0.026		95	b

R	$RH_3P_2O_7$	$RH_2P_2O_7^-$	$RHP_2O_7^{2-}$	$RP_2O_7^{3-}$			
γ-phenylpropyl	1.8	0.50	0.45	0.018		95	296
adenosine		0.54	0.036			95	296

R		$R_2HP_2O_7^-$	$R_2P_2O_7^{2-}$				
sym-di-γ-phenylpropyl			3×10^{-4}			95	296
unsym. diethyl		0.108	4.14			35	294

[a] Some values have been recalculated for consistency in units.
[b] Calculated using diphosphate species rate constants from Ref. 326 and activation energies for pH values 1, 1, 4, and 7, respectively, from Ref. 441.

acid, for example, $(RO)_3POH^+$, is important in hydrolysis of phosphate esters (see Section IV.C).

Over the normal pH range the observed diphosphate hydrolysis rate constant can be described in terms of a weighted average of rate constants for the hydrolysis of the individual protonated species:

$$k_{obs} = k_5x_5 + k_4x_4 + k_3x_3 + k_2x_2 + k_1x_1 + k_0x_0 \qquad (13)$$

In Eq. (13) k represents a first-order rate constant, x represents the mole fraction of the total diphosphate that is in a particular protonated form, and the numerical subscripts 5, 4, 3, . . . , identify the respective protonated forms, $H_5P_2O_7^+$, $H_4P_2O_7$, $H_3P_2O_7^-$, . . . (326). Table XI lists rate constants that have been assigned to each species. Fuchs

(151) and Muus (314) first treated diphosphate hydrolysis kinetics in terms of individual protonated species.

First-order kinetics involving a particular protonated species are mathematically indistinguishable from second-order kinetics involving hydronium ion and the species of one lower protonation. Equation (14) gives the alternate formulations of the rate expression for one level of protonation.

$$\text{Rate} = k_1[H_4P_2O_7] = k_2[H^+][H_3P_2O_7^-] \tag{14}$$

The second-order rate constant, k_2, is simply k_1/K_a, where K_a is the acid ionization constant for $H_4P_2O_7$. There are a number of reasons for preferring the first-order expression. The protonation processes involved in maintaining equilibrium among the species are very fast and cannot be considered kinetically important steps. The rate of hydrolysis of the P—O—P linkage of tetraethyl diphosphate is relatively insensitive to pH over the range 1–7 (230). The rate of hydrolysis of $(RO)_2PO—O—PO_3^{2-}$ is close to that expected for $(RO)(HO)PO—O—PO_3^{2-}$ (294). A second-order formulation based on hydronium ion reaction with the trianion is possible for the latter species but not for the former.

Studies of the hydrolysis of the P—O—P linkage in diphosphate esters are listed in Table XII. These hydrolyses are first order in ester at a given pH (55,191,294,296). The pH dependence of the hydrolysis rate for γ-phenylpropyl diphosphate closely resembles that for inorganic diphosphate (Curve a of Fig. 2). The pattern for this monoester has been fitted with a rate law equivalent to Eq. (13) (296). In Table XI, the rate constants assigned the individual protonated species at 95° can be compared to values calculated for inorganic diphosphate at that temperature. The agreement is close enough to suggest that an alkyl substituent is kinetically equivalent to a proton, except in ability to tautomerize, as will be seen. Curves a and b of Figure 3 show that even the adenosine moiety has no special properties with respect to P—O—P cleavage.

Miller and Westheimer have pointed out that the diester dianion of **5** hydrolyzes much more slowly than does the monoester dianion of **6**.

$$
\begin{array}{ccc}
\text{O O} & \text{O O} & \text{O O} \\
\text{ROPOPOR} & \text{ROPOPOH} & \text{ROPOPO}^- \\
\text{O\_O\_} & \text{O\_O\_} & \text{O O\_} \\
& & \text{H} \\
(\textbf{5}) & (\textbf{6}) & (\textbf{7})
\end{array}
$$

The respective rates for these species ($R = \gamma$-phenylpropyl) at 95° are about 3×10^{-4} and 0.45 hr^{-1} (296). Accepting that a single alkyl

TABLE XII
Diphosphate Ester Hydrolysis Kinetics

Ester	pH	Temp. (°C)	Ref.
Monoesters			
γ-phenylpropyl	0.3–13.1	95, 99	295, 296
adenosine	1.1–9.9	95	296
adenosine	3.5–10.5	80–95	191
adenosine	3.6–9.4	80	389
adenosine	$1N\ H_2SO_4$	100	263
Diesters			
sym. di-γ-phenylpropyl	1.1–13.1	95	296
unsym. diethyl	3–10	20–46	294
Triesters			
trimethyl	8.8		55
Tetraesters			
tetramethyl		25	424
tetraethyl	4.8–9.8	57	55
tetraethyl	8.8	59	356
tetraethyl	1–7	0–60	230
tetraethyl		25–65	36
tetraethyl	3	25, 38	175
tetraethyl	13	30	139
tetraethyl		25	424
tetra-n-propyl		25	424
tetra-isopropyl		25	424, 425
tetra-n-butyl		25	424
tetrabenzyl	(in 1-propanol)	0, 50	136
unsym. dimethyl diethyl		25	425
unsym. dimethyl di-n-propyl		25	425
unsym. dimethyl di-isopropyl		25	425
unsym. diethyl di-n-propyl		25	425
unsym. diethyl di-isopropyl		25	425
unsym. diethyl di-n-butyl		25	425
unsym. di-isopropyl di-n-propyl		25	425
unsym. di-isopropyl di-n-butyl		25	425

substituent is kinetically equivalent to a single bound proton, the kinetic difference for these species can most reasonably be attributed to the rearrangement of **6** to give **7**, a tautomerism that is impossible for **4**.   Although form **7** will be a minor constituent, its kinetic importance can be appreciated from the study by Miller and Ukena (294) of the hydrolysis kinetics for unsymmetrical diethyl diphosphate. Table XI and Figure 3 allow comparison of the hydrolysis rate constant values for this ester at 35° and for γ-phenylpropyl diphosphate at 95°. At the same temperature, the unsymmetrical diester dianion will

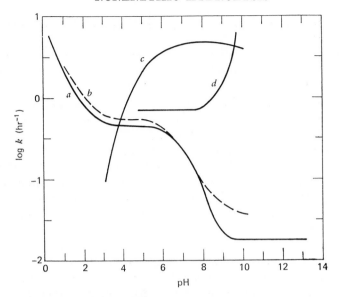

Fig. 3. Diphosphate ester hydrolysis, first-order rate constants for (a) γ-phenyl-propyl diphosphate, 95° (296); (b) adenosine diphosphate, 95°, dashed for clarity (296); (c) unsymmetrical diethyl diphosphate, 35° (294); and (z) tetra-ethyl diphosphate, 57° (55).

hydrolyze far faster than the monoester dianion. The kinetic importance of unsymmetrical diphosphate species was suggested earlier (11,55,356).

From the catalysis of tetraethyl diphosphate hydrolysis by added monophosphate, Brown and Hamer (55) deduced a mechanism involving nucleophilic displacement at phosphorus by HPO_4^{2-}.

$$(EtO)_2PO{-}O{-}PO(OEt)_2 + HPO_4^{2-} \rightarrow$$
$$(EtO)_2PO{-}O{-}PO_3^{2-} + (EtO)_2PO_2^- + H^+ \quad (15)$$

The postulated unsymmetrical diethyl diphosphate intermediate could then undergo P—O—P cleavage by an S_N1 or bimolecular mechanism. The extreme instability of the unsymmetrical intermediate and its phosphorylating ability argued for its hydrolysis by the S_N1 mono-metaphosphate mechanism. Sulfite ion and pyridine were about as effective catalysts as HPO_4^{2-} and could function through related mechanisms. The sterically hindered base lutidine was a very poor catalyst, consistent with nucleophilic attack at phosphorus in the first step. Comparable results have been reported by Westheimer's group (136).

Samuel and Silver (356), using ^{18}O-labeled phosphate ion to catalyze the same reaction, found the enrichment of ^{18}O in the diethyl monophosphate product was that expected on the basis of Eq. (15). Alternate paths for transfer of the label could be excluded. This study also showed that the P—O—P bridge oxygen remained with the substituted phosphate tetrahedron. This supports the monometaphosphate mechanism for the second step as the alternate is the unlikely bimolecular reaction at the unsubstituted phosphorus.

One might ask at this point whether the entire hydrolysis for inorganic diphosphate might not proceed through unsymmetrically protonated $H_2P_2O_7^{2-}$. If this were the case, rate expressions of the form of Eq. (13) would not be found. Since the ratio of the activities of the unsymmetrical and symmetrical forms of $H_2P_2O_7^{2-}$ is independent of pH, the rate expression would involve only the $H_2P_2O_7^{2-}$ activity. All the protonated forms must have kinetic importance in appropriate pH ranges. Around pH 4, where symmetrical $H_2P_2O_7^{2-}$ is the most abundant species, most of the reaction occurs through the unsymmetrical form. This does not tell us anything concerning other pH ranges. In fact, it is likely that in more acid solutions a bimolecular mechanism prevails, with nucleophilic attack by water.

In contrast to the pattern for monoesters (curves a and b of Fig. 3), the rate of hydrolysis of unsymmetrical diethyl diphosphate decreases with decreasing pH through the acid region (curve c). The faster hydrolysis for the dianion than the monoanion is due to the importance of a good leaving group in the monometaphosphate mechanism for species in which proton transfer to the leaving group is impossible (245). Diethyl phosphate is an excellent leaving group. In the monoanion, the proton is presumed to be bound internally in the reactive intermediate. The very low basicity of the phosphoryl oxygen of the diethyl substituted tetrahedron makes the equilibrium concentration of this configuration very low. For a unimolecular mechanism $\Delta S^{\ddagger}$ is rather negative (Table XIII).

The rate increase for tetraethyl diphosphate in alkaline medium (Fig. 3d) was attributed by Brown (55) to an S_N2 displacement by OH^- on phosphorus, followed by loss of a proton. Rapid cleavage of the P—O—P linkage of tetraethyl diphosphate in dilute sodium hydroxide was also noted by Ketelaar (230). A similar acceleration at high pH has been reported for adenosine diphosphate with certain cations and attributed to cation catalysis (Fig. 4) (389). The rate did not increase at high pH for γ-phenylpropyl diphosphate (295,296), for symmetrical di-γ-phenylpropyl diphosphate (296) or for unsymmetrical diethyl diphosphate (294). A steric effect on the hydrolysis of tetraalkyl

TABLE XIII
Activation Parameters for Diphosphate Hydrolysis[a]

Species	pH	E_a (kcal/mole)	Log A (log hr^{-1})	$\Delta S^{\ddagger}$ (eu)	Ref.
$P_2O_7^{4-}$	$3M$ H_2SO_4	15	10.7	-29	67
	$1M$ H_2SO_4	16	10.6	-31	67
	$0.5M$ HCl	21			238
	~ 1	22.8	13.5	-14.5	147
	1	23.3	13.5	-15	441
	1	22.3			89
	4	27.8	15.7	-5	441
	4	28.9			89
	7	30.4	16.6	-1	441
	10	(41.4)	(20.7)	($+18$)	441
ADP	8	24.2			191
unsym. $Et_2P_2O_7^{2-}$	10	18	13.1	-17	294
$Et_4P_2O_7$		10.3	6.6	-47	175
		10.5	6.7	-46	36
		10.7	6.1	-49	230
$H_3P_2O_7^-$	b	30.2	17.7	$+4$	89
$H_2P_2O_7^{2-}$	b	29.0	17.0	$+1$	89

[a] Some values have been calculated from data in the references shown. Other values have been recalculated for consistency in units. The Arrhenius activation values, E_a, have been obtained from $\Delta H^{\ddagger}$ values by means of the relationship for constant pressure solution reactions, $E_a = \Delta H^{\ddagger} + RT$. At 298°K RT amounts to 0.6 kcal/mole. Interconversion of the Arrhenius pre-exponential factor, A, and the activation entropy, $\Delta S^{\ddagger}$, has been accomplished by the relationship for constant pressure solution reactions, $A = (ekT/h) \exp (\Delta S^{\ddagger}/R)$, in which k and h are Boltzmann's and Planck's constants, respectively. Refs 5,150.

[b] For the specific species shown.

diphosphates and unsymmetrical $R_2R_2'P_2O_7$ esters has been noted, with isopropyl groups particularly effective at reducing the rate (424,425).

Activation parameters for diphosphate hydrolysis are reported in Table XIII. Use of the value of $\Delta S^{\ddagger}$ has been suggested by Long as a criterion for the role of water in hydrolysis reactions (273,274,359). For unimolecular mechanisms small $\Delta S^{\ddagger}$ values of either sign are expected, most commonly from 0 to 10 eu. For bimolecular mechanisms large negative $\Delta S^{\ddagger}$ values are expected, generally -15 to -30 eu. On this basis a bimolecular mechanism is indicated for hydrolysis of inorganic diphosphate in strongly acid solution (67), pH ≤ 1. An S_N1, presumably monometaphosphate mechanism is indicated for

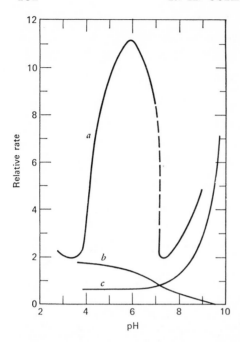

Fig. 4. Metal ion catalysis (389). Hydrolysis of sodium adenosine diphosphate, 20 mM at 80°, with (a) 20 mM Zn^{2+} added; (b) no added ion; and (c) 20 mM Mn^{2+}.

higher pH values. For the diphosphate esters, all the $\Delta S^{\ddagger}$ values listed are very negative, yet there is good evidence for an S_N1 mechanism for hydrolysis of the unsymmetrical diethyl diphosphate. For the three tetraethyl diphosphate studies, the pH values are uncertain and probably changed during the runs. The hydrolysis kinetics for this compound are largely independent of pH, however, and must be bimolecular. With substitution of the phosphoryl oxygen by sulfur there are marked changes in the activation parameters, indicating a change in mechanism. For the hydrolysis in neutral solution of $(EtO)_2P(O)OP(O)(OEt)_2$, $(EtO)_2P(S)OP(O)(OEt)_2$, and $(EtO)_2P(S)OP(S)(OEt)_2$, the activation energies are, respectively, 10.5, 12.4, and 27.9 kcal/mole, and the activation entropies -46, -42, and -5 eu (35,36).

As pointed out in Section II, the role of metal ions is understood only in a general way. Most metal ions inhibit hydrolysis below pH 4, but serve as catalysts at higher pH. Studies of cation effects include Na^+ and tetraalkylammonium salts (89,441); Mg^{2+} and Ca^{2+} salts (97,161,264,315,389); Li^+, Ca^{2+}, Ba^{2+}, and Zn^{2+} (67); Mn^{2+}, Cu^{2+}, Zn^{2+} (55); Be^{2+}, Sr^{2+}, Ba^{2+}, Mn^{2+}, Co^{2+}, Ni^{2+}, Cu^{2+}, Zn^{2+}, Cd^{2+}, Hg^{2+} (389); and salts and complexes of Cu^{2+}, Mn^{2+}, Pb^{2+}, Ce^{3+}, UO_2^{2+}, Mo(VI), Zr(IV), and Th(IV) (189). Many metal hydroxide gels and metal oxides are effective catalysts (20,46). Figure 4 shows typical data. A conductimetric rate study indicated that Cr^{3+} was a potent inhibitor in acid solution (344), but it appears the conductimetric method was misleading (166).

B. Triphosphate

The hydrolysis of triphosphate is first order (147,299,369,373,441). Conditions under which the uncatalyzed (other than by H^+) hydrolysis has been studied are listed in Table XIV. The results of these studies

TABLE XIV

Inorganic Triphosphate Hydrolysis Kinetics Studies

pH	Temp. (°C)	Ref.
2–12	65.5	114
1–13	30–125	165, 440, 441
1.6–10	37–70	299
1–4	39.6–49	147
1–5	65.5	373
5–9	65, 88	161
10, 13	70, 100	31
10	25, 82	342
6M HCl, 8.5	25	226
alk., concd. soln.	70	369
alk., concd. soln.	108–121	83
	60–100	453
	25–35	329

are summarized in Figure 5, from which rate constants and half-lives can be estimated for conditions of interest. Although, as for all the condensed phosphates, the hydrolysis of triphosphate is spontaneous over the entire range of aqueous conditions, the rate is slow near room temperature, especially at pH $\geq$ 7.

The dependence of rate on pH is shown in Figure 6. The rate minimum at pH 10 in curve c was also reported by Watzel (453). In both cases, sodium hydroxide was the base used. In curve b, from the study by Van Wazer and colleagues, with tetramethylammonium ion the only cation present, the rate decreased steadily with increasing pH, showing the catalysis should be attributed to sodium ion in the other studies. Since sodium ion does not catalyze diphosphate hydrolysis, one method of analysis for triphosphate in the presence of diphosphate takes advantage of the much higher rate of hydrolysis for $P_3O_{10}^{5-}$ than for $P_2O_7^{4-}$ in 10–20% NaOH solution (317).

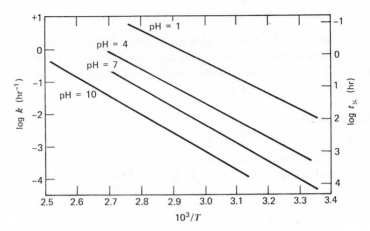

Fig. 5.   First-order rate constants for triphosphate hydrolysis.

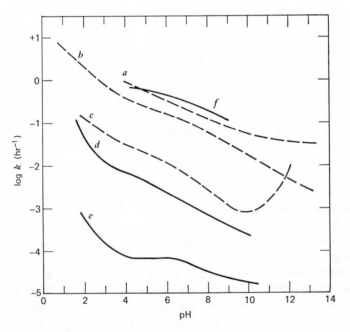

Fig. 6.   First-order rate constants for triphosphate hydrolysis at (a) 90°, Na$^+$ (441); (b) 90°, $(CH_3)_4N^+$ (441); (c) 65.5°, Na$^+$ (114); (d) 50° (299); and (e) adenosine triphosphate, ordinate in arbitrary units (389); and (f) adenosine triphosphate, 95° (296). Dashed lines based on relatively few points.

The pH dependence of triphosphate hydrolysis kinetics has recently been resolved into a dependence on the relative abundance of the various protonated forms of triphosphate (299), Eq. (16).

$$\text{Rate} = (k_5 + k_4'K_5)(H_5P_3O_{10}) + (k_4 + k_3'K_4)(H_4P_3O_{10}^-)$$
$$+ (k_3 + k_2'K_3)(H_3P_3O_{10}^{2-}) + (k_2 + k_1'K_2)(H_2P_3O_{10}^{3-})$$
$$+ (k_1 + k_0'K_1)(HP_3O_{10}^{4-}) + k_5'(H_5P_3O_{10})$$
$$+ k_0(P_3O_{10}^{5-}) \tag{16}$$

where k_5, k_4, . . . are rate constants for reaction with water, k_5', k_4', . . .

TABLE XV

Rate Constants for Individual Species[a] (hr^{-1})

Temp. (°C)	$H_3P_3O_{10}^{2-}$ $(k_3 + k_2'K_3)$[b]	$H_2P_3O_{10}^{3-}$ $(k_2 + k_1'K_2)$	$HP_3O_{10}^{4-}$ $(k_1 + k_0'K_1)$	$P_3O_{10}^{5-}$ (k_0)
37	1.3×10^{-3}	1.9×10^{-3}	9.0×10^{-5}	2.9×10^{-5}
50	1.11×10^{-2}	1.05×10^{-2}	9.0×10^{-4}	2.1×10^{-4}
60	1.67×10^{-2}	2.52×10^{-2}	2.80×10^{-3}	4.6×10^{-4}
70	4.8×10^{-2}	5.9×10^{-2}	7.94×10^{-3}	8.7×10^{-4}

Activation Parameters for Individual Species[c]

$\log A$ (hr^{-1})	13.5	12.8	16.1	20.2
$\Delta H^{\ddagger} = E_a$ (kcal/mole)	23.3	22.0	28.6	36.6
$\Delta S^{\ddagger}$ (eu)	-15	-18	-3	$+16$

[a] From Ref. 299, recalculated.
[b] See Eq. (16).
[c] See footnote (a), Table XIII.

are rate constants for reactions with H_3O^+, and K_5, K_4, . . . are acid ionization constants, with the subscript giving the number of protons in the triphosphate species concerned, $H_5P_3O_{10}$, $H_4P_3O_{10}^-$, ... This is analogous to the treatment earlier accorded diphosphate [Eq. (13)], but mathematical inseparability of the water and hydronium ion reactions is recognized explicitly. Rate constants for the individual species appear in Table XV. The original paper resolves these constants into separate values for reaction with water, k, and reaction with hydronium ion, k', but this evaluation rests on questionable assumptions.

The activation parameters of Table XV are derived from the combined rate constants in that table; they differ from those given in the

TABLE XVI

Triphosphate Ester Hydrolysis Kinetics Studies

Ester	pH	Temp. (°C)	Ref.
adenosine	$\sim$1	40–50	149
	2–10	80	389
	4–9	95	296
	6–12	100	188
	4.2, 8.5		265, 266
	$1N$ H_2SO_4	100	263
γ-phenylpropyl	4–9	95	295, 296

original paper for the separate reactions.   The activation enthalpy and entropy both show minima for $H_2P_3O_{10}{}^{3-}$.   The enthalpy minimum outweighs that for the entropy, and the rate constant is a maximum for this species.   If the $\Delta S^{\ddagger}$ value is used as a criterion of mechanism (273,274,359), a bimolecular mechanism is indicated for $H_3P_3O_{10}{}^{2-}$ and $H_2P_3O_{10}{}^{3-}$, and a unimolecular mechanism for $HP_3O_{10}{}^{4-}$ and possibly $P_3O_{10}{}^{5-}$.

Studies of triphosphate ester hydrolyses in the absence of a catalyst other than hydronium ion are listed in Table XVI.   Miller and Westheimer (296) have shown for triphosphate, as for diphosphate, that the adenosine moiety has no special properties with respect to the uncatalyzed hydrolysis (Table XVII).   The similarity in hydrolysis rate constants for the species with organic substituents and with hydrogen is surprising.   It must arise in part from a cancellation of effects as even the statistical factor of two, favoring inorganic triphosphate hydrolysis, is not evident.   Except at high pH, triphosphate

TABLE XVII

Hydrolysis Rate Constants (hr^{-1}) at 95°

R	$RHP_3O_{10}{}^{3-}$	$RP_3O_{10}{}^{4-}$
γ-phenylpropyl[a]	0.54	0.076
adenosine[a]	0.63	0.12
hydrogen[b]	0.54	0.11

[a] From Ref. 296.
[b] Calculated for 95° from data of Table XV.

monoester hydrolysis occurs almost completely at the terminal P—O—P linkage (142,149,265,296), [Eq. (17)].

$$
\underset{\underset{O\_O\_O\_}{O\ O\ O}}{ROPOPOPOH} + H_2O = \underset{\underset{O\_O\_}{O\ O}}{ROPOPOH} + \underset{\underset{O\_}{O}}{HOPOH} \tag{17}
$$

$$
\underset{\underset{O\_O\_O\_}{O\ O\ O}}{ROPOPOPOH} + H_2O = \underset{\underset{O\_}{O}}{ROPOH} + \underset{\underset{O\_O\_}{O\ O}}{HOPOPOH} \tag{18}
$$

More surprising is the similarity in rate constants for the triphosphate trianions (Table XVII) and for the corresponding diphosphate dianions ($RHP_2O_7{}^{2-}$, R = γ-phenylpropyl, adenosine and hydrogen, Table XI, 95°). The values are alike enough to suggest that the tautomeric mechanism suggested by Miller and Westheimer (296) for dianions of inorganic diphosphate and of diphosphate monoesters applies to the hydrolysis of the trianions of inorganic triphosphate and of triphosphate monoesters. Support for such tautomerism can be found in ^{31}P NMR chemical shift data (117), which show large enough changes in the shift for the middle phosphorus of triphosphate over the pH range 3–10 to strongly suggest a direct interaction of the protons of $H_2P_3O_{10}{}^{3-}$ and $HP_3O_{10}{}^{4-}$ with the central phosphorus tetrahedron, rather than simply inductive effects on the middle phosphorus through a HOPOP (middle) chain. Oxygen isotope studies support this view in that cleavage is at the terminal P—O bridge (214).

An increase in ionic strength (established by tetramethylammonium salt) decreases the rate of triphosphate hydrolysis (441); for example, the rate constant at 50° and pH 4.0 decreased 32% for an increase in ionic strength from 0.1 to 1.0 (299). Because the acid ionization constants increase with increased ionic strength, at least part of the rate shift is accounted for by a shift to less protonated species.

Added metallic cations inhibit inorganic triphosphate hydrolysis in strongly acid solution (147,441), but catalyze it in less acid solution. An increase in the catalytic effect of Mg^{2+} and Ca^{2+} was found as pH was increased from 5 to 9 (161). An increase in the catalytic ability of Na^+ was found from pH 4 to 13 (441). The catalysis by Na^+ resulted from a more favorable $\Delta S^{\ddagger}$ at pH 4 and 7 and from more favorable $\Delta H^{\ddagger}$ and $\Delta S^{\ddagger}$ at pH 10. The smaller effect of cations at lower pH values is to be expected on the basis of limitation of complex formation through competition between protons and metal ions for triphosphate interaction sites. In addition, the reference state in acid solution reflects proton catalysis. To appear to be a catalyst a metal ion must be more effective than the protons it displaces.

Metal-ion inhibition has been reported for ATP hydrolysis in strongly acid solutions (264,265,266), but at pH 1.3 catalysis was found for small additions of NaCl and inhibition for larger additions (149). The association constants for Na^+ and K^+ with ATP (229 and 221 liters/mole, respectively, at pH 9 and 25°) are larger than previously thought (300). Catalytic ability of Cu^{2+} and Zn^{2+} peaks at pH 5–6, with very little dependence on ionic strength over the range 0.1 to 1.0 (363,389). At pH 8.8, Mg^{2+} has been reported to be a catalyst for ATP hydrolysis (264,265,266) and an inhibitor (315).

Miller and Westheimer have investigated the effect of cations on the hydrolysis of ATP and of their ATP model, γ-phenylpropyl triphosphate (297). The formation constants are almost the same for the complexes of Mg^{2+} and Cd^{2+} with both of these triphosphate esters, indicating there is no special interaction of these metal ions with the adenosine residue. The ratio of hydrolysis rates at pH 8.1 for ATP and γ-phenylpropyl triphosphate are affected very little by these cations. On the other hand, Cu^{2+} at pH 5, at one concentration at least, is twenty times as effective a catalyst for hydrolysis of the adenosine ester as for the γ-phenylpropyl. The presence of a Cu^{2+}–adenosine interaction that is lacking for γ-phenylpropyl is supported by a UV spectral study of ATP complexes (362). Coordination with the adenine ring was found to predominate for Cu^{2+}, whereas little such interaction was found for Mg^{2+}, Ca^{2+}, Mn^{2+}, Co^{2+}, Ni^{2+}, or Zn^{2+}. Proton NMR studies indicate that Zn^{2+} interacts with the nitrogen at position 7 of the adenine ring, but that Mg^{2+} and Ca^{2+} do not react with the adenine or pyrimidine portion of ATP (106,178).

In Section II, we called attention to the difficulties with metal-ion-catalysis studies to date. Some of these are emphasized by a relatively thorough study of Tetas and Lowenstein (389). First, metal ions do not catalyze reactions (17) and (18) equally, as is illustrated by Table XVIII. Although reaction (17) remains the more important for every cation tested, for Ca^{2+} 31% of the disappearance of ATP is accounted for by reaction (18). Apparent catalytic effectiveness will depend on how the reaction is followed. Second, pH has a strong influence on metal-ion catalysis. Figure 7 shows the danger of studying these reactions at arbitrarily chosen pH values. The relative catalytic effectiveness of Mn^{2+} and Zn^{2+} strongly depends on whether they are compared at pH 2 or 5 or 8. The relative effectiveness for reactions (17) and (18) must depend on pH, also.

TABLE XVIII
Specificity in Metal-Ion Catalysis of ATP Hydrolysis[a]

Metal ion	Relative rates		Ratio of PO_4^{3-} rate to $P_2O_7^{4-}$ rate
	PO_4^{3-} formation	$P_2O_7^{4-}$ formation	
Ni^{2+}	75	1.0	75
None	39	1.0	39
Mg^{2+}	98	3.3	30
Co^{2+}	137	4.7	29
Mn^{2+}	275	13	21
Ba^{2+}	57	4.1	14
Cu^{2+}	122	8.7	14
Zn^{2+}	170	16	11
Hg^{2+}	79	8.0	10
Cd^{2+}	199	21	9.7
Be^{2+}	162	19	8.5
Ca^{2+}	177	79	2.3

[a] At pH 9 and 80°, Ref. 389.

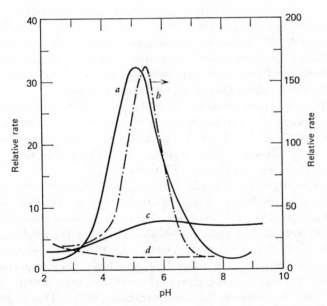

Fig. 7. Metal ion catalysis of hydrolysis of sodium adenosine triphosphate, 20 mM at 80° with (a) 20 mM Zn^{2+} added; (b) 20 mM Cu^{2+} added (ordinate scale at right); (c) 20 mM Mn^{2+} added; and (d) no added ion (389).

139

Speculation on the role of metal ions has assumed that for fully ionized ATP chelation will occur preferentially at the β and γ phosphorus tetrahedra, as shown in **8** instead of as in **9**.

$$\begin{bmatrix} & \text{O} \ \ \text{O} \ \ \text{O} \\ \text{AOPOPOPO} \\ & \text{O} \ \ \text{O} \ \ \text{O} \\ & \qquad \backslash \ / \\ & \qquad \text{M} \end{bmatrix}^{-2} \qquad \begin{bmatrix} & \text{O} \ \ \text{O} \ \ \text{O} \\ \text{AOPOPOPO} \\ & \text{O} \ \ \text{O} \ \ \text{O} \\ & \quad \backslash \ / \\ & \quad \text{M} \end{bmatrix}^{-2}$$

$$\text{(8)} \qquad\qquad\qquad\qquad \text{(9)}$$

This view is generally substantiated by a ^{31}P NMR study of the cation-induced chemical shifts for ATP over the pH range 4.5–9 (106). Interaction with the β and γ phosphate tetrahedra, presumably through chelation, is indicated for Mg^{2+}, Ca^{2+}, Cu^{2+}, and Zn^{2+}. Mn^{2+} and Co^{2+}, on the other hand, interact with all three tetrahedra of the triphosphate chain. It is not clear whether this involves tridentate chelation or averaging over a mixture of complexes. The summary of this article lists Cu^{2+} as interacting at the α and β positions, which has misled some readers; the text and figure make it clear interaction is with the β and γ tetrahedra. No correlation has been established between the structure and properties of these complexes and the catalytic effectiveness or specificity of the metal ions. Perhaps catalytic studies under a wider variety of conditions are needed. Just as kinetic importance was found for unsymmetrical diphosphate species, it is possible that the mechanism of metal-ion catalysis will be found to involve species or configurations other than the most abundant ones.

Alkaline conditions favor the hydrolysis of ATP by reaction (18) to produce inorganic diphosphate and adenosine monophosphate (107,188,229,270,272). Barium hydroxide is commonly employed, as Ba^{2+} is a very effective catalyst (188). It is likely that several points of attack on the triphosphate are involved (270).

In the hydrolysis of inorganic triphosphate, diphosphate and monophosphate are produced in an equimolar ratio. Hydrolysis of the diphosphate product will limit the diphosphate:monophosphate mole ratio to somewhat less than one, although the ratio might approach one in the early stages of the reaction. As Table XIX shows, experimental ratios significantly exceeding one have been reported. The ratio is larger the more concentrated the solution, even if non-phosphate salts are used to increase the concentration (369). This suggests the higher diphosphate yield is favored either by increased Na^+ concentration or by decreased water activity. The increased diphosphate yield cannot be the result of monophosphate recombination nor of a shift in the $P_2O_7^{4-}$–PO_4^{3-} equilibrium. This follows from the disappearance

TABLE XIX

Triphosphate Hydrolysis Products

Solution conc'n (wt %)	Temp. (°C)	Percent hydrolysis	Mole ratio $P_2O_7^{4-}/PO_4^{3-}$	Ref.
1%	100	50	1.32	31
1%, +1% NaOH	100	50	1.19	31
1%	70	20	1.5	31
14%	121	37	1.78	83
5%[a]	70, 100	10	2.0	369

[a] Plus 20% Na_2SO_4 and 10 Na_2SiO_3 for a total of 35 wt % solutes.

of diphosphate with time. It seems likely that phosphorylation of monophosphate by triphosphate is responsible, although the explanations proposed in the literature involve a trimolecular reaction between

$$2 P_3O_{10}^{5-} + H_2O = 2 HP_2O_7^{3-} + P_2O_7^{4-} \tag{19}$$

triphosphate ions and water (83) or a bimolecular reaction between pairs of triphosphate–water complexes (369). An activation energy of 28 ± 5 kcal/mole was found for reaction (19) in 14% $Na_5P_3O_{10}$ solution.

Of importance to practical applications of triphosphate are the observations of Shen and Dyroff (369). Even at high ionic strengths, up to 35 wt % solutes (5 $Na_5P_3O_{10}$, 20 Na_2SO_4, 10 Na_2SiO_3), the hydrolysis of triphosphate is first order. The activation energy in the 35 wt % solution was 19.8 kcal/mole, compared to 22.8 kcal/mole for a 1% solution (1% $Na_5P_3O_{10}$). The first-order rate constant increased exponentially with total solute concentration. An excellent exponential correlation was obtained with Na^+ concentration.

The hydrolysis of triphosphate during the dehydration of $Na_5P_3O_{10}$· $6H_2O$ has been of considerable industrial interest. This hydrate can be stored for long periods without appreciable decomposition, but when dehydration by vacuum and drying agent or thermal dehydration below 150° is attempted, hydrolysis of the triphosphate anion occurs (169). Prolonged grinding at room temperature causes decomposition of the anion (340). Below 80°, dehydration produces amorphous $Na_5P_3O_{10}$ initially, and $Na_4P_2O_7$ as the only crystalline product (370). At 80–120°, the principal reaction appears to be (21) (346,370,410), although $Na_3HP_2O_7$ and Na_2HPO_4 appear to be formed under some circumstances, as well (342,472).

$$Na_5P_3O_{10} \cdot 6 H_2O = Na_5P_3O_{10} + 6 H_2O \text{ (g)} \tag{20}$$

$$Na_5P_3O_{10} \cdot 6 H_2O = Na_4P_2O_7 + NaH_2PO_4 + 5 H_2O \text{ (g)} \tag{21}$$

$$Na_5P_3O_{10} + H_2O \text{ (g)} = Na_4P_2O_7 + NaH_2PO_4 \tag{22}$$

The product often has a diphosphate:monophosphate mole ratio either above or below one (342,370). A value below one can easily be understood through hydrolysis of the diphosphate. The much more common values above one are more difficult to explain. The additional diphosphate might result from phosphorylation of monophosphate by triphosphate, but there would not be much proximity of these possible reactants and diffusion rates would be very small at the temperatures involved. More likely sources of the additional diphosphate are condensation of hydrogen monophosphate ions or a direct reaction between triphosphate ions. It has been reported that at 60° at 40% relative humidity $Na_5P_3O_{10}$ degrades faster than $Na_5P_3O_{10} \cdot 6H_2O$. Furthermore, the anhydrous $Na_5P_3O_{10}$ converts almost completely to $Na_4P_2O_7$ and $Na_3HP_2O_7 \cdot H_2O$, less than 1% monophosphate being formed (431).

Above 120° increasing amounts of anhydrous $Na_5P_3O_{10}$ (form II) appear among the decomposition products (342,410). To remove all of the water from a sample heating to over 165° is required (164). Removal of all the water cannot be accomplished until the mono- and diphosphates formed earlier have condensed to triphosphate.

TABLE XX

Thermodynamics of $Na_5P_3O_{10} \cdot 6 H_2O$ Decomposition at $298°K$[a]

Reaction.	$\Delta H°$ (kcal/mole)	$\Delta S°$ (eu)	$\Delta G°$ (kcal/mole)
(20)	79.8	211.9	16.7
(21)	64.0	174.5	12.0
(22)	−15.8	−37.41	−4.6

[a] Calculated from data of Table III on the basis that species other than water in the reactions are crystalline. Anhydrous $Na_5P_3O_{10}$ was considered to be form II.

Table XX lists values for the thermodynamic changes accompanying some of the reactions that appear to be involved in $Na_5P_3O_{10} \cdot 6 H_2O$ decomposition. If one makes the reasonable assumption that the $\Delta H°$ and $\Delta S°$ values are independent of temperature over the range 25–150°, he can account for several aspects of this decomposition on a thermodynamic basis. One can calculate that $\Delta G°$ is the same for reactions (20) and (21) at 150°. At temperatures below 150° reaction (21) is thermodynamically favored; at temperatures above 150° reaction (20) is favored. This agrees with the observations reported above, that $Na_5P_3O_{10}$-II forms at the higher temperatures. Because of slow diffusion away of water vapor at ordinary pressures, a thermal decomposition that produces water vapor cannot be expected to become rapid in a bulk sample below the temperatures at which $K_p \sim 1$ atm. For the formation of $Na_5P_3O_{10}$-II from $Na_4P_2O_7$ and NaH_2PO_4 previously formed [i.e., the reverse of reaction (22)] this temperature is 150°.

The negative $\Delta G°$ for reaction (22) shows that hydrolysis should be expected in the presence of 1.0 atm of water vapor below 150° for $Na_5P_3O_{10}$-II that might form. The dehydration of $Na_5P_3O_{10} \cdot 6 H_2O$ does not give this P_{H_2O} until 113° however (Table XXI). Combining

TABLE XXI

Dissociation Pressures for $Na_5P_3O_{10} \cdot 6 H_2O$ (atm)[a]

Temp. (°C)	Dehydration [Eq. (20)]	Degradation [Eq. (21)]
25	6×10^{-13}	2×10^{-9}
75		9×10^{-3}
93	4×10^{-2}	1.0
113	1.0	8×10^{1}
120	6.8	4×10^{2}

[a] Calculated from $\Delta H°$ and $\Delta S°$ values of Table XX.

the thermodynamic data for reactions (20) and (22) shows that only above 110° could the partial pressure of water vapor from

$$Na_5P_3O_{10} \cdot 6 H_2O$$

dehydration cause hydrolysis of $Na_5P_3O_{10}$-II by reaction (22). Of course, amorphous $Na_5P_3O_{10}$ would be higher in free energy than $Na_5P_3O_{10}$-II. Amorphous triphosphate might be able to react with the low pressure of water vapor, and this could be the route for reaction (21).

The very low calculated equilibrium dissociation pressures in Table XXI for moderate temperatures explain why a vacuum is commonly used in dehydration at these lower temperatures. Experimental studies of the water vapor pressure in the $Na_5P_3O_{10} \cdot 6 H_2O$ system have recognized that reactions other than dehydration were occurring (370,471). Even the pressures reported by Shen (370) for 99.3% $Na_5P_3O_{10} \cdot 6 H_2O$ are all significantly above those for reaction (20). The pressures listed in Table XXI show a much more marked temperature dependence than the values reported by Shen. In fact, his dissociation pressure–temperature dependence is that for a ΔH of about 19 kcal/mole, less than one-third of the values associated with reactions (20) and (21). Although Shen did not verify that his pressures represented equilibrium, they were reproducible, and suggest that neither reaction (20) or (21) is important in the equilibrium decomposition of $Na_5P_3O_{10} \cdot 6 H_2O$ over the range of temperature studied (60–95°). The formation of $Na_5P_3O_{10}$-amorphous in reaction (20) would increase the ΔH discrepancy.

Much of the behavior of $Na_5P_3O_{10} \cdot 6 H_2O$ on attempted dehydration needs a kinetic explanation. A study by Groves and Edwards (169) involving control of the partial pressure of water vapor showed that the dehydration and the hydrolysis reactions of $Na_5P_3O_{10} \cdot 6 H_2O$ are distinct from one another. Microscopic examination of single crystals showed that in the hydrolysis reaction reacted spots appeared at corners and edges and spread over the crystal; only one-sixth of the water had been lost when the reaction had covered the crystal. For the dehydration reaction, opaque planes appeared and spread through the crystal; all the water had been lost when the crystal was entirely opaque. Microscopic studies have also been carried out by Prodan (338). Studies on powdered material showed that at intermediate partial pressures of water (from about 150 to 500 mm Hg, depending on temperature) only the dehydration reaction occurred. Outside this range the hydrolysis reaction took place. Nucleation is a crucial step in the thermal decomposition of solids that decompose by an interfacial mechanism (152,470). Not only is it prerequisite to reaction, but it can determine which of several reactions occurs. Groves and Edwards report long induction periods in the hydrolytic decomposition of $Na_5P_3O_{10} \cdot 6 H_2O$; for example, at 110° the induction period was about 20 min. This is the time required for growth nuclei to appear. The dehydration reaction evidently normally has longer induction periods, but it was found that with $p_{H_2O} > 210$ mm Hg the induction period for the dehydration reaction could be shortened to 3–7 min in the same temperature range. Controlling the partial pressure of water controlled the kinetics of nucleation and, in turn, controlled which reaction occurred. The presence of an adequate water vapor level is known to aid crystallization, which can be important to the nucleation process.

The hydrolysis reaction kinetics for $Na_5P_3O_{10} \cdot 6 H_2O$ appear to be either diffusion controlled or heat-transfer-rate controlled in the early stages (370) and diffusion controlled in the later stages (339,370). For the initial stages of the reaction, activation energies of 14.0 (370) and 13.4 (339) have been reported. For the later stages, 20.1 (339) and 41.5 kcal/mole (472) have been reported for temperatures below

$110°$ and 18.4 kcal/mole (472) for higher temperatures. For the smaller value below $110°$, the rate of loss of water from the sample was determined, for the larger the rate of disappearance of crystalline hexahydrate. Several reaction steps can separate these. The decomposition of $Na_3NiP_3O_{10}\cdot12\ H_2O$ is faster than that of $Na_5P_3O_{10}\cdot6\ H_2O$ (341).

C. Tetraphosphate

Tetraphosphate hydrolysis kinetics are first order (115,168,329,373, 380). Although it was at first thought that the hydrolysis occurred at the middle P—O—P link (342), it is clear that over a wide range of conditions degradation occurs primarily by end-group clipping (115,168, 380,405,460).

$$
\begin{matrix} O\ \ O\ \ O\ \ O \\ \_OPOPOPOPO\_ \\ O\_O\_O\_O\_ \end{matrix} + H_2O \longrightarrow \begin{matrix} O\ \ O\ \ O \\ \_OPOPOPOH \\ O\_O\_O\_ \end{matrix} + \begin{matrix} O \\ HOPO\_ \\ O\_ \end{matrix} \tag{23}
$$

In addition, metaphosphate abstraction, a reorganization reaction (not a hydrolysis) that becomes more important for longer polyphosphate chains, occurs to a small extent (168,460).

$$
\begin{matrix} O\ \ O\ \ O\ \ O \\ \_OPOPOPOPO\_ \\ O\_O\_O\_O\_ \end{matrix} \longrightarrow \text{(metaphosphate ring)} + \begin{matrix} O \\ \_OPO\_ \\ O\_ \end{matrix} \tag{24}
$$

In contrast to the pattern for longer chains, Eq. 24 appears to have its greatest importance for tetraphosphate in acidic solution (329,460). Evidence for a coiling of the tetraphosphate anion that might be related to Eq. 24 comes from ^{31}P NMR (117) and from acid ionization constant values (452).

Hydrolysis of tetraphosphate is faster than that of triphosphate over the pH range studied. Activation energies at pH 4, 7, and 11 are, respectively, 27.5, 25.2, and 25.3 kcal/mole (168). Acid catalysis is observed but not base catalysis (115,168,460). Although sodium ion catalysis is suggested by a comparison of the curves in Figure 8 for all of which Na^+ is present except curve d, this catalysis must be much less effective than in the case of triphosphate. Reactions (23) and (24) are catalyzed by Mg^{2+} (460). In ammoniacal solution Mg^{2+} catalyzes the phosporylation of monophosphate to diphosphate by tetraphosphate (460), as well as the degradation of tetraphosphate (405). The hydrolysis of adenosine tetraphosphate has been studied at $100°$ in $1N\ H_2SO_4$ (263).

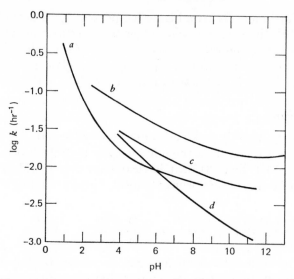

Fig. 8. Tetraphosphate hydrolysis. First-order rate constants at (a) 60° (460);
(b) 65.5° (115); (c) and (d) 60° (168). In study (d), $(CH_3)_4N^+$ was the only
cation; the others had Na^+ present.

D. Other Oligophosphates

The polyphosphate chain can undergo degradation in at least three
ways (276,355,394): (*1*) hydrolytic cleavage of monophosphate from the
chain ends, which has been termed "end-group clipping" (168), (*2*)
splitting out of metaphosphate rings (391,409), and (*3*) random hydrolytic cleavage to produce two chains (276). The metaphosphate
abstraction reaction, which will be discussed more fully in the next
section, is a rearrangement or reorganization process, rather than a
hydrolysis. End-group clipping has been demonstrated for all the
oligophosphates and metaphosphate abstraction for the tetraphosphate
and longer oligomers. Small amounts of random cleavage would be
difficult to detect for the oligophosphates, but this reaction has been
established for longer chains and must have a role in the degradation
of oligomers. Strauss and Day found the rate of random cleavage
negligibly small for oligophosphates through the hexaphosphate
(380).

Studies by Griffith and Buxton (168) and by Strauss and Day (380)
of the aqueous degradation of oligomers through octaphosphate provide
separate rate constants for the end-group clipping and metaphosphate
abstraction reactions. For each example tested, the degradation
kinetics were first order. Some of the results for these sodium salts

are presented in Figure 9. For di- through pentaphosphate the total rate of disappearance of the initial phosphate anion decreases from pH 4 to 11; for hexa- through octaphosphate it is a maximum somewhere between pH 4 and 11. At a given pH, the rate of disappearance of the oligophosphate anion increases with increased chain length. Wieker has shown that at pH 8, after a maximum in the degradation rate constant near a chain length of ten phosphate tetrahedra, a limiting value is reached beyond a chain length of about 100 (459).

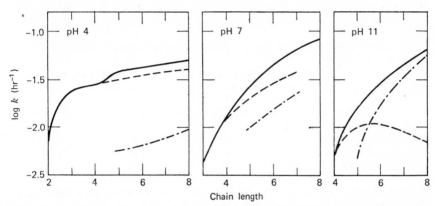

Fig. 9. Sodium oligophosphate degradation, first-order rate constants at 60°. ———— total degradation, ———— end-group clipping, ————— metaphosphate abstraction. Di- and triphosphate data for 1% solutions (441), others for about 5% (168).

Activation energies for end-group clipping decrease with increased pH for a given oligophosphate and with increased chain length for a given pH (Table XXII). Activation energies for triphosphate fit the pattern (441). Activation energies for metaphosphate abstraction increase with increased pH for a given oligophosphate, the opposite of end-group clipping, but are independent of chain length.

In the same work (168), a limited study of the aqueous degradation of tetramethylammonium salts of penta-, hexa-, and heptaphosphate showed virtually no metaphosphate abstraction. Sodium ions must have a role in this mode of degradation. There is also some catalysis of metaphosphate abstraction by hydrogen ion (329,460). Trimethylammonium ion cannot form complexes with polyphosphate chains; both sodium ion and hydrogen ion do. The role of these ions might be to neutralize charges on the polyphosphate chain to permit coiling. In support of this, viscosity measurements show metal ions are more effective than tetramethylammonium ion at inducing coiling of

TABLE XXII

A. Activation Energies (kcal/mole) for Sodium Oligophosphate Degradation[a]

	End-group clipping			Metaphosphate abstraction			Total		
pH	4	7	11	4	7	11	4	7	11
Tetraphosphate	27.5	25.2	25.3				27.5	25.2	25.3
Pentaphosphate	23.4	20.5	11.6	16.6	24.6	25.7	21.8	23.2	26.4
Hexaphosphate	21.5	17.8	12.3	16.4	24.4	25.3	20.3	20.0	27.4
Heptaphosphate	20.1	15.4	12.3	16.2	24.3	25.7	19.5	17.3	26.3

B. Activation Entropies (eu) for Sodium Oligophosphate Degradation[a]

	End-group clipping		Metaphosphate abstraction	
pH	4	11	4	11
Tetraphosphate	−9	−20		
Pentaphosphate	−26	−59	−45	−18
Hexaphosphate	−27	−57	−46	−17
Heptaphosphate	−33	−57	−46	−14

[a] For about 5% solutions, from Ref. 168. See footnote (a), Table XIII. Values are for first-order rate constants.

polyphosphate chains (200). However since the hydrogen ion cannot be chelated, the greater catalytic effectiveness of sodium ion for meta phosphate abstraction might be due to the providing of a favorable configuration in the polyphosphate chain through formation of a chelate. The low rate of metaphosphate abstraction in the absence of metal ion is evidence for this, and ^{31}P NMR provides direct evidence for polyphosphate chelation of Na^+ in solution (118). This topic will be discussed further in the following section.

Metaphosphate abstraction is catalyzed more than end-group clipping by added metal ions (458). In basic solution with sodium ion present, the major degradation route for oligomers longer than pentaphosphate is metaphosphate abstraction. Except for this, end-group clipping is the main mode of degradation for oligomers (168,380).

The catalytic effect of Cu^{2+} and its ethylenediamine and poly-L-lysine complexes on the degradation of tetra-through hexaphosphates has been studied at pH 4 and 8 (305,306). Cupric ion is a weak catalyst for the metaphosphate abstraction reaction. The 1:1 ethylenediamine complex also favored metaphosphate abstraction. Its effectiveness was the same as the water complex at pH 4 but considerably greater at pH 8.   At high concentrations the poly-L-lysine complex is remarkably effective at catalyzing the end-group clipping reaction.

E. Long-Chain Polyphosphates

Degradation of long-chain polyphosphates in aqueous solution occurs in three known ways (276,355,394): (1) end-group clipping, in which monophosphate units are cleaved from the ends of the chains, (2) metaphosphate abstraction, in which metaphosphate units, mostly tribut including some larger rings, are eliminated from the chain, and (3) random cleavage of the chain.   The first and last of these are hydrolytic reactions, but metaphosphate formation requires no water [Eq. (24)] and can be regarded as a reorganization reaction (168,396,416,432).

To some degree these processes are independent of one another (381,392), but the evidence is ambiguous and the completeness of this independence has not been established.   It is quite possible that the degree of independence varies with conditions.   A possible relation between metaphosphate abstraction and chain cleavage will be discussed later in this section.

Long-chain phosphate samples contain a distribution of chain lengths (84,331,434,435) and can most readily be characterized in terms of the average chain length ($\bar{n}$ = average number of phosphorus atoms in the chain).   Typical chain lengths for Graham's salts [glassy $(NaPO_3)_n$] are of the order of 50 to a few hundred phosphate tetrahedra.   Chain lengths can reach into the thousands, especially for KPO_3 samples.   The over-all degradation rate of chain polyphosphates shows a maximum for a chain length of about ten and reaches a limiting value about $\bar{n} = 100$ (459).   Variables investigated in the more important long-chain polyphosphate degradation kinetics studies are listed in Table XXIII.   There are other studies of possible interest (97,135,226,280,281,453).   Correlation of the results of the various studies is complicated by the variety of ways of following the degradation and of expressing the results.   Some first-order rate constants are listed in Table XXIV and over-all activation energies in Table XXV.

The degradation of polyphosphates is cation catalyzed, with the relative effectiveness of a number of cations at pH 8 and 60° shown in Table XXVI.   The catalytic effectiveness of the cations increases with increased charge and with decreased ionic radius, an order that relates to the strength of binding of these cations to the polyphosphate chain.   Only cations bound to the polyphosphate chain are effective.   Consistent with this, the increased rate observed on addition of an organic solvent has been attributed to increased cation catalysis resulting from

TABLE XXIII
Linear Long-chain Sodium Polyphosphate Degradation Studies

pH	Temp. (°C)	Effect of other cation	Ref.
1–12M HCl	0–25		137
1–8	40–80	K^+	418
1–13	50, 100		3
1, 6	25–100	Ca^{2+}	196
4–10	30–90		276
4.5, 13	70, 100		31
4.5, 6	60, 100	NH_4^+	371
5–9	40–80		155
5–9	65, 88		161
5, 8	60	Mg^{2+} (with EDTA)	462
5, 8	60	Mg^{2+}	419
5	65.5		381
8, 9	90	K^+, Mg^{2+}, Ca^{2+}	200
8	60		458
8	60	Variety	459, 461
8.5	27–90	K^+	335

TABLE XXIV

Polyphosphate Degradation Rate Constants[a] (hr^{-1}) at 60°

pH	Metaphosphate abstraction	End-group clipping	Over-all	Ref.
$(NaPO_3)_n$				
1			1.82	418
2			0.103	418
4		9×10^{-4}		276
5	2.6×10^{-3}		6.1×10^{-3}	155
		3.4×10^{-3}	6.48×10^{-3}	381, 418
7	4.9×10^{-4}		1.34×10^{-3}	155
		1.7×10^{-4}		276
8			1.6×10^{-3}	459
			6.4×10^{-4}	418
9	2.8×10^{-4}		4.4×10^{-4}	155
10		1.1×10^{-4}		276
$(NH_4PO_3)_n$				
4.5			2.9×10^{-3}	371
6.0			1.6×10^{-4}	371

[a] Rate constants for expressions first order in (PO_3^-) concentration.

TABLE XXV

Activation Energies for Polyphosphate Degradation[a]

pH	Cation	E_a (kcal/mole)	$\log A$ ($\log hr^{-1}$)	$\Delta S^{\ddagger}$ (eu)	Ref.
Over-all					
Conc. HCl	Na^+	15.6			137
1	Na^+	15.1	10.2	−30	418
3	Na^+	22.8	14.0	−13	418
8	Na^+	25.0	13.3	−16	418
8.5	K^+	25			335
15% NH_4PO_3	NH_4^+	13.8			371
40% ethanol	Na^+	16			43
Hydrolysis alone					
7	Na^+	20.5	11.2	−6	155
Trimetaphosphate abstraction alone					
7	Na^+	23.5	12.2	−7	155

[a] See footnote (a), Table XIII.

increased ion association due to decreased dielectric constant of the medium (462). About the same hydrolysis rate has been found for solvents consisting of 40% methanol, ethanol, acetone, n-propanol, t-butanol, or dioxane, in spite of the range of dielectric constants involved for these mixtures, from 60 to 43 (43). However the kinetic treatment, based on viscosity measurements, assumed that only random cleavage occurred.

Studies of the distribution of initial degradation products as a function of pH (155,276,418) suggest that the relative importance of the three modes of degradation, end-group clipping (a), metaphosphate abstraction (b), and random cleavage (c), based on mole ratios of the appropriate products is $c > a > b$ at very low pH values, $a > b > c$ at pH 3–7 and $a \sim b > c$ at pH 8–10. At pH 10, the principal products of $(NaPO_3)_n$ degradation are mono-, trimeta-, and tetrametaphosphate in mole ratio $1:1:0.25$ (276). At lower pH values the trimeta–tetrameta

TABLE XXVI

Metal-Ion Catalysis of Polyphosphate Degradation[a]

Cation	K^+	Na^+	Li^+	Ba^{2+}	Sr^{2+}	Ca^{2+}	Mg^{2+}	Al^{3+}
k/k_0[b]	~1	~1	1.08	1.56	1.69	2.78	3.52	7.50

[a] At pH 8 and 60° for $\bar{n} = 60$, using 0.164 eq of cation/mole of $(NaPO_3)$, from Refs. 415, 461.

[b] Ratio of rate constant in presence of cation to rate constant in absence.

ratio is larger; a 20:1 ratio in rate of formation of these has been reported for pH 5 (381).

Strauss and Krol have divided the degradation of long polyphosphate chains at pH 5 and 65.5° into end-group clipping, metaphosphate abstraction from chain ends, metaphosphate abstraction from chain middles, and random cleavage (381). Using samples with $\bar{n}$ values of 30, 75, and 170, separate rate constants were obtained for each of these processes. The rate constant for end-group clipping was independent of chain length for a rate expression that was first order in chain ends

TABLE XXVII

$(NaPO_3)_n$ Degradation Rate Constants[a]

$\bar{n}$	End-group clipping k_{EC}	Meta abstraction		Random cleavage k_{RC}
		Ends k_{ME}	Middles k_{MM}	
30			2.4×10^{-4}	
75	3.4×10^{-2}	5×10^{-3}	3.5×10^{-4}	6.5×10^{-4}
170			4.0×10^{-4}	7.3×10^{-4}

[a] First-order rate constants, hr^{-1}, at 65.5° and pH 5 from Ref. 381.

[Eq. (25)]. Rate constants for metaphosphate abstraction from chain ends and chain middles were separated by assuming the reaction at chain ends was insensitive to chain length, which is reasonable and supported by the finding for end-group clipping. The rate constant for random hydrolytic cleavage at other than the terminal P—O—P linkage was dependent on chain length. Rate constant values obtained in this study appear in Table XXVII. The values are not directly comparable to those of Table XXIV because of differences in the rate expressions. The more refined rate expressions used by Strauss and Krol are:

for end-group clipping

$$\frac{d[P_1]}{dt} = k_{EC}[E] \tag{25}$$

for metaphosphate abstraction

$$\frac{d[P_0]}{dt} = k_{MM}[M'] + k_{ME}[E'] - k_0([P_{03}] + [P_{04}]) \tag{26}$$

and for random cleavage

$$\frac{d[E]}{dt} = 2k_{RC}[M] + 2k_0([P_{03}] + [P_{04}]) \tag{27}$$

In these expressions, the square brackets represent moles of the indicated species/100 g-atoms of phosphorus. P_{03} and P_{04} refer to trimeta- and tetrametaphosphate. Since almost all the ring metaphosphate present is trimetaphosphate, the expression ($[P_{03}] + [P_{04}]$) is treated in the rate laws as trimetaphosphate for simplicity. P_1, P_2, P_3, etc., represent mono-, di-, triphosphate, etc. E refers to endgroups other than in monophosphate and is defined by

$$[E] = [E_{total}] - [P_1] \tag{28}$$

M represents middle groups available for random chain cleavage; for this the phosphorus atoms in monophosphate and the ring metaphosphates are eliminated, as are $\frac{3}{2}$ of an atom for each chain end.

$$[M] = 100 - [P_1] - 3[P_{03}] - 4[P_{04}] - \tfrac{3}{2}[E] \tag{29}$$

The reasoning here is that the phosphorus atom next to the end is available to this reaction only through the oxygen bridge—to one side of it. The end and middle phosphorus atoms are redefined slightly for the metaphosphate abstraction reactions:

$$[E'] = [E] - 2\{[P_2] + [P_3] + [P_4]\} \tag{30}$$

$$[M'] = [M] - \tfrac{1}{2}[E] \tag{31}$$

The rate constant for trimetaphosphate ring opening is k_0; its value was determined to be 6.4×10^{-4} hr^{-1} at 65.5° and pH 5. The other rate constants are identified in Table XXVII.

These results provide evidence for random cleavage apart from cleavage that might accompany metaphosphate abstraction; the rate constants for random cleavage are larger than those for metaphosphate abstraction from the middle of the chain. Under the conditions studied hydrolysis of terminal P—O—P linkages is of the order of 50 times as fast as hydrolysis of P—O—P linkages further into the chain (k_{EC} vs. k_{RC}). A factor of 10 to 1000 had been reported earlier (276). The authors suggest that the dependence on chain length of the rate of random cleavage and metaphosphate abstraction from the middle of the chain can be attributed to increased strength of binding of cations (including H$^+$) with increased chain length.

The large amounts of trimetaphosphate formed in long-chain polyphosphate degradation were at one time considered to be evidence that these chains contained trimetaphosphate rings (387). That reorganization must occur during degradation was firmly established by the finding of up to 50% yields of trimetaphosphate in the degradation of Maddrell's salt (409), for the anion of which X-ray diffraction studies confirmed a linear chain structure (134,219,261). The shortest chain for which trimetaphosphate abstraction has been detected is tetraphosphate (168). Hexaphosphate is the shortest for which tetrametaphosphate has been found as a degradation product (418,458). The yield of metaphosphate is dependent on chain length (Table XXVIII).

TABLE XXVIII

Trimetaphosphate Yield in Polyphosphate Degradation[a]

Chain length	2, 3, 4	5	6	7	8	∞
Trimetaphosphate yield (% of P)	0	~7	~16	~25	~38	≥60

[a] Ref. 458.

Although thought at one time to be zero order on the basis of data for only a small extent of reaction (276), metaphosphate abstraction is first order in phosphate formula units (155,380,415,418). The activation energy for metaphosphate abstraction from $(NaPO_3)_n$ at pH 7 is 23.5 ± 1.5 kcal/mole (155). Polyphosphate degradation is not catalyzed by anions but is by metal ions, which increase the rate of metaphosphate abstraction more than the rate of end-group clipping (392,415,419,458,461).

It has been proposed that metaphosphate abstraction results from a spiral coiling of the polyphosphate chains in solution (393,395,416). Metal ion and hydrogen ion catalysis is considered to function, at least in part, through the favoring of a coiled configuration. Neutralization of the charges along the chain by complexing of cations to the chain reduces repulsions between units of the chain and permits the chain to coil. Physical evidence for the coiling of polyphosphate chains in the presence of added salts is available from flow birefringence (442), light scattering (383), and viscosity measurements (382). Viscosity studies also show that tetramethylammonium ion, which does not complex with the chain and which is not effective at catalyzing polyphosphate degradation, is not as effective as metal ions at inducing chain coiling (200). The polyphosphate chains have very little steric hindrance to internal rotation (383).

The bonding of hydrogen ion or metal ions to the chain renders the phosphorus of any tetrahedron involved more positive and more susceptible to nucleophilic attack. Thilo has suggested that the mechanism of metaphosphate abstraction is nucleophilic attack on a relatively electropositive phosphorus atom by a nonbridging oxygen atom of a nearby phosphate tetrahedron (395,416), with the relative stability of a six-membered ring favoring trimetaphosphate formation as opposed to larger rings (**10**).

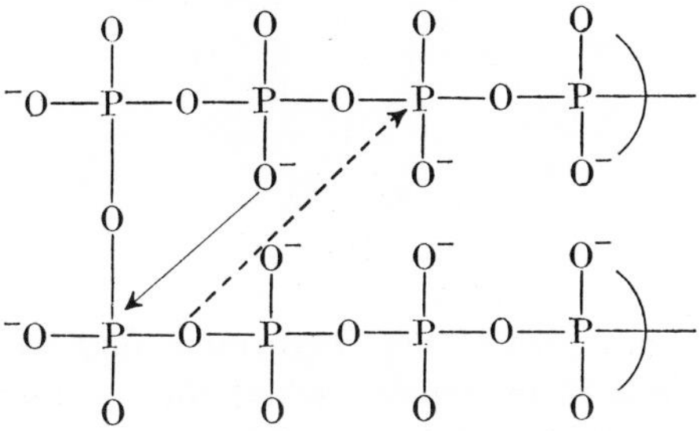

(10)

Chelation of a metal ion to two oxygen atoms of the same tetrahedron (**11**) has been suggested as making that phosphorus particularly susceptible to nucleophilic attack (419).

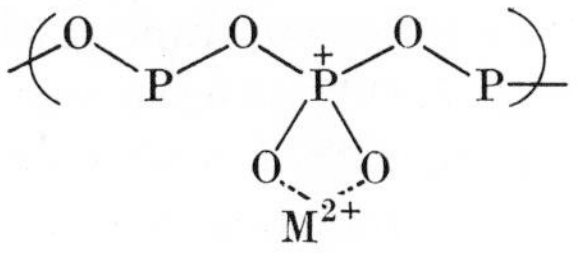

(11)

Because of the statistical difficulties in metaphosphate formation by random coiling (434), one would expect metaphosphate abstraction to be very slow if it proceeded on this basis. The catalytic effect of the metal ion should be equally available to nucleophilic attack by water, demanding that random cleavage be important, which it is not. The four-membered ring of **11** would be highly strained. The nonbridging oxygen atoms of a polyphosphate chain are about 2.5 Å apart, whereas the oxygen atoms in a polyphosphate chelate to Mg^{2+}, for example, should be about 3.0 Å apart (215). This distance would be greater for larger cations. Carrying out the mechanism of **10** with a model suggests that only a miracle would accomplish the repair of the chain indicated by the dotted arrow. The basic suggestion is probably correct, however, that metaphosphate abstraction can occur by an intramolecular mechanism of the type shown by the solid arrow of **10**. Hydrogen bonding of water to the nonbridging oxygen atoms of the polyphosphate chain (**12**) has been suggested as a factor systematizing the coiling of the chain.

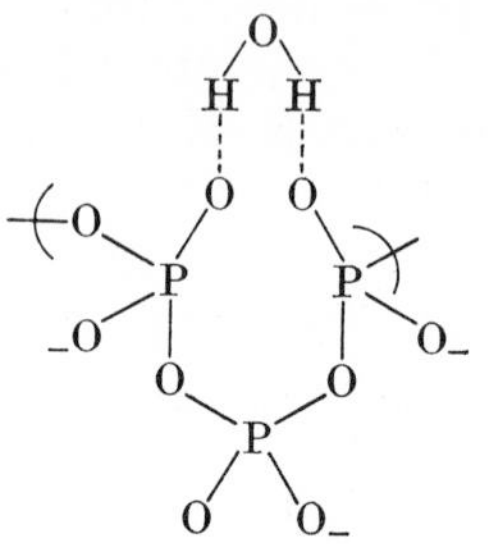

(12)

A spiral, coiled configuration for high-molecular-weight linear polyphosphate anions has been demonstrated for a number of solids, including $[Cd(PO_3)_2]_n$ (423), $(AgPO_3)_n$ (217), and the A and B forms of Kurrol salt, $(NaPO_3)_n$ (218,220,275). Although in each of these there are four phosphate tetrahedra in the repeating pattern of the spiral, because of the extension of the spiral reorganization into trimetaphosphate units is more likely than into tetrametaphosphate units. Other high-molecular-weight chain polyphosphates in the solid state have a zig-zag anion pattern, with a repeat unit of three tetrahedra, as in $(KPO_3)_n$ (221) and $(RbPO_3)_n$ (108); or have a severely flattened spiral, approaching a zig-zag pattern, with a repeat unit of three tetrahedra, as in $[Na_2H(PO_3)_3]_n$ (219), or of four tetrahedra, as in $[Pb(PO_3)_2]_n$ (222) or eight, as in $[Na_3H(PO_3)_4]_n$ (223). An examination of the details of these structures suggests that if reorganization of these zig-zag anions were to occur by nucleophilic attack of an oxygen of one tetrahedron on the phosphorus of another, that trimetaphosphate should be expected from the anion configuration in $(KPO_3)_n$, $(RbPO_3)_n$, and $[Na_2H(PO_3)_3]_n$, while tetrametaphosphate should be expected from the anions of $[Pb(PO_3)_2]_n$ and $[Na_3H(PO_3)_4]_n$. The reasoning here is on a kinetic basis, as the ring phosphate is not the ultimately stable product in solution. Of course, the structures in these solids would not be preserved in solution, but they suggest configurations that might develop over short sections of the chain on chelation in solution.

As noted earlier, both hydrogen ion and metal ions catalyze metaphosphate abstraction. Both permit coiling of the polyphosphate chain, which must be a factor in aiding metaphosphate abstraction. Metal ions are much more effective catalysts for this reaction than is hydrogen, however, and chelation, which occurs with metal ions but not with hydrogen ion, must be an additional catalytic factor.

Octamethylpyrophosphoramide (OMPA) is a bidentate coordinating agent forming complexes containing a chelate ring that should resemble those formed by polyphosphates. Crystal structures have been determined for tris(OMPA) complexes of Mg^{2+}, Co^{2+}, and Cu^{2+} perchlorates (215) and for bis(perchlorato)bis(OMPA) copper(II) (198). In all the tris complexes, the chelate ring was planar. In the bis complex, the ring was slightly puckered. Because of the known effectiveness of Mg^{2+} as a catalyst for metaphosphate abstraction, structural data for the tris (OMPA) Mg^{2+} complex were used to construct a model of Mg^{2+} coordinated to a polyphosphate chain. The dimensions of the chelate ring are given in **13**. For the chain tetrahedral phosphorus atoms, a

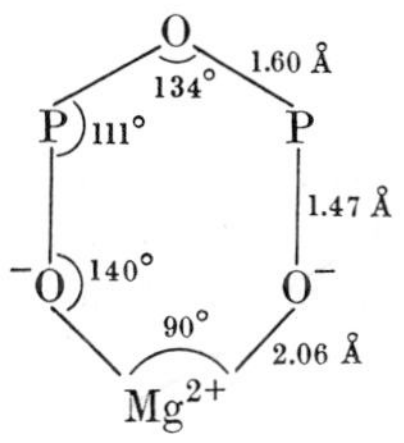

(13)

P—O—P angle of 133° and P—O$_{terminal}$ and P—O$_{bridge}$ distances of 1.50 and 1.60 Å, respectively, were used.

In basic solution, with the chain fully ionized with respect to protons, chelation to two phosphate tetrahedra at the end of the chain is favored

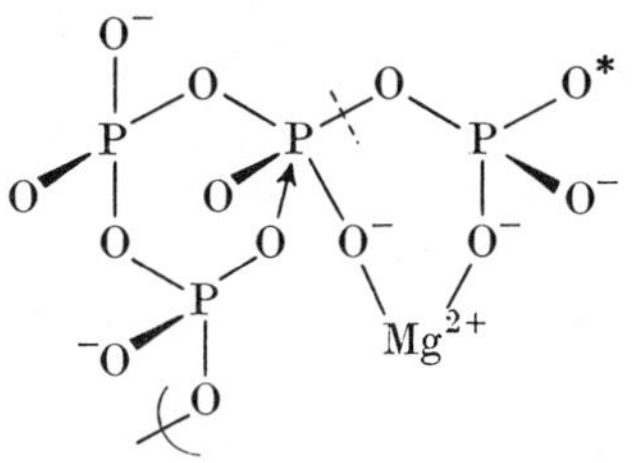

(14)

by the additional charge on the terminal tetrahedron **(14)**. Manipulation of the model suggests that intramolecular attack on the phosphorus of the second tetrahedron by an oxygen of the fourth tetrahedron is the only logical route. The required configuration for this nucleophilic attack is easily achieved. The terminal bridging oxygen would be displaced from the second tetrahedron to liberate the terminal tetrahedron as monophosphate. The experimental finding in a number of studies that monophosphate and trimetaphosphate form in a 1:1 mole ratio over the pH range 8–10 is consistent with this mechanism, as is the observation that tetraphosphate is the shortest chain that produces trimetaphosphate. However in a study by Wieker and Thilo at pH 8 and 60°, the monophosphate–trimetaphosphate mole ratio was 1:1.1 in the presence of Na^+, 1:1.5 with the addition of about 0.5 Mg^{2+} per chain, and about 1:3 with the addition of 5 or more Mg^{2+} per chain (462). Either another mechanism must add to the metaphosphate produced, or the terminal group must rejoin the residue of the chain. The latter is very unlikely for this model, even when tridentate or higher chelation is considered. In a study of oligophosphate degradation (380), Strauss and Day found that the rates of trimetaphosphate abstraction from hexaphosphate and pentaphosphate matched the rates of triphosphate and diphosphate formation, respectively. This finding requires that the metaphosphate ring form from three terminal phosphate tetrahedra or that the chain be repaired after trimetaphosphate abstraction from the middle of the chain. The former seems more likely.

In acid solution, protonation of the end-group makes chelation about equally probable anywhere on the chain. Even in basic solution chelation anywhere would occur with more than two metal ions available per chain. The starred oxygen atom of **14** would then be a bridge oxygen. The mechanism of the previous paragraph would still be possible except that random cleavage instead of end-group clipping would accompany metaphosphate abstraction. It is possible, however, that the cumulative electron-withdrawing effect of chelation and the two adjacent phosphate tetrahedra could lead to breaking of the P—O—P bridge of the chelate, followed by saturation of the affected phosphorus through trimetaphosphate ring formation as in **14**. This intramolecular S_N1 mechanism would be comparable to the "monometaphosphate mechanism" for phosphate ester hydrolysis. Random cleavage should still accompany the metaphosphate abstraction reaction; the model reveals no systematic way for repair of the chain. The rate constants obtained at pH 5 by Strauss and Krol (Table XXVII) show end-group clipping to be faster than metaphosphate abstraction at the chain ends and show random cleavage to be faster than metaphosphate abstraction from chain middles. Because of the availability of water molecules as nucleophiles, ring formation could not be assured in the S_N1 mechanism. The additional rate of end-group clipping and random cleavage found by Strauss and Krol, beyond that matching the metaphosphate abstraction rates, could be attributed to intervention of water in this mechanism. The $\Delta S^{\ddagger}$ value for metaphosphate abstraction suggests a unimolecular mechanism (Table XXV).

Immobilization of a segment of the polyphosphate chain through chelate formation can facilitate ring formation compared to completely free chain rotation. Tetrametaphosphate formation is possible using an oxygen of the fifth tetrahedron. In addition, immobilization of two tetrahedra in the chelate favors attack by an oxygen of the fourth tetrahedron on the phosphorus of the first tetrahedron (15). If the

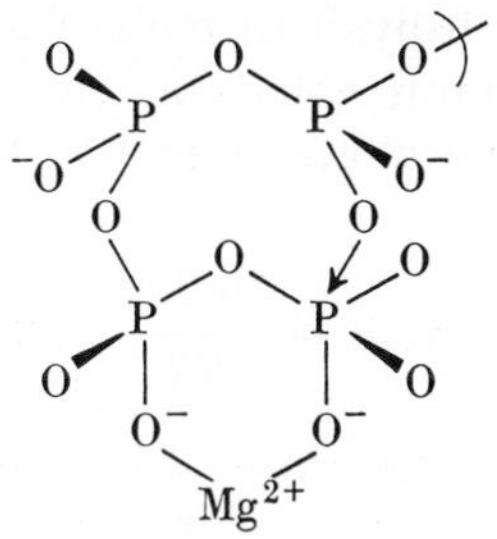

(15)

OMPA chelate structure determinations are disregarded, and the chelate ring allowed to depart from planarity, a number of possible configurations arise involving boat and chair forms of the chelate ring, carrying of the chain from the axial or equatorial oxygen of the second phosphorus tetrahedron, and optical isomerism at the second tetrahedron. For only some of these configurations is metaphosphate abstraction reasonable, but no new avenues appear and no systematic way of repairing the chain after metaphosphate abstraction exists.

Additional study is needed. Additional information on the extent to which the several degradation routes are independent, and the dependence of the several routes on pH, metal ion concentration, metal ion size and charge, and chain lengths, is needed for testing mechanisms. In the metal-ion catalysis of trimetaphosphate abstraction, the chelate might not be the reactive form.

The hydrolysis of polyphosphate in highly acid medium, 1 to 12.6M HCl, is first order in polyphosphate and second order in acid concentration (137). The activated complex proposed to account for the dependence on acidity involved protonation of bridge oxygen atoms in the polyphosphate chain.

Hydrolysis of polyphosphate chains in molten $(NaPO_3)_n$ has been studied (464). For the equilibrium described by Eq. (32), the average

$$n(NaPO_3) \ (1) + H_2O \rightleftharpoons Na_n(H_2P_nO_{3n+1}) \ (1) \tag{32}$$

chain length is given by (33), which can be derived from the mass action law. From the temperature dependence of the value of the equilibrium

$$\bar{n} = \left(\frac{K}{p_{H_2O}}\right)^{\frac{1}{2}} \tag{33}$$

constant of Eq. (33), a value of about 10 kcal/mole was found for the enthalpy of P—O—P hydrolysis in molten long-chain polyphosphate. Using the Clausius-Clapyron equation, the average chain length in molten long-chain polyphosphate from 650 to 850° can be expressed as in Eq. (34).

$$2.303 \log \bar{n} = 8.87 - \frac{10{,}200}{RT} - 2.303 \frac{\log p_{H_2O}}{2} \tag{34}$$

The average chain length decreases with increased temperature and with increased partial pressure of water vapor in the atmosphere surrounding the sample.

F. Trimetaphosphate

The hydrolysis of trimetaphosphate follows the sequence:

$$P_3O_9{}^{3-} + H_2O \rightarrow H_2P_3O_{10}{}^{3-} \tag{35}$$

$$H_2P_3O_{10}{}^{3-} + H_2O \rightarrow H_2P_2O_7{}^{2-} + H_2PO_4{}^- \tag{36}$$

$$H_2P_2O_7{}^{2-} + H_2O \rightarrow 2 \ H_2PO_4{}^- \tag{37}$$

In neutral solution at room temperature trimetaphosphate hydrolysis is very slow. The rates of the above reactions differ in their dependence on temperature and pH. At elevated temperatures in acidic and neutral solution reactions (36) and (37) are enough faster than (35) that appreciable concentrations of tri- and diphosphate do not develop during the hydrolysis (31,418). In strongly basic solution, however, reaction (35) is much the faster, and the alkaline hydrolysis of trimetaphosphate can be made to give almost quantitative yields of triphosphate (31,140,181,404).

The hydrolysis is first order in trimetaphosphate in both acidic and basic solution (51,183). The value of the hydrolysis rate constant depends strongly on hydrogen ion activity (Figs. 10 and 11). Trimetaphosphoric acid is often described as a strong acid because its titration curve resembles that for a strong acid. A single sharp break

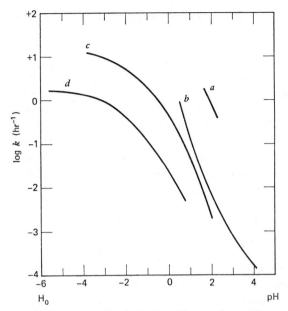

Fig. 10. Trimetaphosphate hydrolysis. First-order rate constants at (a) 69.40° (183); (b) 39.50° (328); (c) 25.03° (328); and (d) 0.00° (328).

is observed when three equivalents of base have been added per mole of the acid (443). Nonetheless trimetaphosphoric acid is not truly strong in the sense that its degree of ionization can be assumed to be 1. The ionization constant for its final ionization [Eq. (38)] is 0.0083 (119,302), and the single sharp break in the titration curve for $H_3P_3O_9$ means that the first and second ionization constant values cannot be

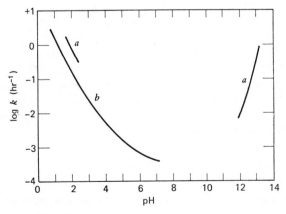

Fig. 11. Trimetaphosphate hydrolysis. First-order rate constants at (a) 69.40° (183) and (b) 60° (418).

much larger than this. This observation is consistent with the

$$HP_3O_9{}^{2-} = H^+ + P_3O_9{}^{3-} \tag{38}$$

generalization that there is one acidic proton with $pK_a \sim 2$ for each phosphorus atom of any condensed phosphoric acid. Since the acid is not truly strong, changes in pH should affect the hydrolysis kinetics through changes in the distribution of trimetaphosphate among the various protonated forms: $H_3P_3O_9$, $H_2P_3O_9{}^-$, $HP_3O_9{}^{2-}$, and $P_3O_9{}^{3-}$. In addition, ^{31}P NMR of various phosphates in 100% sulfuric acid, chlorosulfuric acid, and oleums up to 65% sulfur trioxide gives evidence for the presence of more highly protonated species, including $H_4P_3O_9{}^+$ and $H_5P_3O_9{}^{2+}$ (130).

In a study of the hydrolysis kinetics over the relatively narrow pH range from 1.7 to 2.3 (Fig. 10a), Healy and Kilpatrick attributed the observed rate primarily to reactions of $HP_3O_9{}^{2-}$ and $P_3O_9{}^{3-}$, but found evidence of $H_2P_3O_9{}^-$ involvement, as well (183). Experiments carried into more acidic solution (Fig. 10b–d) require the involvement of more highly protonated species in the kinetic treatment (328). As in the case of the chain phosphates, the influence on kinetics of hydrogen ion concentration can be formulated in terms of an involvement of hydrogen ion in a second-order reaction or in terms of a hydrogen ion induced shifting of the equilibrium concentrations of the various protonated forms of trimetaphosphate, as shown for one protonation level in Eqs. (39) and (40). In these equations $k' = kK_3$, where K_3

$$\text{Rate} = k[H^+][P_3O_9{}^{3-}] \tag{39}$$

$$\text{Rate} = k'[HP_3O_9{}^{2-}] \tag{40}$$

is the ionization constant for $HP_3O_9{}^{2-}$. Both descriptions of the kinetics have been used. Because the equilibrium shifts are known to occur and can be separately evaluated, and because the treatment is simpler on that basis, the first-order formulation appears preferable. Later in this section $\Delta S^{\ddagger}$ values are cited as evidence for a unimolecular mechanism in acid solution. Equations (39) and (40) consider only one of the several levels of protonation that would have to be considered in describing the observed rate of hydrolysis in acidic solution. The rate law should be of the form:

$$-\frac{dc}{dt} = k_0[P_3O_9{}^{3-}] + k_1[HP_3O_9{}^{2-}] + k_2[H_2P_3O_9{}^-] + \cdots. \tag{41}$$

where c represents the total concentration of all trimetaphosphate species.

The hydrolysis in basic solution is first order in hydroxide ion, as well as in trimetaphosphate (183,293).

$$\frac{d[P_3O_9{}^{3-}]}{dt} = k[OH^-][P_3O_9{}^{3-}] \qquad (42)$$

The right hand side of Eq. (42) should be added to (41) to obtain the complete rate law. The addition of this dependence of the kinetics on OH^- concentration leads to a rate minimum around pH 9–10, as can be seen from Figure 11.

An increase in ionic strength established by an increase in tetra-n-propylammonium perchlorate concentration decreases the rate of trimetaphosphate hydrolysis in acid solution (183). Presumably there is no complex formation between the tetralkylammonium ion and trimetaphosphate. This kinetic influence of ionic strength is of the direction expected from the Bronsted-Bjerrum rule for a reaction between oppositely charged ions, consistent with Eq. (39). On the other hand, increased ionic strength will increase the acid ionization constant values for the protonated trimetaphosphate species, and the shifting concentrations of these species will also explain a decrease in rate [Eq. (41)]. There are no unambiguous data in the literature for assessing the effect of ionic strength changes on the kinetics of the hydrolysis reaction in basic solution in the absence of metal ions. An increase in ionic strength decreases the rate of hydrolysis in the presence of metal ions, but this is best interpreted as being largely due to an increase in dissociation constant for the complex of the catalytic metal ion with trimetaphosphate (183).

The influence of temperature on trimetaphosphate hydrolysis is summarized in the activation parameters listed in Table XXIX. A change in mechanism with pH is clearly indicated by the $\Delta S^{\ddagger}$ values (273,274,359). The small values for the low pH range suggest an S_N1 mechanism, whereas the large negative values for the high pH range suggest a bimolecular mechanism.

As is true for the chain phosphates, metal ions have a greater catalytic effect on trimetaphosphate hydrolysis in alkaline solution than in acidic. The metal ion effect is limited in acid solution by competition from hydrogen ion for coordination sites on the condensed phosphates. In acidic solution added Na^+, Ca^{2+}, and Ba^{2+} cause small decreases in the rate of trimetaphosphate hydrolysis (183). This effect must be the result of (1) a decrease through an ionic strength effect due to both ions of the salt, (2) a decrease through displacement of catalytic hydrogen ions from sites on the trimetaphosphate, and (3) an increase through catalysis by metal ions newly bound to the trimetaphosphate. Uranyl

TABLE XXIX

Activation Parameters for Sodium Trimetaphosphate Hydrolysis[a]

pH	Added cation	E_a (kcal/mole)	$\log A$ ($\log \text{hr}^{-1}$)	$\Delta S^{\ddagger}$ (eu)	Ref.
1.7[b]		24.0	15.5	$+2$	183, 184
2.0[b]		22.8	14.8	-0.9	183, 184
2.2[b]		23.4	15.1	$+0.5$	183, 184
3		21.8	10.9	-27	418
7		19.6	7.7	-42	418
~12[b]		18.2	~11	-9	201
12.7[b]		15.5	8.4	-30	183, 184
13		20.0	12.9	-18	51
13.1[b]		17.5	10.0	-23	183, 184
13.2[b]		16.1	9.1	-27	183, 184
~2	UO_2^{2+}	17			206
~12[b]	Ba^{2+}	15.5	15.52	-5.8	202
~12[b]	Ca^{2+}	11.5	14.02	-13	202
~12	Na^+	16.1–16.8			204

[a] See footnote (a), Table XIII. Values are for first-order rate constants except where indicated.

[b] For second-order rate constants. A in liters/(mole) (hr).

ion, however, has been shown to be catalytically effective at pH 2 (206).

In basic solution (pH 11–13), the rate of trimetaphosphate hydrolysis is lowest if the only cation present is a tetraalkylammonium ion. The catalytic effectiveness of metal ions increases in the order $Na^+ < Ba^{2+} < Ca^{2+}$, with even Na^+ a very effective catalyst (183,202,204). The rate constant is directly proportional to the concentration of each cf these ions (201,204). The rate expression for the alkaline hydrolysis of trimetaphosphate in the presence of metal ions should contain a term for each complex-forming cation present [Eq. (43)]. It has been estimated that the rate constants at 69° for hydrolysis of the Ca^{2+} and

$$-\frac{dc}{dt} = k_1[OH^-][P_3O_9^{3-}] + k_2[OH^-][M^{II}P_3O_9^{1-}] + \cdots \quad (43)$$

Ba^{2+} complexes are of the order of one thousand times the rate constant for the Na^+ complex (183). Using dissociation constant values of 6.8×10^{-2}, 3.4×10^{-4}, and 4.5×10^{-4}, respectively, for $NaP_3O_9^{2-}$, $CaP_3O_9^-$, and $BaP_3O_9^-$ (119,216); and concentrations representative of those employed in the experiments on which the above estimate of catalytic effectiveness is based (1.0×10^{-4} M metal ion, 6.0×10^{-6} M $P_3O_9^{3-}$), one calculates that about one hundred times as much of the trimetaphosphate is in the form of metal-ion complex for the Ca^{2+} and Ba^{2+} solutions as for the Na^+ solution (23, 18, and 0.15% of the trimetaphosphate complexed, respectively). If these dissociation constants are reliable, much of the greater effectiveness of the divalent cations results from the smaller dissociation constants for their complexes. It must be noted that the dissociation constant reported by Indelli (205) for $NaP_3O_9^{2-}$, 7.6×10^{-4}, is much smaller than that of Davies and Monk used above. At a given pH both the energy and entropy of activation appear to be made more favorable by the introduction of catalytically effective metal ions (Table XXIX).

Catalysis of trimetaphosphate hydrolysis has also been observed with gelatinous precipitated hydroxides of a variety of metals (19). Those of La^{3+}, Ce^{4+}, Pr^{3+}, Nd^{3+}, Sm^{3+}, Yt^{3+}, Zr^{4+}, and Th^{4+} were particularly effective; those of Ti^{4+}, Pb^{2+}, and Mn^{2+} moderately effective; and those of Al^{3+}, Cu^{2+}, Mg^{2+}, Ba^{2+}, Zn^{2+}, Fe^{3+}, Co^{2+}, and Ni^{2+} only slightly effective. Healy and Kilpatrick felt that chloride ion was as effective as hydroxide ion at accelerating trimetaphosphate hydrolysis in alkaline solution (183). It is difficult to separate the effects of Cl^- and Na^+ with their data, however, and Na^+ is known to be effective. More highly charged anions, SO_4^{2-} and $Fe(CN)_6^{3-}$, have an inhibiting effect on the alkaline hydrolysis, attributed to an ionic strength effect (183). A contributing factor would be a reduction in metal ion available to the trimetaphosphate caused by ion-pair association with the sulfate and ferricyanide.

The total enthalpy of hydrolysis of trimetaphosphate to orthophosphate, determined for the enzymatic hydrolysis at pH 7 and 33°, is -18.6 kcal/mole (288). The average value per P—O—P linkage in the trimetaphosphate anion is -6.2 kcal/mole. This is of the order of the values listed in Table I for disphosphate hydrolysis. Of course, this average value is not necessarily the value for reaction (35) or for any other formulation of the breaking of the trimetaphosphate ring to form triphosphate.

Other studies of the aqueous hydrolysis have been relatively limited (31,161,257,293). The investigations by Kiehl and co-workers (29,239) of the reversion kinetics for "sodium monometaphosphate" are of limited value. Although one might believe this to have been the trimetaphosphate, it evidently was a mixture of trimetaphosphate with long-chain polyphosphate (119). These studies were further handicapped because the investigators were unaware of triphosphate as an intermediate in the hydrolysis.

The cleavage of the trimetaphosphate ring in aqueous ammonia at pH ≥ 9 involves ammonolysis rather than hydrolysis (144,343). The product is monoamidotriphosphate [Eq. (44)]. The reaction is fast

$$P_3O_9^{-3} + NH_3 = \overset{O\ \ O\ \ O}{\underset{O\_O\_O\_}{HOPOPOPNH_2}} \tag{44}$$

at pH 12, and the product can be isolated readily by precipitation as $Ba_2(P_3O_9NH_2)$. On acidification of a monoamidotriphosphate solution the trimetaphosphate ring is reformed. Similar reactions have been observed with methylamine and ethylamine to give N-alkyl-amidotriphosphate (143). The reforming of trimetaphosphate in acid solution is a more pronounced property of these compounds than of the simple amidotriphosphate. Fluoride ion cleaves the trimetaphosphate ring in solution to form monofluorotriphosphate (141). Some tendency to reform trimetaphosphate is exhibited, but in alkaline solution. The trimetaphosphate ring is cleaved in alkaline solution by phenol; m- and p-$C_6H_4(OH)_2$; and o-, m-, and p-cresol to give the monophenyl-triphosphate anion (142). This anion persists in aqueous solution at room temperature, but it undergoes hydrolysis to monophosphate and phenyldiphosphate in hot solution. Reaction of trimetaphosphate with o-$C_6H_4(OH)_2$ in alkaline solution rapidly produced diphosphate and phenylmonophosphate.

The alkaline hydrolysis of sodium trimetaphosphate has been studied in concentrated solutions (up to 30 wt % solids), such as are encountered in the slurries employed in the manufacture of detergents (368). The reaction is second order, showing dependence on $P_3O_9^{3-}$ and OH^- concentrations. Hydrolysis of the trimetaphosphate to triphosphate is fast. Up to 20% NaOH, further degradation of the triphosphate is slow, but beyond this concentration the rate of $P_3O_{10}^{5-}$ hydrolysis increases, until it is quite fast at 50% NaOH. The hydrolysis of $(NaPO_3)_3$ in the slurry appears to be a practical source of $Na_5P_3O_{10}$ for the resulting detergent composition. A ΔH of -21.5 kcal/mole was measured for reaction (45).

$$Na_3P_3O_9 + 2\,NaOH + 5\,H_2O = Na_5P_3O_{10}{\cdot}6\,H_2O \tag{45}$$

This heat effect and the take-up of water in the hydrated product help with the drying of the slurry. Activation energies from 17.5 to about 16 kcal/mole were obtained for slurries from 5 to 20% solids, respectively. These values are comparable to those for more dilute alkaline solutions (Table XXIX).

Hydrolysis has been noted during the dehydration of $Li_3P_3O_9{\cdot}3\,H_2O$ (171) and of both forms of $Na_3P_3O_9{\cdot}H_2O$ (400,401,414). In each case, acid polyphosphates form. For the lithium salt the anion chain length gradually increases with increased temperature. Heating the sodium salt to 350–625° produces the anhydrous trimetaphosphate from the acid polyphosphates first formed.

G. Tetrametaphosphate

The first step in tetrametaphosphate hydrolysis is cleavage of the ring to form the tetraphosphate chain. Further degradation of the

$$P_4O_{12}^{4-} + H_2O = H_2P_4O_{13}^{4-} \qquad (46)$$

tetraphosphate is primarily by end-group clipping (Section III-C). As is the case for trimetaphosphate hydrolysis, in acid solution the later steps are faster than the first, and the solution does not accumulate appreciable concentrations of the species intermediate between the metaphosphate and monophosphate (115,418). In basic solution the later steps involving chain phosphate hydrolysis are much slower than the metaphosphate ring cleavage, with the result that appreciable concentrations of intermediates, especially of tetraphosphate, are obtained (33,342,405,456,457). In fact, the alkaline hydrolysis of tetrametaphosphate affords a convenient preparation of the tetraphosphate ion.

The hydrolysis in acid solution is first order in tetrametaphosphate (51,115,418). Although the rate studies have not been interpreted in this way, by analogy with trimetaphosphate hydrolysis the dependence of the first-order rate constant on pH in the acid region should be attributed to the effect of hydrogen ion activity on the relative abundances of the protonated tetrametaphosphate species: $P_4O_{12}^{4-}$, $HP_4O_{12}^{3-}$, $H_2P_4O_{12}^{2-}$, . . . , with separate, larger rate constants applicable to the successively more highly protonated species.

As is the case with trimetaphosphate, the alkaline hydrolysis of tetrametaphosphate is first order in tetrametaphosphate and in hydroxide ion (203,293,405). The rate expression for decrease in total tetrametaphosphate concentration, c, should be of the form of (47), paralleling the rate law for trimetaphosphate hydrolysis.

$$-\frac{dc}{dt} = k_{OH}[OH^-][P_4O_{12}^{4-}] + k_0[P_4O_{12}^{4-}]$$
$$+ k_1[HP_4O_{12}^{3-}] + k_2[H_2P_4O_{12}^{2-}] + \cdots \qquad (47)$$

The rate increase in alkaline solution through the relatively rapid bimolecular reaction of hydroxide ion with $P_4O_{12}^{4-}$, along with hydrogen-ion catalysis in acid solution through increasing abundances of the more highly protonated species, results in a minimum in the rate of hydrolysis of tetrametaphosphate. From Figure 12 we can estimate that at about 60° this minimum occurs at pH 7–9, but this figure is based on a very small number of experimental points.

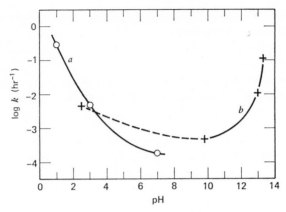

Fig. 12. Tetrametaphosphate hydrolysis. First-order rate constants at (a) 60° (418) and (b) 65.5° (115).

Figure 12 closely resembles in form the corresponding Figure 11 for trimetaphosphate. The rate constants for trimetaphosphate hydrolysis at a given temperature and pH are from 2 to 10 times larger, however, than those for tetrametaphosphate. For example, at pH 3 and 60°, Thilo and Wieker found the rate constants for hydrolysis of trimetaphosphate and tetrametaphosphate to be, respectively, 1.73×10^{-2} and 4.35×10^{-3} hr^{-1}, a ratio of about 4 (418). In the following section, it will be seen that this trend toward slower hydrolysis rate continues through the larger rings for one pH, at least.

It is of interest to compare rates for comparable forms of the metaphosphates. Davies and Monk have reported ionization constant values of 0.0083 and 0.0018 for $HP_3O_9^{2-}$ and $HP_4O_{12}^{3-}$, respectively (119,302). From these values, one can calculate for pH 3 that 11% of the trimetaphosphate is present as $HP_3O_9^{2-}$ and about 36% of the tetrametaphosphate as $HP_4O_{12}^{3-}$, with the remainder of each completely ionized. The ratio of these percentages is 3.3, in the opposite direction to the ratio of rates. This suggests that the ratio of hydrolysis rates for comparable species, say for the monoprotonated species, is greater than the ratio observed by Thilo and Wieker. The sodium ion present in the solutions studied is also a catalyst (203). Using dissociation constants for the sodium complexes from the above references to Davies and Monk, one can estimate that there would be about twice as much of the tetrametaphosphate as $NaP_4O_{12}^{3-}$ as there would be trimetaphosphate as $NaP_3O_9^{2-}$ in the 1% solution used for the rate comparisons. Again, the greater amount of catalysis is at work in the slower observed hydrolysis. The ring is more susceptible to hydrolysis in trimetaphosphate than in tetrametaphosphate, not only for similar solution conditions, but even more markedly for analogous species.

The reported activation energies for tetrametaphosphate hydrolysis appear in Table XXX, along with $\Delta S^{\ddagger}$ values calculated by us from the same data. The poor agreement among the values forbids discussion. The values at pH 1 and 3 are based on rate constants for only two temperatures. The other values are derived from relatively poor Arrhenius plots for which we estimate activation energy uncertainties of about ± 4 kcal/mole.

TABLE XXX

Activation Parameters for Sodium Tetrametaphosphate Hydrolysis[a]

pH	E_a (kcal/mole)	log A (hr^{-1})	$\Delta S^{\ddagger}$ (eu)	Ref.
0.8	25.3	15.7	-5	51,52
1	20.0	12.5	-19	418
3	20.0	10.8	-28	418
~12	21			203
13	26	15.3	-7	51,52

[a] See footnote (a), Table XIII. Values are for first-order rate constants.

As with trimetaphosphate, aqueous ammonia cleaves the tetrametaphosphate ring. The product in the latter case is monoamidotetraphosphate. On acidification of a solution containing this ion some tetrametaphosphate is reformed, but other products include $P_3O_9^{3-}$, NH_4^+, $P_4O_{13}^{6-}$, $P_3O_{10}^{5-}$, and $PO_3NH_2^-$ (393).

The thermal decomposition of lithium tetrametaphosphate–tetrahydrate has been reported to involve the loss of two molecules of water at 60°, with hydrolysis of the anion to form tetra- and octaphosphate occurring at higher temperatures (171). Dehydration of $Pb_2(P_4O_{12})\cdot$ $2 H_2O$ at 110° gives a quantitative yield of $Pb(H_2PO_4)_2$ (365). When heated in humid atmospheres $Na_4P_4O_{12}$ and $K_4P_4O_{12}$ are converted to the long-chain Maddrell's salts (171).

H. Higher Metaphosphates

The literature contains many references to "hexametaphosphate" in describing sodium phosphate glasses of composition approaching $NaPO_3$. These are known to be primarily mixtures of polyphosphates of various chain lengths (435), and hydrolysis of material of this type is covered in Sections III-E. and III-I. The "hexametaphosphate" misnomer can be attributed to Fleitmann, who was able to prepare double salts of the phosphate glass anions with cation ratios as high as $1:5$ (145). Although Fleitmann stated more work was necessary to establish the complexity of the anion, the name came into common usage. Now that true hexametaphosphates have been characterized it is particularly important to avoid the incorrect application of this nomenclature.

Interestingly, metaphosphate rings of various sizes, including the hexametaphosphate, are present in the sodium phosphate glasses

TABLE XXXI
Sodium Hexametaphosphate Hydrolysis, First-Order
Rate Constants (10^5 hr^{-1})[a]

Temp.	pH		
(°C)	4	7	11
30	2.1 ± 1.0	2.2 ± 0.7	1.9 ± 0.4
60	7.6 ± 0.7	5.4 ± 1.8	5.8 ± 1.4
E_a	9	6	6

[a] From Ref. 167.

referred to above (444). Although they are only minor constituents in these glasses, pentameta- and hexametaphosphate have been isolated by precipitation fractionation from this source and from material prepared by brief heating of sodium trimetaphosphates (407,408). Lithium phosphate systems with a $Li_2O-P_2O_5$ ratio near $7:5$ prepared below 300° contain 5–20% hexametaphosphate, providing another source (167). The octametaphosphate ion has been prepared by controlled heating of lead tetrametaphosphate (365).

For all the metaphosphates studied, the first step in the hydrolysis is the cleavage of a P—O—P linkage in the ring to produce the corresponding polyphosphate chain. The hydrolyses are first order in metaphosphate (167,408). The kinetics of hydrolysis of hexametaphosphate have been studied by Griffith and Buxton (167) as a function of pH and temperature (Table XXXI). In spite of the low reliability of the data, which was limited by the analytical procedure, two surprising features are evident. In contrast to all other phosphate hydrolyses studied, the kinetics are independent of pH within experimental error, and the activation energy is half or less of the usual value. The lack of dependence of the rate constant on pH should not be extrapolated to pH values outside the range studied. Reference to Figures 11 and 12 will show that for the trimetaphosphate and tetrametaphosphate there are relatively small changes in rate constant between pH 4 and 11. In addition, the rate constant value at pH 13.6 in Table XXXII shows an increase in rate in alkaline solution, but one must recognize that rate constants from different sources are not necessarily comparable.

The small activation energies in conjunction with the small rate constants require that the Arrhenius A value be very small or, equivalently, that the entropy of activation be very negative. The data of Table XXXI lead, in fact, to $\Delta S^{\ddagger}$ values of the order of -70 to -80 eu

TABLE XXXII

First-Order Hydrolysis Rate Constants for $(PO_3^-)_n$ (hr^{-1}) at $30°$

		n			
pH	3	4	5	6	Ref.
4				2.1×10^{-5}	167
7				2.2×10^{-5}	167
11				1.9×10^{-5}	167
13.6	1.5×10^{-1}	4.4×10^{-3}	2.9×10^{-3}	6.9×10^{-4}	408

and to log A values of 0 to -2 log hr^{-1}, which are extremely far removed from the usual values. Values of this magnitude would indicate that large charges are produced in the transition state or that the $\Delta S^{\ddagger}$ value is a cumulative one, including entropy changes for several steps preceding the rate-determining step. Thilo and Schülke have shown (408) that the rate constant for hydrolysis at pH 13.6 decreases with increased ring size (Table XXXII). Because of the limited number of studies of the activation parameters, other than for trimetaphosphate, and the inconsistency of those that exist, it is impossible to relate this to a trend in activation enthalpy or entropy.

The alkaline hydrolysis of the larger metaphosphate rings does not serve as a method of preparation for the longer oligophosphates. This is a result of the decrease in alkaline hydrolysis rate with increased size for metaphosphates, coupled with the increase in alkaline hydrolysis rate with increased length for polyphosphate chains (Section III-D). Because of these two trends, significant concentrations of the primary hydrolysis product from the rings do not accumulate in the solution. Griffith and Buxton reported that in the hydrolysis of hexametaphosphate at pH 4, 7, and 11, the concentration of none of the degradation products other than monophosphate exceeded 1% (166). The alkaline hydrolysis of octametaphosphate yields octaphosphate and its degradation products, but large concentrations of the intermediates do not develop (365). The kinetics of this hydrolysis were not studied.

Thermal dehydration of sodium hexametaphosphate–hydrate is accompanied by cleavage and reorganization of the metaphosphate ring, with the formation of trimetaphosphate, amorphous long-chain polyphosphate and some anhydrous hexametaphosphate (408). Concentrated aqueous ammonia cleaves the pentameta- and hexametaphosphate rings to produce amidopenta- and amidohexaphosphate in reactions parallel to those occurring with trimeta- and tetrametaphosphate (408).

I. Ultraphosphate

Under this heading we are interested in the hydrolysis occurring at branch points in the oxygen-bridged phosphate structure. Branch points are those phosphate tetrahedra in which three oxygen atoms are shared with other phosphate tetrahedra. For the metaphosphate anion composition, with the general formula $(PO_3^-)_n$, there is a deficiency of one oxygen atom per phosphorus atom relative to the unbridged monophosphate anion composition, PO_4^{3-}. One can easily establish that for this deficiency each tetrahedron must be involved in an average of two bridges. For this case, the simplest structure type is chains of infinite length, such as rings are and high molecular weight polyphosphates approach. Phosphate anions of lower O to P ratio than 3:1 must average more than two bridges on each phosphate tetrahedron. An O to P ratio smaller than 3:1 corresponds to an

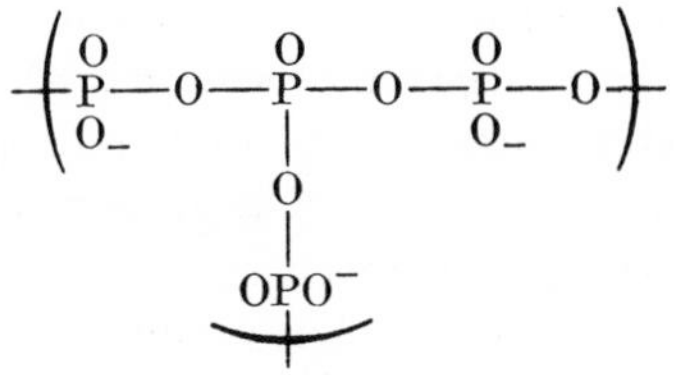

(16) 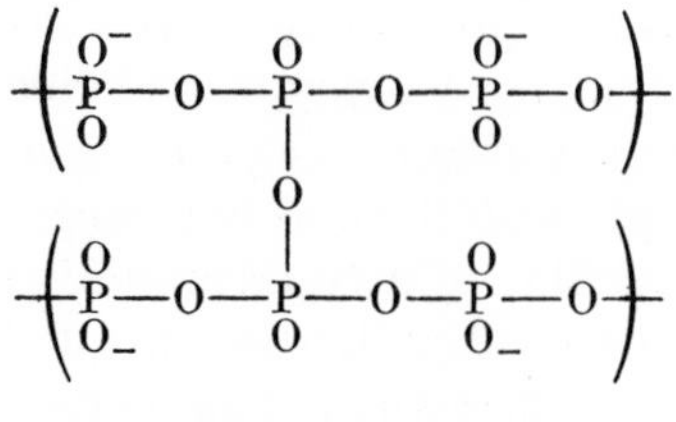   (17)

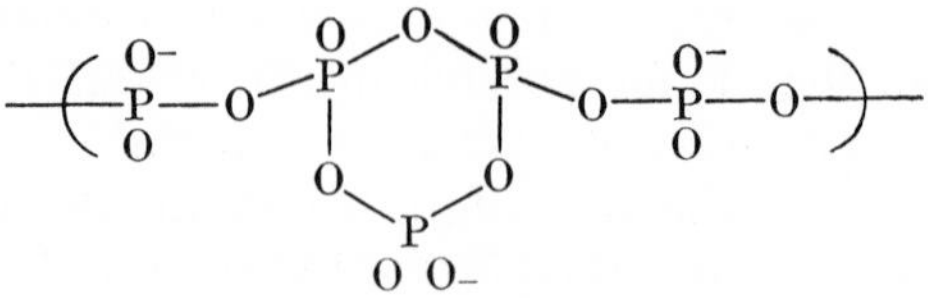

(18)

$M_2O-P_2O_5$ or $MO-P_2O_5$ ratio smaller than 1:1. Phosphates in this composition range are called ultraphosphates. They are generally noncrystalline; most are high-molecular-weight glasses.

Possible ultraphosphate structural elements include simple branch points (16), cross-linkages between chains (17), and the incorporation of rings in the chains (18). A useful generic term for the branch points of any type is "tertiary tetrahedra." The ultraphosphates are sometimes referred to as "network phosphates."

Thilo and Sonntag (411) considered the possible presence of structures 16 and 17 in Graham's salt type ultraphosphate, high-molecular-weight glasses prepared by dehydration of partially neutralized phosphoric acid solutions (Na to P ratio less than 1). They reasoned that hydrolysis of the first structure produces a strongly acidic proton and a weakly acidic proton [Eq. (48)], whereas hydrolysis of the cross-linkage of the second structure produces only strongly acidic protons [Eq. (49)].

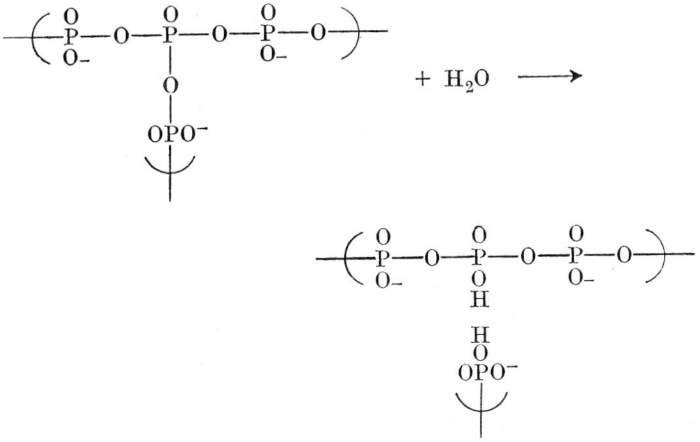

$$+ H_2O \longrightarrow \tag{48}$$

$$+ H_2O \longrightarrow \tag{49}$$

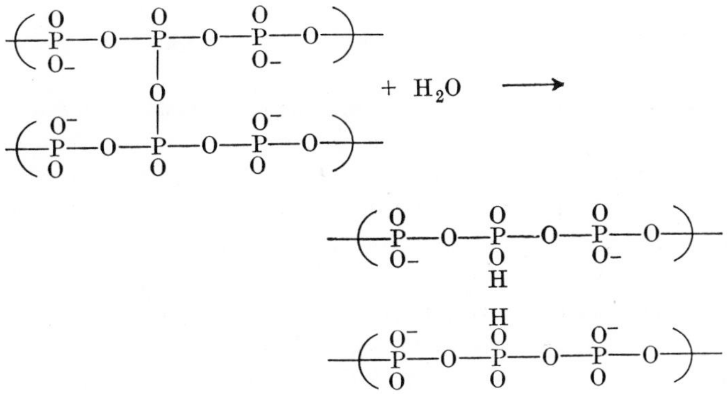

Determination of changes during hydrolysis in the strong acid function (titratable to about pH 4.5) and of the weak acid function (titratable roughly from pH 4.5 to 9.0) allows assessment of the amount of each structural element. On this basis, they found that cross-linked ultra-phosphates form first during the dehydration. Longer heating or heating at higher temperatures increases the number of branch points markedly at the expense of cross-link points. Their final data for a sample with an Na to P ratio of 0.79, heated 40 hr at 650° showed about 40% cross-link points and 60% branch points. In this study a cross-link consists of a single oxygen atom bridge; a cross-linked structure in which one or more phosphate tetrahedra were present in the link between the chains would be counted as two branch points. This study did not consider the possibility that the crosslink might hydrolyze as in Eq. (50).

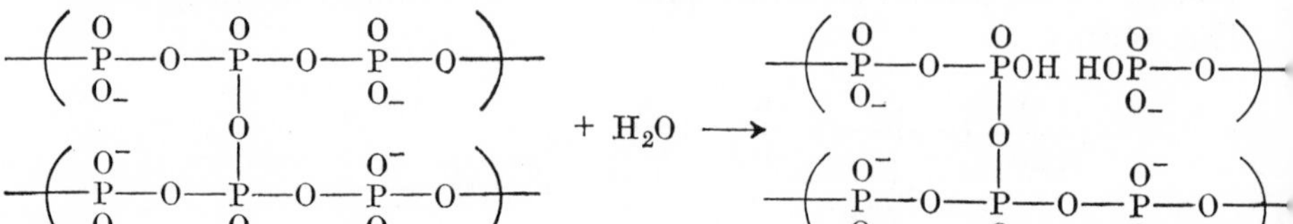

$$(50)$$

This reaction would cause the cross-link to be counted as a branch point. If we assume the mechanism of hydrolysis at tertiary tetrahedra to be attack by water on the phosphorus atom, for which the evidence will be cited later, there is a statistical factor of two favoring reaction (50) over (49). On the other hand, there should be a sizable steric factor favoring reaction (49). Assuming predominance of the steric factor, it would be reasonable to ignore reaction (50). The possibility of rings in the chain was not considered (18, for example). Hydrolysis of the tertiary tetrahedra involved in the ring structure would cause these sites to be counted as cross-links (19) or as branch-points (20 and 21).

Aiken and Gill have suggested that six-membered rings are an important structural feature in ultraphosphates (4). The basis for this is somewhat ambiguous, namely the finding of trimetaphosphate after hydrolysis of the tertiary tetrahedra in ultraphosphate samples. Gill and Riaz, in a study of sodium ultraphosphate glasses for which the Na to P ratio ranged from 1.0 to 0.7, found up to 18% of the phosphorus present as trimetaphosphate in the primary hydrolysis products. Since the degradation of polyphosphate chains to trimeta-phosphate (Section III-E) is orders of magnitude slower than the

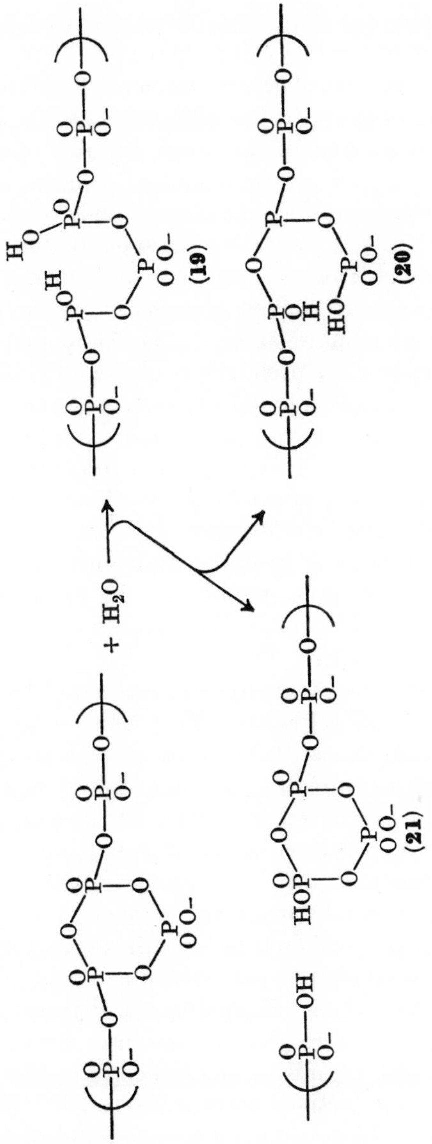
+ H_2O
(19)
(20)
(21)

hydrolysis of ultraphosphates, they concluded that trimetaphosphate rings must be a part of the chain (18), and must be freed from the chain by hydrolysis at the two tertiary tetrahedra involved (21). This conclusion was consistent with the occurrence in the IR spectra of all the ultraphosphate glasses of a broad doublet corresponding to a doublet in the spectrum of sodium trimetaphosphate, where it has been attributed to deformation of the P—O—P angle of the ring.

It is not clear that Gill and Riaz considered the possibility of the formation of 19 or 20 occurring, the possibility of trimetaphosphate formation in hydrolysis at tertiary tetrahedra, or the possible presence of discrete trimetaphosphate anions as a component of the glasses. It is difficult to establish a preferred route among 19, 20, or 21. At pH 7, which was used in this study, hydrolyses at the secondary tetrahedra of trimetaphosphate and long-chain polyphosphate occur at comparable rates. For a given tertiary tetrahedron there is no statistical advantage to any of the three reactions, and for a bimolecular mechanism probably little steric differentiation.

If one assumes the rings to be in the chair form, a serious steric problem arises for attack of H_2O opposite to the equatorial position, for which the chain connection might be expected to show a preference, but this preference is small. The free energy difference for axial and equatorial chain attachments is probably less than 1 kcal/mole, based on data for cyclohexane substituents. This free energy difference corresponds to an equilibrium constant of about 5 relating the two forms. One concludes that rupture of the rings should not be ignored. If this is the case, a much higher proportion of the phosphate tetrahedra of ultraphosphates would have to be in rings, and the chain lengths would have to be unreasonably short in view of the viscosity data for ultraphosphates. No doubt rings are present in the chains, but the extent of their occurrence should be examined further.

Elucidation of the structural patterns of ultraphosphates is difficult. The materials we have been discussing are glasses and are not susceptible to X-ray structure determinations. Studies of ultraphosphates in solution can be misleading concerning the original glasses. Significant numbers of tertiary tetrahedra can be hydrolyzed during the solution process. In fact, for ultraphosphates of low Na to P ratio such hydrolysis is probably a necessary feature of the solution process. Tertiary phosphate tetrahedra have been detected in ultraphosphate glasses by their IR spectra (30). They have been detected by ^{31}P NMR spectra in solution in tetramethylurea (160) and in chlorosulfuric acid and sulfuric acid oleums (130). X-ray studies have not detected tertiary phosphate tetrahedra in crystalline, cation-deficient KPO_3

(335). Presumably, this is a result of the relative insensitivity of the X-ray method.

Evidence for the presence of tertiary tetrahedra in phosphates and for the very rapid hydrolysis of these sites comes primarily from solution studies, such as investigations of viscosity (276,335,383,384,436, 464), changes in end-group titration values (411), and changes in solution pH (383,384). It was recognized earlier by Van Wazer (443) that rapid hydrolysis of tertiary phosphate tetrahedra is to be expected on the basis of the lengthening of the P—O—P bridge in P_4O_{10}. Pauling's rules concerning sharing of oxygen atoms in oxyanions (332) predict that the stability of a structure is decreased by an increase in the number of oxygen atoms shared by a given pair of central atoms.

Strauss and Treitler examined alternatives to attributing the initial pH and viscosity decreases in ultraphosphate solutons to hydrolysis at tertiary tetrahedra (384). Hydrolysis at secondary tetrahedra (tetrahedra in linear portions of the chains) is easily excluded. As pointed out below, the hydrolysis rate at such points is known and is orders of magnitude too small. Furthermore there would be no reason for hydrolysis at secondary tetrahedra to stop while the anion molecular weights are still large. Thilo and co-workers have shown that incorporation of arsenic in a polyphosphate chain, in place of some of the phosphorus atoms, causes rapid hydrolysis at those sites (402,403). Analysis of the samples studied by Strauss and Treitler showed the levels of arsenic (and of silicon) were at least an order of magnitude too small to account for the number of fast hydrolysis sites and could not account for a dependence of the number of sites on the Na to P ratio.

Based on an extrapolation to zero solution time of the initial rapid viscosity change in ultraphosphate solutions, Strauss and Treitler estimated the number of tertiary tetrahedra in the ultraphosphate sample before solution (384). A sample with an Na to P ratio of 0.97, heated 12 hr at about 940°, had of the order of 4 tertiary tetrahedra/100 phosphate tetrahedra. The fraction of tertiary tetrahedra decreased with increased Na to P ratio, but $NaPO_3$ glasses with an Na to P ratio of 1 had about 3/1000, and those with Na to P ratios slightly greater than 1 still had about 1 tertiary tetrahedra/1000. The samples studied had chain lengths of about 200, so that for Na:P = 1 about half the anions had tertiary tetrahedra, on the average.

The hydrolysis at tertiary phosphate tetrahedra is first order in tertiary tetrahedra (384,411). The kinetics do not appear to depend on the percentage of tertiary tetrahedra. Thilo and Sonntag obtained first-order hydrolysis rate constants of 0.28 and 0.27 hr^{-1} for samples with 1.9 and 9.0%, respectively, of tertiary tetrahedra, using 1%

solutions at pH 5 and 21° (411).   Strauss and Trietler obtained rate constants of $0.48 \pm 0.24\,hr^{-1}$ at 25° for samples with from 1 to about 40 tertiary tetrahedra/1000 phosphorus atoms (384).   The rate is unaffected by the addition of up to $0.35M$ NaBr or by the lowering of the dielectric constant of the medium from 82 to 70 by addition of organic solvents (385).

In contrast to the pattern for polyphosphate hydrolysis, the kinetics are independent of hydrogen ion activity between pH 3.4 and 11 (384, 385).   Most of this range is covered in Table XXXIII.   The lack of

TABLE XXXIII

Hydrolysis at Tertiary Tetrahedra, First-Order Rate Constants (hr^{-1})[a]

pH	Temperature (°C)						
	0	21	25	28	32	41	45
4		0.25					
5		0.28			0.89	1.59	
6		0.38					
7		0.38					
7.1	0.025		0.56				3.9
8		0.36					
9		0.28					
11				0.59			

[a] From Ref. 411, except data at pH 7.1 and 11 which are from Ref. 385.

dependence on pH is not surprising over most of this range.   Protonation of the tertiary phosphate tetrahedra cannot be expected until a significantly lower pH than 3.4, as this protonation step would be comparable to creating $H_4PO_4^+$.   Hydroxide-ion catalysis is not important until about pH 9–11 (cf. Figs 3, 6, and 11).

The rate constants for hydrolysis at tertiary phosphate tetrahedra are much larger than those for the chain and metaphosphates   The first-order rate constant for hydrolysis of diphosphate can be estimated from Fig. 1 to be about $2 \times 10^{-6}\,hr^{-1}$ at pH 7 and 21°.   For the over-all rate of degradation of long-chain polyphosphate at the same conditions one obtains a value of $1.2 \times 10^{-5}\,hr^{-1}$, using data from Tables XXIV and XXV.   The value for tertiary tetrahedra hydrolysis at these conditions is $0.38\,hr^{-1}$, a value 2×10^5 to 4×10^4 times as large.

Activation parameters for hydrolysis at tertiary tetrahedra are given in Table XXXIV.   The activation energies are distinctly lower than for hydrolysis of polyphosphates at the same pH values.   The activation energy differences are sufficient to account for the observed rate

TABLE XXXIV

Activation Parameters for Tertiary Tetrahedron Hydrolysis[a]

pH	E_a (kcal/mole)	$\log A$ (log hr^{-1})	$\Delta S^{\ddagger}$ (eu)	Ref.
5	15.4	10.8	-28	411
7.1	18.9	13.5	-15	385

[a] See footnote (a), Table XIII. Values are for first-order rate constants.

differences. For example, for rounded-off activation ·energies of 18 and 24 kcal/mole at pH 7 for tertiary tetrahedra and polyphosphate, respectively, the ratio of rates at 21° would be 3 × 10^4.

Thilo has pointed out that the activation energies for tertiary tetrahedra hydrolysis at pH values 5 or 7 are similar to the values for polyphosphate hydrolysis at pH 1 (411). At pH 1 the polyphosphates are largely protonated. A protonated tetrahedron in the middle of a polyphosphate chain is similar to a tertiary tetrahedron in several respects. Hydrogen and phosphorus have the same Pauling electronegativity value, for which reason they might be similar in their effects on the electron density in the oxygen bridges to the tertiary tetrahedron. On this basis, they would have similar effects on the strengths of those bridges and on the attractiveness to nucleophilic attack of the phosphorus atom of the tertiary tetrahedron. From the valence bond viewpoint, protonation and branching both reduce the resonance at the phosphate tetrahedron involved and thus destabilize that tetrahedron. From a molecular orbital viewpoint, hydrogen and phosphorus addition are less closely equivalent in nature of bonding if the participation of phosphorus d orbitals in π dp bonding is allowed. The importance of this interaction has not yet been established (2,26,48,285,286,298).

From pH 1.7 to about 4.5, the same ^{31}P NMR chemical shift, about 11 ppm upfield relative to 85% H_3PO_4, is observed for the α and γ phosphorus atoms of adenosine triphosphate at an ionic strength of 1.0 with tetramethylammonium ion the only cation present (105). Over most of this pH range the principal species in solution is the monoprotonated anion, **22**. Evidently protonation is closely equivalent to

$$
\begin{array}{c}
\text{O O O} \\
\text{A—OPOPOPOH} \\
\text{O\_O\_O\_} \\
\text{(22)}
\end{array}
$$

esterification with respect to influence on the environment of the phosphorus atom of a phosphate tetrahedron. At pH values above 4.5, with the removal of the proton from the terminal phosphate tetrahedron, the chemical shift for its phosphorus atom decreases markedly. Over the entire pH range studied, 1.7 to about 9, the central phosphorus atom shows a much greater chemical shift, about 22 ppm relative to 85% H_3PO_4. From this evidence it would appear that protonation of a phosphate tetrahedron is not closely equivalent to the involvement of the tetrahedron in an additional oxygen bridge to phosphorus.

The $\Delta S^{\ddagger}$ values of Table XXXIV suggest that the hydrolysis of tertiary tetrahedra is bimolecular. Strauss and Treitler came to the same conclusion on other bases (385). The lack of kinetic dependence on the dielectric constant of the medium eliminates the S_N1 mechanism, for which a charge separation would be required in the transition state. Electrophilic attack by water on a bridging oxygen atom at the tertiary tetrahedon was ruled out on the basis of the absence of a pH effect on the kinetics. The hydronium ion should be more effective than a water molecule for this route. Nucleophilic attack by water on the phosphorus atom of the tertiary tetrahedron was viewed as most likely, with the formation of a bond between the oxygen of the water and the phosphorus selected as the rate-determining step. For this mechanism the large rate increase accompanying formation of a third P—O—P link must be attributed to changes in the electronic environment of the phosphorus, rather than to changes in the strength of the P—O—P bridges.

Two small anions containing tertiary phosphate tetrahedra have been reported. Study of these should be particularly fruitful in developing an understanding of the characteristics of branch points. The isotetrametaphosphate anion shown in **23** is strictly not an ultraphosphate as O:P = 3, but it does contain a tertiary tetrahedron.

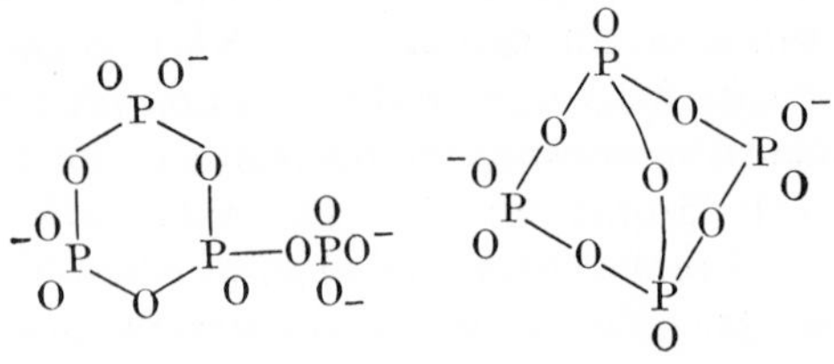

(23)   (24)

Although they did not study the kinetics, Rätz and Thilo found that the tetraethyl ester of isotetrametaphosphate hydrolyzes essentially instantaneously at room temperature (350). The structure of the ester was established on the basis of the molar proportion of ultimate hydrolysis products: 2 moles of monoethyl phosphate to 1 of diethyl phosphate to 1 of inorganic phosphate. Evidence from ^{31}P NMR suggests the formation of the anion 24 in the condensation of H_3PO_4 by dicyclohexylcarbodiimide in tetramethylurea and in the reaction of P_4O_{10} with diisopropylurea in tetramethylurea (160). Structure 24 is the first product in the hydrolysis of P_4O_{10} (Section III-J).

J. Phosphorus(V) Oxide

Phosphorus(V) oxide is the ultimate ultraphoshate. The common volatile form of this compound has the molecular formula P_4O_{10} and the structural formula shown in Figure 13. The four phosphorus atoms are at the corners of a regular tetrahedron. The P—O bond length in the P—O—P linkages is 1.60 Å, well within the 1.55–1.67 Å range for such bonds in chain and ring phosphates (109). The P—O bond length for the phosphoryl oxygen atoms is 1.40 Å, shorter than typical. This shortening is to be expected, as each phosphate tetrahedron of P_4O_{10} is tertiary. The average of all the P—O distances in any phosphate anion can be expected to be 1.54 $\pm$ 0.02 Å (Table IV.1 of Ref. 109). Since a tertiary tetrahedron has three of the longer bridge bonds, a complementary shortening of the phosphoryl bond is to be expected. From the Schomaker–Stevenson relationship, one would predict P—O single and double bond lengths of 1.67 and 1.53 Å, respectively (364). The average P—O bond in phosphates should be considered a double bond, presumably involving delocalized $dp\pi$ bonding (48,262,285,298). In a tertiary tetrahedron the lower electron density in the three bridge bonds evidently is compensated for by an increase in electron density in the phosphoryl bond. The phosphate tetrahedron in P_4O_{10} is somewhat distorted, with the OPO angle 101.5° for the angle formed with two bridge oxygen atoms and 116.5° for the angle formed with a bridge oxygen and the phosphoryl oxygen (121,179).

Two high-molecular-weight orthorhombic forms of phosphorus(V) oxide are known. One has an extended three-dimensional structure (121) and the other an extended sheet structure (277). These highly polymerized forms dissolve in water slowly. They hydrolyze as they dissolve, or more accurately, dissolve as they hydrolyze, as hydrolysis is necessary to set free units capable of passing into solution. Only the hydrolysis of the more fully studied, volatile, rhombohedral P_4O_{10} will be considered here.

A priori, one can say that the successive cleavage of the oxygen bridges in the hydrolysis of P_4O_{10} will follow the pattern of Figure 13. Reasoning from the P_4O_{10} structure in this way facilitated understanding

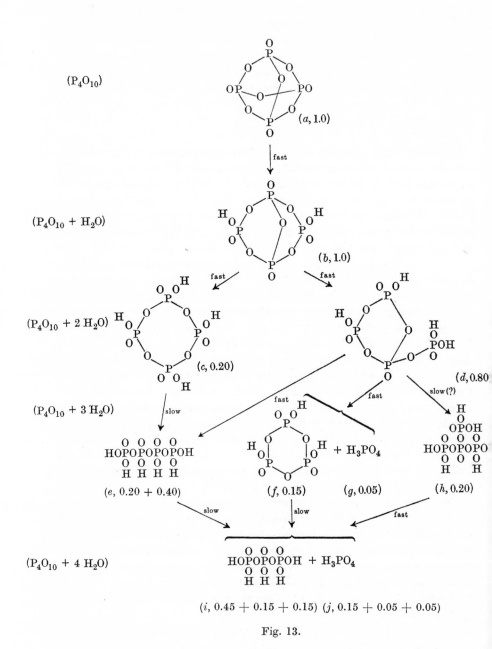

(P_4O_{10})

$(P_4O_{10} + H_2O)$

$(P_4O_{10} + 2 H_2O)$

$(P_4O_{10} + 3 H_2O)$

$(P_4O_{10} + 4 H_2O)$

Fig. 13.

182

of the route of hydrolysis and lead to a method of preparation for tetra-metaphosphate, which is the principal product when P_4O_{10} is dissolved in excess water at 0° (33,345,346,456).

The parentheses associated with each species of Figure 13 give a letter designation for the species and the portion of the phosphorus that would be found in that form at that level of hydrolysis, assuming that the reaction products are statistically determined. For example, at hydrolysis level (P_4O_{10} + 2 H_2O), only the breaking of the cross-link of species (b) will produce tetrametaphosphate, but the breaking of any of the other four P—O—P linkages will produce isotetrametaphoshate, leading to the respective amounts, 0.20 and 0.80. Where a sum of figures appears in the parentheses, each figure refers to the product of one route. The possibility of tetraphosphate cleaving at the central P—O—P linkage need not be considered (cf. Section III-C). The pattern of principal hydrolysis products for P_4O_{10} in both acidic and basic solution is (353):

$$P_4O_{10} \longrightarrow H_4P_4O_{12} \longrightarrow H_6P_4O_{13} \longrightarrow$$
$$H_5P_3O_{10} + H_3PO_4$$
$$\longrightarrow H_4P_2O_7 + H_3PO_4$$
$$\longrightarrow 2\ H_3PO_4$$

Species (b), (d), and (h) have not been isolated from the P_4O_{10} hydrolysis products. Apparently they are too reactive to persist in aqueous medium. Evidence has been presented for the cross-linked structure (b), however, as the product of condensation of monophosphoric acid by dicyclohexylcarbodiimide in tetramethylurea solution (160), and for the ethyl and isopropyl esters of (d) as products of the reaction of P_4O_{10} with diethyl and diisopropyl ethers (350,420).

Average distributions of hydrolysis products obtained in two studies by dissolving P_4O_{10} in an excess of water at 0° are shown in Table XXXV. For comparison, the expected distribution of products is given on several bases in the same table. For Case I, the assumption is made that only those reactions eliminating a tertiary tetrahedron will be fast enough to be considered. These reactions are labeled, "fast" in Figure 13, and lead to tetrametaphosphate, tetraphosphate and monophosphate as products. Assuming that the amount of material hydrolyzing by alternate routes is statistically determined on the basis of the number of bonds that can be broken to give that route, one obtains the distribution of products listed as Case I. Tetraphosphate has been considered to appear analytically as long-chain polyphosphate, $O(PO_3)_n^{-(n+2)}$. There is little resemblance between Case I

TABLE XXXV

P_4O_{10} Hydrolysis Products, % of Total P

PO_4^{3-}	$P_2O_7^{4-}$	$P_3O_{10}^{5-}$	$O(PO_3)_n^{-(n+2)}$	$P_3O_9^{3-}$	$P_4O_{12}^{4-}$	Ref.
4.4	1	15.0	3.6	2.3	72.1	417
2.1	2.1	4.2	16.4	14.5	60.0	372
Statistically determined:						
7	0	0	53	20	20	Case I
3	0	0	27	10	60	Case II
7	0	21	0	0	72	Case III

and the distributions found experimentally. The experimental yields of PO_4^{-3}, $O(PO_3)_n^{-(n+2)}$, and $P_3O_9^{3-}$ are lower than predicted, whereas the yields of $P_3O_{10}^{5-}$ and $P_4O_{12}^{4-}$ are significantly higher than expected.

It is evident that the alternate modes of hydrolysis for the cross-linked species (b) are not statistically controlled. Cleavage of the cross-link to form tetrametaphosphate must be at least twelve times as fast as cleavage of any particular one of the other four P—O—P linkages. If, as Case II, one assumes this factor of twelve but continues to allow statistics to determine the products of hydrolysis of the isotetrametaphosphate, the second statistical distribution of Table XXXV is obtained. This distribution agrees reasonably well with the data of Shima, et al. (372). The data of Thilo (417) show lower yields of long-chain polyphosphate and trimetaphosphate, with a higher yield of triphosphate. This can be accommodated within the hydrolysis scheme of Figure 13 by revising the rate assignments to the three hydrolysis paths for the isotetrametaphosphate. If the hydrolysis of this species to isotetraphosphate (species h) is assumed to be much the fastest of the three, further hydrolysis of isotetraphosphate is assumed fast because of its tertiary tetrahedron, and the relative rates for hydrolysis of the cross-link and of a particular noncross-link P—O—P bridge of species (b) are set at 3.6 (to match Thilo's $P_4O_{12}^{4-}$ yield); the product distribution for Case III is obtained. This distribution agrees satisfactorily with Thilo's results. The agreement could be improved easily by allowing small amounts of the hydrolysis of the isotetrametaphosphate to tetraphosphate and to trimetaphosphate and monophosphate. Thus both sets of data are consistent with the hydrolysis scheme of Figure 13 if an explanation can be found for alteration of the pattern of product distribution within the hydrolysis scheme.

Thilo and Wieker observed that some trimetaphosphate is always produced in P_4O_{10} hydrolysis, with the proportion dependent on conditions. Data for their experiments at 0° appear in Table XXXVI. The solutions for the experiments at uncontrolled pH would, of course, become acidic. On this basis, the decrease in trimetaphosphate yield down Table XXXVI could be associated with decreasing hydrogen ion activity in the solution during the P_4O_{10} hydrolysis. Since protonation of polyphosphate species affects their hydrolysis rates, it is very likely that the relative rates for alternate routes of hydrolysis will be affected by the level of protonation. The data of Table XXXVI

TABLE XXXVI
Trimetaphosphate Yield from P_4O_{10} Hydrolysis[a]

% of P as $P_3O_9{}^{3-}$	pH	P_4O_{10} used (mg/40 ml H_2O)
8	Uncontrolled	2302[b]
5.2	Uncontrolled	839[b]
2.0	Uncontrolled	428
1.5	Neutral	490
1.5	Alkaline	673
1.1	Alkaline	609

[a] At 0°, data from Ref. 417.
[b] Turbidity observed in these solutions, taken as indication of the formation of high molecular weight material.

suggest that under acidic conditions, with some degree of protonation of the isotetrametaphosphate species, the hydrolysis to trimetaphosphate and monophosphate is favored, whereas under alkaline conditions, with at least the more strongly acidic protons ionized, hydrolysis to the isotetraphosphate is favored. On this basis it would have to be said that conditions were more acidic during the hydrolyses of reference (372). It is true that larger $P_4O_{10}:H_2O$ ratios were used in this study. Of interest would be a study of the hydrolysis in a definitely acidic solution, in which the isotetrametaphosphate could be kept essentially un-ionized. It is significant in this regard that the principal route of hydrolysis of the ethyl and isopropyl esters of isotetrametaphosphate is to trimetaphosphate and monophosphate (420), corresponding to the route suggested above for the un-ionized acid.

Thilo and Wieker associated the formation of larger amounts of trimetaphosphate with the appearance of turbidity in the solution during the P_4O_{10} hydrolysis (417). The trimetaphosphate formed was attributed to the trimetaphosphate abstraction degradation reaction observed with the longer chain polyphosphates (cf. Section III-E).

The same view has been taken in a more recent study (186), in which anion-exchange chromatography demonstrated the presence of up to 10% long-chain phosphate in the P_4O_{10} hydrolysis products. A quantity of the long-chain product was isolated and found to resemble octaphosphate in that (1) degradation of the product for 90 min at 100° yielded 38% trimetaphosphate (cf. Table XXVIII) and (2) the ratio 1:4.2 found for the weak acid function before and after total hydrolysis is close to that expected for octaphosphate. Uniformity of chain length in the sample was not demonstrated. This oligophosphate material cannot be a hydrolysis product of P_4O_{10} but must be a product of phosphorylation reactions occurring during the hydrolysis.

Trimetaphosphate abstraction from long-chain polyphosphate must be responsible for some of the $P_3O_9{}^{3-}$ observed, but it is questionable whether this reaction is fast enough to account for all of the $P_3O_9{}^{3-}$. Furthermore, relatively large amounts of triphosphate are present in the P_4O_{10} hydrolysis products. On the basis of known and presumed rates, the formation of this through hydrolysis of isotetrametaphosphate is more logical than through hydrolysis of trimeta- and tetrametaphosphate. The complementary percentages of triphosphate and trimetaphosphate in the hydrolysis products found by Thilo and by Shima (Table XXXV) are consistent with a hydrolysis scheme in which both are derived from the isotetrametaphosphate. Alternatively, the complementary percentages of tri- and trimetaphosphate could be attributed to accelerated hydrolysis of the trimetaphosphate under more acidic conditions. This would require in Table XXXV that the solutions used by Thilo (417) be the more acidic, which is inconsistent with the experimental conditions. Furthermore, there should be significant hydrolysis of the tetrametaphosphate as well as of the trimetaphosphate, which is inconsistent with the high yields of tetrametaphosphate found by Thilo. Again, a study of the hydrolysis of P_4O_{10} in distinctly acidic solution would be of interest.

Although isotetrametaphosphate is at most a minor product in the hydrolysis of P_4O_{10} through the stage ($P_4O_{10} + 2 H_2O$) and cannot be isolated there, in the reaction of P_4O_{10} with ethers this route is the more important, and the resulting isotetrametaphosphate ester persists among the reaction products. By the reaction in chloroform of diethyl ether with P_4O_{10}, Rätz and Thilo prepared the ethyl ester of isotetrametaphosphate (350). Although it was preserved in the nonaqueous medium, it hydrolyzed very rapidly. The structure was established on the basis of the ultimate hydrolysis products, which were a 2:1:1 mole ratio of RPO_4H_2, R_2PO_4H, and H_3PO_4. Thilo and Woggon found in the reaction of P_4O_{10} with excess isopropylether about 70%

of the product was isotetrametaphosphate and the remainder normal tetrametaphosphate (420). This is the reverse of the weighting of these reaction paths for the hydrolysis of P_4O_{10}.

The principal early hydrolysis product for both the ethyl and isopropyl esters is the diester of monophosphoric acid, resulting from cleavage of the isotetrametaphosphate side chain. The hydrolysis is very rapid at 100°, but not instantaneous. At 0° hydrolysis is very much slower.

Thilo and Wieker have pointed out that the proportion of isotetrametaphosphate and normal tetrametaphosphate produced in the reaction of P_4O_{10} with two moles of ether is close to that expected on a statistical basis (417). There must be no differentiation of possible reaction sites in the ester form of the cross-linked species (b) of Figure 13. The marked deviation from the statistical expectation in the case of P_4O_{10} hydrolysis can then be attributed to ionization of the acid groups that are present in species (b) for the hydrolysis but not for the etherolysis. With the ionization of these protons inductive effects which are absent in the ester remove the chemical equivalence of the P—O—P linkages of species (b). Thilo also attaches some importance to the mutual repulsion of the ionized tetrahedra, tending to open the structure to the tetrametaphosphate ring. Increasing the acidity of the water used to hydrolyze P_4O_{10} should shift the route from normal tetrametaphosphate to isotetrametaphosphate. This is again consistent with a more acidic medium in the hydrolysis studies of Shima, et al., in Table XXXV (372).

Figure 14 summarizes what appear to be the principal routes of P_4O_{10} solvolysis through the very rapid stages. The H^+ associated with two of the arrows shows the route favored by protonation or esterification of the species. In addition to the reactions shown, condensation reactions occur to produce species with more than the four phosphorus atoms of a P_4O_{10} molecule.

Hydrolysis of P_4O_{10} under conditions restricting the rate of access of water to the P_4O_{10} has the effect of increasing the yield of long-chain phosphate. Condensation through phosphorylation reactions is favored under these conditions. The suspension of P_4O_{10} in ether or benzene containing low levels of water leads to higher levels of long-chain phosphate when excess water is subsequently added (372). Surprisingly, ensuring that a large excess of water is available at initial contact also increases the long-chain phosphate yield, as when P_4O_{10} vapor in a stream of dry oxygen is passed over water (186). The solvolysis of P_4O_{10} in acetone, in ethanol, and in mixtures of these with water has been studied (372), as well as the reaction of P_4O_{10} with fused H_3PO_4,

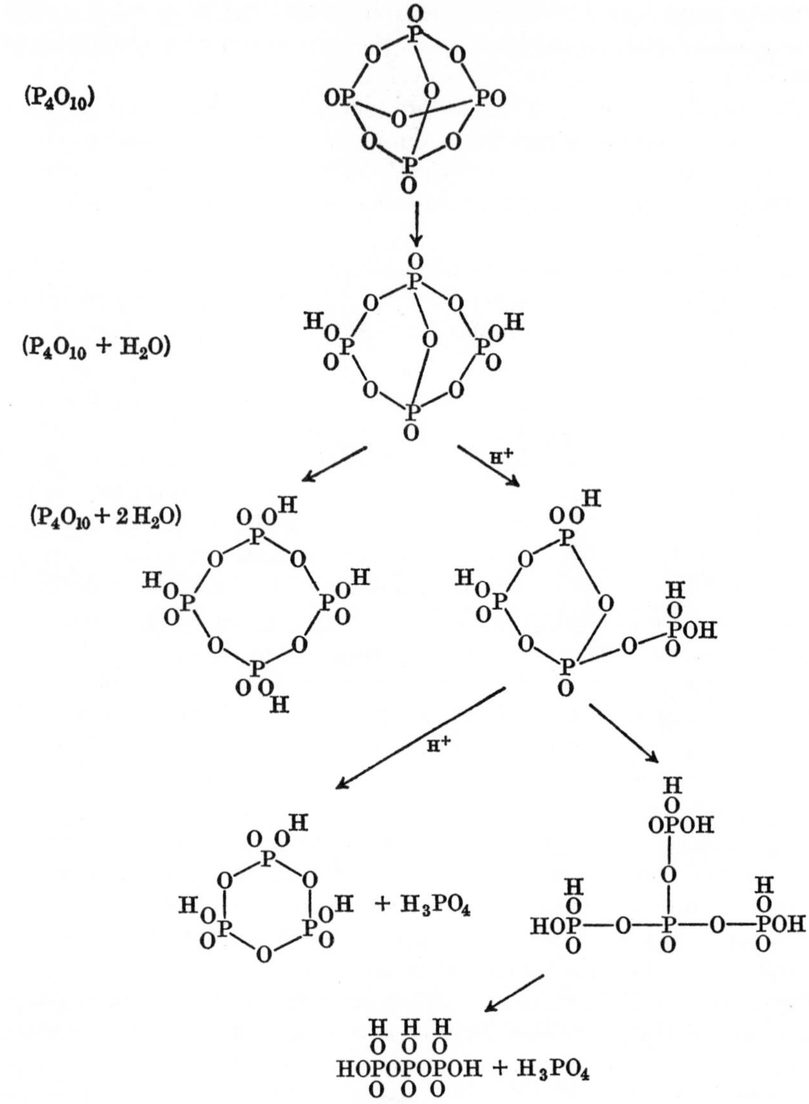

Fig. 14

with $(EtO)_3PO$, with low-molecular-weight primary and secondary alcohols, with ethylene glycol, and with Et_3N, $PhNH_2$, and p-$ClC_6H_4NH_2$ (96).

NMR studies of solutions of P_4O_{10} in 100% sulfuric acid and in 20% and 65% sulfuric acid oleums show that an equilibrium set of products is formed which is the same as for various phosphates in these solvents.

Typical species are $P(OH)_4^+$ and $(HO)_3POP(OH)_3^{2+}$. In P_4O_{10} solutions in chlorosulfuric acid branched phosphate structures are present (130).

IV. P—O Hydrolysis in Linkages to Other Elements

A. Phosphato-Metal Complexes

In aqueous solution the complex ions $Co(NH_3)_5OP(OCH_3)_3^{3+}$, $Co(NH_3)_5OP(O\text{-}t\text{-butyl})_3^{3+}$, $Co(NH_3)_5OPO(OCH_3)_2^{2+}$, $Co(NH_3)_5OP(OH)_3^{3+}$, and $Co(NH_3)_5OPO(OH)_2^{2+}$ hydrolyze to the aquopentammine cobalt(III) complex (360). In the case of the phosphate ester complexes, very little cleavage of the ester linkage accompanies the Co—O—P rupture. In spite of this, the hydrolysis of $Co(NH_3)_5$-$OP(COH_3)_3^{3+}$ in ^{18}O-enriched water showed 20% of the Co—O—P cleavage occurred at the P—O bond. In alkaline solution 12% of the cleavage was at this bond. Rate constant values of 0.90 and 0.50 hr^{-1} were determined for the hydrolysis of the trimethyl- and tri-t-butylphosphato complexes, respectively, at 25° in acidic solution. For the trimethylphosphato complex, the rate constant was the same in 0.1 and 0.4M $HClO_4$. In alkaline solution, pH about 12, the hydrolysis was very fast. Hydrolysis was slower for the other complexes listed above.

Lincoln and Stranks have studied the kinetics and equilibria involved in the hydrolysis of a number of phosphatocobalt complexes. The phosphate ligand can be bidentate or monodentate. It is to be expected that hydrolysis of a bidentate complex must proceed through a monodentate intermediate [Eq. (51)].

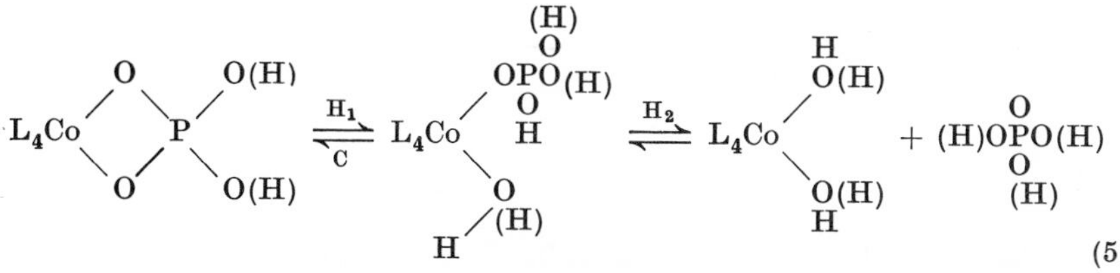

$$\tag{51}$$

In this general equation, L represents a ligand atom occupying an octahedral position not involved in the reaction. The presence or absence of the parenthetical protons will be determined by the solution pH. Ionization constants for these protons were determined by Lincoln and Stranks (Table XXXVII). UV spectra of phosphatocobalt complexes in 10–11.7 M $HClO_4$ gave evidence of an additional protonation of the species of Table XXXVII.

TABLE XXXVII

pK$_a$ Values for Phosphato Complexes[a]

Species	Temp. (°C)	pK$_1$	pK$_2$	pK$_3$	pK$_{aq}$	Ref.
OP(OH)$_3$	19.7	1.72	5.92	12.3		360
Co(NH$_3$)$_5$OP(OH)$_3$$^{3+}$		−0.67				267
cis-Co(NH$_3$)$_4$(OH)$_2$OP(OH)$_3$$^{3+}$	5	−0.34 ± 0.05	3.60 ± 0.05	8.50 ± 0.05		267
Coen$_2$(OH)$_2$OP(OH)$_3$$^{3+}$	5		3.2 ± 0.1	9.2 ± 0.1	6.70 ± 0.05	267
Coen$_2$(OH)$_2$OP(OH)$_3$$^{3+}$	5		3.30 ± 0.05	9.75 ± 0.05	7.25 ± 0.05	267
Coen$_2$O$_2$P(OH)$_2$$^{2+}$ c	5	0.00 ± 0.14^b	4.25 ± 0.05			267

[a] pK$_1$, pK$_2$, and pK$_3$ are for the successive acid ionizations of the phosphate portion of the species. pK$_{aq}$ is for the acid ionization of the bound water.

[b] Ref. 268.

[c] Bidentate phosphate. In the other complexes phosphate is monodentate.

It is interesting, but not surprising, that coordination enhances the acidity of each of the protons of H_3PO_4. This does not mean that Co is more effective than H in increasing the acidity of protons on other oxygen atoms of the phosphate tetrahedron. In fact, it is less effective. Thus for the series of tertiary phosphate tetrahedra in the series H_3PO_4, $Coen_2(OH)_2OPO(OH)_2{}^{2+}$, and $Coen_2O_2PO(OH)^+$, the pK values for the ionization of the most acidic proton present are 1.72, 3.30, and 4.25, respectively.

The rates of the opposing reactions constituting the first equilibrium of Eq. (51) have been studied for $Coen_2O_2P(OH)_2{}^{2+}$ over the range from pH 11 to $2.5M$ $HClO_4$ (268). The dependence of rate constant on pH in the range 3–11 is shown in Figure 15 for reactions H_1 (hydrolysis 1) and C (chelation) of Eq. (51). The kinetics were first order in complex ion at all pH values. From the relative rates it can be seen that over the approximate pH range 5 to 10 the bidentate complex is the more stable; above and below this range the monodentate form is favored.

The observed rate constant for the hydrolysis reaction [H_1, Eq. (51)] was resolved into separate values for the following hydrolysis reactions:

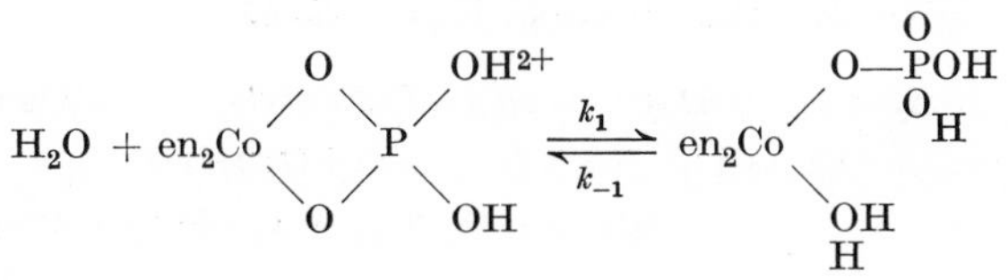

$$H_2O + en_2Co\diagdown\!\!\!\diagup{}^{O}_{O}\diagdown P\diagup{}^{OH^{2+}}_{OH} \underset{k_{-1}}{\overset{k_1}{\rightleftharpoons}} en_2Co\diagup{}^{O-POH}_{OH}{}^{O}_{H} \qquad (52)$$

$$H_2O + en_2Co\diagdown\!\!\!\diagup{}^{O}_{O}\diagdown P\diagup{}^{OH^+}_{O} \underset{k_{-2}}{\overset{k_2}{\rightleftharpoons}} \left[en_2Co\diagup{}^{O-PO}_{OH}{}^{O}_{H} \right]^+ \qquad (53)$$

$$H_2O + en_2Co\diagdown\!\!\!\diagup{}^{O}_{O}\diagdown P\diagup{}^{O}_{O} \underset{k_{-3}}{\overset{k_3}{\rightleftharpoons}} en_2Co\diagup{}^{O-PO}_{O}{}^{O}_{H}{}_{H} \qquad (54)$$

$$OH^- + en_2Co\diagdown\!\!\!\diagup{}^{O}_{O}\diagdown P\diagup{}^{O}_{O} \underset{k_{-4}}{\overset{k_4}{\rightleftharpoons}} \left[en_2Co\diagup{}^{O-PO}_{O}{}^{O}_{H} \right]^- \qquad (55)$$

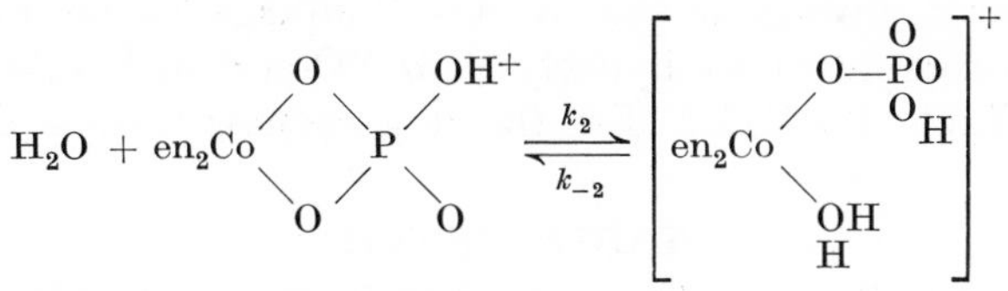

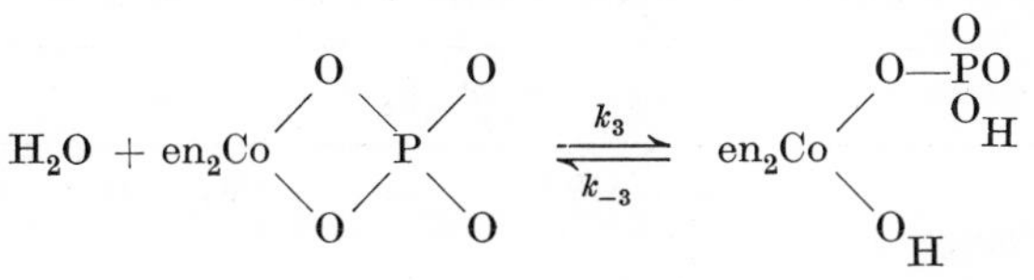

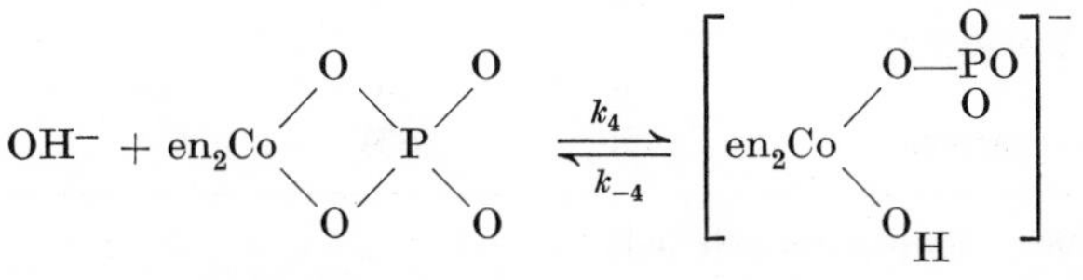

Equations (52–55) define the rate constants in the rate expression below. Values for these rate constants appear in Table XXXVIII.

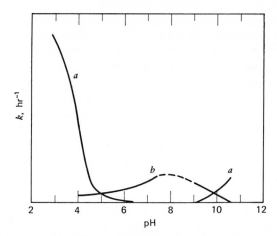

Fig. 15. Phosphatocobalt(III) complex reactions (268). First-order rate constants at 22.5° for (a) opening bidentate phosphate ring, H_1 of Eq. (51); (b) forming bidentate ring, C of Eq. (51).

In the same study, the separate rate constants were also determined for the chelation reactions that reverse Eqs. (53–55).

$$\text{Rate } H_1 = k_{obs}[\text{en}_2\text{Co O}_2\text{P(OH)}_2{}^{2+} + \text{en}_2\text{Co O}_2\text{PO(OH)}^+ + \text{en}_2\text{Co O}_2\text{PO}_2]$$
$$= k_1[\text{en}_2\text{Co O}_2\text{P(OH)}_2{}^{2+}] + k_2[\text{en}_2\text{Co O}_2\text{PO(OH)}^+]$$
$$+ k_3[\text{en}_2\text{Co O}_2\text{PO}_2] + k_4[\text{OH}^-][\text{en}_2\text{Co O}_2\text{PO}_2] \quad (56)$$

The point of cleavage of the Co—O—P linkage has been determined for several complexes by hydrolysis in ^{18}O-enriched water. Results appear in Table XXXIX. For the ring-opening process discussed in

TABLE XXXVIII

Kinetics of Bidentate Phosphatocobalt Hydrolysis at 22.5°, hr^{-1} [a]

Rate constant:	k_1	k_2	k_3	k_4
	1.5×10^3 [b]	20	9×10^{-2}	6.3×10^3 [c]
Approx. pH range of dominance	<2	2–6	6–10	>10
% P—O cleavage at 60°:[d]	0	0	40	35
Chemical equation:	(52)	(53)	(54)	(55)

[a] Ref. 268.   Ionic strength >1.
[b] At 10°.
[c] k_4 is a second-order constant, liters/(mole)(hr).
[d] From Table XXXIX.

TABLE XXXIX

Phosphatocobalt Hydrolysis, % P—O Cleavage at 60°

Complex[a]	Medium				
	11.4M HClO$_4$	pH ~ 0	pH ~ 7	pH ~ 13	Ref.
Monodentate:					
Co(NH$_3$)$_5$OP(OCH$_3$)$_3$$^{3+}$		20		12	360
Co(NH$_3$)$_5$OP(OH)$_3$$^{3+}$	0	0	0	0	267
cis-Coen$_2$(OH$_2$)OP(OH)$_3$$^{3+}$	0	0		0	267
Bidentate:					
Coen$_2$O$_2$P(OH)$_2$$^{2+}$ [b]	0	0	40	35	267

[a] Principal species at pH about 0 shown.
[b] % P—O cleavage in the chelate ring-opening step.

the preceding paragraphs, Eqs. (52) and (53) occur with no P—O cleavage. Equations (54) and (55) involve only 35–40% P—O cleavage. The marked increase in rate constant value and the decrease in importance of P—O cleavage with protonation of the phosphate ligand can be attributed to the improvement of phosphate as a leaving group with protonation. It can also be said that protonation of the phosphate lowers the electron density in the Co—O bonds through an inductive effect. For Eqs. (54) and (55) rate constant values of 4×10^{-2} hr^{-1} and 2×10^3 liter/(mole)(hr), respectively, can be estimated for the P—O rupture route at 22.5°.

The P—O cleavage data for Co(NH$_3$)$_5$OP(OH)$_3$ and Co(NH$_3$)$_5$-OP(OCH$_3$)$_3$ in Table XXXIX appear to be inconsistent. At pH 0, OP(OH)$_3$ and OP(OCH$_3$)$_3$ should be about equally good leaving groups. At pH 13, OP(OCH$_3$)$_3$ should be much better than OPO$_3$$^{3-}$, which the inorganic phosphate would have become. The P—O cleavage route should be as important for the inorganic phosphate at pH 0 as for the ester, and it should be more important for the inorganic phosphate at pH 13. For the inorganic phosphate complex, the incorporation of ^{18}O in the liberated phosphate was determined. For the ester complex, the ^{18}O content in the Co(NH$_3$)$_5$(OH$_2$)$^{3+}$ product was determined. Both procedures should be valid.

The kinetics of the hydrolysis of the monodentate complexes Co(NH$_3$)$_5$OPO$_3$, cis-Co(NH$_3$)$_4$(OH$_2$)OPO$_3$ and cis-Coen$_2$(OH)$_2$OPO$_3$, and of the bidentate complexes Co(NH$_3$)$_4$O$_2$PO$_2$ and Coen$_2$O$_2$PO$_2$ have been studied at HClO$_4$ concentrations from 10^{-3} to 11.4M at 45–70 (269). These reactions will not be discussed as the bidentate phosphate ring-opening reaction at these acidities and the monodentate phosphate hydrolysis at all acidities are by Co—O cleavage alone.

B. Other Inorganic Oxygen Bridges

Studies of mixed polyanions in which the phosphate tetrahedron is oxygen bridged to other atoms have been reviewed by Shigeru Ohashi in Volume 1 of this series. We shall minimize overlap with that chapter.

A random distribution of the alien central atoms and rapid, essentially exclusive, cleavage at those atoms has been demonstrated for copolymers of phosphate tetrahedra with arsenate (170,399,402,403, 467), vanadate (323), or silicate tetrahedra (138,324). Aqueous solutions of the copolymers contain monoarsenate, monovanadate or monosulfate, and a mixture of polyphosphate chains. In the $NaPO_3$–$NaAsO_3$ system both glassy and crystalline (Maddrell salt structure) linear copolymers can be prepared (402,403). Samples in the KPO_3–$KAsO_3$ system are primarily crystalline and include a trimeric ring structure as well as several crystalline long-chain modifications (170, 399,467). If the glassy $NaPO_3$–$NaVO_3$ copolymers are crystallized by tempering them, most of the phosphate is segregated as crystalline trimetaphosphate. Preparations in the $NaPO_3$–Na_2SiO_3, $NaPO_3$–SiO_2, KPO_3–SiO_2, and CaO–P_2O_5–SiO_2 systems appear to have a pattern of largely separated phosphate and silicate structures in crystalline samples, but random copolymers in glasses. The review by Ohashi can be consulted for further information.

In a study of the Na_2O–SO_3–P_2O_5 system, Shaver and Stites found that sulfatophosphates form at 400° or higher in melts with NaO: $(P_2O_5 + SO_3)$ ratios smaller than 1 (367). Thilo and Blumenthal discovered that two hydrolysis reactions with significantly different rates occurred on dissolving sulfatophosphate melts (398). For several reasons the faster reaction was attributed to hydrolysis of $S_2O_7^{2-}$: (1) at the end of the faster reaction the solution had lost its previous ability to react with pyridine or aniline in a reaction characteristic of $S_2O_7^{2-}$ (28); (2) the faster hydrolysis had the same rate and activation energy as $S_2O_7^{2-}$ hydrolysis; and (3) as for the hydrolysis of $S_2O_7^{2-}$, the faster reaction showed no dependence of rate on pH between 4.5 and 9.5. Comparative data for the $S_2O_7^{2-}$ hydrolysis were obtained from studies by Thilo and von Lampe (412,413).

That the slower reaction was due to hydrolysis of P—O—S linkages was concluded on the following bases: (1) the rate and activation energy for the slower reaction were intermediate between those for P—O—P and S—O—S cleavage; (2) solutions of the melts contained ordinary SO_4^{2-} which is easily eluted from an anion-exchange column and, in addition, another source of sulfate which yielded sulfate and polyphosphate together on elution with $3N$ HCl; and (3) the amount of ordinary SO_4^{2-} obtained in the initial elution for a melt of known $SO_3:P_2O_5$ ratio was consistent with the presence of sulfatophosphates in which sulfate tetrahedra occur as endgroups on polyphosphate chains. Further evidence for the presence of sulfate tetrahedra only as end-groups was the observation that only one strongly acidic proton was produced per sulfur atom on hydrolysis of the sulfatophosphate. Two would be expected if sulfate tetrahedra were present as middle groups. A rate constant of 0.0925 hr^{-1} was found for the hydrolysis of P—O—S linkages at pH 4.5 and 60°. For comparison, the rate constants for hydrolysis of $S_2O_7^{2-}$ and $P_2O_7^{4-}$ under the same conditions are 3.5×10^2 and 0.01 hr^{-1}, respectively.

Assuming the existence of an equilibrium between S—O—S (in $S_2O_7^{2-}$) and P—O—P linkages on the one hand and P—O—S linkages on the other [Eq. (57)], Thilo and Blumenthal evaluated an equi-

$$-\overset{|}{\underset{|}{P}}-O-\overset{|}{\underset{|}{P}}- \; + \; -\overset{|}{\underset{|}{S}}-O-\overset{|}{\underset{|}{S}}- \; = \; 2(-\overset{|}{\underset{|}{P}}-O-\overset{|}{\underset{|}{S}}-) \qquad (57)$$

librium constant [Eq. (58)] for the reaction. The value of [SOS] was

$$K = \frac{[POS]^2}{[POP][SOS]} \qquad (58)$$

taken as one-half the molar amount of SO_4^{2-} derived from the rapid hydrolysis, [POS] as the molar amount of SO_4^{2-} from the slow hydrolysis, and [POP] as $P(\bar{n} - 1)/\bar{n}$, where P is the moles of phosphate tetrahedra present and $\bar{n}$ the average polyphosphate chain length. The concept of such an equilibrium was supported by the finding that for six melts with S:P atom ratios from 0.9 to 69 the equilibrium constant value calculated from (58) remained within the range 0.60 to 1.18 at 700°.

Since the equilibrium constant value is about 1.0, the standard free energy change for reaction (57) at 700° is about zero. Since $\Delta S°$ must be positive for this reaction, $\Delta H°$ must also be positive. The fraction of sulfur in the sulfatophosphate form must increase with increasing temperature of preparation. Consistent with this, Thilo and Blumenthal found values of 0.92, 1.15, and 2.59 for the ratio of sulfur in sulfatophosphate to sulfur in disulfate at 400, 600, and 800°, respectively, in melts of $Na_2O:P_2O_5:SO_3$ mole ratio about 2.0:1.0:1.9.

Pure samples of the simplest inorganic sulfatophosphate known thus far, disulfatomonophosphate (25), have been prepared as the sodium and

$$
\begin{array}{c}
\text{O O O} \\
\text{-OSOPOSO-} \\
\text{O O\_O}
\end{array}
$$
(25)

potassium salts by von Lampe (445). Data on the kinetics of hydrolysis of this anion, as well as a disulfatooligophosphate appear in Table XL.

TABLE XL

Hydrolysis of Disulfatophosphates First-Order Rate Constants, hr^{-1} [a]

Anion	pH	Temperature (°C)			E_a (kcal/mole)
		20	60	100	
Disulfato-mono-phosphate	3.8	6.25×10^{-3}	1.05	7×10^1	25
	12	$\sim 9 \times 10^{-5}$	6.36×10^{-3}	1.7×10^{-1}	20.4
Disulfato-oligo-phosphate[b]	3.8	1.79×10^{-2}	1.05	26	19.5
	12	$\sim 2 \times 10^{-4}$	0.11	2.64	15

[a] Ref. 445. The separate activation energy values do not appear in the original paper.
[b] Average length of the polyphosphate part of the chain is 5.

Benkovic and Hevey found that the P—O—S hydrolysis in phenyl phosphosulfate, $ROPO_3SO_3^{2-}$, is primarily by S—O cleavage (40). Hydrolysis in ^{18}O-enriched water showed only 7% and 9% P—O cleavage, respectively at pH 1–2 and pH 3.71. In this pH range, the S—O cleavage is of the order of twelve times as fast as P—O cleavage. An acid-catalyzed unimolecular mechanism, involving elimination of SO_3, was assigned on the basis of the predominant S—O cleavage, a small negative $\Delta S^{\ddagger}$ (−9.5 eu), sulfating properties of the monoanion in mixed alcohol–water solvents, and a deuterium oxide solvent effect typical of a A-1 mechanism. At a pH of about 3, Al^{3+}, Mg^{2+}, and Ca^{2+} catalyzed the hydrolysis, and Mn^{2+}, Co^{2+}, Cu^{2+}, and Zn^{2+} inhibited it. The influence of these cations on the point of cleavage was not determined. The hydrolysis rate constant for phenyl phosphosulfate was independent of pH from pH 10 to 6, and then rose steadily as the pH was reduced to 0.

C. Phosphate Esters

The hydrolysis of orthophosphate esters and of mixed anhydrides, such as acetyl phosphate, has been studied extensively. For simplicity in this discussion, we shall use the general term ester to include the anhydrides. Considering the existence of mono-, di- and triesters of H_3PO_4, the levels of protonation possible for each, the choice of alkyl, aryl, or acyl organic groups, and the possibilities for steric, strain and electronic effects in the organic groups, hydrolysis of P—O—C linkages easily could present a hopelessly confusing picture. Instead, there are well-defined kinetic and mechanistic patterns.

TABLE XLI

% P—O Cleavage in Phosphate Triester Hydrolyses

| | Reactive species | | |
Ester	R_3PO_4	$R_3PO_4{}^+OH^-$	Ref.
acetyl dibenzyl phosphate		~5	41
2-(2,4-dinitrophenoxy)-2-oxo-1,3,2-dioxaphosphorinan	100	100	233
triphenyl phosphate	100[a]		23
triphenyl phosphate		~100[b]	25
Ac(Me)CHOPO(OR)$_2$		97	54
Ac(Me)$_2$COPO(OR)$_2$		77	54
PhCO(Ph)CHOPO(OR)$_2$		85	54
trimethyl phosphate	30?	~100	45
triethyl phosphate	~0		35

[a] In 75% dioxane–25% water. The hydrolysis of the conjugate acid must also be by P—O cleavage.

[b] In 60% dioxane–40% water.

For the purposes of this chapter, an encyclopediac compilation of phosphate ester hydrolysis studies is inappropriate. We are concerned here with cleavage of the P—O bond of the ester linkage at a PO_4 tetrahedron. Mechanistic ideas concerning P—O cleavage can be developed more fully in ester hydrolysis than in polyphosphate hydrolysis because of the additional study methods available. Additional reviews, considering rupture of the C—O bond as well as the P—O, are available elsewhere (60,65,66,111,193,207,235,308,455).

1. POINT OF CLEAVAGE

Studies of the point of cleavage of phosphate ester linkages by ^{18}O-tracer methods are summarized in Tables XLI–XLIV. The esters are

TABLE XLII
% P—O Cleavage in Phosphate Diester Hydrolyses

	Reactive species				
Ester	$R_2PO_4H_2^+$	R_2PO_4H	$R_2PO_4^-$	$R_2PO_4^-$ $+ OH^-$	Ref.
bis(2,4-dinitrophenyl)-phosphate	~100	~100	~100	~60?	69
methyl 2,4-dinitrophenyl-phosphate		~95	~95	~50	249
diphenyl phosphate		100			23
dimethyl phosphate	~11	~22		~16	79
dimethyl phosphate		31	11		172
dimethyl phosphate			35[a]		172
cyclohexyl 3-hydroxy-2-butyl phosphate			<10		57
acetyl phenyl			0[b]		131

[a] Catalyzed by Ba^{2+}.

[b] Determined by products of hydrolysis in 90% methanol.

arranged generally in each table in order of increasing pK_a of the organic leaving group that would be formed by P—O rupture. In general, this is seen to list the compounds in order of decreasing tendency for such cleavage. Table XLV summarizes present views on the point of cleavage in these hydrolyses. We shall be concerned with the examples exhibiting predominantly P—O cleavage.

2. TRIESTERS

For the trialkyl phosphates, reaction with water involves C—O cleavage, whereas attack by OH^- occurs at the phosphorus atom. This pattern was deduced earlier on a kinetic basis (195), and it is consistent with Pearson's "hard" and "soft" reagent principle (334); harder nucleophiles, such as OH^-, are expected to favor reaction at the hard phosphorus (V) site. For the triaryl phosphates, reaction occurs at phosphorus with H_2O as well as with OH^-. Triaryl phosphates have the advantage (for P—O cleavage) that phenoxide is a more stable anion than alkoxide, as indicated by a larger pK_a for ionization of the protonated species.

The hydrolysis of phosphate triesters in definitely basic solution is second order, the rate dependent on ester and hydroxide concentrations (78). These reactions have been assigned S_N2 mechanisms for

TABLE XLIII

% P—O Cleavage in Phosphate Monoester Hydrolyses

Ester	Reactive species					Ref.
	$RPO_4H_3^+$	RPO_4H_2	RPO_4H^-	RPO_4^{2-}	RPO_4^{2-} + OH^-	
salicyl phosphate				100		39
succinyl phosphate			~0	~60		449
3,3-dimethylbutyryl	~0					336
acetyl phosphate	<20?	~20?	>89		0	330
acetyl phosphate				100	0	41
2,4-dinitrophenyl phosphate				100	~100	74
2,4-dinitrophenyl phosphate				100		245
p-nitrophenyl phosphate	100	100	100			23
p-nitrophenyl phosphate				≥96		242
phenyl phosphate		100	100			23
p-tolyl phosphate			100			23
α-D-glucoside-4,6-phosphate					~100	187
α-D-glucose-1-phosphate		~0	~100			77, 104
phenyl-β-D-glucoside-4,6-phosphate					~100	187
glucose-6-phosphate						68
1-methoxypropyl-2-phosphate			100	100		82
butanediol-1,3-phosphate					~100	187
methyl phosphate	27	~0	~100			76
methyl phosphate	27					253
methyl phosphate	0					75
methyl phosphate		63				172
ethyl phosphate	48	100				253
isopropyl phosphate	1		100			253
t-butyl phosphate	0	0	75			260

TABLE XLIV

% P—O Cleavage in Hydrolysis of Cyclic Phosphate Esters

Ester	Reactive species		Ref.
	Neutral	Monoanion + OH^-	
ethyl β-hydroxy-*trans*-cinnamic acid ester of H_3PO_4[a]	0		284
ethylene phosphate[b]	100	100	172

[a] Six-membered ring. Data for ring-opening reaction.

[b] Five-membered ring.

TABLE XLV

Generalizations Regarding Point of Cleavage in Phosphate Ester Hydrolysis

Ester type	Species involved[a]	Possible leaving group		
		Acyl	Aryl	Alkyl
R_3PO_4	$+1$	N.E.[b]	P—O	N.E.
	0	N.E.	P—O	C—O
	$0 + OH^-$	C—O	P—O	P—O
R_2HPO_4	$+1$	N.E.	P—O	Predominantly C—O
	0	N.E.	P—O	Predominantly C—O
	-1	C—O	P—O[c]	Predominantly C—O
	$-1 + OH^-$	N.E.	Mixed	Predominantly C—O
RH_2PO_4	$+1$	C—O	P—O	C—O
	0	Predominantly C—O	P—O	Mixed
	-1	P—O	P—O	P—O
	-2	P—O	P—O	P—O
	$-2 + OH^-$	C—O	P—O	N.E.

[a] The symbols $+1$, 0, -1, and -2 refer to the conjugate acid, the neutral molecule, the monoanion, and the dianion, respectively. One of these symbols "$+ OH^-$" refers to the bimolecular reaction with hydroxide ion.

[b] N.E. means not established.

[c] It is questionable whether this reaction has, in fact, been observed (248).

both the aryl and the alkyl esters (25,45,70,232,430). However the activation energies for these reactions are generally somewhat low and the activation entropies too negative for typical S_N2 reactions (Table XLVI). For this reason, an addition–elimination mechanism has been considered (25). For this mechanism exchange of the phosphoryl oxygen with the solvent is to be expected if equilibria (a) and (b) of Eq. (59) are fast compared to the elimination reaction (d),

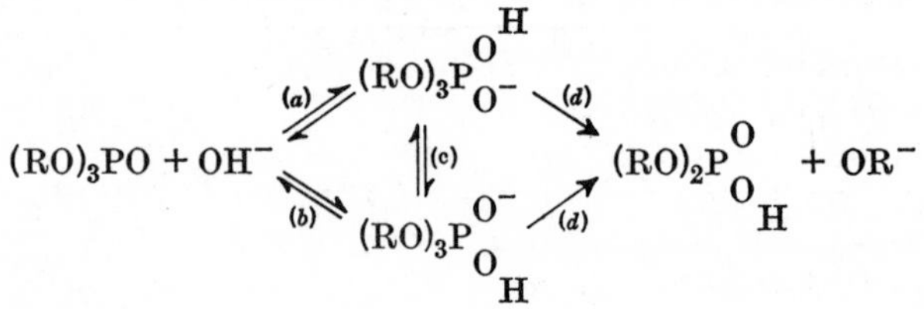

$$(RO)_3PO + OH^- \rightleftharpoons \cdots \rightarrow (RO)_2P\overset{O}{\underset{O}{}}H + OR^- \qquad (59)$$

and if the proton transfer step (c) is fastest of all. Since these are

TABLE XLVI

Activation Parameters for Triester–OH$^-$ Reaction[a]

Ester	E_a (kcal/mole)	log A [log (liter)(mole)$^{-1}$ (hr)$^{-1}$]	$\Delta S^{\ddagger}$ (eu)	Ref.
In water				
dimethyl				
p-nitrophenyl				
phosphate	12.20 $\pm$ 0.20	11.11	-26.0 ± 0.7	158
diethyl				
p-nitrophenyl				
phosphate	13.52 $\pm$ 0.21	11.51	-24.1 ± 0.7	158
diisopropyl				
p-nitrophenyl				
phosphate	14.51 $\pm$ 0.21	11.20	-25.5 ± 0.6	158
di-n-propyl				
p-nitrophenyl				
phosphate	13.03 $\pm$ 0.36	10.99	-26.5 ± 1.2	158
di-t-butyl				
p-nitrophenyl				
phosphate	13.25 $\pm$ 0.19	11.08	-26.1 ± 0.9	158
di-sec-butyl				
p-nitrophenyl				
phosphate	14.34 $\pm$ 0.30	10.76	-27.6 ± 1.0	158
di-n-butyl				
p-nitrophenyl				
phosphate	13.19 $\pm$ 0.38	11.11	-26.0 ± 1.2	158
trimethyl				
phosphate	16.2	11.7	-23	25
triethyl				
phosphate	15.0	9.72	-32.4	231
triethyl				
phosphate	15.0	11.50	-24.2	38
In mixed solvents				
tris(p-nitrophenyl)				
phosphate[b]	4.1	8.12	-39.7	232
methyl				
bis(p-nitrophenyl)				
phosphate[b]	13.4	13.03	-17.2	232
ethyl				
bis(p-nitrophenyl)				
phosphate[b]	14.3	13.74	-13.9	232

[a] See footnote (a), Table XIII. A 1 molar solution at 25° is the standard state for $\Delta S^{\ddagger}$ in these second order reactions.

[b] In 50% acetone–50% water.

TABLE XLVI (cont.)

Ester	E_a (kcal/mole)	log A [log (liter)(mole)$^{-1}$ (hr)$^{-1}$]	$\Delta S^{\ddagger}$ (eu)	Ref.
diphenyl p-nitrophenyl phosphate[c]	10.5	9.9	−28	70
diethyl p-nitrophenyl phosphate[d]	12.4	10.42	−29.2	70
triphenyl phosphate[e]	10.2	8.5	−38	25
diethyl phenyl phosphate[d]	15.0	10.54	−28.1	231
diethyl [m-$(CH_3)_3N^+$ phenyl] phosphate[d]	15.8	13.68	−14.2	231
triethyl phosphate[f]	16.1	17.13	+1.6	231
triethyl phosphate[g]	14.9	9.4	−34	390

[c] In 5% dioxane–95% water.
[d] In 50% ethanol–50% water.
[e] In 60% dioxane–40% water.
[f] In 75% ethanol–25% water.
[g] In 50% dioxane–50% water.

reasonable assumptions, the lack of any detectable exchange argues against the addition–elimination route (25). Also in favor of the concerted S_N2 (P) mechanism is the demonstration of inversion of configuration for some nucleophilic displacements at phosphorus (194).

Evidence for pseudorotation in pentacovalent phosphorus species (42) calls for reexamination of the addition–elimination mechanism. For pentacovalent phosphorus, pseudorotation is the process by which the trigonal bipyramidal structure by deformation of bond angles and bond lengths appears to have been rotated 90° about one of the phosphorus-substituent bonds. In Eq. (60) phosphorus is located at the

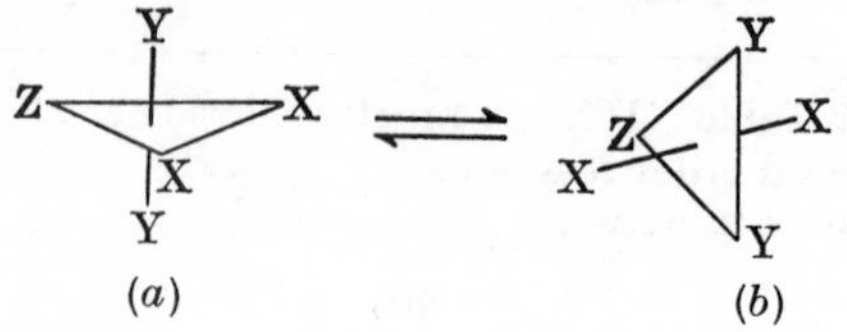

$$(60)$$

$$(a) \qquad (b)$$

intersection of the axis with the equatorial plane of the trigonal bipyramid. Form (a) can be transformed into (b) by warping the Y substituents to the right and the X substituents to the left in the figure. The apparent rotation of the trigonal bipyramid 90° about the P–Z axis has been accomplished by deformation instead of by rotation. Atom Z can be referred to as the pivot atom. Of course, any substituent atom in the equatorial plane could serve as a pivot. The transition state for the pseudorotation can be seen to be a tetragonal pyramid with the pivot atom at the apex.

The concept of pseudorotation in pentacovalent phosphorus compounds has recently been reviewed by Westheimer (455), with the advancing of several mechanistically important principles. The addition of a nucleophile to tetrahedral phosphorus to produce a pentacovalent addition product occurs at an axial (apical) position, as does the elimination of a group from the trigonal bipyramidal intermediate (123,177). If the substituent atoms differ appreciably in polarity the more polar substituents occupy the apical positions. Substituents forming five-membered rings with phosphorus bridge between an apical and an equatorial position to take advantage (for strain relief) of the 90° phosphorus bond angle for these positions.

The availability of d orbitals for phosphorus makes it difficult to carry over mechanistic generalizations from carbon chemistry. Evidence for this is available in a comparison of the inertness of bridgehead carbon atoms to substitution (27) with the substitution reactivity of bridgehead silicon atoms (375,376). Like phosphorus, silicon has valence d orbitals.

On the basis of electrostatics and the rule requiring that entering and leaving groups occupy axial positions, nucleophilic attack by hydroxide ion on the phosphorus atom of a triester would be expected to produce the intermediate species (a) of Eq. (61). Through pseudorotation the

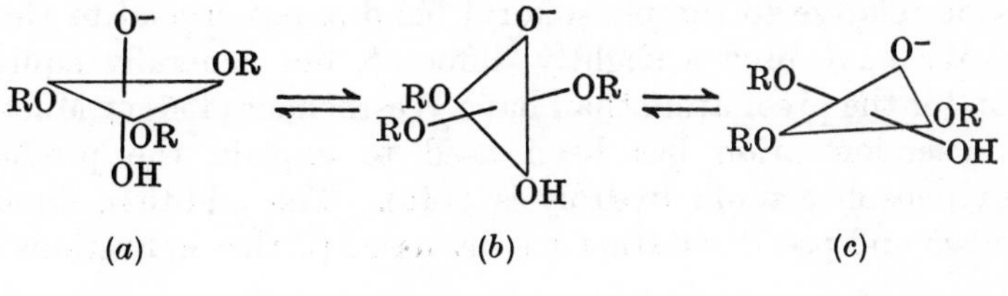

$$\text{(a)} \qquad\qquad \text{(b)} \qquad\qquad \text{(c)} \tag{61}$$

initially equatorial ester substituents will move into the axial positions necessary for elimination (form b). With further pseudorotation, the establishment of an equilibrium among the several forms is to be expected.

Shortness of the bond to a given substituent will favor its holding an equatorial position and acting as pivot atom (455). A useful analogy can be drawn with ClF_3, which has a T-shaped structure, bond angle 87.5°, in which F occupies one equatorial position and the two axial positions of a trigonal bipyramid, with lone pairs presumably occupying the remaining two equatorial positions. Considering the 120° angles between bonds to equatorial positions and the 90° bond angles for axial–equatorial combinations, the equatorial positions are electrostatically advantageous for substituent atoms that concentrate electron density close to the phosphorus atom. SF_4 has, in effect, a trigonal bipyramidal structure with the single lone pair occupying an equatorial position.

Corbridge reports average bond lengths in a number of structures of 1.48, 1.56, and 1.59 Å for the phosphoryl, hydroxy, and ester P—O bonds, respectively (109). For seven compounds containing both POH and POR linkages, the PO bond lengths in the former are, on the average, 0.04 Å shorter. Of necessity, these data are for tetrahedral phosphorus species. The shortness of the phosphoryl bond would favor the phosphoryl oxygen maintaining an equatorial position and acting as pivot atom. An equatorial position would be favored, but to a lesser degree, for the hydroxy group, as well. Extended Hückel MO calculations for possible trigonal bipyramidal structures for the methyl ethylene phosphate–water addition product place the phosphoryl oxygen atom at an equatorial position in all the lower energy forms and predict the smallest energy barrier for pseudorotation if the phosphoryl oxygen atom serves as pivot (47). On these bases, we can conclude that in Eq. (61) form (b) will be most abundant and form (a) least. This pattern of abundance favors ester hydrolysis and, at the same time, inhibits oxygen exchange. In addition, phenoxide is a better leaving group than hydroxide. The data, then, are consistent with an addition–elimination mechanism. The direction of attack by hydroxide relative to the phosphoryl bond is not crucial to the argument. We have used a slightly different, but generally equivalent, criterion for the pivot atom than have Westheimer (455) or Muetterties (307). Pseudorotation has been used to explain the products of dimethylphosphoacetoin hydrolysis (146). The addition–elimination mechanism and pseudorotation are discussed further in Sections II and IV-C-5.

The importance of the leaving tendency of the eliminated ion as an influence on rate is brought out for aryl substituents in Table XLVII, taken from a study by van Hooidonk and Ginjaar. The order of rates

TABLE XLVII
Substituent Effect on Triester-Hydroxide Reaction Rates[a]

X	k^b [(liter)/(mole)(hr)]	$pK_a{}^c$	$\nu_{P-O-C(Ar)}$ (cm^{-1})
o-NO$_2$	79.8 ± 1.8^d	7.21 ± 0.02^d	1242
p-NO$_2$	39.4 ± 2.1	6.99 ± 0.02	1234
m-NO$_2$	23.0 ± 0.4	8.39 ± 0.01	1225
p-CN	22.9 ± 0.3	7.85 ± 0.02	1230
p-COCH$_3$	11.6 ± 0.1	8.01 ± 0.02	1213
o-Cl	9.37 ± 0.05	8.46 ± 0.02	1230
o-Br	8.59 ± 0.12	8.33 ± 0.02	1232
o-I	6.30 ± 0.01	8.44 ± 0.02	1227
m-Br	5.95 ± 0.01	9.06 ± 0.01	1210
m-Cl	5.62 ± 0.05	9.13 ± 0.01	1214
p-I	4.64 ± 0.01	9.21 ± 0.01	1216
p-Br	4.54 ± 0.01	9.27 ± 0.01	1219
p-Cl	4.15 ± 0.04	9.35 ± 0.01	1218
p-COO$^-$	3.31 ± 0.04	9.32 ± 0.02	
p-SCH$_3$	2.67 ± 0.01	9.53 ± 0.01	1216
m-OCH$_3$	2.24 ± 0.01	9.65 ± 0.01	
H	1.67 ± 0.02	9.89 ± 0.02	1212
o-OCH$_3$	1.32 ± 0.01	9.90 ± 0.01	1211
m-CH$_3$	1.27 ± 0.02	10.09 ± 0.01	1209
p-isoC$_3$H$_7$	1.21 ± 0.02	10.04 ± 0.02	1219
p-C$_2$H$_5$	1.17 ± 0.01	10.18 ± 0.01	1212
p-OCH$_3$	1.01 ± 0.03	10.12 ± 0.01	1208
p-N(CH$_3$)$_2$	1.00 ± 0.04	10.08 ± 0.02	1221
p-NH$_2$	0.876 ± 0.04	10.44 ± 0.01	1194
o-C$_2$H$_5$	0.690 ± 0.02	10.27 ± 0.01	1226
o-isoC$_3$H$_7$	0.618 ± 0.01	10.31 ± 0.01	1227
o-N(CH$_3$)$_2$	0.534 ± 0.01	10.62 ± 0.03	1221

[a] Ref. 430. Data for diethyl-substituted phenyl phosphates of the general formula $(C_2H_5O)_2P(O)OC_6H_4X$.

[b] Second-order rate constants for the alkaline hydrolysis at 25° ± 0.05°.

[c] pK_a values determined for 0.1M KCl solution at 25 ± 0.05°.

[d] Standard deviation.

is close to the order of pK_a values for the parent phenols of the eliminated phenoxides. There is some relation between the reaction rate and the stretching frequency of the P—O bond in the phosphorus–phenoxide linkage. The Hammett relationship gave fairly satisfactory correlations with the rates, with a slope of −1.1 for the log k_{rate} versus pK_a plot (430). In a study of fewer examples but involving elimination of phenoxides of more acidic phenols, Khan and Kirby obtained a slope of about -0.4 for the same relationship (233). This low value for the slope, which appeared to be a limiting value of the slope for a series of nucleophiles of increasing effectiveness (water, acetate, HPO_4^{2-}, CO_3^{2-}, OOH^-, $CF_3CH_2O^-$, and OH^-, with respective slopes -0.99, -0.88, -0.65, -0.54, -0.35, -0.34, and -0.4), suggested an addition–elimination mechanism in which, for the better nucleophiles, the rate was primarily determined by the addition step rather than by the leaving tendency of the eliminated phenoxide. The source of the difference in slope in the two studies is not clear. Two possible reasons are that (1) the second study used better leaving groups, for which the kinetics by an addition–elimination mechanism would be more likely to be principally determined by the addition process, and (b) the pK_a values used in the two studies were not perfectly consistent.

In contrast to the other triesters that have been studied, hydroxide ion does not catalyze the hydrolysis of tri-t-butyl phosphate, and the rate is greatly decreased if the polarity of the solvent is lowered (112). The mechanism must be different for this hydrolysis. Ionization to give a t-butyl carbonium ion has been suggested. Such a reaction is common in acid solution for monoesters whose alkyl groups readily ionize as carbonium ions.

The dependence of rate constant on pH is shown in Figure 16 for several triesters whose general formula appears in **26**. The rate constants for the reaction of the neutral triesters with water is generally

(26)

much smaller than those for the reactions with OH^- and with H_3O^+, so much so that for the poorer leaving groups the rate over virtually the entire pH range is controlled by the OH^- and H_3O^+ reactions. Barnard et al. (25) have pointed out that at $100°$ in 60% dioxane–40% water the respective specific second-order rate constants for reaction of triphenyl phosphate with hydroxide ion and with water are 1.5×10^3 and 1×10^{-5} hr^{-1}, respectively, giving a ratio of about 10^8. The corresponding factor for attack on a saturated carbon atom is about 10^4. This is a further reason for the respective preferences of hydroxide ion and water for attack at phosphorus and carbon in triester hydrolysis.

Khan and Kirby found the rate of the reaction of neutral triester with water is more sensitive to the leaving group than is the OH^-

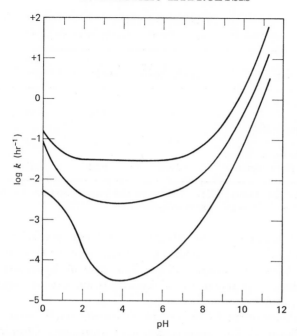

Fig. 16. Phosphate ester hydrolysis (233). First-order rate constants at 39.0° for (a) 2,4-dinitrophenyl phosphate; (b) 4-acetyl-2-nitrophenyl phosphate; and (c) 4-nitrophenyl phosphate.

reaction (233). A linear free energy plot of the log of the first-order rate constant versus the pK_a of the leaving group gave a line described by the equation $\log k = -0.82–0.99\ pK_a$. The slope, -0.99, is considerably greater than the value of -0.4 found by the same authors for the reaction with OH⁻. On the basis of the above value for the slope and its relation to that for diester hydrolysis, the very negative entropy of activation (Table XLVIII), a significant solvent deuterium isotope effect $(k_H/k_D = 2.0)$ and general base catalysis by 2,6-lutidine, a S_N2 (P) mechanism was assigned. The possibility was recognized that the reaction might not be concerted in nature but have a somewhat stable pentacovalent intermediate, as is likely for the OH⁻ reaction already discussed. It was suggested that in addition to the water molecule directly involved, additional water provides general base catalysis. Barnard et al., before the development of the concept of pseudorotation in pentacovalent intermediates, assigned a concerted S_N2 mechanism to the neutral hydrolysis (25).

A study of the kinetics of neutral and acid trimethyl phosphate hydrolysis as a function of solvent composition (water–dimethyl

TABLE XLVIII

Activation Parameters for Triester–H$_2$O Reaction[a]

Ester	E_a (kcal/mole)	log A (log hr^{-1})	$\Delta S^{\ddagger}$ (eu)	Ref.
tris(p-nitrophenyl) phosphate	4.1	3.62	−60	232
ethyl bis(p-nitrophenyl) phosphate	14.3	8.51	−38	232
2-(2,4-dinitrophenoxy)-2-oxo-1,3,2-dioxaphosphorinan	15.0	9.01	−35.6	233

[a] See footnote (a), Table XIII. Values of the parameters are for first-order rate constants for hydrolysis of aryl substituent.

sulfoxide and water–ethylene glycol), of solvent deuterium isotope effects, and of salt effects supports C—O cleavage for trimethyl phosphate under these conditions (279). The rate of hydrolysis of trimethylphosphate is independent of hydrogen ion activity from pH 7 to 3M HClO$_4$ (24).

The dependence of the hydrolysis rate constant on H$^+$ concentration is given in Figure 17 for concentrated acid solutions. Closely similar results have been reported for the triphenyl phosphate hydrolysis,

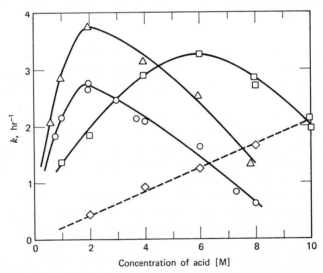

Fig. 17. Hydrolysis of p-nitrophenyl diphenyl phosphate at 100° in concentrated acids (70). Symbols are △ for H$_2$SO$_4$, ○ for HClO$_4$, □ for HCl, and ◇ for HCl + LiCl at ionic strength 10.0.

which shows a maximum rate at about $1.5M$ $HClO_4$ in 60% dioxane (23). The triesters involved are not fully protonated to the conjugate acid at the rate maximum. For triphenyl phosphate, for example, the pK_a for the conjugate acid is -2.77 (23).

Using an expression for the rate constant based on a Bronsted-Bjerrum treatment of an A2 reaction with the assumptions that only one water molecule is forming a new covalent bond in the transition state, that the log of the collected activity coefficient terms varies linearly with electrolyte concentration, and that each proton is bonded to four water molecules, Bunton and co-workers obtained a good fit to the solid curves of Figure 17, except at the highest acid concentrations, where the crucial assumption that each proton removes four water molecules from availability breaks down (70). Although the treatment was highly simplified, it demonstrated one way in which rate maxima might arise in concentrated acid hydrolysis. In effect, the acid "salts-in" the reaction substrate but "salts-out" the transition state because of its strong requirement for hydration arising from protonation of the substrate. The rate maximum at $1.5M$ $HClO_4$ in the acid-catalyzed triphenyl phosphate hydrolysis was earlier explained by the same treatment by Barnard et al. (23). In the case of aryl esters with good electron-withdrawing groups, the rate in strong acids might be limited by the rate of pseudorotation (250).

Lithium chloride and hydrogen chloride have similar effects on water activity. If they also have similar (or compensatory) effects on the activity coefficients of the other species involved, the water activity and the collected activity coefficient terms of the Bronsted-Bjerrum rate expression can be kept constant by maintaining the ionic strength constant with LiCl while the acidity is adjusted with HCl. When this was done, the dotted curve of Figure 17 was obtained, providing further evidence for activity effects as the source of the rate maximum in the earlier curves.

Linear relationships were found in plots of $\log k + H_0$ versus $\log c_{H^+} + H_0$ for the several acids (70). For $HClO_4$, H_2SO_4, and HCl the slopes were 1.25, 1.48, and 1.09, respectively. Bunnett and Olsen consider slopes in this range to be indicative of the involvement of water as a proton transfer agent in the rate-limiting reaction step (64). Satisfactory fits could not be obtained to Bunnett's earlier equations (62,63).

The alkaline hydrolysis of p-nitrophenyl diphenyl phosphate is strongly catalyzed by cationic micelles of cetyltrimethylammonium bromide (80). The rate rises abruptly as the concentration of detergent goes above the critical micelle concentration. The improved rate results from a much more favorable entropy of activation, about -8.5 eu, which overcomes a less favorable activation energy, about 14.0 kcal/mole, relative to reaction in the absence of the micelles (cf. Table XLVI). The ester is solubilized by the micelles, implying the ester must then enter the reaction from a lower enthalpy state. The more favorable $\Delta S^{\ddagger}$ might be the result of a lower degree of hydration of the transition state in association with the micelles. The uncharged detergent, Igepal, strongly inhibits the reaction, even at submicellar concentrations. Sodium lauryl sulfate also inhibits the reaction strongly, but only above the critical micelle concentration. The inhibition is related to the extent to which the ester is incorporated into the micelles or is associated with the detergent species. Decyltrimethylammonium bromide has very little catalytic effect (81); apparently micelles of this detergent do not incorporate the ester effectively enough.

A comparison of the catalytic effects of two cationic detergents (27; I:R $= $ H, II:R $= $ CH$_3$, R$' = $ n-C$_{10}$H$_{21}$ and C$_{12}$H$_{25}$) showed the

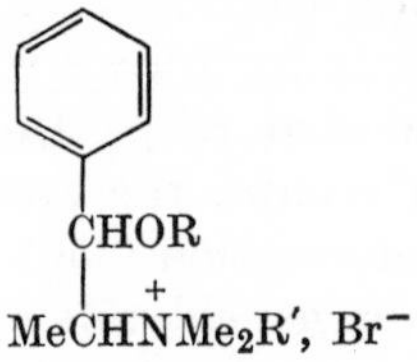

$$
\begin{array}{c}
\text{C}_6\text{H}_5 \\
| \\
\overset{|}{\text{C}}\text{HOR} \\
| \quad \quad + \\
\text{MeCHNMe}_2\text{R}', \ \text{Br}^-
\end{array}
$$

(27)

hydroxyl group in the polar head of I could serve as a nucleophilic reagent, particularly when ionized at high pH (81). For a given chain length of R$'$, I is a much more effective catalyst than is II. Racemic I is more effective than $(-)$-I. At 25° the factor by which the maximum rate is increased for R$' = $ C$_{12}$H$_{25}$ is 26, 22, and 6, respectively, for $(\pm)$-I, $(-)$-I, and $(-)$-II.

3. DIESTERS

Dialkyl monophosphate esters under all conditions studied hydrolyze predominantly at the C—O bond (Table XLV). The hydrolysis of aryl groups of diesters occurs essentially completely at the P—O linkage, except perhaps for the reaction with hydroxide ion. Because the remaining proton in a diester of phosphoric acid has a pK_a of about 1, over most of the pH range diesters exist in solution primarily as the anion. As a group the diesters are the least reactive of the phosphate esters. They are relatively reactive only for aryl esters carrying strongly electron-withdrawing substituents, which makes the phenoxide an excellent leaving group.

There have been very few studies of the kinetics of hydrolysis of simple dialkyl esters. Rate constants for individual reactions are listed in Table XLIX. These constants describe the over-all observed kinetics when inserted into an equation of the form of (62). The acid catalysis represented by the first term of Eq. (62) continues to $5M$ $HClO_4$, at least, for dimethyl phosphate (79). The rate of

$$\text{Rate} = k_1[H^+][R_2HPO_4] + k_2[R_2HPO_4]$$
$$+ k_3[R_2PO_4^-] + k_4[R_2PO_4^-][OH^-] \quad (62)$$

dimethyl phosphate hydrolysis in basic solution is low (about 0.02

TABLE XLIX

Rate Constants for R_2HPO_4 Hydrolysis

Reacting species	R=CH$_3$[a]		R=C$_6$H$_5$CH$_2$[b] combined
	P—O	C—O	
$R_2HPO_4 + H^+$ [c]	4×10^{-4}	3.3×10^{-3}	1.63
R_2HPO_4[d]	4×10^{-3}	0.12	0.11
$R_2PO_4^-$[d]			1.5×10^{-4}

[a] At 100°. K_a for dimethyl hydrogen phosphate is 0.025. Ref. 79.
[b] At 75.6° and unit ionic strength. Predominantly C—O cleavage. K_a is 0.09 for dibenzyl hydrogen phosphate. Ref. 255.
[c] Second-order rate constants, liters/(mole)(hr).
[d] First-order rate constants, hr^{-1}.

liters/(mole)(hr) in $1M$ OH$^-$ at 125°), but there might be slight catalysis with increasing base strength (254).

There is enhanced alkaline reactivity for phosphate diesters containing a hydroxyl group close to the phosphate (56). Two mechanisms compete in the alkaline hydrolysis of 2-hydroxyalkyl esters [Eq. (63)]: (a) cleavage of a P—O bond to produce alkoxide coupled with the formation of a cyclic phosphate intermediate and (b cleavage of a C—O bond to produce a glycol via an epoxide intermediate (57). The relative importance of these routes is strongly influenced by the leaving ability of RO$^-$. As could be expected, changing R makes little difference in the rate of the epoxide route, but illustrates the great importance of a good leaving group for route (a). The data of Table L gave a very satisfactory linear free energy relationship of slope -0.56 between the log of the rate constant for route (a) and the

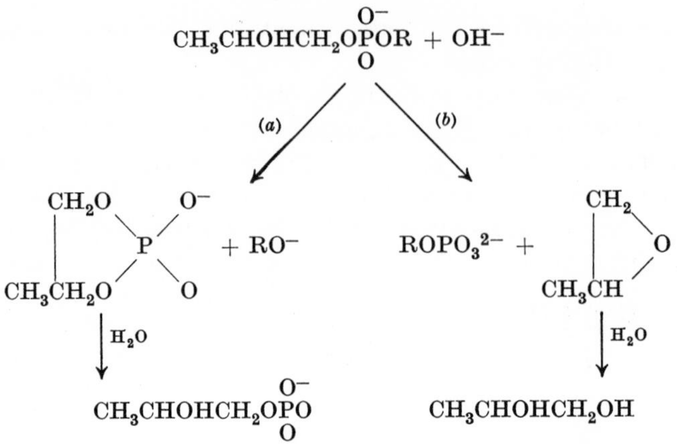

$$\begin{matrix} & O^- & \\ & | & \\ CH_3CHOHCH_2OPOR & + & OH^- \\ & || & \\ & O & \end{matrix}$$

$$\overset{CH_2O}{\underset{CH_3CH_2O}{\big|}}\overset{O^-}{\underset{O}{P}} + RO^- \qquad ROPO_3^{2-} + \overset{CH_2}{\underset{CH_3CH}{\big|}}O \qquad (63)$$

with H_2O routes leading to:

$$\begin{matrix} & O^- & \\ & | & \\ CH_3CHOHCH_2OPO & & CH_3CHOHCH_2OH \\ & || & \\ & O & \end{matrix}$$

pK_a of the alcohol or phenol form of the leaving group (58).   The rate differences for route (a) with change in R are primarily due to activation energy differences.

The hydrolysis of aryl diesters has been studied more fully than that of dialkyl esters, perhaps because of a more convenient (greater) rate in the former case.   Except in the reaction with hydroxide ion, for which the point of cleavage is mixed, aryl groups are removed from diesters by P—O scission.   The dependence of the hydrolysis rate constant on pH is presented in Figure 18 for several diaryl esters.   The

TABLE L

Dependence of Alkaline Hydrolysis Rate on Leaving Group[a]

	Rate constant[c] for route		% by route	
R[b]	(a)	(b)	(a)	(b)
p-nitrophenyl	9.4×10^3		100	
p-chlorophenyl	9.0×10^3		100	
phenyl	2.66×10^2		100	
2-methoxyethyl	0.45	0.28	94	6
2-hydroxyethyl	0.85	0.49	63	37
methyl	0.46	0.35	93	7
ethyl	0.47	0.24	70	30
isopropyl	2.4×10^{-3}	0.11	19	81
cyclohexyl	2.0×10^{-3}	0.11	16	84

[a] Ref. 58.   Conditions were 80° and 0.05M NaOH.
[b] R appears in Eq. (63).
Second-order rate constant, liters/(mole)(hr).

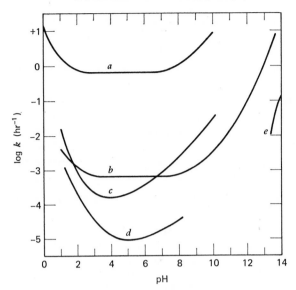

Fig. 18. Diaryl ester hydrolysis. First-order rate constants for (a) bis(2,4-dinitrophenyl) phosphate, 100° (248); (b) bis(2,4-dinitrophenyl) phosphate, 25° (69); (c) bis(4-nitrophenyl) phosphate, 100° (248); (d) diphenyl phosphate, 100° (23); and (e) diphenyl phosphate, 100° (224).

order of rates at 100° (curves a, c, d, and e) is consistent with the leaving abilities of the phenoxides. Kirby and Younas found an excellent linear free energy relationship between the log of the first-order rate constants for hydrolysis of a number of phosphate diester anions and the pK_a values of the corresponding phenols (Table LI). The slope of the log k versus pK_a line was -0.97, showing a strong rate sensitivity to the leaving ability of the phenoxide. The Hammett equation

TABLE LI

Substituent Effects on $Ar_2PO_4^-$ Hydrolysis[a]

Ar	Rate constant (hr^{-1})	pK_a
2,4-dinitrophenyl	0.54	4.07
4-acetyl-2-nitrophenyl	2.57×10^{-2}	5.09
4-chloro-2-nitrophenyl	1.47×10^{-3}	6.36
4-nitrophenyl	2.27×10^{-4}	7.15
2-nitrophenyl	2.02×10^{-4}	7.23
3-nitrophenyl	1.81×10^{-5}	8.35

[a] At 100° and ionic strength 1.0 (KCl). Ref. 248.

derived from the data for the 4-substituted 2-nitrophenyl esters predicted a hydrolysis rate constant for the diphenyl phosphate anion a 100° of 4.4 × 10⁻⁹ hr⁻¹. This is the basis for footnote (c) to Table XLV. Under conditions for which hydrolysis of diphenyl phosphate anion is expected, the observed hydrolysis is probably a composite of the neutral ester hydrolysis and the anion-hydroxide reaction.

Other curves might have been added to Figure 18. The log k versus pH profile for di(p-ethoxy) phosphate is close to that for diphenyl phosphate in shape and position (289). That for bis(3-nitrophenyl) phosphate is slightly below the curve for bis(4-nitrophenyl) phosphate (248). The shape of the curve for dibenzoyl phosphate is close to that for the bis(2,4-dinitrophenyl) ester, including the pronounced

TABLE LII

Rate Constants for Bis(2,4-dinitrophenyl) Phosphate Hydrolysis[a]

Reaction	Rate constant
$(ArO)_2PO(OH) + H^+$	1.7×10^{-2} liters/(mole)(hr)
$(ArO)_2PO(OH)$	$\sim 6 \times 10^{-3}$ hr⁻¹
$(ArO)_2PO_2^-$	6.5×10^{-4} hr⁻¹
$(ArO)_2PO_2^- + OH^-$	$\sim 1 \times 10^1$ liters/(mole)(hr)

[a] At 25.0°, using $pK_a = 1$ for $(ArO)_2PO(OH)$. Ref. 69.

flat portion, which extends from pH 2.2 to 6.3 for the dibenzoyl ester (251).

Over the usual pH range the rate pattern is determined by the relative importance of the last three terms of Eq. (62). For bis(2,4-dinitrophenyl) phosphate these terms are important, respectively, over the pH ranges <2, 3–7, and >8 (cf. Figure 18). Values of the rate constants are listed in Table LII. For esters with poorer phenoxide leaving groups, the anion hydrolysis is very slow, making the $k_3[R_2PO_4^-]$ term of Eq. (62) negligible. For these esters the rate constant–pH profile is determined by the R_2HPO_4–H_2O and $R_2PO_4^-$–OH^- reactions. This explains the absence of flatness at the minima of curves c and d of Figure 18.

Activation parameters for diester hydrolysis appear in Table LIII. Considering these, in particular the $\Delta S^\ddagger$ values, all the reactions appear to be bimolecular except that of dimethyl phosphate with water, and this is a case of predominantly C—O cleavage. The mechanism of the bis(2,4-dinitrophenyl) phosphate anion hydrolysis has been considered

TABLE LIII
Activation Parameters for R_2HPO_4 Hydrolysis[a]

Reactive species	R	E_a (kcal/mole)	$\log A$ ($\log hr^{-1}$)	$\Delta S^{\ddagger}$ (eu)	Ref.
$R_2PO_4H_2{}^+$	2,4-dinitrophenyl	19.1	13	-19	69
R_2PO_4H	p-ethoxyphenyl	21.4	10.79	-27.4	290
R_2PO_4H	p-methoxyphenyl	20.1	8.65	-37.2	290
R_2PO_4H	o-methoxyphenyl	19.3	10.23	-30.0	290
R_2PO_4H	methyl[b]	25.5	16.6	-0.7	79
$R_2PO_4{}^-$	2,4-dinitrophenyl	19.4	11	-26	69
$R_2PO_4{}^-$	2,4-dinitrophenyl	19.6	11.2	-25.5	248
$R_2PO_4{}^-$	acetyl phenyl[b]			-28.8	132
$R_2PO_4{}^-$ + OH^-	2,4-dinitrophenyl	17.2	14	-14	69
$R_2PO_4{}^-$ + OH^-	p-nitrophenyl	18.3	11.8	-23	231
$R_2PO_4{}^-$ + OH^-	phenyl	16	13.6	-15	224
$R_2PO_4{}^-$ + OH^-	methyl[b]	$\sim$28	10.2	-30	254

[a] See footnote (a), Table XIII.
[b] Predominantly C—O fission.

in detail (69,248). A unimolecular mechanism analogous to the mono-metaphosphate mechanism for monoesters [Eq. (64)] is unlikely for

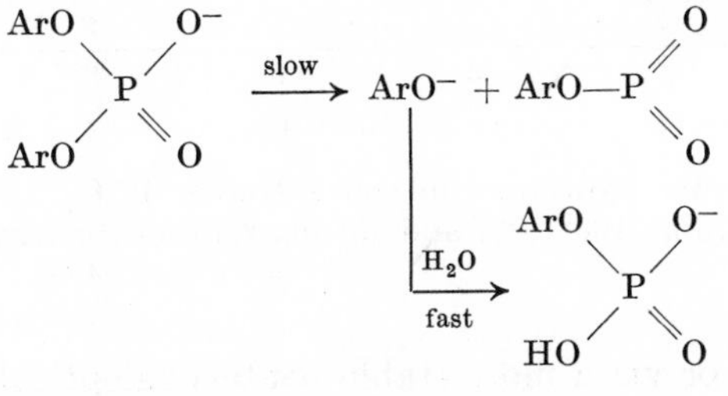

$$(64)$$

several reasons: (*1*) the kinetic parameters are inconsistent with a unimolecular mechanism, (*2*) the small negative charge of the diester anion provides little driving force for an elimination reaction, (*3*) there is only a moderate sensitivity of the rate to the leaving group (slope of $\log k$ versus pK_a relationship is -0.97 for diester anions versus -1.23 for monoester dianions, for which the monometaphosphate mechanism applies), (*4*) the solvent deuterium isotope effect, k_H/k_D, is about 1.5 for the diester anion reaction as opposed to 1.0 for the mono-ester dianion hydrolysis, and (*5*) the addition of organic solvents decreases the rate of diester anion reaction, but not the rate for the monoester mono- or dianion.

A bimolecular mechanism, Eq. (65) is consistent with the $\Delta S^{\ddagger}$ value, the solvent deuterium isotope effect, the moderate leaving group sensitivity, and the decrease in rate through the lowering of water activity

$$\begin{array}{c} ArO \\ \diagdown \\ \diagup P \diagdown \\ ArO \quad O \end{array} + OH_2 \rightleftharpoons ArO{-}\underset{\underset{O_-}{|}}{\overset{\overset{OAr}{|}}{P}}{-}OH_2^+ \longrightarrow ArOH + \begin{array}{c} ArO \quad OH \\ \diagdown \diagup \\ P \\ \diagup \diagdown \\ O \quad O^- \end{array} \quad (65)$$

in mixed solvents. Whether the bimolecular reaction is by a concerted

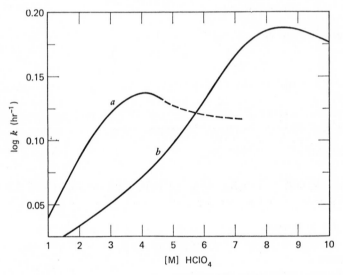

Fig. 19. Diaryl ester hydrolysis in concentrated $HClO_4$. Curves are for (a) diphenyl phosphate at $100°$ (23) and (b) bis(2,4-dinitrophenyl) phosphate, $25°$ (69).

S_N2 mechanism or via a more stable pentacovalent intermediate is an open question. Bimolecular mechanisms are probably valid also for the hydrolyses of the neutral bis(2,4-dinitrophenyl)phosphate and its conjugate acid and for the anion-hydroxide reaction (69). If an S_N1 mechanism is not found with the excellent leaving group 2,4-dinitrophenoxide, it should not be expected for other diesters. Salt effects on the hydrolysis kinetics of bis(2,4-dinitrophenyl) phosphate are small (69).

The dependence of the rate constant on acid concentration in more acidic solutions is shown in Figure 19 for two diesters. Considering the temperature difference, the better leaving group again gives the greater hydrolysis rate. In similar plots for di-o-methoxy, di-p-methoxy-, and di-p-ethoxyphenyl phosphates rate maxima appear at about $4M$ $HClO_4$ (290). Bis(2,4-dinitrophenyl) phosphate shows a rate maximum at about $4M$ sulfuric acid but shows none in hydrochloric acid up to $10M$ (69). Presumably, the explanation for these rate maxima is the same as for the similar maxima observed for triaryl and monoaryl esters, discussed in the preceding and following sections. However the only hydrolysis kinetics study at constant ionic strength for diaryl esters in concentrated acid solutions appears to have been that of Mhala and co-workers with the alkoxyphenyl esters (290). In this study no acid catalysis was observed at constant ionic strength, and the maxima were considered to result from a positive salt effect combined with changing water activity.

Intramolecular catalysis has been observed for a number of diesters with a neighboring hydroxyl group on one of the organic substituents. This catalysis can be highly effective, as very reactive five-membered cyclic phosphate intermediates are possible [cf. Eq. (63) and Section IV-C-5]. Intramolecular catalysis by a carboxyl group can occur through the reactive five-membered ring formation when the carboxyl is properly placed in the structure, or it can occur through intramolecular proton transfer from the carboxyl group to the C—O—P bridge. These modes of catalysis have been reviewed (60). Intramolecular nucleophilic catalysis at phosphorus is also possible, as in the formation of a six-membered ring with a neighboring carboxylate group (234). In this case, the minimum in the log k versus pH profile occurs at the pH of neutral species abundance, rather than of anion abundance.

The anion hydrolysis of diphenyl phosphate is catalyzed by Ba^{2+} (185) and that of acetyl phenyl phosphate by Mg^{2+} and Ca^{2+} (322). In the absence of divalent cation, the latter reaction, which involves loss of acetyl, is principally by C—O cleavage, yet no change in the C=O stretching frequencies was noted on addition of the divalent cation to an acetyl phosphate solution. The point of cleavage for the metal-ion catalyzed reaction was not determined.

The reaction of bis(2,4-dinitrophenyl) phosphate with hydroxide ion is catalyzed up to 30-fold by cationic micelles of cetyltrimethylammonium bromide, unaffected by anionic micelles of sodium lauryl sulfate, and inhibited by uncharged micelles of the nonionic detergent Igepal (61). Catalysis is attributed to the ability of the cationic micelle to bring together the two anionic reagents and to stabilize the transition state relative to the initial state.

4. MONOESTERS

Phosphate monoesters are diprotic acids with pK_a values of about 1 and 6, the exact values depending on the organic function. Over the usual pH range three species are encountered: the neutral molecule, the monoanion, and the dianion. In addition to the direct hydrolyses of each of these, the acid-catalyzed reaction of the neutral molecule (or alternatively, the reaction of its conjugate acid) and the hydroxide-catalyzed reaction of the dianion must be considered at and beyond the limits of the usual pH range. These reactions are entirely or predominantly by P—O scission for all of the monoaryl species, for the monoalkyl monoanion, and for the monoacyl mono- and dianions (Table XLV). A summary of monoester hydrolysis studies appears in Table LIV. The listing is not exhaustive, but an attempt was made to include all studies of any magnitude involving P—O cleavage, plus the major studies involving C—O cleavage.

The influence on rate due to the change in solution species with pH is shown in Figures 20 and 21 for a variety of esters. The pH dependence of rate was studied early and attributed to changing solution species by Bailly (12–15) and by Desjobert (124–128). The figures have been labeled to show the approximate pH range of importance for each species. Because of differences in the activation parameters for the reactions of the separate species, the form of the rate constant-pH profile is not necessarily independent of temperature (Fig. 22). The rate constant dependence on strong acid concentration is presented in Figure 23 for several monoalkyl phosphates and in Figure 24 for several monoaryl esters.

Over the entire range of acidity/basicity conditions, the observed rate of hydrolysis of a phosphate monoester can be described by Eq. (66), each term representing a separate possible reaction. In general,

$$\text{Rate} = k_1[\text{RH}_3\text{PO}_4^+] + k_2[\text{RH}_2\text{PO}_4] + k_3[\text{RHPO}_4^-]$$
$$+ k_4[\text{RPO}_4^{2-}] + k_5[\text{RPO}_4^{2-}][\text{OH}^-] \quad (66)$$

over a moderate range of acidity, only one, two, or three of the terms need be considered. Over most of the usual pH range only the third and fourth terms are important. The rate constant versus pH profiles, such as appear in Figures 20–22, are determined almost entirely by the rate constants k_3 and k_4 of Eq. (66) and by the ionization constant for the monoanion, i.e., K_2 for the ester molecule. For most monoesters k_3 is larger than k_4, resulting in a rate maximum around pH 4, near which the monoanion peaks in abundance (Figures 20a–c and 21c). The pattern is reversed for those aryl and acyl esters for which the

TABLE LIV
Studies of Monoester Hydrolysis

R	Temp. (°C)	pH	Concd. acid, M HCl	Concd. acid, M HClO$_4$	Concd. acid, M H$_2$SO$_4$	Catalysts and inhibitors	Ref
Acyl							
acetyl	0–50	3–9					132
acetyl	25		1–5				336
acetyl	~25	4, 14	0–3			py	330
acetyl	39	0–12				Mg^{2+}	252
acetyl	25	7–8				a	49
acetyl	25–39	6–9				b	320
acetyl	24	1–11				c	283
acetyl	0–40					Li$^+$	256
fluoroacetyl	24	1–11				c	283
propionyl	37	1–13					449
succinyl	37	1–14					449
glutaryl	37	1–13					449
maleyl	37	1–14					449
isobutyryl	25		2–5				336
isovaleryl	25		2–5				336
trimethyl-acetyl	25		1–5				336
3,3-dimethyl-butyryl	25		1–5	1–4			336
carbamyl	26–37	7–9				d	321
choline	100	0–13				e	8,9
choline	100		0–9	0–8	0–7		10
benzoyl	37	2–13					251
p-methoxy-benzoyl	39	1–7					132
p-nitrobenzoyl	39	1–8					132
3,5-dinitro-benzoyl	39	1–7					132
Aryl							
2,4-dinitro-phenyl	25, 73	0–14	0–10	0–8	3–8		71
2,4-dinitro-phenyl	25	0–14					69
2,4-dinitro-phenyl	27–100	2–12	1–3				245
2,4-dinitro-phenyl	39	1–13					244
2,5-dinitro-phenyl	25, 73	1–14		0–8			71
2,6-dinitro-phenyl	73	1–14		0–8			71
p-nitrophenyl	100	0–8		0–7			292
p-nitrophenyl	25, 73	1–13	0–8	0–11	1–7		23
p-nitrophenyl	25	0–5					72

219

TABLE LIV (cont.)

R	Temp. (°C)	pH	HCl	HClO$_4$	H$_2$SO$_4$	Catalysts and inhibitors	Ref.
			Concd. acid, M				
Aryl (cont.)							
p-nitrophenyl	39	7–14					242
p-nitrophenyl	68–82	2–9					192
p-nitrophenyl	37	1–13					129
o-nitrophenyl	25–100	1–14		0–8			72
m-nitrophenyl	25–100	1–14	1–8	0–8	1–4		72
salicyl	30	4–7				f	310
salicyl	30	3–7				g	190
salicyl	27–47	2–10					95
salicyl	25	1–7					39
m-carboxy-phenyl	70, 80	1–10					92
p-carboxy-phenyl	70, 80	1–10					92
1,3-dicarboxy-phenyl-2-	35	2–9				f,h	311
1-methoxy-carbonyl-3-carboxy-phenyl-2-	35	2–8				f,i	311
8-carboxy-α-naphthyl	70, 80	1–10					92
2-carboxy-1-naphthyl	31, 37	2–10					94
3-carboxy-2-naphthyl	31, 37	2–10					94
1-carboxy-2-naphthyl	31, 37	2–10					94
p-acetylphenyl	73–101			0–8			23
p-bromophenyl	80, 98	0–8		0–7			292
p-chlorophenyl	80, 98	0–8		0–7			292
p-chlorophenyl	100	2–8	2–8				72
o-chlorophenyl	73, 100	0–8	1–8				72
p-chloro-m-tolyl	80, 97	0–8		0–7			291
phenyl	100	0–10		0–7			24
phenyl	70, 80	1–10					92
phenyl	80	1–5					313
phenyl	73–100	2–7		0–8			23
α-naphthyl	70, 80	1–10					92
β-naphthyl	70, 80	1–10					92
p-tolyl	73–100	2–6		0–5			23
p-tolyl	80–100	0–8		0–7			291
2,6-dimethyl-phenyl	100		3–7				23
o-t-butyl-phenyl	100	0–8	1–7				72

TABLE LIV (cont.)

R	Temp. (°C)	pH	HCl	HClO$_4$	H$_2$SO$_4$	Catalysts and inhibitors	Ref.
Aryl (cont.)							
p-t-butyl-phenyl	100	3–8					72
2,6-t-butyl-4-methylphenyl	100	0–8	1–7				72
o-methoxyphenyl	90	0–8		0–8			289
p-methoxyl-phenyl	99	0–8		0–8			289
p-ethoxyphenyl	90	0–8		0–8			289
Alkyl							
methyl	100	0–7		1–10		I$^-$	76
methyl	85–111			4			253
methyl	100			1–3		k	75
methyl	101						366
methyl	100	0–10		1–5			24
ethyl	100			0–8			253
aminoethyl	78	9				La(OH)$_3$	82
hydroxyethyl	78	9				La(OH)$_3$	82
isopropyl	100	1–9		1–6			253
1-methoxy-propyl-2-	78	0–8			0–5	La(OH)$_3$	82
t-butyl	44–74	0–7		0–3			260
neopentyl	100	2–6	0–8		4		75
benzyl	77	0–7					255
2-pyridyl-methyl	70, 90	1–9					313
3-pyridyl-methyl	70, 90	1–9					313
4-pyridyl-methyl	90	1–9					313
3-hydroxy-2-pyridylmethyl	60–80	1–10					312
glucose-1-	25–100	1–8		0–3			77
glucose-6-	45–100	0–12	2–6^m	0–6	5–6		68
fructose-6-	40, 50						148

a Li$^+$, Na$^+$, K$^+$, NH$_4^+$, Mg^{2+}, Ca^{2+}, Sr^{2+}, Ba^{2+}, La^{3+}.
b Mg^{2+}, Ca^{2+}, Mn^{2+}, Co^{2+}, Ni^{2+}, Zn^{2+}.
c Mg^{2+}, Ca^{2+}, pyridine.
d Mg^{2+}, Ca^{2+}, Ba^{2+}.
e Be^{2+}, Mg^{2+}, Zn^{2+}.
f Cu^{2+} chelates.
g Fe^{3+}, Co^{2+}, Ni^{2+}, Zn^{2+}, Cd^{2+}, ZrO^{2+}, VO^{2+}, UO^{2+}.
h VO^{2+}, Ni^{2+}, Cu^{2+}, dipyridyl Co(II).
i VO^{2+}, Cu^{2+}.
j Mg^{2+}, Co^{2+}, Ni^{2+}, Cu^{2+}, Zn^{2+}.
k Cl$^-$, Br$^-$, I$^-$.
l Ni^{2+}, Cu^{2+}.
m And 5–6M HBr.

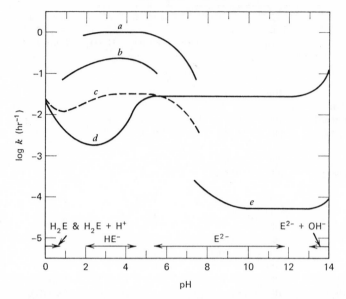

Fig. 20. Monoester hydrolysis. First-order rate constants for (a) phenyl phosphate, $100°$ (23); (b) p-nitrophenyl phosphate, $73°$ (23); (c) methyl phosphate, $100°$, dashed for clarity (76); (d) 2,4-dinitrophenyl phosphate, $25°$ (69); and (e) p-nitrophenyl phosphate, $39°$ (242).

phenoxide or carboxylate anion is particularly stable, as measured by a large ionization constant for the corresponding phenol or acid (Figures 20d and 21a and b). The reason for this will be discussed in connection with the mechanisms for mono- and dianion hydrolysis. Since alkoxide is a poor leaving group, monoalkyl esters characteristically show rate maxima near pH 4.

Except with phosphate esters for which the alkyl groups easily ionize as carbonium ions or oxocarbonium ions, the un-ionized ester molecule is relatively unreactive, less reactive than the monoanion. For monoaryl esters with strongly electron-withdrawing substituents and for alkyl esters, there is pronounced acid catalysis in concentrated acid [the first term of Eq. 66]. In general, there is some acceleration of rate in strongly alkaline solutions. Table LV summarizes the patterns.

Table XLV shows the point of cleavage in the hydrolysis of each phosphate ester species. Our attention will be given to cleavage of the P—O bond as we consider in turn the reactions represented by the terms of Eq. (66).

Activation parameters for the hydrolysis of the conjugate acid species are listed in Table LVI. Scission of the C—O bond is largely

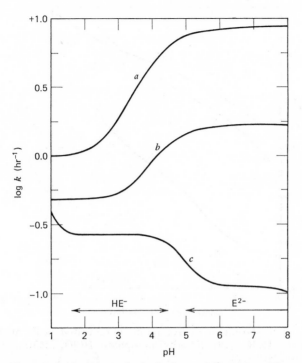

Fig. 21. Monoacyl phosphate hydrolysis at 39° (132). First-order rate constants for (a) 3,5-dinitrobenzoyl phosphate; (b) p-nitrobenzoyl phosphate; and (c) p-methoxybenzoyl phosphate.

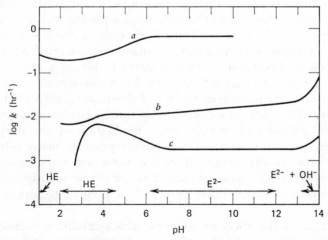

Fig. 22. Effect of temperature (71). First-order rate constants for hydrolysis of 2,5-dinitrophenyl phosphate at (a) 73.0°; (b) 45.0°; and (c) 25.0°.

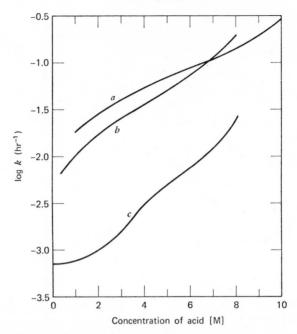

Fig. 23. Monoalkyl phosphate hydrolysis in concentrated acids at 100°, (a) methyl phosphate, $HClO_4$ (76); (b) ethyl phosphate, $HClO_4$ (253); and (c) neopentyl phosphate, HCl (75).

responsible for acyl and alkyl ester hydrolyses, and at least two mechanisms apply to the latter (253). The $\Delta S^{\ddagger}$ values for hydrolysis of the aryl ester conjugate acid species are in the range for a bimolecular mechanism. An A2 mechanism at phosphorus has been tentatively assigned to the acid-catalyzed hydrolysis of the P—O bond of neopentyl phosphate (75), although at present the concerted A2 mechanism cannot be distinguished from a route involving a more or less stable pentacovalent intermediate. An A1 mechanism at carbon has been assigned to the hydrolysis of t-butyl phosphate (260) and glucose-1-phosphate (77). An A2 mechanism at carbon has been assigned in the case of glucose-6-phosphate (68). Evaluation of the conjugate acid and the neutral ester kinetics is handicapped for many esters by overlap of the acidity ranges in which these are important. The pK_a values for the conjugate acid and the ester are not greatly different. Unless the rate constants differ greatly, it is difficult to separate the kinetics.

In contrast to the alkyl monoesters, which exhibit a normal acid catalysis (Fig. 23), and the phenyl ester, which shows no acid catalysis

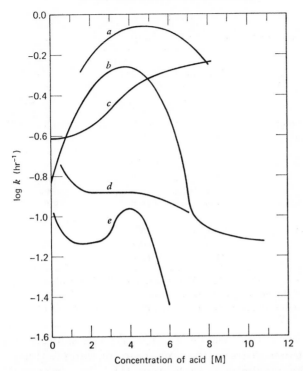

Fig. 24. Monoaryl phosphate hydrolysis in concentrated acids, (a) p-nitro-phenyl phosphate, HCl, 73° (23); (b) p-nitrophenyl phosphate, HClO$_4$, 73° (23); (c) p-nitrophenyl phosphate, HCl + LiCl, ionic strength 8.0, 73° (23); (d) phenyl phosphate, HClO$_4$, 100° (23); and (e) p-methoxyphenyl phosphate, HClO$_4$, 99° (289).

(Fig. 24d), aryl esters carrying strongly electron-withdrawing sub-stituents have a maximum rate of hydrolysis at 3–5M hydrogen ion, as shown in Figure 24a and b (72). The maximum is not the result of complete protonation of the ester to the conjugate acid. As shown for p-nitrophenyl phosphate in Figure 24c, if the ionic strength is held constant with LiCl, continuous acid catalysis is exhibited, with no

TABLE LV

Principal Reactive Species in Monoester Hydrolysis

Alkyl: [a]		$RH_3PO_4^+$	$RHPO_4^-$	
Aryl:	ordinary		$RHPO_4^-$	
	good-leaving	$RH_3PO_4^+$	$RHPO_4^-$	
	excellent-leaving	$RH_3PO_4^+$		RPO_4^{2-}

[a] Hydrolysis of the neutral alkyl ester becomes important if a carbonium ion can be formed.

TABLE LVI
Activation Parameters for $RH_3PO_4^+$ Hydrolysis[a]

R	E_a (kcal/mole)	log A (log hr^{-1})	$\Delta S^\ddagger$ (eu)	Ref.
acetyl	21.8 ± 0.1	15.7 ± 0.1[b]	−5.1 ± 0.5[b]	336
trimethylacetyl	21.8 ± 0.3	14.7 ± 0.2[b]	−9.7 ± 1.2[b]	336
3,3-dimethylbutyryl	23.8 ± 0.2	16.9 ± 0.1[b]	+0.6 ± 0.7[b]	336
2,4-dinitrophenyl	18	9.0	−23	71
o-nitrophenyl	20	11.8	−23	72
m-nitrophenyl	20	11.3	−25	72
methyl	25.2 ± 1.2	13.2	−16.3	253
ethyl	29.3 ± 1.0	15.5	−6.0	253
isopropyl	31.0 ± 0.2	18.6	+8.2	253
3-hydroxy-2-pyridylmethyl	25.1	15	−5	312
glucose-1-	28.5	20.1	+14.9	77
glucose-6-	24	13.4	−19	68

[a] See footnote (a), Table XIII. Values are based on first-order rate constants for hydrolysis of the conjugate acid species.
[b] Based on second-order constants. Units of log A are log liters/(mole)(hr).

maximum. Since the effect on water activity is similar for Li$^+$ and H$^+$, this result demonstrates the importance of water activity changes in shaping the maximum. In Section III-C-2 a treatment of the similar rate maximum for triphenyl phosphate was described. Presumably the same explantion, in terms of the conflict between increasing acid-catalysis and decreasing water activity, applies here. The rate maximum observed in concentrated $HClO_4$ solution for p-methoxy- (Fig. 24e), o-methoxy-, and p-ethoxyphenyl phosphates does not involve acid catalysis, as the rate is independent of $HClO_4$ concentration if the ionic strength is kept constant (289). Neopentyl phosphate, an alkyl ester for which nucleophilic attack at the α carbon is inhibited, shows acid catalysis without a rate maximum (75).

Except for hydrolyses involving ready carbonium ion formation (76,260), neutral ester hydrolysis is relatively slow for alkyl esters. Although concurrent hydrolysis by other routes makes it difficult to establish the point of cleavage, rupture appears to occur principally at the C—O bond for this category. For monoaryl esters, not only is nucleophilic substitution at carbon inhibited, but also the neutral ester hydrolysis at the P—O link is accelerated Barnard has given the first-order rate constants at 100° for the P—O cleavage route for methyl, p-tolyl, phenyl, and p-nitrophenyl phosphates as 4 × 10^{-4}, 0.066, 0.110, and 1.08 hr^{-1}, respectively (23). The order of values correlates well with the leaving abilities of the organic substituent. The molecularity of the reaction at phosphorus has not yet been established, but the data of Table LVII suggest there is more than one mechanism, even for the aryl esters.

Most thoroughly studied of the monoester hydrolysis reactions are those of the mono- and dianion. Except for the monoanions of phosphorylcholine and of p-bromophenyl, o-methoxyphenyl, p-methoxyphenyl, and p-ethoxyphenyl phosphates, the activation parameters of

TABLE LVII

Activation Parameters for RH_2PO_4 Hydrolysis[a]

R	E_a (kcal/mole)	log A (log hr^{-1})	$\Delta S^{\ddagger}$ (eu)	Ref.
o-methoxyphenyl	19.4	11.12	−25.9	289
p-ethoxyphenyl	20.2	10.76	−27.6	289
p-chlorophenyl	28.46	16.11	−3.1	292
p-chloro-m-tolyl	27.46	16.45	−1.6	291
3-hydroxy-2-pyridylmethyl	30.6	15	+6	312
2-pyridylmethyl	28.2			313
3-pyridylmethyl	29.2			313
glucose-1-	31.0	20.19	+15.6	77

[a] See footnote (a), Table XIII. Values are based on first-order rate constants for the hydrolysis of the neutral monoester molecule.

Tables LVIII and LIX are consistent with a unimolecular mechanism generally applicable to hydrolyses of monoester mono- and dianions. The common feature is believed to be cleavage of the P—O bond to eliminate a somewhat hypothetical monomeric metaphosphate [Eq. (67)].

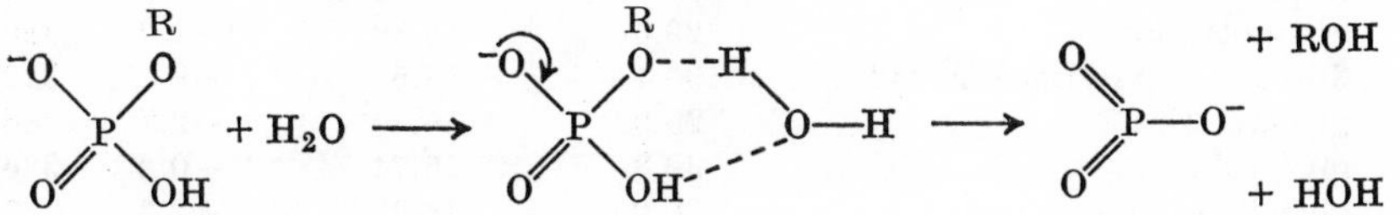

$$(67)$$

This mechanism was initially suggested by Westheimer (82,255), Vernon, Bunton, Barnard (24) and co-workers for the hydrolysis of monoanions of phosphate monoesters. Without direct evidence, it is now generally accepted for the hydrolysis of monoanions of acyl, aryl, and alkyl monoesters and for certain classes of dianions (65,66,111).

TABLE LVIII
Activation Parameters for $RHPO_4^-$ Hydrolysis[a]

R	E_a (kcal/mole)	$\log A$ ($\log hr^{-1}$)	$\Delta S^{\ddagger}$ (eu)	Ref.
Acyl				
acetyl	22.5	16.00	−3.6	132
choline	24.3	13.3	−16	8
Aryl				
2,4-dinitrophenyl	24.8	14.5	−6.0	245
p-nitrophenyl	26.0			192
p-nitrophenyl	27	16.6	−1	72
p-nitrophenyl	30.0	18.12	+6.1	23
o-nitrophenyl	28	18	+5	72
m-nitrophenyl	31	17.7	+4	72
p-carboxyphenyl	28.3	17.0	+0.9	92
m-carboxyphenyl	27.9	13.7	−1.4	92
salicyl[b]	23.0			95
salicyl[c]	24.1	16.53	−1.2	95
phenyl	29.0	16.88	+0.4	23
phenyl	29.0	17.0	+0.9	92
8-carboxy-α-naphthyl	31.0	17.88	+5.0	92
2-carboxy-1-naphthyl[c]	24.5	17.36	+2.6	94
3-carboxy-2-naphthyl[c]	25.4	17.75	+4.4	94
1-carboxy-2-naphthyl[c]	28.6	18.2	+6.5	94
α-naphthyl	30.6	17.69	+4.1	92
β-naphthyl	29.5	18.0	+5.5	92
p-tolyl	29.0	16.82	+0.1	23
2,6-dimethylphenyl	30	17.34	+2.5	72
o-chlorophenyl	27	15.81	−4.5	72
p-chlorophenyl	27.46	15.92	−4.0	292
p-bromophenyl	21.70	12.31	−20.5	292
p-chloro-m-tolyl	27.46	15.94	−3.9	291
o-methoxyphenyl	21.4	12.77	−18.4	290
p-methoxyphenyl	22.9	13.60	−14.6	290
p-ethoxyphenyl	19.4	11.01	−26.4	290
Alkyl				
methyl	30.6	16.37	−1.9	76
isopropyl	32.1	17.31	+2.4	253
ethanolamine	29.6	16.38	−1.9	386
3-hydroxy-2-pyridylmethyl	30.4	18.5	+8	312
glycerol-1-	29.9	16.19	−2.7	386
glycerol-2-	30.3	16.74	−0.2	386
glucose-1-	30.0	16.26	−2.3	77
glucose-6-	30.5	16.8	0.0	68
2-pyridylmethyl[d]	28.2	15.91	−4.0	313
3-pyridylmethyl[d]	29.3	16.17	−2.8	313

[a] See footnote (a), Table XIII. Values are based on first-order rate constants for the hydrolysis of the monoanion, i.e., for the monoprotonated phosphate tetrahedron. Probably all the reactions listed involve P—O cleavage.

[b] Carboxyl un-ionized.

[c] Carboxyl ionized.

[d] For hydrolysis of zwitterion, $(C_5H_5NH^+)CH_2OP(OH)O_2^-$.

TABLE LIX

Activation Parameters for RPO_4^{2-} Hydrolysis[a]

R	E_a (kcal/mole)	$\log A$ ($\log hr^{-1}$)	$\Delta S^{\ddagger}$ (eu)	Ref.
Acyl				
acetyl	26.6	18.1	5.8	320
acetyl	25.4	17.60	+3.7	132
acetyl	22.7	16.0[b]	−3.7[b]	256
carbamyl	25.2	17.28	+2.3	321
Aryl				
2,4-dinitrophenyl	25.5	17.34	+2.5	71
2,5-dinitrophenyl	28	17.0	+1	71
2,6-dinitrophenyl	26	18.21	+6.5	71
p-nitrophenyl	31.2	17.55	+3.5	242
p-carboxyphenyl	27.4	16.11	−3.1	92
m-carboxyphenyl	27.7	16.22	−2.6	92
8-carboxy-α-naphthyl	30.9	19.11	+10.6	92
Alkyl				
glucose-6-	33	18.9	+9	68
3-hydroxy-2-pyridylmethyl	29.1	18	+6	312

[a] See footnote (a), Table XIII. Values based on first-order rate constants for the hydrolysis of the dianion.

[b] Recalculated from second order to first order basis.

Equation (67) has been written to show transfer of the monoanion proton to the leaving group occurring through water. This cannot be considered to be established as added organic solvents have little influence on the rate of hydrolysis of phosphoramidic acid, for which the same mechanism is thought to apply (93). Postulating the transfer through water avoids the energetically unfavorable formation of a four-membered ring, which would be required for direct proton transfer, trading this for a less favorable entropy of activation. The proton transfer is invoked to avoid the elimination of an energetically unfavorable alkoxide or aryloxide ion   It is a key part of the mechanism for this reason and because the unavailability of a proton for this transfer in a monoester dianion or a diester monoanion explains the generally slower hydrolysis rates for these species. A linear free energy plot of the log of the rate constant for monoester monoanion hydrolysis versus the pK_a of the leaving group has a slope of about −0.3 for a wide variety of alkyl and aryl esters (72,245). This relative insensitivity to the leaving group is evidence for protonation of the leaving group before its elimination, as the innately better leaving groups are less receptive to the proton transfer.

Although the generally preferred formulation of the mechanism shows the proton transfer concerted with elimination, it is not clear that this is the case. The deuterium solvent isotope effect is smaller, in most cases, than would be expected; and the $\Delta S^{\ddagger}$ values are less negative than expected for the transition state of Eq. (67). Kirby and Varvoglis (245) have suggested that the proton transfer occurs before the rate-limiting step in what amounts to an A1 mechanism [Eq. (68)]

$$
\begin{matrix}
& \text{R} & & & \text{R} & & \\
^-\text{O} & \text{O} & & ^-\text{O} & ^+\text{OH} & & \text{O} \\
& \diagdown \diagup & & & \diagdown \diagup & & \diagdown \diagup \\
& \text{P} & \rightleftharpoons & & \text{P} & \longrightarrow & \text{P—O}^- + \text{ROH} \\
& \diagup \diagdown & & & \diagup \diagdown & & \diagup \\
\text{O} & \quad \text{OH} & & \_\text{O} & \text{O} & & \text{O}
\end{matrix}
\qquad (68)
$$

Consistent with this, the deuterium isotope effect, k_H/k_D, which is 0.87 for the methyl phosphate monoanion hydrolysis at 100°, is 1.45 for the hydrolysis of the 2,4-dinitrophenyl phosphate monoanion at 39°. Kirby and Varvoglis argue that for methyl phosphate the proton transfer occurs earlier than the rate-limiting step, whereas for the excellent leaving group 2,4-dinitrophenol, the rate is so fast that the equilibrium of Eq. (68) cannot be maintained, and the proton transfer has become concerted with the elimination, as revealed by the isotope effect.

The monometaphosphate species eliminated is coordinately unsaturated and cannot be isolated   It undergoes rapid combination with a nucleophile, usually with water to produce monophosphate. The reactivity of monometaphosphate should be such that it would be highly unselective in this further reaction. Ester hydrolysis in mixed solvents in some cases produces suitably mixed products (93) and in others shows a high degree of selectivity (71). It appears now that mixed solvents provide a poor test of the mechanism. The reactivity of the monometaphosphate intermediate is such that it never has a free existence, but reacts with a species in the solvation sphere of the transition state. Apparent selectivity of the intermediate is attributed to selectivity in the transition state solvation sphere (66). If the solvent does not provide a suitable nucleophile the intermediate phosphorylates a solute, producing condensed phosphates in some cases (55,99,100,132).

A bimolecular mechanism for the monoanion hydrolysis can be eliminated on the basis of the $\Delta S^{\ddagger}$ values, the smaller than expected steric effects from bulky ortho substituents in monoaryl phosphates (72), and the greater hydrolysis rate for monoanion hydrolysis than for the neutral ester, in spite of the expected inhibition of nucleophilic attack in the case of the monoanion due to its negative charge.

Since internal proton transfer to produce a more favorable leaving group is not possible for monoester dianions, slower hydrolysis of the dianions by an analogous route is to be expected generally [Eq. (69)].

$$^-O \diagdown \diagup OR \qquad O \diagdown \qquad \qquad$$

$$P \longrightarrow P-O^- + {}^-OR \qquad (69)$$

$$O \diagup \diagdown O^- \qquad O \diagup$$

Kirby and Varvoglis have shown that exceptions will arise for leaving groups that are readily eliminated as the anion (245). Comparative mono- and dianion hydrolysis rate constants appear in Table LX

TABLE LX

ArH_2PO_4 Hydrolysis Rate Constants[a]

Ar	pK_a[b]	Monoanion	Dianion
phenyl	9.99	3.42×10^{-4}	very slow
2-nitrophenyl	7.23	3.22×10^{-3}	2.42×10^{-4}
4-nitrophenyl	7.14	3.85×10^{-3}	5.6×10^{-5}
3,5-dinitrophenyl	6.68	3.57×10^{-3}	1.0×10^{-4}
2-nitro-4-chloro	6.46	2.21×10^{-3}	1.18×10^{-3}
2,4,6-trichloro	6.1	1.32×10^{-3}	1.6×10^{-4}
2-chloro-4-nitro	5.45	8.77×10^{-4}	8.16×10^{-4}
2-nitro-4-COMe	5.09	8.28×10^{-3}	3.70×10^{-2}
2,4-dinitro	4.07	1.82×10^{-2}	6.3×10^{-1}

[a] First-order rate constants, hr^{-1}, at 39° and unit ionic strength (245).

[b] pK_a for the corresponding phenol.

for a number of monoaryl esters. For 2-chloro-4-nitrophenyl phosphate, the rate constants are about the same for the two anions. A rate constant versus pH profile for this ester shows a very small step at about pH 5. The esters listed above this one in Table LX (pK_a of phenol > 5.5) have rate maxima about pH 2–4; those lower in the table (pK_a of phenol > 5.5) have rate minima in this region.

Because the stability of the carboxylate ion makes it an excellent leaving group, some monoacyl phosphate dianions hydrolyze more rapidly than the monoanions (Fig. 21). As for the monoaryl phosphates, the rate relationship depends on the pK_a of the parent organic acid, although the critical pK_a value for equality of rate constants is about 4.0 for the monobenzoyl phosphates (Table LXI). The mechanism of hydrolysis for an acyl phosphate dianion is directly analogous

TABLE LXI

AcH_2PO_4 Hydrolysis Rate Constants[a]

Ac	Monoanion	Dianion
p-methoxybenzoyl	0.29	0.11
p-methylbenzoyl	0.34	0.16
benzoyl	0.37	0.19
p-chlorobenzoyl[b]	0.34	0.39
m-nitrobenzoyl	0.49	1.8
p-nitrobenzoyl	0.49	1.7
3,5-dinitrobenzoyl	1.02	9.0

[a] First-order rate constants, hr^{-1}, at $39°$ (132).
[b] pK_a for p-chlorobenzoic acid is 3.98.

to Eq. (69). For the monoanion there is no need to postulate the involvement of a water molecule in the proton transfer step:

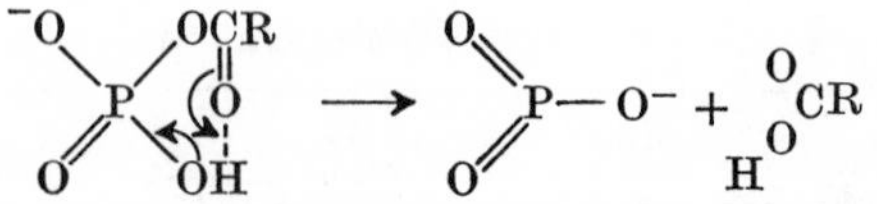

$$\begin{array}{c}\overset{-}{O}\diagdown\underset{O}{\overset{OCR}{\diagup}}O\diagdown\\P\longrightarrowP-O^- + \overset{O}{\underset{O}{CR}}\\O\diagupOHO\diagupH\end{array}\tag{70}$$

Westheimer considers the monometaphosphate mechanism to have utility as a predictive device, in that significantly rapid hydrolysis of phosphates, phosphonates, etc., is to be expected whenever a reasonable reaction can be written based on a hypothetical monometaphosphate intermediate (426).

For a variety of acyl, aryl, and alkyl monoesters the slope was -1.23 for a plot of the log of the dianion hydrolysis rate constant against the pK_a of the acid, phenol, or alcohol parent to the leaving group (245). This is much larger than the slope of about -0.3 found for monoanion hydrolysis and is consistent with the elimination of an organic anion in the dianion hydrolysis. The greater sensitivity of rate to substituents for the dianion is evident in Figure 21.

Simple aryl and alkyl dianions are unreactive at high pH unless a neighboring group is present to provide intramolecular catalysis (111) or unless an E2 elimination breaks the C—O bond (259). For nitrophenyls, studies carried past pH 12 show increased hydrolysis rates. Presumably, a bimolecular dianion–hydroxide reaction becomes important at high hydroxide concentrations, although electrolyte effects are possibly responsible. Studies in this region have been limited. The sensitivity of the rate constant to pK_a of the phenol is greater than for the spontaneous dianion reaction (71). For the reaction of acetyl phosphate dianion with hydroxide ion $\Delta H^{\ddagger}$ and $\Delta S^{\ddagger}$ values of 27.3

kcal/mole and $+27.9$ eu have been reported (256). Although an [18]O study has shown that P—O cleavage occurs in the reaction of 2,4-dinitrophenyl phosphate dianion with hydroxide (74), Kirby and Jencks have suggested for the p-nitrophenyl phosphate dianion reaction with hydroxide that C—O cleavage occurs, based on a similarity of rate to that of the reaction of OH⁻ with 1-chloro-4-nitrobenzene (242,243). The hydrolysis of p-nitrophenyl phosphate in strongly basic solution is several orders of magnitude slower than that of 2,4-dinitrophenyl phosphate.

A variety of types of catalysis for monoester hydrolysis have been observed (111,308). Intramolecular catalysis often results from neighboring groups facilitating the transfer of the proton to the leaving group and/or the stabilization of the association of the proton with the leaving group in the transition state (39,60,309,312), or through the formation of a highly reactive five-membered cyclic phosphate (56). Solutes that have been evaluated as catalysts appear in conjunction with Table LIV. Halide ions, particularly iodide, catalyze the acid hydrolysis of monoalkyl phosphates (75), while metal ions, especially Li⁺ and the polyvalent ions or their hydroxides, catalyze a variety of hydrolyses (17,18). The role of the metal ion is not completely clear. Possible contributions include: (1) partially neutralizing the charge on the ester anion, making it more approachable to a nucleophile, (2) making favorable changes in the transition state through coordination and chelation, and (3) forming a metal–hydroxy complex which might more readily enter the dianion–hydroxide reaction (49). It has been established that the degree of catalysis by a given metal ion is related to the concentration of metal–phosphate complex, but there has been a failure to determine either the molecularity of the ester hydrolysis reaction for the complex or the point of cleavage. The possibility appears not to have been considered yet that in a metal chelate the metal ion, with relatively long M—O bonds, might function as an equatorial bridging group to lower the activation energy for formation of a trigonal bipyramidal transition state and thus accelerate ester hydrolysis (cf. following section). Hydrolysis of the dianions of aryl phosphates is catalyzed by cationic micelles of cetyltrimethylammonium bromide, but not by nonionic or anionic micelles (73). In contrast to the tri- and diesters, the reaction with hydroxide ion is not catalyzed. The hydrolysis of nitrophenyl dihydrogen phosphates is accelerated by light (182,245). As in the nonphotochemical reaction, it is the P—O linkage that is cleaved (246). The ester behaves in the photochemical reaction as though it were a derivative of a much more acidic phenol, probably an excited singlet state. Photochemical acceleration of hydrolysis of nitrophenyl groups from di- and triesters involves C—O cleavage (247).

5. CYCLIC ESTERS

Five-membered cyclic esters of phosphoric acid hydrolyze to open-chain monoesters much more rapidly than the hydrolysis of noncyclic diesters to monoesters (172,194,254). Considering P—O cleavage (the cyclic ester undergoes about 100% P—O rupture), the ratio of the rates of hydrolysis for ethylene phosphate and dimethyl phosphate is about 10^8 at 25° in alkaline solution and greater than that at 100° in acidic solution (172). The alkaline hydrolysis of o-phenylene phosphate is about 6×10^6 times as fast as that of diphenyl phosphate (224).

During acidic hydrolysis (but not alkaline), oxygen exchange between ethylene phosphate and the solvent is also extremely fast, exchange occurring at about one-fifth the rate of hydrolysis (172). From this it is evident that breaking of the five-membered ring is not responsible for the rapid reaction, but instead, relief of strain in the transition state. Furthermore, in a close parallel to the oxygen exchange, the alkaline hydrolysis of methyl ethylene phosphate involves only P—O cleavage in the ring, but the acidic hydrolysis includes loss of methoxyl (5–30%), as well as ring opening (110). The hydrolysis rate for methyl ethylene phosphate is about 10^6 times that of trimethyl phosphate. Evidence for strain in the five-membered ring is found in the enthalpy of hydrolysis for methyl ethylene phosphate, -28.5 kcal/mole, which is about 6 kcal/mole more exothermic than that of dimethyl hydroxyethyl phosphate or of trimethyl phosphate (113,225). The O—P—O bond angle in the ring is 99° (377).

Other very rapid phosphate triester hydrolyses in which the five-membered ring present is preserved are those of phenyl ethylene phosphate (122) and of ethyl, propyl, butyl, and isobutyl dimethylvinylene phosphates (37). The double bond in the ring of the latter compounds increases the strain and the rate of reaction relative to the ethylene ester. The dimethylvinylene phosphate anion hydrolyzes about ten times as fast at 25° as the ethylene phosphate anion (34).

Activation parameters for the neutral hydrolysis of several cyclic dimethylvinylene triesters appear in Table LXII. The small activation energies are consistent with relief of the ring strain in the transition state, and in view of the unfavorable $\Delta S^\ddagger$ values, are responsible for the great speed of these reactions. The 99° O—P—O bond angle in the ring is somewhat greater than that required for minimizing strain in the rest of the ring (429); a decrease in this angle is required to relieve

TABLE LXII

Activation Parameters for Cyclic Phosphate Esters[a]

R	E_a (kcal/mole)	$\log A$ (log hr^{-1})	$\Delta S^{\ddagger}$ (eu)
ethyl	8.4 ± 1.2	7.7 ± 0.9	-42 ± 4
propyl	10.8 ± 0.5	9.45 ± 0.36	-34 ± 2
butyl	9.8 ± 1.0	8.53 ± 0.76	-38 ± 3
iso-butyl	8.3 ± 0.9	7.35 ± 0.69	-43 ± 3

[a] See footnote (a), Table XIII. Values are based on first-order rate constants for hydrolysis of the neutral ester. R applies to the general formula of **30**. Ref. 37.

the strain. This observation and the very negative $\Delta S^{\ddagger}$ values suggest an addition–elimination mechanism with a trigonal bipyramidal transition state. With ethylene bridging between apical and equatorial positions, an OPO angle in the ring of 90° would result [Eq.

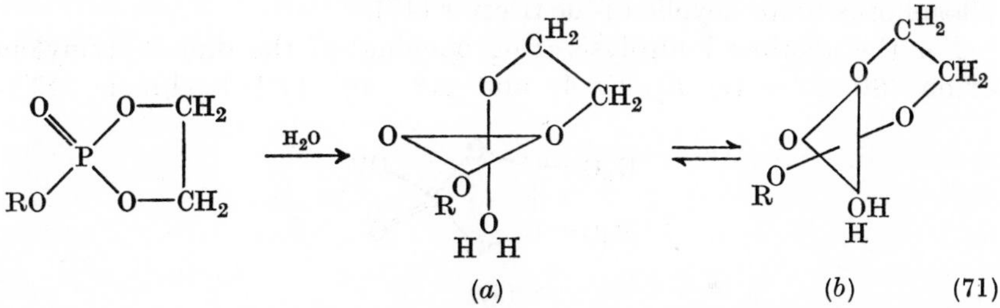

$$ \text{(a)} \qquad\qquad \text{(b)} \qquad\qquad (71) $$

(71a)]. The activation energy for the reaction would then be reduced by the amount of the strain energy, relative to the activation energy for an acyclic ester. Since addition and elimination operations are restricted to apical positions, ring opening would involve cleavage of the upper P—O bond of (a). MO calculations predict that approach by water is favored opposite to one of the P—O bonds of the ring (47).

With pseudorotation (see **26** and associated text), oxygen exchange in ethylene phosphate (R = H) or loss of methoxyl from methyl ethylene phosphate (R = CH₃) become possible, as shown in the equilibrium between forms (a) and (b). This alteration of the transition state occurs without change in the ring strain. The phosphoryl oxygen can be expected to serve as pivot atom (Section IV-C-2). Thus pseudorotation of a pentacovalent transition state explains how ring strain in the initial reactant accelerates both ring and nonring hydrolysis. Presumably proton transfer occurs within the transition state to the oxygen atom of the leaving group. MO calculations support this view (47), predicting the lowest energy configurations for the transition state to be **28**, then **29**.

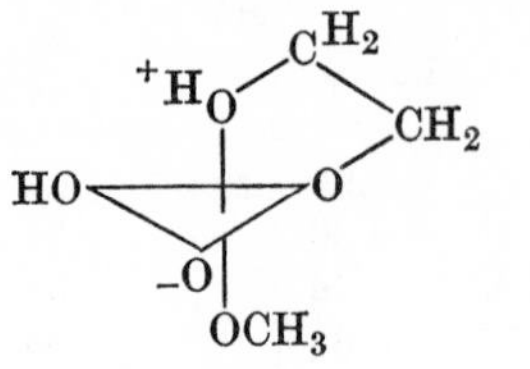

(28) 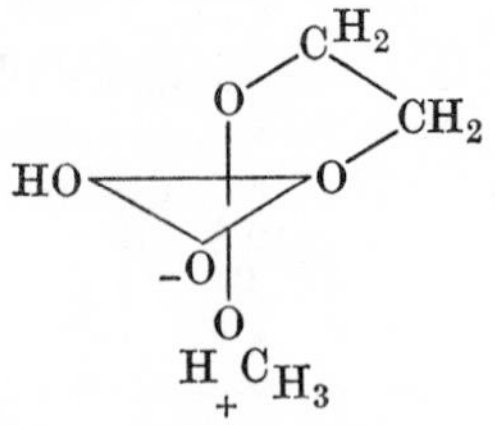   (29)

Estimates of the strain energy in the ethylene phosphate ring account for only part of the increased rate. Inhibition of dp π bonding in five-membered ring phosphates has been credited with contributing through the making available of a phosphorus $3d$ orbital as a site for nucleophilic attack (318). According to MO calculations, the phosphorus atom of methyl ethylene phosphate is 0.04 more positive than the phosphorus of an acyclic di- or triester (47).

For the alkaline hydrolysis (ring opening) of the dimethylvinylene anion (**30**, R = H) E_a, log A, and $\Delta S^{\ddagger}$ are 17.1 kcal/mole, 13.74

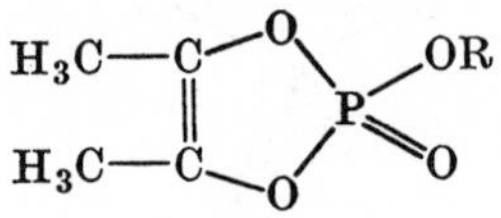

(30)

[log liters/(mole)(hr)] and -3.05 eu (standard state $1M$ at $25°$), respectively, based on second-order rate constants (34). Since the ring opening of ethylene phosphate is entirely by P—O cleavage (172), we may assume with the unsaturated carbon here that P—O cleavage also occurs. The moderate $\Delta S^{\ddagger}$ value suggests that the mechanism is unimolecular, rather than addition–elimination. With ring strain favoring ring opening, this would be consistent wth the lack of oxygen exchange for ethylene phosphate in alkaline solution (172) and the limitation of hydrolysis of methyl ethylene phosphate in alkaline solution to ring opening (110), although pseudorotation accommodates the latter observations in a bimolecular mechanism.

The lability of five-membered cyclic phosphate esters was first deduced from the unusual reactivity of 2-hydroxyalkyl esters (56). The hydrolysis in dilute alkali of methyl 2-hydroxyethyl phosphate is much faster than that of methyl 2-methoxyethyl phosphate (16); that of the 2-keto ester dimethyl phosphoacetoin is 2×10^6 times as fast as that of trimethyl phosphate (348). Other examples of rapid hydrolysis through five-membered cyclic intermediates formed with appropriate neighboring groups include P,P-dimethyl phosphoenol pyruvate (101), the phenyl ester of cis-tetrahydrofuran-3,4-diol phosphate (319), dimethylphosphoacetoin (146), and several α-keto triesters (54). Some understanding of the stereochemistry of the pentacovalent transition state can be obtained from studies on the pentacovalent phosphorus in oxyphosphoranes (347). Oxyphosphoranes containing a five-membered ring hydrolyze extremely rapidly with P—O cleavage (348,349):

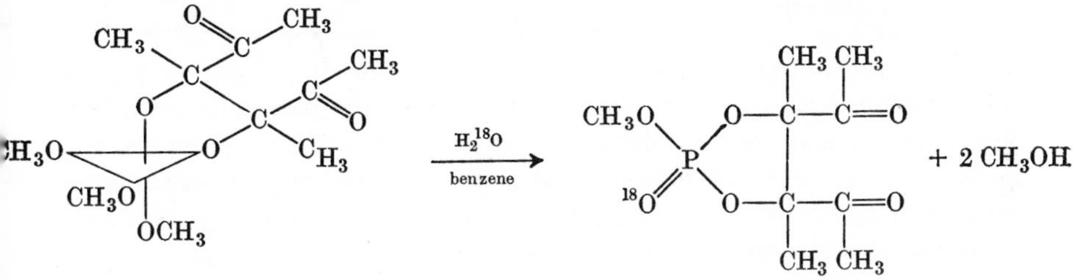

(72)

The hydrolysis of six- and seven-membered cyclic phosphates, for which ring strain is lacking, is comparable in rate to that of the simple dialkyl phosphates, much slower than for the five-membered ones (98,236).

V. Oxygen Exchange

The exchange of oxygen atoms between a phosphate species and water is a hydrolysis reaction which can be followed only by using isotopic tracer methods. The early, more-or-less qualitative studies of this exchange are marked by contradictions (44,174,421,422,466). Even the more recent kinetic studies show poor agreement (Table LXIII).

Exchange rates have been studied over the range from $8.93M$ $HClO_4$ to pH 8.61 by Bunton, et al. (78). The kinetics are first order in phosphate. The rate constant shows a maximum at pH 4.5–5.5, the peak abundance range for the monoanion, $H_2PO_4^-$; shows a minimum at a pH of about 1, where the phosphate is primarily un-ionized; and increases steadily as the acidity increases further (Fig. 25). Presumably the conjugate acid is involved at high acidities. In spite of the close similarity of this pattern to that for the hydrolysis of phosphate

TABLE LXIII

H_3PO_4–$H_2{}^{18}O$ Exchange, First-Order Rate Constants, hr^{-1} at $100°$

$H_4PO_4^+$	H_3PO_4	$H_2PO_4^-$	Ref.
	2.3×10^{-4}	a	50
$1.97 \times 10^{-2}K_a{}^b$	4.61×10^{-3}	1.45×10^{-2}	78
		3×10^{-4}	465

[a] No detectable exchange.
[b] For rate expression first order in $H_4PO_4^+$, Eq. (74). K_a is acid ionization constant for $H_4PO_4^+$.

monoesters, the mechanisms can only be the same for the monoanions. For oxygen exchange with $H_2PO_4^-$, by arguments paralleling those for the ester monoanion hydrolysis (76), it was concluded that the reaction is unimolecular. The rate constants at $100°$ for these reactions are similar, 1.45×10^{-2} hr^{-1} and 2.96×10^{-2} hr^{-1}, respectively.

Mechanisms were not established for oxygen exchange with the neutral and conjugate acid species. The mechanism for exchange with the neutral H_2PO_4 molecule cannot be analogous to that for $CH_3OPO_3H_2$ hydrolysis, as the latter occurs principally by C—O cleavage. For the respective conjugate acid species the mechanisms must also be different; the exchange kinetics for $H_4PO_4^+$ show a positive salt effect, whereas

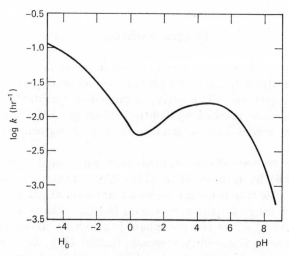

Fig. 25. Solvent oxygen exchange with H_3PO_4, first-order rate constants at $100°$ (78).

there is no salt effect for the hydrolysis of $CH_3OPO_3H_3^+$. A second order rate expression, Eq. (73) was used to describe the kinetics in highly acid solution in place of the mathematically equivalent Eq. (74).

$$\text{Rate} = k[H^+][H_3PO_4] \tag{73}$$
$$= k(K_a[H_4PO_4^+]) = k'[H_4PO_4^+] \tag{74}$$

Others have employed the species $H_4PO_4^+$ in interpretation of exchange kinetics (388). The complete rate law for oxygen exchange with monophosphate over the acidity range studied by Bunton is given in Eq. (75).

$$\text{Rate} = k_1[H_4PO_4^+] + k_2[H_3PO_4] + k_3[H_2PO_4^-] \tag{75}$$

The values of k_2 and k_3 at 100° appear in Table LXIII. From Eqs.

TABLE LXIV

H_3PO_4–$H_2^{18}O$ Exchange Rate in Concentrated H_3PO_4[a]

H_3PO_4 (f)	Temperature (°C)					E_a (kcal/mole)
	40	60	80	100	110	
5.9				0.0400		
8.8				0.123	0.28	26.5
12.2		0.0103	0.0796	0.411	0.944	24.0
15.1	0.00494	0.0382	0.242	1.24	3.12	23.1
17.8	0.025	0.166	0.739	3.41		19.7
18.35				4.82		

[a] Rate in g-atom/(liter)(hr).   Ref. 227.

(73) and (74), one can see that k_1 at 100° is equal to the $1.97 \times 10^{-2} K_a$ value of Table LXIII and, in turn, that the first term of Eq. (75) can be replaced by $1.97 \times 10^{-2} [H^+][H_3PO_4]$.

Keisch, Kennedy, and Wahl determined the water–H_3PO_4 exchange rates in concentrated H_3PO_4 solutions, 5.9 to 18.35 f in H_3PO_4 (227) Concentrations were expressed in formality because of the formation of $H_4P_2O_7$ in the more concentrated solutions. A number of their results appear in Table LXIV. The values shown are rates (not rate constants) for exchanges of oxygen atoms, both tracer and nontracer, occurring per unit time (379). A change in mechanism with phosphoric acid concentration is suggested by the trend in activation energy values. The following rate law was deduced:

$$\text{Rate} = k_1 a_{H_3PO_4} a_{H_2O} + k_2 a_{H_3PO_4}^2 + k_3 a_{H_3PO_4}^3 a_{H_2O}^{-1} \tag{76}$$

The activity of water was taken to be 1.00 in pure water and the activity of H_3PO_4 to be 0.05 at 0.05 mole fraction. Activities for water and H_3PO_4 at concentrations of interest were then derived from vapor pressure data. The first two terms predominate at H_3PO_4 formalities below and above 14, respectively. The third term becomes more important than the first only above about 17.5 f H_3PO_4. The respective values assigned the three rate constants at 100° are 0.16, 8.6 × 10⁻⁴, and 1.7 × 10⁻⁸ g-atom/(liter)(sec).

The first term of Eq. (76) relates to the bimolecular reaction of H_3PO_4 and H_2O. At the H_3PO_4 concentrations employed the water activity is significantly diminished, and a pseudo first order expression can no longer be used. The water activity in 6.0 f H_3PO_4, for example, is 0.770 (relative to pure water as 1.00).

TABLE LXV

Comparison of $H_4P_2O_7$ Hydrolysis and H_3PO_4–H_2O Exchange[a]

Conditions	Hydrolysis rate [mole/(liter)(hr)]	Exchange rate [g-atom/(liter)(hr)]
25°, 18.0 f H_3PO_4	5.4 × 10⁻³	6.8 × 10⁻³
60°, 17.8 f H_3PO_4	1.7 × 10⁻¹	1.7 × 10⁻¹

[a] Ref. 227.

It is interesting to compare Bunton's kinetic term for H_3PO_4, 4.61 × 10⁻³ [H_3PO_4] hr⁻¹ [Eq. (75)], with the first term of Eq. (76), 0.16 $a_{H_3PO_4} a_{H_2O}$, by evaluating rates from both for a $0.4M$ H_3PO_4 solution, the approximate concentration used by Bunton. In such a solution a_{H_2O} is about 0.98 and $a_{H_3PO_4}$ about 0.007 (mole fraction). The respective rates are then 1.8 × 10⁻³ (Bunton) and 1.1 × 10⁻³ hr⁻¹ (Keish). Even without considering the differences in conditions for the two studies, the agreement is good; it suggests that the same mechanism is involved.

The second term of Eq. (76), $k_2 a_{H_3PO_4}^2$ is attributed to exchange occurring through the reversible formation and hydrolysis of $H_4P_2O_7$ [Eq. (77)]. To test this, $H_4P_2O_7$ hydrolysis rates were determined and

$$2\,H_3PO_4 = H_4P_2O_7 + H_2O \qquad (77)$$

compared with H_3PO_4–H_2O oxygen exchange rates (Table LXV). The rates are nearly the same, consistent with the exchange occurring through the $H_4P_2O_7$ hydrolysis. Statistically, if the $H_4P_2O_7$ hydrolysis and the H_3PO_4 exchange reactions simply occurred by the same mechanisms, the H_3PO_4–H_2O exchange should occur at four times the rate of $H_4P_2O_7$ hydrolysis. The activation energy for the exchange process in about 14–16 f H_3PO_4 (Table LXIV) is close to the about 22 kcal/mole value found for $H_4P_2O_7$ at a pH of about 1 (Table XIII). From 8.8 to 17.8 f H_3PO_4, $\Delta S^{\ddagger}$ is -10 to -21 eu.

At the highest H_3PO_4 concentrations the first two terms of Eq. (76) do not account for all of the oxygen atom exchange rate. The third term assumes oxygen exchange occurring through the reversible formation of triphosphoric acid [Eq. (78)]. Because of the narrow concen-

$$H_4P_2O_7 + H_3PO_4 = H_5P_3O_{10} + H_2O \qquad (78)$$

tration range over which this term is kinetically important and the lower reliability of the exchange data at the highest H_3PO_4 concentrations, the form of this term is not as well tested as for the first two. Although the concentration added was relatively small (0.3 f), KH_2PO_4 does not appear to affect the exchange rate in concentrated H_3PO_4. For this reason, no term in $H_2PO_4^-$ was included in the rate expression. The addition of 0.9 f $HClO_4$ to 13.4 f H_3PO_4 increased the rate less than half as much as did a 0.9 f increase in the H_3PO_4 concentration. The addition of finely divided TiO_2 increased the rate slightly.

In a study of oxygen exchange at 120 and 160° Garus, et al. (153), found for Na_2HPO_4 and K_2HPO_4 solutions that the rates were constant for water–phosphate mole ratios from 12:1 to 36:1. For NaH_2PO_4 and KH_2PO_4, the rate increased significantly over about the same range of solution composition. Plots of the log of the rate constant versus the square root of the ionic strength were linear for all four phosphates. The plots for the dihydrogen phosphates converged at an ionic strength of zero, while those for the monohydrogen phosphates were approximately parallel. The authors suggested that in the monohydrogen phosphate solutions there is considerable ion-pair association. An activation energy of about 25 kcal/mole was obtained for oxygen exchange with the dihydrogen phosphates.

Oxygen exchange occurring during the hydrolysis of ethylene phosphate, dimethyl phosphate, and monomethyl phosphate has been important in deducing mechanisms for these hydrolyses (122,172). The exchange with monomethyl phosphate in concentrated acid solution (76) is acid-catalyzed (357).

Winter and Briscoe found no detectable exchange in 41 hr at 100° in $1.88M$ $Na_4P_2O_7$ (465). With the addition of $0.83M$ NaOH, they found 2.5% exchange in 41 hr, and with the addition of $0.13M$ H_2SO_4 15% exchange occurred in 64 hr. These results correspond to first-order rate constants of 6.2×10^{-4} and 2.6×10^{-3} hr^{-1}, respectively. No exchange between $POCl_3$ and H_2O was detected in 5 hr at ambient temperatures (176), but 5 hr is not a significant period of time at ordinary temperatures.

REFERENCES

1. Abbott, G. A., *J. Amer. Chem. Soc.*, **31**, 763 (1909).
2. Adams, C. J., A. J. Downs, and S. Cradock, *Ann. Repts. Chem. Soc.*, **65**, 216 (1968).
3. Aghileri, L. J., *Intern. J. Appl. Radiation Isotopes*, **15**, 549 (1964).
4. Aiken, S. M. and J. B. Gill, *J. Inorg. Nucl. Chem.*, **28**, 2460 (1966).
5. Amdur, I. and G. G. Hammes, *Chemical Kinetics*, McGraw-Hill, New York, 1966, pp. 55–56.
6. Andon, R. J. L., J. F. Counsell, J. F. Martin, and C. J. Mash, *J. Appl. Chem.*, **17**, 65 (1967).
7. Arlman, E. J., *Rec. Trav. Chim. Pays-Bas*, **56**, 919 (1937).
8. Attias, J., *J. Chim. Phys.*, **58**, 310 (1961).
9. Attias, J., *Compt. Rend.*, **254**, 1166 (1962).
10. Attias, J., *Chimia (Aarau)*, **20**, 17 (1966).
11. Avison, A. W. D., *J. Chem. Soc.*, **1955**, 732.
12. Bailly, M. C., *Bull. Soc. Chim. Fr.* (5) **9**, 314 (1942).
13. Bailly, M. C., *Bull. Soc. Chim. Fr.* (5) **9**, 340 (1942).
14. Bailly, M. C., *Bull. Soc. Chim. Fr.* (5) **9**, 405 (1942).
15. Bailly, M. C., *Bull. Soc. Chim. Fr.* (5) **9**, 421 (1942).
16. Bailly, O. and J. Gaumé, *Bull. Soc. Chim. Fr.* (5) **3**, 1396 (1936).
17. Bamann, E. and M. Meisenheimer, *Ber.*, **71B**, 1980 (1938).
18. Bamann, E. and M. Meisenheimer, *Ber.*, **71B**, 1711 (1938).
19. Bamann, E. and M. Meisenheimer, *Ber.* **71B**, 2086 (1938).
20. Bamann, E. and M. Meisenheimer, *Ber.*, **71B**, 2233 (1938).
21. Banks, B. E. C., *Chem. Brit.*, **5**, 514 (1969).
22. Banks, B. E. C. and C. A. Vernon, *Chem. Brit.*, **6**, 541 (1970).
23. Barnard, P. W. C., C. A. Bunton, D. Kellerman, M. M. Mhala, B. Silver, C. A. Vernon, and V. A. Welch, *J. Chem. Soc.* (*B*), **1966**, 227.
24. Barnard, P. W. C., C. A. Bunton, D. R. Llewellyn, K. G. Oldham, B. L. Silver, and C. A. Vernon, *Chem. Ind.*, **1955**, 760.
25. Barnard, P. W. C., C. A. Bunton, D. R. Llewellyn, C. A. Vernon, and V. A. Welch, *J. Chem. Soc.*, **1961**, 2670.
26. Bartell, L. S., L. S. Su, and H. Yow, *Inorg. Chem.*, **9**, 1903 (1970).
27. Bartlett, P. D. and E. S. Lewis, *J. Amer. Chem. Soc.*, **72**, 1005 (1950).
28. Baumgarten, P., *Chem. Ber.*, **67**, 1100 (1934).
29. Beans, H. T. and S. J. Kiehl, *J. Amer. Chem. Soc.*, **49**, 1878 (1927).
30. Bekturov, A. B., S. I. Kalmkov, K. V. Khon, and E. V. Poletaev, *Izv. Akad. Nauk Kaz. SSR, Ser. Khim.*, **20**, 1 (1970).
31. Bell, R. N., *Ind. Eng. Chem.*, **39**, 136 (1947).
32. Bell, R. N., *Ind. Eng. Chem.*, **40**, 1464 (1948).
33. Bell, R. N., L. F. Audrieth, and O. F. Hill, *Ind. Eng. Chem.*, **44**, 568 (1952).
34. Bel'skii, V. E., N. N. Bezzubova, and I. P. Gozman, *Zh. Obshch. Khim.*, **38**, 1330–1334 (1968); *J. Gen. Chem. USSR*, **38**, 1281 (1968).
35. Bel'skii, V. E. and M. V. Efremova, *Izv. Akad. Nauk SSSR, Ser. Khim.*, **1968**, 409.

36. Bel'skii, V. E., M. V. Efremova, and Z. V. Lustina, *Izv. Akad. Nauk SSSR, Ser. Khim.*, **1967**, 1236.
37. Bel'skii, V. E. and I. P. Gozman, *Zh. Obshch. Khim.*, **37**, 2730 (1967); *J. Gen. Chem. USSR*, **37**, 2599 (1967).
38. Bel'skii, V. E., A. N. Pudovik, M. V. Efremova, V. N. Eliseenokov, and A. R. Panteleeva, *Dokl. Akad. Nauk SSSR*, **180**, 351 (1968).
39. Bender, M. L. and J. M. Lawlor, *J. Amer. Chem. Soc.*, **85**, 3010 (1963).
40. Benkovic, S. J. and R. C. Hevey, *J. Amer. Chem. Soc.*, **92**, 4971 (1970).
41. Bentley, R., *J. Amer. Chem. Soc.*, **71**, 2765 (1949).
42. Berry, R. S., *J. Chem. Phys.*, **32**, 933 (1960).
43. Bhargava, H. N. and D. C. Srivastava, *J. Phys. Chem.*, **74**, 36 (1970).
44. Blumenthal, E. and J. B. M. Herbert, *Trans. Faraday Soc.*, **33**, 849 (1937).
45. Blumenthal, E. and J. B. M. Herbert, *Trans. Faraday Soc.*, **41**, 611 (1945).
46. Boyd, C. M., H. P. House, and O. Menis, *U.S. At. Energy Comm.* **ORNL-2893** (1960), *Chem. Abstr.*, **54**, 14901i (1960).
47. Boyd, D. B., *J. Amer. Chem. Soc.*, **91**, 1200 (1969).
48. Boyd, D. B., *Theor. Chim. Acta*, **18**, 184 (1970).
49. Briggs, P. J., D. P. N. Satchell, and G. F. White, *J. Chem. Soc. (B)*, **1970**, 1008.
50. Brodskii, A. I. and L. V. Sulima, *Dokl. Adak. Nauk SSSR*, **92**, 589 (1953).
51. Brovkina, I. A., *Zh. Obshch. Khim.*, **22**, 1917 (1952).
52. Brovkina, I. A., *Trudy Moskov Aviatsion Inst.*, **1955**, No. 52, 53; *Chem. Abstr.*, **53**, 10919a (1959).
53. Brown, D. M., J. A. Flint, and N. K. Hamer, *J. Chem. Soc.*, **1964**, 326.
54. Brown, D. M. and M. J. Frearson, *Chem Commun.*, **1968**, 1342.
55. Brown, D. M. and N. K. Hamer, *J. Chem. Soc.*, **1960**, 1155.
56. Brown, D. M. and A. R. Todd, *J. Chem. Soc.*, **1952**, 52.
57. Brown, D. M. and D. A. Usher, *J. Chem. Soc.*, **1965**, 6547.
58. Brown, D. M. and D. A. Usher, *J. Chem. Soc.*, **1965**, 6558.
59. Brown, I. D. and C. Calvo, *J. Solid State Chem.*, **1**, 173 (1970).
60. Bruice, T. C. and S. J. Benkovic, *Bioorganic Mechanisms*, Benjamin, New York, 1966.
61. Buist, G. J., C. A. Bunton, L. Robinson, and L. Sepulveda, *J. Amer. Chem. Soc.*, **92**, 4072 (1970).
62. Bunnett, J. F., *J. Amer. Chem. Soc.*, **82**, 499 (1960).
63. Bunnett, J. F., *J. Amer. Chem. Soc.*, **83**, 4956 (1961).
64. Bunnett, J. F. and F. P. Olsen, *Can. J. Chem.*, **44**, 1917 (1966).
65. Bunton, C. A., *J. Chem. Educ.*, **45**, 21 (1968).
66. Bunton, C. A., *Accts. Chem. Res.* **3**, 257 (1970).
67. Bunton, C. A. and H. Chaimovich, *Inorg. Chem.*, **4**, 1763 (1965).
68. Bunton, C. A. and H. Chaimovich, *J. Amer. Chem. Soc.*, **88**, 4082 (1966).
69. Bunton, C. A. and S. J. Farber, *J. Org. Chem.*, **34**, 767 (1969).
70. Bunton, C. A., S. J. Farber, and E. J. Fendler, *J. Org. Chem.*, **33**, 29 (1968).
71. Bunton, C. A., E. J. Fendler, and J. H. Fendler, *J. Amer. Chem. Soc.*, **89**, 1221 (1967).
72. Bunton, C. A., E. J. Fendler, E. Humeres, and K. U. Yang, *J. Org. Chem.*, **32**, 2806 (1967).
73. Bunton, C. A., E. J. Fendler, G. L. Sepulveda, and K. U. Yang, *J. Amer. Chem. Soc.*, **90**, 5512 (1968).

74. Bunton, C. A. and J. M. Hellyer, *J. Org. Chem.*, **34**, 2798 (1969).
75. Bunton, C. A., D. Kellerman, K. G. Oldham, and C. A. Vernon, *J. Chem. Soc.* (*B*), **1966**, 292.
76. Bunton, C. A., D. R. Llewellyn, K. G. Oldham, and C. A. Vernon, *J. Chem. Soc.*, **1958**, 3574.
77. Bunton, C. A., D. R. Llewellyn, K. G. Oldham, and C. A. Vernon, *J. Chem. Soc.*, **1958**, 3588.
78. Bunton, C. A., D. R. Llewellyn, C. A. Vernon, and V. A. Welch, *J. Chem. Soc.*, **1961**, 1636.
79. Bunton, C. A., M. M. Mhala, K. G. Oldham, and C. A. Vernon, *J. Chem. Soc.*, **1960**, 3293.
80. Bunton, C. A. and L. Robinson, *J. Org. Chem.*, **34**, 773 (1969).
81. Bunton, C. A., L. Robinson, and M. Stam, *J. Amer. Chem. Soc.*, **92**, 7393 (1970).
82. Butcher, W. W. and F. H. Westheimer, *J. Amer. Chem. Soc.*, **77**, 2420 (1955).
83. Buyers, A. G., *J. Phys. Chem.*, **66**, 939 (1962).
84. Callis, C. F., J. R. Van Wazer, and P. G. Arvan, *Chem. Revs.*, **54**, 785 (1954).
85. Calvo, C., *Can. J. Chem.*, **43**, 1139 (1965).
86. Calvo, C., *Can. J. Chem.*, **43**, 1147 (1965).
87. Calvo, C., *Acta Cryst.*, **23**, 289 (1967).
88. Calvo, C., *Inorg. Chem.*, **7**, 1345 (1968).
89. Campbell, D. O. and M. L. Kilpatrick, *J. Amer. Chem. Soc.*, **76**, 893 (1954).
90. Chambers, R. W. and H. G. Khorana, *Chem. Ind.*, **1956**, 1022.
91. Chambers, R. W. and H. G. Khorana, *J. Amer. Chem. Soc.*, **80**, 3749 (1958).
92. Chanley, J. D. and E. Feageson, *J. Amer. Chem. Soc.*, **77**, 4002 (1955).
93. Chanley, J. D. and E. Feageson, *J. Amer. Chem. Soc.*, **85**, 1181 (1963).
94. Chanley, J. D. and E. M. Gindler, *J. Amer. Chem. Soc.*, **75**, 4035 (1953).
95. Chanley, J. D., E. M. Gindler, and H. Sobotka, *J. Amer. Chem. Soc.*, **74**, 4347 (1952).
96. Cherbuliez, E., J. P. Leber, and M. Schwarz, *Helv. Chim. Acta*, **36**, 1189 (1953).
97. Cherbuliez, E., J. P. Leber, and P. Stucki, *Helv. Chim. Acta*, **36**, 537 (1953).
98. Cherbuliez, E., H. Probst, and J. Rabinowitz, *Helv. Chim. Acta*, **42**, 1377 (1959).
99. Clark, V. M., D. W. Hutchinson, and A. Todd, *J. Chem. Soc.*, **1961**, 722.
100. Clark, V. M., D. W. Hutchinson, G. W. Kirby, and A. Todd, *J. Chem. Soc.*, **1961**, 715.
101. Clark, V. M. and A. J. Kirby, *J. Amer. Chem. Soc.*, **85**, 3705 (1963).
102. Clark, V. M., A. J. Kirby, and A. Todd, *J. Chem. Soc.*, **1957**, 1497.
103. Clark, V. M. and S. G. Warren, *Proc. Chem. Soc.*, **1963**, 178.
104. Cohn, M., *J. Biol. Chem.*, **180**, 771 (1949).
105. Cohn, M. and T. R. Hughes, *J. Biol. Chem.*, **235**, 3250 (1960).
106. Cohn, M. and T. R. Hughes, *J. Biol. Chem.*, **237**, 176 (1962).
107. Cook, W. H., D. Lipkin, and R. Markham, *J. Amer. Chem. Soc.*, **79**, 3607 (1957).
108. Corbridge, D. E. C., *Acta Cryst.*, **9**, 308 (1956).

109. Corbridge, D. E. C., in *Topics In Phosphorus Chemistry*, M. Grayson and E. J. Griffith, Eds., Interscience, New York, 1966, Vol. 3.
110. Covitz, F. and F. H. Westheimer, *J. Amer. Chem. Soc.*, **85**, 1773 (1963).
111. Cox, J. R., Jr. and O. B. Ramsay, *Chem. Revs.*, **64**, 317 (1964).
112. Cox, J. R., Jr., O. B. Ramsay, and M. G. Newton, unpublished work cited in Ref. 111.
113. Cox, J. R., Jr., R. E. Wall, and F. H. Westheimer, *Chem. Ind.* **1959**, 929.
114. Crowther, J. P. and A. E. R. Westman, *Can. J. Chem.*, **32**, 42 (1954).
115. Crowther, J. and A. E. R. Westman, *Can. J. Chem.*, **34**, 969 (1956).
116. Cruickshank, D. W. J., *J. Chem. Soc.*, **1961**, 5486.
117. Crutchfield, M. M., C. F. Callis, R. R. Irani, and G. C. Roth, *Inorg. Chem.*, **1**, 813 (1962).
118. Crutchfield, M. M. and E. R. Irani, *J. Amer. Chem. Soc.*, **87**, 2815 (1965).
119. Davies, C. W. and C. B. Monk, *J. Chem. Soc.*, **1949**, 413.
120. de Decker, H. C. J. and C. H. MacGillavry, *Rec. Trav. Chim. Pays-Bas*, **60**, 153 (1941).
121. de Decker, H. C. J., *Rec. Trav. Chim. Pays-Bas*, **60**, 413 (1941).
122. Dennis, E. A. and F. H. Westheimer, *J. Amer. Chem. Soc.*, **88**, 3431 (1966).
123. Dennis, E. A. and F. H. Westheimer, *J. Amer. Chem. Soc.*, **88**, 3432 (1966).
124. Desjobert, A., *Bull. Soc. Chim. Fr.*, **1947**, 809.
125. Desjobert, A., *C. R. Acad. Sci.*, **224**, 575 (1947).
126. Desjobert, A., *Bull. Soc. Chim. Fr.*, **1951**, 42.
127. Desjobert, A., *Bull. Soc. Chim. Biol.*, **36**, 475 (1954).
128. Desjobert, A., *Bull. Soc. Chim. Biol.*, **37**, 683 (1955).
129. Desjobert, A., *Bull. Soc. Chim. Fr.*, **1963**, 683.
130. Dillon, K. B. and T. C. Waddington, *J. Chem. Soc. (A)*, **1970**, 1146.
131. DiSabato, G. and W. P. Jencks, *J. Amer. Chem. Soc.*, **83**, 4393 (1961).
132. DiSabato, G. and W. P. Jencks, *J. Amer. Chem. Soc.*, **83**, 4400 (1961).
133. Doremieux-Morin, C., J. M. Verdier, and R. Vincent, *Bull. Soc. Chim. Fr.*, **1967**, 1628.
134. Dornberger-Schiff, K., F. Liebau, and E. Thilo, *Acta Cryst.*, **8**, 752 (1955).
135. Drucker, C., *Acta Chem. Scand.*, **1**, 221 (1947).
136. Dudek, G. O. and F. H. Westheimer, *J. Amer. Chem. Soc.*, **81**, 2641 (1959).
137. Dunlap, P. M. and W. E. Groves, *Texas J. Sci.*, **14**, 151 (1962).
138. du Plessis, D. J., *Angew. Chem.*, **71**, 697 (1959).
139. Dvornikoff, M. N. and H. L. Morrill, *Anal. Chem.*, **20**, 935 (1948).
140. Faber, W., Ger. Pat. 734,511 (April 17, 1943).
141. Feldmann, W., *Z. Anorg. Allgem. Chem.*, **338**, 235 (1965).
142. Feldmann, W., *Chem. Ber.*, **99**, 3251 (1966).
143. Feldmann, W. and E. Thilo, *Z. Anorg. Allgem. Chem.*, **327**, 159 (1964).
144. Feldmann, W. and E. Thilo, *Z. Anorg. Allgem. Chem.*, **328**, 113 (1964).
145. Fleitmann, T., *Poggendorff's Ann. Phys. Chem.*, **78**, 233, 361 (1849).
146. Frank, D. S. and D. A. Usher, *J. Amer. Chem. Soc.*, **89**, 6360 (1967).
147. Friess, S. L., *J. Amer. Chem. Soc.*, **74**, 4027 (1952).
148. Friess, S. L., *J. Amer. Chem. Soc.*, **74**, 5521 (1952).
149. Friess, S. L., *J. Amer. Chem. Soc.*, **75**, 323 (1953).
150. Frost, A. A. and R. G. Pearson, *Kinetics and Mechanism*, Wiley, New York, 1953, p. 97.
151. Fuchs, N., *J. Russ. Phys. Chem. Soc.*, **61**, 1035 (1929); *Chem. Abstr.*, **24**, 543 (1930).

152. Garner, W. E., *Chemistry of the Solid State*, Butterworths, London, 1955.
153. Garus, L. I., M. O. Tereshkevich, and O. K. Skarre, *Zh. Fiz. Khim.*, **40**, 2222 (1966).
154. Gilham, P. T. and H. G. Khorana, *J. Amer. Chem. Soc.*, **80**, 6212 (1958).
155. Gill, J. B. and S. A. Riaz, *J. Chem. Soc.* (*A*), **1969**, 843.
156. Gillespie, R. J., R. Kapoor, and E. A. Robinson, *Can. J. Chem.*, **44**, 1203 (1966).
157. Ging, N. A. and J. M. Sturtevant, *J. Amer. Chem. Soc.*, **75**, 2087 (1953).
158. Ginjaar, L., and S. Vel, *Rec. Trav. Chim. Pays-Bas*, **77**, 956 (1958).
159. Giran, H., *Ann. Chim. Phys.* (7), **30**, 203 (1903).
160. Glonek, T., T. C. Meyers, P. Z. Han, and J. R. Van Wazer, *J. Amer. Chem. Soc.*, **92**, 7214 (1970).
161. Green, J., *Ind. Eng. Chem.*, **42**, 1542 (1950).
162. Greenwood, N. N. and A. Thompson, *J. Chem. Soc.*, **1959**, 3485.
163. Greenwood, N. N. and A. Thompson, *J. Chem. Soc.*, **1959**, 3493.
164. Griffith, E. J., *Anal. Chem.*, **29**, 198 (1957).
165. Griffith, E. J., *Ind. Eng. Chem.*, **51**, 240 (1959).
166. Griffith, E. J., *J. Inorg. Nucl. Chem.*, **27**, 1172 (1965).
167. Griffith, E. J. and R. L. Buxton, *Inorg. Chem.*, **4**, 549 (1965).
168. Griffith, E. J. and R. L. Buxton, *J. Amer. Chem. Soc.*, **89**, 2884 (1967).
169. Groves, W. O. and J. W. Edwards, *J. Phys. Chem.*, **65**, 645 (1961).
170. Grunze, I., K. Dostál, and E. Thilo, *Z. Anorg. Allgem. Chem.*, **302**, 221 (1959).
171. Grunze, H. and E. Thilo, *Z. Anorg. Allgem. Chem.*, **281**, 284 (1955).
172. Haake, P. C. and F. H. Westheimer, *J. Amer. Chem. Soc.*, **83**, 1102 (1961).
173. Hagman, L. O., I. Jansson, and C. Magnéli, *Acta Chem. Scand.*, **22**, 1419 (1968).
174. Hall, N. F. and O. R. Alexander, *J. Amer. Chem. Soc.*, **62**, 3455 (1940).
175. Hall, S. A. and M. Jacobson, *Ind. Eng. Chem.*, **40**, 694 (1948).
176. Halmann, M. and L. Kugel, *J. Chem. Soc.*, **1964**, 3733.
177. Hamer, N. K., *J. Chem. Soc.* (*C*), **1966**, 404.
178. Hammes, G. G., G. E. Maciel, and J. S. Waugh, *J. Amer. Chem. Soc.*, **83**, 2394 (1961).
179. Hampson, G. C. and A. J. Stosick, *J. Amer. Chem. Soc.*, **60**, 1814 (1938).
180. Hartley, S. B., W. S. Holmes, J. K. Jacques, M. F. Mole, and J. C. McCoubrey, *Quart. Rev., Chem. Soc.*, **17**, 204 (1963).
181. Hatch, G. B., U.S. Pat. 2,365,190 (December 19, 1944).
182. Havinga, E., R. O. DeJongh, and W. Dorst, *Rec. Trav. Chim. Pays-Bas*, **75**, 378 (1956).
183. Healy, R. M. and M. L. Kilpatrick, *J. Amer. Chem. Soc.*, **77**, 5258 (1955).
184. Healy, R. M. and M. L. Kilpatrick, *J. Amer. Chem. Soc.*, **79**, 6575 (1957).
185. Helleiner, C. W. and G. C. Butler, *Can. J. Chem.* **33**, 705 (1955).
186. Henry, W. M., G. Nickless, and F. H. Pollard, *J. Inorg. Nucl. Chem.*, **29**, 2479 (1967).
187. Hetzer, C., *Publ. Sci. Tech. Min. Air. Fr. No.* **153**, pp. 94 (1966).
188. Hock, A. and G. Huber, *Biochem. Z.*, **328**, 44 (1956).
189. Hofstetter, R. and A. E. Martell, *J. Amer. Chem. Soc.*, **81**, 4461 (1959).
190. Hofstetter, R., Y. Murakami, G. Mont, and A. E. Martell, *J. Amer. Chem. Soc.*, **84**, 3041 (1962).
191. Holbrook, K. A. and L. Ouellet, *Can. J. Chem.*, **35**, 1496 (1957).

192. Holbrook, K. A. and L. Ouellet, *Can. J. Chem.*, **36**, 686 (1958).
193. Hudson, R. F., *Structure and Mechanism in Organo-Phosphorus Chemistry*, Academic, New York, 1965.
194. Hudson, R. F. and M. Green, *Angew. Chem., Intern. Ed.*, **2**, 11 (1963).
195. Hudson, R. F. and D. C. Harper, *J. Chem. Soc.*, **1958**, 1356.
196. Huffman, E. O. and J. D. Fleming, *J. Phys. Chem.*, **64**, 240 (1960).
197. Huhti, A. L. and P. A. Gartaganis, *Can. J. Chem.*, **34**, 785 (1956).
198. Hussain, M. S., M. D. Joesten, and P. G. Lenhert, *Inorg., Chem.*, **9**, 162 (1970).
199. Huxley, A. F., *Chem. Brit.*, **6**, 477 (1970).
200. Iler, R. K., *J. Phys. Chem.*, **56**, 1086 (1952).
201. Indelli, A., *Ann. Chim. (Rome)* **46**, 367 (1956).
202. Indelli, A., *Ann. Chim. (Rome)*, **46**, 717 (1956).
203. Indelli, A., *Ann. Chim. (Rome)*, **47**, 586 (1957).
204. Indelli, A., *Ann. Chim. (Rome)*, **48**, 332 (1958).
205. Indelli, A., *Ahn. Chim. (Rome)*, **48**, 345 (1958).
206. Indelli, A. and G. Saglietto, *J. Inorg. Nucl. Chem.*, **25**, 1259 (1963).
207. Ingold, C. K., *Structure and Mechanism in Organic Chemistry*, 2nd ed., Cornell Univ. Press, Ithaca, N.Y., 1969.
208. Irani, R. R., *J. Phys. Chem.*, **65**, 1463 (1961).
209. Irani, R. R. and C. F. Callis, *J. Phys. Chem.*, **64**, 1398 (1960).
210. Irani, R. R. and T. A. Taulli, *J. Inorg. Nucl. Chem.*, **28**, 1011 (1966).
211. Irving, R. J. and H. McKerrell, *Trans. Faraday Soc.*, **63**, 2913 (1967).
212. Irving, R. J. and H. McKerrell, *Trans. Faraday Soc.*, **64**, 875 (1968).
213. Irving, R. J. and H. McKerrell, *Trans. Faraday Soc.*, **64**, 879 (1968).
214. Jencks, W. P., *Survey of Progress in Chemistry*, Academic, New York, 1963, Vol. 1, p. 249.
215. Joesten, M. D., M. S. Hussain, and P. G. Lenhert, *Inorg. Chem.*, **9**, 151 (1970).
216. Jones, H. W., C. B. Monk, and C. W. Davies, *J. Chem. Soc.*, **1949**, 2693.
217. Jost, K. H., *Acta Cryst.*, **14**, 779 (1961).
218. Jost, K. H., *Acta Cryst.*, **14**, 844 (1961).
219. Jost, K. H., *Acta Cryst.*, **15**, 951 (1962).
220. Jost, K. H., *Acta Cryst.*, **16**, 428 (1963).
221. Jost, K. H., *Acta Cryst.*, **16**, 623 (1963).
222. Jost, K. H., *Acta Cryst.*, **17**, 1539 (1964).
223. Jost, K. H., *Acta Cryst., Sec. (B)*, **24**, 992 (1968).
224. Kaiser, E. T. and K. Kudo, *J. Amer. Chem. Soc.*, **89**, 6725 (1967).
225. Kaiser, E. T., M. Panar, And F. H. Westheimer, *J. Amer. Chem. Soc.*, **85**, 602 (1963).
226. Kawabe, M., O. Ohashi, and I. Yamaguchi, *Bull. Chem. Soc. Jap.*, **43**, 3705 (1970).
227. Keisch, B., J. W. Kennedy, and A. C. Wahl, *J. Amer. Chem. Soc.*, **80**, 4778 (1958).
228. Kerker, M. and W. F. Epenscheid, *J. Amer. Chem. Soc.*, **80**, 776 (1958).
229. Kerr, S. E., *J. Biol. Chem.*, **139**, 131 (1941).
230. Ketelaar, J. A. A., and A. H. Bloksma, *Rec. Trav. Chim. Pays-Bas*, **67**, 665 (1948).
231. Ketelaar, J. A. A., and H. R. Gersmann, *Rec. Trav. Chim. Pay-Bas*, **77**, 973 (1958).

232. Ketelaar, J. A. A., H. R. Gersmann, and K. Koopmans, *Rec. Trav. Chim. Pays-Bas*, **71**, 1253 (1952).
233. Khan, S. A. and A. J. Kirby, *J. Chem. Soc.* (*B*), **1970**, 1172.
234. Khan, S. A., A. J. Kirby, M. Wakselman, D. P. Horning, and J. M. Lawlor, *J. Chem. Soc.* (*B*), **1970**, 1182.
235. Khorana, H. G., *Some Recent Developments in the Chemistry of Phosphate Esters of Biological Interest*, Wiley, New York, 1961.
236. Khorana, H. G., G. M. Tener, R. S. Wright, and J. G. Moffatt, *J. Amer. Chem. Soc.*, **79**, 430 (1957).
237. Knorana, H. G. and J. P. Vizsolyi, *J. Amer. Chem. Soc.*, **81**, 4660 (1959).
238. Kiehl, S. J. and E. Claussen, Jr., *J. Amer. Chem. Soc.*, **57**, 2284 (1935).
239. Kiehl, S. J. and H. P. Coats, *J. Amer. Chem. Soc.*, **49**, 2180 (1927).
240. Kiehl, S. J. and W. C. Hansen, *J. Amer. Chem. Soc.*, **48**, 2802 (1926).
241. Kiehl, S. J. and G. H. Wallace, *J. Amer. Chem. Soc.*, **49**, 375 (1927).
242. Kirby, A. J. and W. P. Jencks, *J. Amer. Chem. Soc.*, **87**, 3209 (1965).
243. Kirby, A. J. and W. P. Jencks, *J. Amer. Chem. Soc.*, **87**, 3217 (1965).
244. Kirby, A. J. and A. G. Varvoglis, *J. Amer. Chem. Soc.*, **88**, 1823 (1966).
245. Kirby, A. J. and A. G. Varvoglis, *J. Amer. Chem. Soc.*, **89**, 415 (1967).
246. Kirby, A. J. and A. G. Varvoglis, *Chem. Commun.*, **1967**, 405.
247. Kirby, A. J. and A. G. Varvoglis, *Chem. Commun.*, **1967**, 406.
248. Kirby, A. J. and M. Younas, *J. Chem. Soc.* (*B*), **1970**, 510.
249. Kirby, A. J. and M. Younas, *J. Chem. Soc.* (*B*), **1970**, 1165.
250. Kluger, R. and F. H. Westheimer, *J. Amer. Chem. Soc.*, **91**, 4143 (1969).
251. Koefoed, J. and A. H. Jensen, *Acta Chem. Scand.*, **5**, 23 (1951).
252. Koshland, D. E., Jr., *J. Amer. Chem. Soc.*, **74**, 2286 (1952).
253. Kugel, L. and M. Halmann, *J. Org. Chem.*, **32**, 642 (1967).
254. Kumamoto, J., J. R. Cox, Jr., and F. H. Westheimer, *J. Amer. Chem. Soc.*, **78**, 4858 (1956).
255. Kumamoto, J. and F. H. Westheimer, *J. Amer. Chem. Soc.*, **77**, 2515 (1955).
256. Kurz, J. L. and C. D. Gutsche, *J. Amer. Chem. Soc.*, **82**, 2175 (1960).
257. Kuz'michev, S. I., *Trudy Moskov Aviatsion Inst.*, **1955**, (52), 36; *Chem. Abstr.*, **52**, 19665h (1958).
258. Lambert, S. N. and J. I. Watters, *J. Amer. Chem. Soc.*, **79**, 4262 (1957).
259. Lapidot, A., D. Samuel, and B. Silver, *Chem. Ind.*, **1963**, 468.
260. Lapidot, A., D. Samuel, and M. Weiss-Broday, *J. Chem. Soc.*, **1964**, 637.
261. Liebau, F., *Acta Cryst.*, **9**, 811 (1956).
262. Liebau, F., *Angew. Chem., Intern. Ed.*, **2**, 562 (1963).
263. Liébecq, C., *Arch. Intern. Physiol. Biochem.*, **65**, 141 (1957).
264. Léibecq, C. and M. Jacquemotte-Louis, *Bull. Soc. Chim. Biol.*, **40**, 67 (1958).
265. Liébecq. C. and M. Jacquemotte-Louis, *Bull. Soc. Chim. Biol.*, **40**, 759 (1958).
266. Liébecq, C. and M. Jacquemotte-Louis, *Arch. Intern. Physiol. Biochim.*, **66**, 72 (1958).
267. Lincoln, S. F. and D. R. Stranks, *Aust. J. Chem.*, **21**, 37 (1968).
268. Lincoln, S. F. and D. R. Stranks, *Aust. J. Chem.*, **21**, 57 (1968).
269. Lincoln, S. F. and D. R. Stranks, *Aust. J. Chem.*, **21**, 67 (1968).
270. Lipkin, D. R. Markham, and W. H. Cook, *J. Amer. Chem. Soc.*, **81**, 6075 (1959).

NONENZYMIC HYDROLYSIS 249

271. Lipmann, F., *Adv. Enzymol.*, **1**, 99 (1941).
272. Lohmann, K., *Biochem. Z.*, **233**, 460 (1931).
273. Long, F. A. and R. Bakule, *J. Amer. Chem. Soc.*, **85**, 2313 (1963).
274. Long, F. A., J. G. Pritchard, and F. E. Stafford, *J. Amer. Chem. Soc.*, **79**, 2362 (1957).
275. McAdam, A., K. H. Jost, and B. Beagley, *Acta Cryst., Sec. (B)*, **24**, 1621 (1968).
276. McCullough, J. F., J. R. Van Wazer, and E. J. Griffith, *J. Amer. Chem. Soc.*, **78**, 4528 (1956).
277. MacGillavry, C. H., H. C. J. de Decker, and L. M. Nijland, *Nature*, **164**, 448 (1949).
278. McGilvery, J. D. and J. P. Crowther, *Can. J. Chem.*, **32**, 174 (1954).
279. McTique, P. T. and P. V. Renowden, *Aust. J. Chem.*, **23**, 297 (1970).
280. Malmgren, H., *Acta Chem. Scand.*, **2**, 147 (1948).
281. Mal'tseva, I. M., I. N. Shokin, E. L. Yakhontova, and T. A. Shirokova, *Tr. Mosk. Khim.-Tekhnol. Inst.*, **1967**, 221; *Chem. Abstr.*, **70**, 14793t (1969).
282. March, J., *Advanced Organic Chemistry: Reactions, Mechanisms, and Structure*, McGraw-Hill, New York, 1968.
283. Marcus, A. and W. B. Elliott, *J. Amer. Chem. Soc.*, **80**, 4287 (1958).
284. Marecek, J. F. and D. L. Griffith, *J. Amer. Chem. Soc.*, **92**, 917 (1970).
285. Marsmann, H., L. C. D. Groenweghe, L. J. Schaad, and J. R. Van Wazer, *J. Amer. Chem. Soc.*, **92**, 6107 (1970).
286. Marsmann, H., J. R. Van Wazer, and J. B. Robert, *J. Chem. Soc. (A)*, **1970**, 1566.
287. Meadowcroft, T. R. and F. D. Richardson, *Trans. Faraday Soc.*, **59**, 1564 (1963).
288. Meyerhof, O., R. Shatas, and A. Kaplan, *Biochim. Biophys. Acta*, **12**, 121 (1953).
289. Mhala, M. M., C. P. Holla, G. Kasturi, and K. Gupta, *Indian J. Chem.*, **8**, 51 (1970).
290. Mhala, M. M., C. P. Holla, G. Kasturi, and K. Gupta, *Indian J. Chem.*, **8**, 333 (1970).
291. Mhala, M. M. and M. D. Patwardhan, *Indian J. Chem.*, **6**, 704 (1968).
292. Mhala, M. M., M. D. Patwardhan, and T. R. Kasturi, *Indian J. Chem.*, **7**, 145 (1969).
293. Mil'chenko, V. V., *Trudy Moskov Aviatsion Inst.*, **1955** (52), 47; *Chem. Abstr.*, **53**, 10919c (1959).
294. Miller, D. L. and T. Ukena, *J. Amer. Chem. Soc.*, **91**, 3050 (1969).
295. Miller, D. L. and F. H. Westheimer, *Science*, **148**, 667 (1965).
296. Miller, D. L. and F. H. Westheimer, *J. Amer. Chem. Soc.*, **88**, 1507 (1966).
297. Miller, D. L. and F. H. Westheimer, *J. Amer. Chem. Soc.*, **88**, 1514 (1966).
298. Mitchell, K. A. R., *Chem. Revs.*, **69**, 157 (1969).
299. Mitra, R. P. and B. R. Thurkal, *Indian J. Chem.*, **8**, 350 (1970).
300. Mohan, M. S. and G. A. Rechnitz, *J. Amer. Chem. Soc.*, **92**, 5839 (1970).
301. Monk, C. B., *J. Chem. Soc.*, **1949**, 423.
302. Monk, C. B., *J. Chem. Soc.*, **1949**, 427.
303. Mooney, R. W. and M. A. Aia, *Chem. Rev.*, **61**, 433 (1961).
304. Morey, G. W., *J. Amer. Chem. Soc.*, **80**, 775 (1958).
305. Moriguchi, Y. and M. Miura, *Bull. Chem. Soc. Jap.*, **38**, 678 (1965).

306. Moriguchi, Y., *Bull. Chem. Soc. Jap.*, **39**, 2656 (1966).
307. Muetterties, E. L. and R. A. Schunn, *Quart. Revs.*, **20**, 245 (1966).
308. Murakami, Y., *Topics in Chelate Chemistry and Biochemistry*, Special Publication No. 79 of Kagaku no Ryoiki, A. Nakahara, Ed., Nankodo, Tokyo, 1967, pp. 153–188.
309. Murakami, Y., *Nippon Kagaku Zasshi*, **91**, 185 (1970).
310. Murakami, Y. and A. E. Martell, *J. Phys. Chem.*, **67**, 582 (1963).
311. Murakami, Y. and A. E. Martell, *J. Amer. Chem. Soc.*, **86**, 2119 (1964).
312. Murakami, Y., J. Sunamoto, and H. Ishizu, *J. Chem. Soc. (D)*, **1970**, 1665.
313. Murakami, Y. and M. Takagi, *J. Amer. Chem. Soc.*, **91**, 5130 (1969).
314. Muus, J., *Z. Phys. Chem.*, **159A**, 268 (1932).
315. Nanninga, L. B., *J. Amer. Chem. Soc.*, **79**, 1144 (1957).
316. Nelson, A. K., *J. Chem. Eng. Data.* **9**, 357 (1964).
317. Netherton, L. E., A. R. Wreath, and D. N. Bernhart, *Anal. Chem.*, **27**, 860 (1955).
318. Newton, M. G., J. R. Cox, Jr., and J. A. Bertrand, *J. Amer. Chem. Soc.*, **88**, 1503 (1966).
319. Oakenfull, D. G., D. I. Richardson, and D. A. Usher, *J. Amer. Chem. Soc.*, **89**, 5491 (1967).
320. Oestreich, C. H., and M. M. Jones, *Biochemistry*, **5**, 2926 (1966).
321. Oestreich, C. H. and M. M. Jones, *Biochemistry*, **5**, 3151 (1966).
322. Oestreich, C. H. and M. M. Jones, *Biochemistry*, **6**, 1515 (1967).
323. Ohashi, S. and T. Matsumura, *Bull. Chem. Soc. Jap.*, **35**, 501 (1962).
324. Ohashi, S. and F. Oshima, *Bull. Chem. Soc. Jap.*, **36**, 1489 (1963).
325. Ohlmeyer, P. and R. Shatas, *Arch. Biochem. Biophys.*, **36**, 411 (1953).
326. Osterheld, R. K., *J. Phys. Chem.*, **62**, 1133 (1958).
327. Osterheld, R. K. and J. P. Lampi (in preparation).
328. Osterheld, R. K. and M. M. Lardy (in preparation).
329. Otani, S., M. Miura, and T. Doi, *Kogyo Kagaku Zasshi*, **66**, 593 (1963).
330. Park, J. H. and D. E. Koshland, *J. Biol. Chem.*, **233**, 986 (1958).
331. Parks, J. R. and J. R. Van Wazer, *J. Amer. Chem. Soc.*, **79**, 4890 (1957).
332. Pauling, L., *The Nature of the Chemical Bond*, 3d ed., Cornell Univ. Press, Ithaca, N.Y., 1960, pp. 559–562.
333. Pauling, L., *Chem. Brit.*, **6**, 468 (1970).
334. Pearson, R. G. and J. Songstad, *J. Amer. Chem. Soc.*, **89**, 1827 (1967).
335. Pfanstiel, R. and R. K. Iler, *J. Amer. Chem. Soc.*, **74**, 6059 (1952).
336. Phillips, D. R. and T. H. Fife, *J. Amer. Chem. Soc.*, **90**, 6803 (1968).
337. Postnikov, L. M., *Vestnik Moskov. Univ.*, **5**, *Ser. Fiz-Mat. i Estest. Nauk* No. 3, 63 (1950); *Chem. Abstr.*, **45**, 4594d (1951).
338. Prodan, E. A., M. Pavlyuchenko, and V. A. Budnikova, *Vestsi Akad. Navuk Berlaruss. SSR, Ser. Khim. Navuk*, **1967**, 95.
339. Prodan, E. A., M. M. Pavlyuchenko, and V. A. Budnikova, *Vestsi Akad. Navuk Belaruss. SSR, Ser. Khim. Navuk*, **1968**, 53.
340. Prodan, E. A., M. M. Pavlyuchenko, L. A. Lesnikovich, and V. A. Sotnikova-Yuzhi, *Dokl. Akad. Nauk Beloruss. SSR*, **14**, 905 (1970).
341. Prodan, E. A., M. M. Pavlyuchenko, Yu. M. Sotnikov-Yuzhik, Yu. G. Zonor, and V. A. Budnikova, *Dokl. Akad. Beloruss. SSR*, **11**, 708 (1967).
342. Quimby, O. T., *J. Phys. Chem.*, **58**, 603 (1954).
343. Quimby, O. T. and T. J. Flautt, *Z. Anorg. Allgem. Chem.*, **296**, 220 (1958).

344. Rainey, J. M., M. M. Jones, and W. L. Lockhart, *J. Inorg. Nucl. Chem.*, **26**, 1415 (1964).
345. Raistrick, B., *Disc. Faraday Soc.*, **5**, 234 (1949).
346. Raistrick, B., *Roy. Coll. Sci. J.*, **19**, 9 (1949).
347. Ramierz, F., *Accts. Chem. Res.*, **1**, 168 (1968).
348. Ramirez, F., B. Hansen, and N. B. Desai, *J. Amer. Chem. Soc.*, **84**, 4588 (1962).
349. Ramirez, F., O. P. Madan, N. B. Desai, S. Meyerson, and E. M. Banas, *J. Amer. Chem. Soc.*, **85**, 2681 (1963).
350. Rätz, R. and E. Thilo, *Liebig's Ann. Chem.*, **572**, 173 (1951).
351. Robertson, B. E. and C. Calvo, *Acta Cryst.*, **22**, 665 (1967).
352. Robertson, B. E. and C. Calvo, *Can. J. Chem.*, **46**, 605 (1968).
353. Rodionova, N. I. and J. W. Khodakov, *J. Gen. Chem. USSR*, **20**, 1401 (1950).
354. Ross, R. A. and C. A. Vernon, *Chem. Brit.*, **6**, 539 (1970).
355. Saini, G. and L. Trossarelli, *Ann. Chim. (Rome)*, **46**, 243 (1956).
356. Samuel, D. and B. Silver, *J. Chem. Soc.*, **1961**, 4321.
357. Samuel, D. and B. Silver, unpublished results cited in *Advan. Phys. Org. Chem.*, **3**, 123 (1965).
358. B. Sansoni, *Angew. Chem.*, **67**, 327 (1955).
359. Schaleger, L. L. and F. A. Long, *Advances in Physical Organic Chemistry*, V. Gold, Ed., Academic, New York, 1963, Vol. 1, p. 1.
360. Schmidt, W. and H. Taube, *Inorg. Chem.*, **2**, 698 (1963).
361. Schmulbach, C. D., J. R. Van Wazer, and R. R. Irani, *J. Amer. Chem. Soc.*, **81**, 6347 (1959).
362. Schneider, P. W. and H. Brintzinger, *Helv. Chim. Acta*, **47**, 992 (1964).
363. Schneider, P. W. and H. Brintzinger, *Helv. Chim. Acta*, **47**, 1717 (1964).
364. Schomaker, V. and D. P. Stevenson, *J. Amer. Chem. Soc.*, **63**, 37 (1941).
365. Schülke, U., *Z. Anorg. Allgem. Chem.*, **360**, 231 (1968).
366. Selim, M. and P. Leduc, *Compt. Rend.*, **248**, 1187 (1959).
367. Shaver, K. J. and J. G. Stites, Jr., 131st National Meeting, American Chemical Society, Miami, Florida, April, 1957.
368. Shen, C. Y., *Ind. Eng. Chem., Prod. Res. Develop.*, **5**, 272 (1966).
369. Shen, C. Y. and D. R. Dyroff, *Ind. Eng. Chem., Prod. Res. Develop.*, **5**, 97 (1966).
370. Shen, C. Y., J. S. Metcalf, and E. V. O'Grady, *Ind. Eng. Chem.*, **51**, 717 (1959).
371. Shen, C. Y., N. E. Stahlheber, and D. R. Dyroff, *J. Amer. Chem. Soc.*, **91**, 62 (1969).
372. Shima, M., K. Hamamoto, and S. Utsumi, *Bull. Chem. Soc. Jap.*, **33**, 1386 (1960).
373. Smith, M. J., *Can. J. Chem.*, **37**, 1115 (1959).
374. Sommer, L. H., *Stereochemistry, Mechanism, and Silicon*, McGraw-Hill, New York, 1965.
375. Sommer, L. H. and O. F. Bennett, *J. Amer. Chem. Soc.*, **79**, 1008 (1957).
376. Sommer, L. H. and O. F. Bennett, *J. Amer. Chem. Soc.*, **81**, 251 (1959).
377. Steitz, T. A. and W. N. Lipscomb, *J. Amer. Chem. Soc.*, **87**, 2488 (1965).
378. Stiller, M., T. Diamondstone, R. Witonsky, D. Baltimore, R. J. Rutman, and P. George, *Fed. Proc.*, **24** (2), Abstract No. 1300 (1965).
379. Stranks, D. R. and R. G. Wilkins, *Chem. Revs.*, **57**, 743 (1957).

380. Strauss, U. P. and J. W. Day, *J. Polym. Sci.*, *Part C*, **16**, 2161 (1967).
381. Strauss, U. P. and G. J. Krol, *J. Polym. Sci.*, *Part C*, **16**, 2171 (1967).
382. Strauss, U. P. and E. H. Smith, *J. Amer. Chem. Soc.*, **75**, 6186 (1953).
383. Strauss, U. P., E. H. Smith, and P. L. Wineman, *J. Amer. Chem. Soc.*, **75**, 3935 (1953).
384. Strauss, U. P. and T. L. Treitler, *J. Amer. Chem. Soc.*, **77**, 1473 (1955).
385. Strauss, U. P. and T. L. Treitler, *J. Amer. Chem. Soc.*, **78**, 3553 (1956).
386. Swoboda, P. A. T. and E. M. Crook, *Biochem. J.*, **59**, xxiv (1955).
387. Teichert, W. and K. Rinman, *Acta Chem. Scand.*, **2**, 414 (1948).
388. Tereshkevich, M. O., L. I. Garus, A. F. Kulish, E. S. Varenko, and V. P. Galushko, *Teor. i Eksperim. Khim.*, *Akad. Nauk Ukr. SSR*, **2**, 213 (1966).
389. Tetas, M. and J. M. Lowenstein, *Biochemistry*, **2**, 350 (1963).
390. Thain, E. M., *J. Chem. Soc.*, **1957**, 4694.
391. Thilo, E., *Chem. Technik*, **4**, 345 (1952).
392. Thilo, E., *Chem. Soc.* (*London*) *Spec. Publ. No. 15*, 33 (1961).
393. Thilo, E., *Adv. Inorg. Chem. Radiochem.*, **4**, 1 (1962).
394. Thilo, E., *Colloq. Intern. Centre Natl. Rech. Sci.* (*Paris*), *No. 106*, 491 (1962).
395. Thilo, E., *Rev. Chim.*, *Acad. Rep. Populaire Roumaine*, **7**, 585 (1962); *Chem. Abstr.*, **59**, 2381b (1963).
396. Thilo, E., *Angew. Chem.*, *Intern, Ed.*, **4**, 1061 (1965).
397. Thilo, E., *Pure Appl. Chem.*, **12**, 463 (1966).
398. Thilo, E. and G. Blumenthal, *Z. Anorg. Allgem. Chem.*, **348**, 77 (1966).
399. Thilo, E. and K. Dostál, *Z. Anorg. Allgem. Chem.*, **298**, 100 (1959).
400. Thilo, E., I. Grunze, and H. Grunze, *Monatsber. Deut. Akad. Wiss. Berlin*, **1**, 40 (1959).
401. Thilo, E. and U. Hauschild, *Z. Anorg. Allgem. Chem.*, **261**, 324 (1950).
402. Thilo, E. and L. Kolditz, *Z. Anorg. Allgem. Chem.*, **278**, 122 (1955).
403. Thilo, E. and I. Plaetsche, *Z. Anorg. Allgem. Chem.*, **260**, 297 (1950).
404. Thilo, E. and R. Rätz, *Z. Anorg. Allgem. Chem.*, **258**, 33 (1949).
405. Thilo, E. and R. Rätz, *Z. Anorg. Chem.*, **260**, 255 (1949).
406. Thilo, E. and R. Sauer, *J. Prakt. Chem.* (4), **4**, 324 (1957).
407. Thilo, E. and U. Schülke, *Angew. Chem.*, *Intern. Ed.*, **2**, 742 (1963).
408. Thilo, E. and U. Schülke, *Z. Anorg. Allgem. Chem.*, **341**, 293 (1965).
409. Thilo, E., G. Schulz, and E. M. Wichmann, *Z. Anorg. Allgem. Chem.*, **272**, 182 (1953).
410. Thilo, E. and H. Seeman, *Z. Anorg. Allgem. Chem.*, **267**, 65 (1951).
411. Thilo, E. and A. Sonntag, *Z. Anorg. Allgem. Chem.*, **291**, 186 (1957).
412. Thilo, E. and F. von Lampe, *Z. Anorg. Allgem. Chem.*, **319**, 387 (1963).
413. Thilo, E. and F. von Lampe, *Z. Anorg. Allgem. Chem.*, **349**, 1 (1967).
414. Thilo, E. and M. Wallis, *Chem. Ber.*, **86**, 1213 (1953).
415. Thilo, E. and W. Wieker, *J. Polym. Sci.*, **53**, 55 (1961).
416. Thilo, E. and W. Wieker, *Mezhdunarod. Simpozium po Makromol. Khim.*, *Doklady, Moscow*, **1960**, Sektsiya 3, 399; *Chem. Abstr.* **56**, 6694d (1962).
417. Thilo, E. and W. Wieker, *Z. Anorg. Allgem. Chem.*, **277**, 27 (1954).
418. Thilo, E. and W. Wieker, *Z. Anorg. Allgem. Chem.*, **291**, 164 (1957).
419. Thilo, E. and W. Wieker, *Z. Anorg. Allgem. Chem.*, **313**, 296 (1961).
420. Thilo, E. and H. Woggon, *Z. Anorg. Allgem. Chem.*, **277**, 17 (1954).
421. Titano, T. and K. Goto, *Bull. Chem. Soc. Jap.*, **13**, 667 (1938).
422. Titano, T. and K. Goto, *Bull. Chem. Soc. Jap.*, **14**, 77 (1939).

423. Tordjman, I., M. Beucher, J. C. Guitel, and G. Bassi, *Bull. Soc. Fr. Mineral. Crystallogr.*, **91**, 344 (1968).
424. Toy, A. D. F., *J. Amer. Chem. Soc.*, **70**, 3882 (1948).
425. Toy, A. D. F., *J. Amer. Chem. Soc.*, **72**, 2065 (1950).
426. Traylor, P. S. and F. H. Westheimer, *J. Amer. Chem. Soc.*, **87**, 553 (1965).
427. Tukaszewicz, K. and R. Smajkioewicz, *Roczniki Chem.*, **35**, 741 (1961).
428. Tukaszewicz, K., *Bull Acad. Polon. Sci. Ser. Sci. Chem.*, **15**, 47 (1967).
429. Usher, D. A., E. A. Dennis, and F. H. Westheimer, *J. Amer. Chem. Soc.*, **87**, 2320 (1965).
430. van Hooidonk, C. and L. Ginjaar, *Rec. Trav. Chim. Pays-Bas*, **86**, 449 (1967).
431. Vanstrom, R. E., F. McCullough, and L. E. Netherton, *Abstract 15, Div. Inorg. Chem.*, American Chemical Society Meeting, Boston, Mass., April 1959.
432. Van Wazer, J. R., *Phosphorus and Its Compounds*, Interscience, New York, 1958, Vol. 1.
433. Van Wazer, J. R., *Phosphorus and Its Compounds*, Interscience, New York, 1958, Vol. 1, p. 490.
434. Van Wazer, J. R., *J. Amer. Chem. Soc.*, **72**, 644 (1950).
435. Van Wazer, J. R., *J. Amer. Chem. Soc.*, **72**, 647 (1950).
436. Van Wazer, J. R., *J. Amer. Chem. Soc.*, **72**, 906 (1950).
437. Van Wazer, J. R. and C. F. Callis, *Chem. Revs.*, **58**, 1011 (1958).
438. Van Wazer, J. R. and D. A. Campanella, *J. Amer. Chem. Soc.*, **72**, 655 (1950).
439. Van Wazer, J. R. and E. J. Griffith, *J. Amer. Chem. Soc.*, **77**, 6140 (1955).
440. Van Wazer, J. R., E. J. Griffith, and J. F. McCullough, *J. Amer. Chem. Soc.*, **74**, 4977 (1952).
441. Van Wazer, J. R., E. J. Griffith, and J. F. McCullough, *J. Amer. Chem. Soc.*, **77**, 287 (1955).
442. Van Wazer, J. R., M. Goldstein, and E. Farber, *J. Amer. Chem. Soc.*, **75**, 1563 (1953).
443. Van Wazer, J. R. and K. A. Holst, *J. Amer. Chem. Soc.*, **72**, 639 (1950).
444. Van Wazer, J. R. and E. Karl-Kroupa, *J. Amer. Chem. Soc.*, **78**, 1772 (1956).
445. von Lampe, F., *Z. Anorg. Allgem. Chem.*, **367**, 170 (1969).
446. Wagman, D. D., W. H. Evans, V. B. Parker, I. Halow, S. M. Bailey, and R. H. Schumm, N.B.S. Technical Note, *TN 270-3* (1968).
447. Wagman, D. D., W. H. Evans, V. B. Parker, I. Halow, S. M. Bailey, and R. H. Schumm, N.B.S. Technical Note, *TN 270-4* (1969).
448. Wall, F. T. and R. H. Doremus, *J. Amer. Chem. Soc.*, **76**, 868 (1954).
449. Walsh, C. T., J. G. G. Hildebrand, and L. B. Spector, *J. Biol. Chem.*, **245**, 5699 (1970).
450. Watters, J. I., S. M. Lambert, and E. D. Loughran, *J. Amer. Chem. Soc.*, **79**, 3651 (1957).
451. Watters, J. I. and S. Matsumoto, *J. Inorg. Nucl. Chem.*, **29**, 2955 (1967).
452. Watters, J. I., P. E. Sturrock, and R. E. Simonaitis, *Inorg. Chem.*, **2**, 765 (1963).
453. Watzel, R., *Die Chemie*, **55**, 356 (1942).
454. Webb, N. C., *Acta Cryst.*, **21**, 942 (1966).
455. Westheimer, F. H., *Accts. Chem. Res.*, **1**, 70 (1968).

456. Westman, A. E. R. and A. E. Scott, *Nature*, **168**, 740 (1951).
457. Westman, A. E. R., A. E. Scott, and J. T. Pedley, *Chem. in Can.*, **4**, 35 (1952).
458. Wieker, W., *Z. Elektrochem.*, **64**, 1047 (1960).
459. Wieker, W., *Z. Anorg. Allgem. Chem.*, **313**, 309 (1961).
460. Wieker, W., *Z. Anorg. Allgem. Chem.*, **355**, 20 (1967).
461. Wieker, W. and E. Thilo, *Z. Anorg. Allgem. Chem.*, **306**, 48 (1960).
462. Wieker, W. and E. Thilo, *Z. Anorg. Allgem. Chem.*, **313**, 296 (1961).
463. Wilkie, D., *Chem. Brit.*, **6**, 472 (1970).
464. Winkler, A. and E. Thilo, *Z. Anorg. Allgem. Chem.*, **298**, 302 (1959).
465. Winter, E. R. S. and H. V. A. Briscoe, *J. Chem. Soc.*, **1942**, 631.
466. Winter, E. R. S., M. Carlton, and H. V. A. Briscoe, *J. Chem. Soc.*, **1940**, 131.
467. Wodtcke, F. and E. Thilo, *Monatsber. Deut. Akad. Wiss. Berlin*, **1**, 508 (1959).
468. Wolhoff, J. A. and J. Th. G. Overbeek, *Rec. Trav. Chim. Pays-Bas*, **78**, 759 (1959).
469. Wu, C. H., R. J. Witonsky, P. George, and R. J. Rutman, *J. Amer. Chem. Soc.*, **89**, 1987 (1967).
470. Young, D. A., *Decomposition of Solids*, Pergamon, New York, 1966.
471. Zettlemoyer, A. E. and C. H. Schneider, *J. Amer. Chem. Soc.*, **78**, 3870 (1956).
472. Zettlemoyer, A. C., C. H. Schneider, H. V. Anderson, and R. J. Fuchs, *J. Phys. Chem.*, **61**, 991 (1957).

Cyclophosphates

S. Y. KALLINEY

Schering Corporation, Bloomfield, New Jersey

CONTENTS

I. Introduction

The continued widespread applications of phosphorous compounds resulted in considerable expansion of the literature on phosphorous chemistry. The role of phosphate in biological activities has long been recognized. The use of condensed phosphates as food additives gave them importance in the field of toxicology.

Since early in this century, most attention has been focused on the chain phosphates; it is only in recent decades that an interest in the cyclic phosphates has developed. This may be attributed to modern analytical techniques for the structural characterization of condensed phosphates; this in turn led to the acceptance of a general system for classification of condensed phosphates (284,327).

This chapter is concerned with the known members of inorganic cyclophosphates which are heterocycles in which phosphorus(V) and oxygen are alternating around the ring. They are made exclusively of middle PO_4 groups. They have the general formula $M_n^I(PO_3)_n$

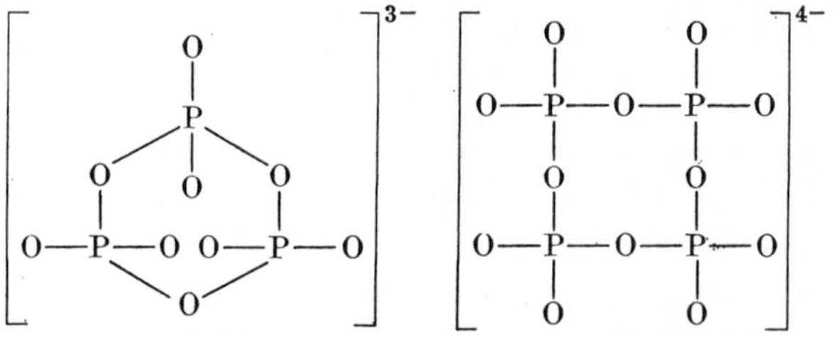

trimetaphosphate
(cyclotriphosphate)
$P_3O_9^{3-}$

tetrametaphosphate
(cyclotetraphosphate)
$P_4O_{12}^{4-}$

where $n = 3, 4, 5, \ldots$, etc. The most common members are the tri- and tetrametaphosphate anions. Haiduc (119) has recently introduced the term cyclophosphate (or cyclophosphoxanate) as a replacement for the metaphosphate terminology. For example, cyclotriphosphate would substitute for trimetaphosphate and cyclotetraphosphate for tetrametaphosphate. The cyclophosphate terminology is preferred since it is more descriptive of the structure than metaphosphate, which originated from Berzelius' double oxide notation as will be described in the historical section. The cyclophosphate terminology will be used in most of the text.

Chain or linear phosphates will refer to "polyphosphates"; this terminology is used frequently in the literature. The prefix "poly" is meaningless for this particular series since all condensed phosphates are poly. Perhaps it is possible to do away with this prefix. For instance, diphosphate could be used to describe pyrophosphate and triphosphate to describe tripolyphosphate.

Because of the wide differences encountered in the nomenclature of the condensed phosphates, it would appear desirable to conduct an international meeting for naming them. The need for systematic nomenclature cannot be overstressed.*

Two extensive reviews on the cyclophosphates have been made by Van Wazer (321) and Haiduc (119). Others have also been written in connection with other types of condensed phosphates (145,176,228, 284–295,322). Most of the literature has appeared under the term "metaphosphate." Special emphasis will be placed on the syntheses and thermochemistry of the cyclophosphates, the discovery of new members of the cyclophosphate family, the application of chemical analysis to cyclophosphate determination, and the interaction of cyclophosphates with metal ions in aqueous solution. Recent developments in this area will receive particular attention.

II. History

The syntheses and properties of cyclophosphates (metaphosphates) and the other condensed phosphates have been a subject of study since the early days of the 18th century. Although Graham (98) is credited with the discovery of cyclotriphosphate, earlier work by Berzelius and Proust must also be acknowledged. Berzelius (23) prepared monosodium orthophosphate and described heating it to redness prior to Na_2O and P_2O_5 analysis. Proust (227) described in some detail his preparation of glassy sodium phosphate by heating $NaNH_4HPO_4$.

Graham's discovery of cyclotriphosphate and other long-chain phosphates came during the course of his investigation of the function of water of crystallization in phosphate salts in order to solve the conflict between Lavoisier's theory and Clark's discovery of sodium pyrophosphate. Clark (41) prepared sodium pyrophosphate by heating $Na_2HPO_4 \cdot 12\,H_2O$. He demonstrated that the new compound,

$$2\,Na_2HPO_4 \cdot 12\,H_2O \xrightarrow[\text{bath}]{\text{sand}} 2\,Na_2HPO_4 + 24\,H_2O \xrightarrow[\text{heat}]{\text{red}} Na_4P_2O_7 + 25\,H_2O$$

* After this chapter was written, the IUPAC Commission on the Nomenclature of Inorganic Chemistry designated cyclic and chain structures in a similar manner used in this chapter [IUPAC Commission on the Nomenclature of Inorganic Chemistry, *Pure and Appl. Chem.*, **28**, 27 (1971), Rule 4.14].

pyrophosphate, was a chemical entity possessing different properties than disodium orthophosphate which has merely lost its water of crystallization. One of the differences he found between the two compounds was that the former gave a white precipitate with silver nitrate while the latter gave a yellow one. However according to Lavoisier's theory salts were compounds derived from basic and acid oxides and were characterized by the ratio of base to acid they contained. All water present in a salt was considered as water of crystallization and nonconstitutional. Accordingly, the three salts in the chemical equation presented above were considered to be the same.

Graham observed that $NaH_2PO_4 \cdot H_2O$, $Na_2HPO_4 \cdot 12 H_2O$, and $Na_3PO_4 \cdot 12 H_2O$ behaved differently on heating. Heating the third salt resulted merely in the loss of the water of crystallization with no further change. When disodium phosphate was heated, it gave pyrophosphate as described by Clark. Heating the monosodium phosphate gave an acid pyrophosphate as the first product; on further heating a new salt was formed which he named metaphosphate and is now known as Graham's salt. He observed that there were three kinds

$$2 NaH_2PO_4 \xrightarrow{190°} Na_2H_2P_2O_7 + H_2O \xrightarrow[\text{heat}]{\text{red}} 2 NaPO_3 + 2 H_2O$$

of metaphosphates. One, now known as sodium cyclotriphosphate, $Na_3P_3O_9$, was soluble in water; the second, known as Maddrell's salt, was insoluble in water; and the third, now known as Graham's salt, was a water-soluble glass which was formed from rapidly cooling molten NaH_2PO_4. It became apparent to Graham that water may have different functions in salts, particularly phosphate salts. He explained his experimental results according to Lavoisier's theory by introducing the assumption that water can act as a base and replace other bases in a salt. A salt is characterized by the ratio of acid to base (sum of both water and basic oxides). According to this ratio, he classified phosphates into ortho-, pyro-, and metaphosphates. Orthophosphates contain three molecules of base per molecule of P_2O_5, the pyrophosphates two, and metaphosphate one. This was a demonstration of Berzelius' law of multiple proportion. All water other than that mentioned in the composition of the salt was considered water of crystallization.

The concept that phosphoric acid molecules can condense to form a polymer was developed by Liebig (358,292). Gay-Lussac and Dulong believed that acids were hydrogen compounds from which salts were formed when hydrogen was replaced by metals, and also that acid oxides alone could not be considered as acids. These views were extended by Liebig to polybasic acids and particularly phosphoric acid. His concept was that different phosphoric acids which can be transferred to acids or salts by partial or complete replacement of hydrogen ions can also be transformed into compounds with a greater number of phosphorous atoms by elimination of water. Thus,

phosphoric acid: $\quad\quad H_3PO_4$
pyrophosphoric acid: $\quad 2\,H_3PO_4 - H_2O = H_4P_2O_7$
triphosphoric acid: $\quad\,\, 3\,H_3PO_4 - 2\,H_2O = H_5P_3O_{10}$
metaphosphoric acid: $\,\, nH_3PO_4 - (n-1)\,H_2O = (HPO_3)_n$

Cyclophosphates have often been referred to as "salts of phosphoric acid poorer in water of constitution than orthophosphoric acid" (228).

Fleitmann and Henneberg (88) studied the differences between the metaphosphate salts obtained by Graham. Liebig's assumption that the differences between these salts were due to the degree of polymerization of phosphoric acid was adopted. They investigated the water-soluble salts now known as cyclotriphosphate and obtained the silver, lead, and barium salts and two double salts of sodium and barium. These compounds have the formulas $Ag_3P_3O_9$, $Pb_3(P_3O_9)_2\cdot3\,H_2O$, $Ba_3(P_3O_9)_2$, $NaBaP_3O_9$, and $NaBaP_3O_9\cdot4\,H_2O$.

Fleitmann and Henneberg concluded that the metaphosphate anion must be trimeric in order to account for the double salt and the quantity of water of hydration of the other salts. Although the conclusion was correct in this case, they arrived at the wrong ones concerning other salts. The simple whole number atomic ratio in a salt is not a proof of the degree of polymerization of the anion which it contains. This is evident from the double salt $Na_9K_3(P_3O_9)_4$, discovered by Griffith and Van Wazer (108) in which the ratio of sodium to potassium is $3:1$. The old idea of obtaining structural data from the composition of double salts would indicate that the salt is a cyclotetra- and not a cyclotriphosphate.

Fleitmann (86,87) also investigated the other two salts described by Graham in an analogous manner. He observed that when alcohol was added to NH_4Cl dissolved in a solution of Graham's salt, a syrupy mixture was separated in which the ratio of sodium to ammonium in equivalents varied from $1:1$ to $1:5$. Addition of $CaCl_2$ to this syrupy product produced a cheese-like precipitate with a composition approximating $5\,CaO\cdot(NH_4)_2O\cdot6\,P_2O_5$. From this observation he concluded that the hexamer was the most probable form. In accordance with this work Graham's salt was called a hexametaphosphate. The structure of this salt was later proven to be incorrect by several investigators (248,250,314–317). However this incorrect name persisted in the literature of the technological and scientific fields for over a century. Graham's salt is neither a cyclic nor a hexameric anion. It is a mixture of salts consisting mainly of anions of high molecular weight in the chain form rather than the cyclic form. A true cyclohexaphosphate was, however, prepared by Griffith and Buxton (106).

The copper cyclotetraphosphate was discovered by Maddrell (174). Fleitmann (86,87) crystallized copper cyclotetraphosphate from the melt of monobasic orthophosphate of copper or another heavy metal with excess phosphoric acid at about 300° or higher. The resulting mass was then washed with water to leave the copper or heavy metal cyclotetraphosphate insoluble salts. By metathesis crystalline double salts could be obtained in which the equivalent ratio of one metal to another was always 1:1. Fleitmann concluded that these salts must therefore be dimetaphosphates. However the anion of these salts was later proven by Warschauer (337) to be a cyclotetraphosphate.

The insoluble alkali-metal metaphosphates did not form double salts, and therefore Fleitmann (86,87) decided that they must be monometaphosphate. Structural studies on Maddrell's salt (58,289) indicated that it consists of long chains made up of interconnected PO_4 groups.

The evidence on which Fleitmann and Henneberg based their conclusion about the structure was weak by today's standards. This was primarily due to the lack of analytical tools available at that time. A state of confusion in establishing the identity of the molecular specie and in naming it grew up for a period of a century and is well documented in the literature. Different names were used to refer to a single salt prepared in a definite way, according to the structural hypothesis of the author. This arose from misinterpretation of data on colligative properties obtained by classical techniques of a large number of impure preparations which contained high proportions of ring phosphates. The degree of ionization of the cations in aqueous solutions of these salts in various preparations were not known. In computing the molecular weights the authors usually assumed either no ionization whatsoever, or complete ionization of the cations—using no activity corrections in either case. Some of the preparations and names of cyclotri- and cyclotetraphosphate salts were reviewed by Karbe and Jander (142) and Yost and Russell (356).

Partridge (208) attracted attention to the chaotic numerical terminology in the old literature. He proposed another system of terminology which would differentiate between the different physically distinguishable phosphates by their physical properties. He described the cyclophosphates by empirical formulas and Roman numerals. Partridge numbered the crystal forms of the sodium salts which resulted from dehydrating NaH_2PO_4 downward from the form stable at the melting point. He designated sodium cyclotriphosphate as $NaPO_3$-I, because it is stable at its melting point of 628°. Polymorphic forms of this compound were designated $NaPO_3$-I' and $NaPO_3$-I". Partridge reported the preparation of two different insoluble salts by heating NaH_2PO_4 at temperatures ranging between 300 and 475°. He designated one salt as $NaPO_3$-II and the other as $NaPO_3$-III. Finally, phosphate glasses were described in terms of the relative amounts of constituent oxides.

The use of modern analytical tools such as X-ray, paper chromatography, and nuclear magnetic resonance helped in establishing the identity and degree of ionization of different phosphates. Using paper chromatography, Ebel (64) demonstrated the erroneous nature of a number of phosphate salts claimed to be cyclic monomeric and dimeric species. Similarly, Van Wazer (321) showed that an ammonium-dimetaphosphate reported by Pepinsky et al. (217) is really an ortho-phosphate. As a result of the advancement in establishing the identity and the degree of polymerization, phosphates were classified by both Thilo (284) and Van Wazer and Griffith (327) according to their structure into three groups: poly-, meta-, and ultraphosphates.

The existence of a monomeric PO_3^- or a dimeric anion in the cyclophosphate series has not been established. In the monomer anion, phosphorus(V) would be three-coordinated using sp^2 hybridization of phosphorus with π-bonding. The sp^2 hybridization is uncommon for phosphorus(V) or any other element below the first row in groups IV and V of the periodic table. Such an anion would have a high degree of coordinative unsaturation and would consequently be reactive chemically. This would lead to polymerization of PO_3^- to achieve the coordination number of four which is the most stable for phosphorus(V). However such a monomer anion might occur in high-temperature transformations of condensed phosphates (112,113,119,352).

The existence of a cyclodiphosphate, $P_2O_6^{2-}$, is improbable. It would be a four-membered ring with two PO_4 tetrahedra joined by sharing two oxygen atoms. The structural strain as well as electrostatic repulsion would make such a structure highly unlikely.

The existence of cyclophosphates with more than four phosphorus atoms in the anion ring was first observed by Van Wazer and Karl-Kroupa (330) in the chromatograms of the hydrolysate of Graham's salt. Cyclopenta- and cyclohexaphosphates of Na, Ag, and Ba containing water of crystallization have been prepared in gram quantities by Thilo and Schülke (304,305) from fractional crystallization of Graham's salt via acetone, silver nitrate, and hexaminocobalt(III) chloride. The Ag and Ba salts were isolated by precipitation with $AgNO_3$ and $BaCl_2$. Kura and Ohashi (156) have also isolated several grams of cyclopenta- and cycloheptaphosphate from Graham's salt by means of an anion-exchange Dextran gel column. The syntheses of both cyclohexa- and cyclooctaphosphate were carried out from starting materials other than Graham's salt. Griffith and Buxton (106) prepared the anhydrous and hexahydrates of both sodium and lithium cyclohexaphosphates. More recently Schülke reported the preparation of the cyclooctaphosphate ($M_8P_8O_{24}$) salts (259,260).

III. Preparation of Cyclophosphates

A. Cyclotriphosphate Salts

Although numerous direct methods for the synthesis of cyclophosphate salts, particularly $Na_3P_3O_9$, exist, the mechanism of many reactions is still not understood. Direct methods refer to methods of preparation of cyclotriphosphate salts from starting materials containing phosphate salts excluding any of the cyclic trimer. In most cases, these methods involve condensation polymerization of hydrogen phosphates, or depolymerization of long-chain phosphates. Unless mentioned, the term "quantitative yield" refers to yields of 90% or more based on P_2O_5 content.

Most commercial methods for preparing $Na_3P_3O_9$, from which all other cyclotriphosphate salts are made, are based on the thermal dehydration of NaH_2PO_4 under controlled conditions as discovered by Graham. The condensation polymerization is carried out between 500 and 600° for about 5 hr, followed by slow cooling (18). The reaction (162) can be formulated as:

$$NaH_2PO_4 \cdot 2\,H_2O \xrightarrow[-2\,H_2O]{50-110°} NaH_2PO_4 \xrightarrow[-\frac{1}{2}\,H_2O]{170-200°} Na_2H_2P_2O_7 \xrightarrow[-\frac{1}{2}\,H_2O]{250-320°}$$

$$(NaPO_3)_n \xrightarrow[500-600°]{} Na_3P_3O_9\text{-I}$$

Shaffery and Strumpf's (264) process is a typical example for the synthesis of $Na_3P_3O_9$. Their process involves continuous feeding of 70–73% NaH_2PO_4 solution at 100° into one end of an inclined rotary kiln. Combustion gas at 800–850° is fed into the opposite end. Feed rates and retention time (about 3 hr) are adjusted to provide complete dehydration and a product discharge temperature of 520–580°.

The form and yield of $Na_3P_3O_9$ obtained by thermal dehydration of the hydrated NaH_2PO_4 are dependent on temperature, rate of cooling, water vapor pressure, sample thickness, and time of roasting. The first two factors are the most critical. Cooling the melt of NaH_2PO_4·H_2O slowly gives three different forms of $Na_3P_3O_9$ depending on the temperature at which the supercooled melt is tempered (166). $Na_3P_3O_9$-I is stable over 443° under an equilibrium pressure of water vapor. This product melts at 625° and is formed exclusively when NaH_2PO_4·H_2O is heated between 550 and 628° (187). Two other unstable polymorphic forms, $Na_3P_3O_9$-II and $Na_3P_3O_9$-III, are obtained by controlled tempering of the supercooled melt at about 525° (292). Both change into the stable $Na_3P_3O_9$-I form if cooled too slowly after crystallization.

Heating NaH_2PO_4 above 625° and quenching it yields the glassy, soluble Graham's salt. Cooling the melt very slowly yields $Na_3P_3O_9$-I, while cooling the melt moderately slowly between 630 and 500° yields a small portion of form A of Kurrol's salt (brittle fibrous crystalline variety) along with both $Na_3P_3O_9$-I and Graham's salt. The crystallization of form A of Kurrol's salt could be induced by nucleating the surface of the melt between 600 and 550° with fragments of preformed form A, silica, carborundum, and any infusable foreign substance. These act as a nuclei for the growth of Kurrol's salt (form A) (313). Form B of Kurrol's salt is produced from form A by mechanical treatment or by exposure to moist air.

By heating NaH_2PO_4·H_2O between 250 and 280° Maddrell's salt is obtained mixed with $Na_3P_3O_9$-I and pyrophosphate. By heating NaH_2PO_4·H_2O between 350 and 400° Maddrell's salt-II is obtained along with $Na_3P_3O_9$-I (313). The interrelation between these condensed phosphates is illustrated in diagrams given by Thilo (292) and Haiduc (119).

The water vapor pressure has an important effect on the yield of $Na_3P_3O_9$. Under an equilibrium pressure of water vapor, the thermal dehydration of NaH_2PO_4 could be represented (292) as:

$$NaH_2PO_4 \underset{}{\overset{169°}{\rightleftharpoons}} \begin{pmatrix} \text{Solid } Na_2H_2P_2O_7; \\ \text{melted } Na_2H_2P_2O_7 \\ \text{and } NaH_2PO_4; \\ \text{where } n = 1\text{--}7; \text{ vapor} \end{pmatrix} \underset{}{\overset{375°}{\rightleftharpoons}}$$

$$\begin{pmatrix} \text{Solid Maddrell's } (h); \\ \text{melted } Na_nH_2P_nO_{3n+1} \\ \text{where } n = 1\text{--}7; \text{ vapor} \end{pmatrix} \underset{}{\overset{443°}{\rightleftharpoons}} \begin{pmatrix} \text{Solid } Na_3P_3O_9; \\ \text{melted } Na_nH_2P_nO_{3n+1} \\ \text{where } n = 1\text{--}7; \text{ vapor} \end{pmatrix}$$

An increase in water vapor pressure at constant temperature increases the yield of Maddrell's salt under static conditions (180) and has the opposite effect under dynamic conditions (162). During the calcination process moisture is evolved. When the hydrated NaH_2PO_4 is heated a hard crust forms on the surface and a considerable amount of moisture is entrapped inside. The thicker the sample, the longer it takes the water vapors to escape. The yield of $Na_3P_3O_9$ is proportional to sample thickness, providing the other factors are held constant (162). On the other hand, if the thermal dehydration of $NaH_2PO_4 \cdot H_2O$ is carried out under dry conditions, products are obtained in addition to $Na_3P_3O_9$. The phase diagram of this system has been developed by Morey and Ingerson (190).

At any temperature, an increased time of roasting increases the yield of $Na_3P_3O_9$ as long as the other factors are constant (162).

The condensation polymerization of the dihydrogen phosphate salts of sodium and other alkali metals ($M^IH_2PO_4$) by organic acid anhydrides has been achieved at temperatures ranging from 40 to 120°. Verdier and Boulle (332) prepared $Na_3P_3O_9 \cdot 1\frac{1}{2} H_2O$ from the condensation of $Na_2HPO_4 \cdot 12 H_2O$ with acetic anhydride. Thilo et al. (113–115) used acetic acid and acetic anhydride to condense NaH_2PO_4 and KH_2PO_4 into $Na_3P_3O_9$ and $K_3P_3O_9$ respectively at temperatures ranging between 80 and 90°. It is of interest to note that thermal dehydration of KH_2PO_4 or any monoalkali dihydrogen phosphate, with the exception of sodium, does not yield cyclotriphosphate.

As mentioned, the thermal dehydration of $NaH_2PO_4 \cdot H_2O$ yields long-chain phosphates. However these polymeric chains with $n > 5$ can depolymerize and rearrange to give mainly $P_3O_9^{3-}$. This can occur in the melt, in the solid state, or in the solution.

All sodium dihydrogen chain phosphates are converted into $Na_3P_3O_9$ when tempered within the last 50° below the melting point. The transition follows Stranski and Kaischew's theory of crystallization (277). The activation energy is 62.9 kcal/mole (117). The kinetics of crystallization and crystal decomposition of $Na_3P_3O_9$ in the range between 270 and 625° was investigated. The dependence of the rate on the temperature was evaluated. The crystal growth followed Frank's screw-dislocation mechanism (116).

Small amounts of noble metals such as Ir, Pt, Rh, Pd, Au, and Ag cause depolymerization and rearrangement of Graham's salts into orthorhombic $Na_3P_3O_9$. The metals are arranged above in the order of their decreasing catalytic activity (118).

The depolymerization and rearrangement of long-chain phosphates can also occur in solution to give $Na_3P_3O_9$. The yield is higher in the presence of added cations. In the presence of Mg^{2+}, $P_3O_9^{3-}$ yield, with a small amount of $P_4O_{12}^{4-}$, is about 70% of Maddrell's salt, compared to a 50% yield in the absence of Mg^{2+} (292).

The formation of $P_3O_9^{3-}$ from chain phosphates involves the breaking of two P—O—P bonds along the chain followed by cyclization. The cleavage of this bond was assumed earlier to be hydrolytic in nature (19,309,329). However experimental evidence suggests that rearrangement rather than hydrolysis is involved (292,294). $H_2^{18}O$ was used during the degradation of chain phosphates. ^{18}O was not found in the formed $P_3O_9^{3-}$ whose oxygen does not exchange with chain phosphates (333). Further evidence was obtained from depolymerization of chain phosphates to form $P_3O_9^{3-}$ quantitatively under completely anhydrous conditions. This was the case when chain phosphates of certain amines dissolved in benzene were heated at 60° (350).

The formation of $P_3O_9^{3-}$ from chain phosphates is currently regarded as a simple transformation. Chain phosphates with $n > 5$ are probably present as a spiral. The repeating unit is on the average 3 PO_4 from which $P_3O_9^{3-}$ is formed by transformation, as shown below. This is

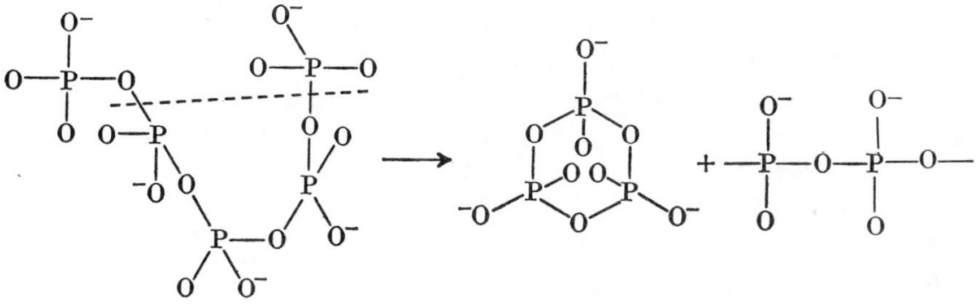

supported from structural analysis of the crystals of chain phosphates (292,295). The acceleration by cations of the formation of $P_3O_9^{3-}$ from chain phosphates in solution can accordingly be explained. Cations are known to form complexes with chain phosphates. Chelate formation helps in promoting coiling of the chains which facilitates the transformation into cyclic structures (292,295).

Lee and Bond (162) pointed out that the greater yield of $Na_3P_3O_9$ with the increase in water retention associated with the increased NaH_2PO_4 thickness might possibly be due to hydrolytic cleavage of P—O—P linkages, facilitation of ionic diffusion, or the retardation of Maddrell's salt. The argument against the hydrolytic cleavage has already been presented. The role of the other two alternatives perhaps is the most important. Water helps the diffusion and crystallization of $Na_3P_3O_9$, thus facilitating the change from Maddrell's salt to $Na_3P_3O_9$.

The other possibility is that water retards the formation of Maddrell's salt when samples are heated up to the selected temperature. This leads to the conversion of pyrophosphate into $Na_3P_3O_9$ directly without formation of significant amounts of the intermediate Maddrell's salt (162).

Sodium phosphate salts other than NaH_2PO_4 may be thermally dehydrated to prepare $Na_3P_3O_9$. The conditions under which $Na_3P_3O_9$ is formed from $Na_2H_2P_2O_7$ were determined (55). Knorre's method (149) in which Na_2HPO_4 or $Na_4P_4O_7$ are dehydrated in the presence of ammonium nitrate is well known.

$$6\ Na_2HPO_4 + 6\ NH_4NO_3 \xrightarrow{\Delta} 2\ Na_3P_3O_9 + 6\ NH_3 + 6\ NaNO_3 + 6\ H_2O$$

Other ammonium salts have been used with different phosphate mixtures in which the over-all Na_2O/P_2O_5 ratio is close to unity. Moore and Shen (186) obtained $Na_3P_3O_9$ quantitatively by calcinating $Na_3P_3O_9$ at 450–620° in an atmosphere containing greater than 95% water vapor by volume. Another process used a mixture of NaH_2PO_4 and 2–6% ammonium salts of mono- or dibasic mineral acids (NH_4NO_3, NH_4Cl, $(NH_4)_2SO_4$, or $(NH_4)_2S_2O_8$) or a compound which forms NH_3 at reaction temperatures (urea or guanidine) (39). The reaction mechanism has not been studied yet.

In addition to the phosphate salts mentioned above, sodium phosphite and some sodium alkyl phosphates have successfully yielded $Na_3P_3O_9$. Examples reported include pyrolysis of methyl and ethyl

$$NaO-\overset{\displaystyle O}{\underset{\displaystyle OH}{\overset{\|}{P}}}-OR \qquad (R = Me\ or\ Et)$$

phosphate, and the oxidation of disodium phosphite with bromine or silver salts (152,153,211,213). The hydrolysis of hexachlorocyclotriphosphazene and octahalogenocyclotetraphosphazenes into cyclotri- and cyclotetraphosphates could also be included in this category (194,231,233,267). The hydrolysis of the former salt with sodium acetate was found to occur in two steps (302).

If Na_2O/P_2O_5 is less than 1:1, other cations are incorporated and double salts are obtained. Griffith (101) obtained $Na_2HP_3O_9$ by evaporating an aqueous orthophosphate liquor having a Na_2O/P_2O_5 mole ratio of $\frac{2}{3}$ and allowing crystallization to proceed at 300° where the Na_2O/P_2O_5 mole ratio reaches $\frac{1}{2}$. Other examples of double salts which have been prepared are $Na_9K_3(P_3O_9)_4$ and $CaNa_4(P_3O_9)_2$ (108, 187–189).

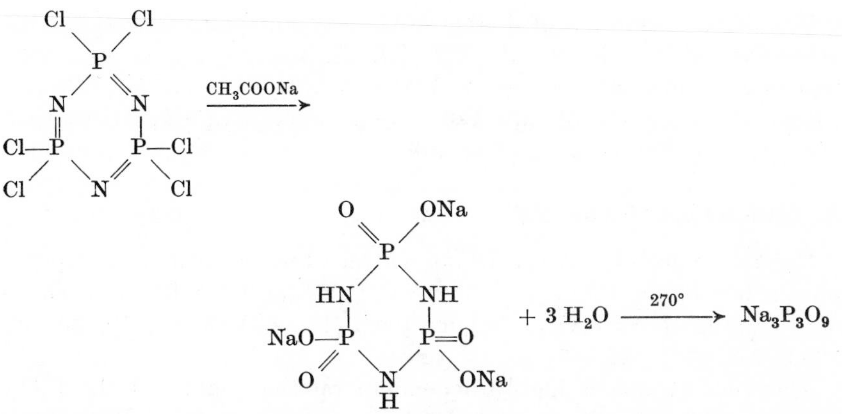

The degree of polymerization of K, Rb, and Cs dihydrogen ortho-phosphates could be influenced to stop at the trimer, producing the corresponding cyclophosphate salts during the thermal dehydration. This could be accomplished by the chemical action of HCl, NH_4NO_3, or by conducting the reaction in a low-melting alkali-metal nitrate solvent (226). K, Rb, and Cs nitrate melts were used as the media to dehydrate the corresponding dihydrogen phosphate salts to obtain the respective cyclotriphosphate salts (226). $(NH_4)_3P_3O_9$ was prepared by treating H_3PO_4 with urea in a mole ratio of N to P ranging from 6:1 to 2:1 at 200–300° (240). The preparation of Mn^{II} cyclotri- and cyclo-tetraphosphates is another example of the above method (163). Manganese violet when heated gives both products in the following sequence:

$$2\ Mn^{III}NH_4P_2O_7 \xrightarrow[120–340°]{-H_2O} [Mn_2^{III}P_4O_{13}(NH_4)_2]$$

(manganese violet)   (blue, unstable)

$$340–460° \left| \begin{array}{l} -NH_3 \\ -NH_2OH \rightarrow N_2 + H_2O \end{array} \right.$$

$$Mn_3^{II}(P_3O_9)_2 \xleftarrow{\text{heat}} Mn_2^{II}(P_4O_{12})$$

(manganous cyclotri-   (manganous cyclotetraphosphate, pink)   phosphate, white)

Conversion of some chain phosphates into ring structures via reactions not involving condensation or depolymerization is known to occur. This is exemplified in the closure of the triphosphate anion or its derivatives into $P_3O_9^{3-}$. Heating $Na_5P_3O_{10}\cdot6\ H_2O$ with acetic acid at 60° yields $Na_3P_3O_9\cdot H_2O$ (301). Acidification of amido-, N-alkyl-amido-, and fluorotriphosphates of Na, K, NH_4, Ag, and Ba results in quantitative formation of the cyclic trimer of the corresponding cation (77,78,82,83,230,233,292).

The direct formation of double salts of cyclotriphosphate of cations other than sodium by the thermal dehydration method has also been reported. Durif and co-workers (178) were able to obtain $NH_4M^{II}P_3O_9$, where $M^{II} = Mg$, Co, Zn, Mn, Cd, or Ca, by heating $(NH_4)_2HPO_3$ with the corresponding carbonates at 300°.

B. Cyclotetraphosphate Salts

Careful hydrolysis of the hexagonal crystalline form of tetraphosphorus decaoxide, P_4O_{10}, with iced NaOH, or Na_2CO_3 solution has been used to prepare $Na_4P_4O_{12}$ in a good yield (11,19,34,46,145,150,236,239, 308,309,313,315,334,347).

The yield, which is highly dependent on the quality of the P_4O_{10} (292), is in the neighborhood of 75% for a pure sample. Although all P—O—P linkages in P_4O_{10} are susceptible to cleavage, the breaking of the two internal cage anhydride P—O—P linkages occurs preferentially.

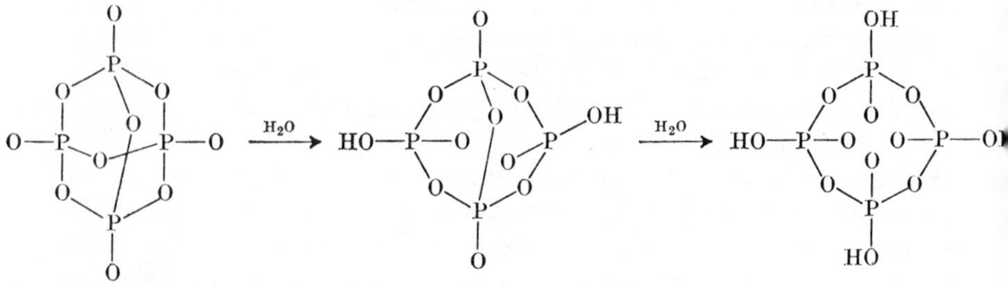

Paper chromatography was used by Thilo and Wieker (308) to study the reaction. The mechanism was also discussed by Van Wazer (321).

Until very recently, the classical methods for preparing cyclotetraphosphate salts consisted mainly of the thermal dehydration of dihydrogen monophosphate at temperatures ranging between 400° and 500°, alone or mixed with a small excess of phosphoric acid. This is similar to the preparation of $Na_3P_3O_9$ from NaH_2PO_4. With the cyclotetraphosphate the radii of the cations present during the thermal dehydration lie between 0.57 and 1.03 Å. They include Ba^{2+}, Pb^{2+}, Cu^{2+}, Zn^{2+}, Mg^{2+}, Cd^{2+}, Fe^{2+}, Fe^{3+}, Co^{2+}, Ni^{2+}, Mn^{2+}, and Al^{3+} (87,88, 95,157,298,299,300,337). In the case of cyclotriphosphate heating NaH_2PO_4 alone gives only $Na_3P_3O_9$. However heating an equimolar mixture of NaH_2PO_4 and H_3PO_4 at 400° yields $Na_2H_2P_4O_{12}$ (101,102) according to this reaction:

$$2\,NaH_2PO_4 + 2\,H_3PO_4 \xrightarrow{400°} Na_2H_2P_4O_{12} + 4\,H_2O$$

Cations whose radii are not in the 0.57–1.03 Å range, yield high-molecular-weight chain phosphates (119,292,297).

A mixture of 2–4 equivalents of H_3PO_4 with metal oxides, hydroxides, or carbonates may also be used in the preparation of $P_4O_{12}^{4-}$ salts (300). Treating the insoluble salts of the mentioned cations with Na_2S yields the sodium cyclotetraphosphate. The type of cation has a considerable influence on the products obtained by thermal dehydration. This might be attributed to their influence in promoting coiling of the chains and influencing the spatial arrangements of the PO_4 tetrahedra so as to change the number of PO_4 repeating units along the spiral. The radius of the cation, its coordination number, and the character of the metal–O bond are influencing factors in this arrangement (16,271,295). Highly ionic bonds favor structures with two PO_4 tetrahedra repeating units while less ionic bonds favor those with four (271). Also, the P—O bonds are shorter and P—O—P angles are larger in compounds with covalent metal–O bonds in comparison to those which are mainly ionic.

The thermal dehydration of monohydrogen orthophosphates can also be used to prepare cyclotetraphosphate salts. Durif et al. (59) prepared four double salts as follows:

$$4\ (NH_4)_2HPO_4 + SrCO_3 + M_2^ICO_3 \rightarrow SrM_2^I(P_4O_{12}) + 2\ CO_2 + 8\ NH_3 + 6\ H_2O$$

where $M^I = NH_4^+$, K^+, Rb^+, or Tl^+. A large excess of $(NH_4)_2HPO_4$ was used and the temperature of the reaction mixture was raised steadily from ambient to 450°. After cooling, the reaction products were tempered at 400°. The thermal dehydration of manganese violet to product $Mn_2^{II}P_4O_{12}$ has already been mentioned.

The depolymerization of high-molecular-weight linear phosphates of sodium produces mainly $Na_3P_3O_9$. In contrast, the depolymerization of linear phosphates in the presence of complex cations such as $[Ni(NH_3)_6]^{2+}$ at 400°, and $[Co(NH_3)_6]^{2+}$ or $[Cu(NH_3)_4]^{2+}$ at 540° yields mainly the cyclotetraphosphate of the corresponding cation. Ammonia is eliminated during the reaction (148).

C. Cyclopenta-, -hexa-, -hepta-, and -octaphosphates

Nearly 90% of Graham's salt is a mixture of linear phosphates with average n of about 100 units. The remaining 10% consists of cyclic structure phosphates of 3–10% trimer, 2–4% tetramer, and less than 1% pentamer, hexamer, and higher cyclic structures. The concentration of the individual cyclic structures decreases exponentially as the number of atoms in the ring increases (293). Thilo and Schülke (304–306) prepared Graham's salt richer in higher cyclic structures by rapid heating of anhydrous $Na_3P_3O_9$. The stepwise addition of acetone to this sample resulted in the precipitation of high-molecular-weight chain phosphates. Subsequent treatment of the mother liquor with a large amount of acetone leads to the separation of an oily substance in which the cyclophosphate fraction is high. Fractional precipitation of the oil with silver nitrate and hexaminocobalt(III) chloride resulted in the separation of cyclopenta- and cyclohexaphosphate, the cyclohexaminocobalt(III) barium and silver salts, and pure $Na_6(P_6O_{18})$ and its hexahydrate.

Fractionation of Graham's salt has also been achieved recently by the use of a QAE-Sephadex A-25 gel column (156). The separated cyclic phosphates ranged from the trimer to the octamer (see the section on column chromatography). Although cyclopenta- and cyclohepta- have been previously isolated on a microscale (349), this procedure allows their separation on a macroscale.

The lithium salt of cyclohexaphosphate was obtained by Griffith and Buxton (106) as a result of heating lithium carbonate with H_3PO_4 at 200–275°. The salt constitutes 5–20% of the lithium phosphate system with an Li_2O/P_2O_5 ratio near 7:5.

The octamer has also recently been synthesized by Schülke (259,260) in 70% yield by heating $Pb_2(P_4O_{12})\cdot 4\ H_2O$. The course of the reaction was followed by chromatography, Guinier photographs, rate of water loss, and differential thermal analysis. The starting material loses two molecules of water of crystallization. The residual water of crystallization causes the hydrolysis of the ring to give lead mono-, di-, and triphosphates. The linear phosphates are condensed to high-molecular-weight chain polymers and lead cyclotetraphosphate which is converted into $Pb_4^{II}(P_8O_{24})$. No rates were given. $Pb_4^{II}(P_8O_{24})$ also occurs in a poor yield when $Pb(H_2PO_4)_2$ is thermally dehydrated (259). $Na_8P_8O_{24}$ is prepared by the metathesis of the lead salt with Na_2S solution.

A large number of cyclophosphate salts have been obtained by indirect methods. This is accomplished by exchanging the cations in the starting material either totally or partially with other cations. Precipitation and metathesis exchange were used earlier for this purpose. The former involves the mixing of soluble sodium salts of cyclotri- and cyclotetraphosphate in particular, with a solution containing a multiple charged or a heavy-metal salt. The solution is evaporated to allow gradual crystallization (see section on chemical analysis by precipitation).

In the metathesis exchange a solution containing a soluble salt, such as Na_2S, K_2S, H_2S, $CaCl_2$, or Na_2CO_3, is added to a multiple charged or heavy-metal cyclophosphate precipitate such as $Ag_3P_3O_9$ or $Cu_2P_4O_{12}$ (86,87). In addition cation exchange has been widely used recently as an indirect method for the synthesis of new cyclophosphate salts. The soluble salt of a cyclophosphate is passed through the resin which is charged by a different metal ion or in the form of H^+ (19,54,106,270, 282,329). In the latter case the free acid can be obtained; this is then neutralized immediately with the desired base. Most of the cyclophosphate salts prepared by the different methods were tabulated by Van Wazer (321) and more recently by Haiduc (119).

IV. Structure

Lindboom (167) in 1875 and Glatzel in 1880 (95) proposed the cyclic structures of the cyclotri- and cyclotetraphosphates respectively. Proofs of the cyclic structure are based on X-ray analysis, cryoscopic determination of molecular weights, titration curves, measurements of conductance of dilute solutions, NMR, paper chromatography and alkaline ring cleavage. This evidence has been discussed in this series as well as in other reviews (43,44,119,292,321).

The ring conformation of both cyclotri- and cyclotetraphosphates has been studied in both solid state and in solution. From X-ray analysis, the solid $Na_3P_3O_9$ has a chair conformation (C_{3v} symmetry) with a P—O(ring) distance of 1.620 ± 0.0004 Å (205). Infrared spectra also supported this conformation (109). In aqueous solution, the $P_3O_9{}^{3-}$ ring exhibits a planar conformation (D_{3h} symmetry) (110,270, 273,274).

In aqueous solution, $P_4O_{12}{}^{4-}$ anion has a puckered structure with C_{2h} symmetry (110). The crystalline monoclinic $Na_4P_4O_{12}\cdot4\,H_2O$ and $(NH_4)_4P_4O_{12}$ have the same structure (X-ray). Both crystals have alternating P—O (endocyclic) bonds whose lengths in the former salt are 1.635 and 1.584 ± 0.016 Å. In the latter salt, the lengths of the bonds are 1.58 and 1.65 ± 0.03 Å (50,206). Different conformations for the tetramer ring in various salts were also found (3,4,50,58,135,136,172, 173,204,206,215,241,242).

V. Analysis and Characterization of Condensed Phosphates

Because of the importance of phosphates, analysis of phosphate mixtures is essential to many chemical, biological, and nuclear investigators. An important feature in connection with this analysis is the ability of different phosphate ions, which are similar in their physical and chemical properties, to exist simultaneously in solution. For this reason the characterization and analysis of cyclophosphates cannot be discussed with the exclusion of other condensed phosphates. Methods of analysis should be able to distinguish between the different phosphates and should be able to quantize each in the mixture.

Many wet analytical procedures have prevailed for a long period of time for the detection, estimation, and separation of condensed phosphates. These include colorimetry, titrimetry, and precipitation, but they are not widely applicable to complex mixtures, particularly to those containing interfering substances. Methods which use physical properties, such as X-ray diffraction, viscosity, optical measurements, melting point, and ultracentrifugation, also do not permit resolution of the mixtures, and very little information about the unknown or minor constituents of mixtures could be obtained.

However the analysis and characterization of condensed phosphates has undergone important evolution in the last twenty years. The application of chromatographic techniques and spectroscopic tools such as NMR and Raman spectroscopy has resulted in considerable progress not only in the mentioned fields, but also in the classification of phosphates, in the discovery of new members, and in the reactions of condensed phosphate anions themselves. General reviews (144,147) on the analysis of cyclophosphates have appeared in connection with other types of condensed phosphates. An authoritative review by Ebel (66) has recently been published.

Most of the known methods of analysis and fractionation of cyclophosphates and other condensed phosphates can be artificially classified into two categories: methods requiring separation and those which do not (see below). Any particular procedure may use one or more of these methods.

Analytical and Fractionation Methods

No Separation	Separation
Titrations	Precipitation
Acid–base titration	Inorganic and
(pH titration)	organic cations
Titration in molten	Organic solvents
salts	Chromatography
Spectroscopy	Paper chromatography
NMR	Column chromatography
Ultraviolet	TLC
X-ray Diffraction	Electrophoresis
Vibrational Spectrum	

A. Methods Requiring No Separation

1. TITRATIONS

a. Acid–Base Titration. Different phosphate groups have different capacities in acid–base reactions. Orthophosphate can associate with three hydrogen ions in solution. The end group —O—PO_3^{2-} can associate with two hydrogen ions, while middle groups, —O—PO_2^{-}—, can associate with one. All three groups are weakly basic with respect to the association of one hydrogen ion. Consequently, the fully protonated groups are strong acids for which the pK_a is of the order of 2.

All phosphoric acids when titrated with a strong base have an inflection point around pH 4.3 (321,329). Protons of cyclophosphoric acids are strong and are equivalents. One inflection point has been observed in the course of the titration. At the equivalence point the number of equivalents of a base is equal to the number of middle PO_4 tetrahedra.

Terminal phosphate groups with $n > 3$ and HPO_4^{2-} are weak acids with pK_a's of about 8. Both are titrated simultaneously with a strong base and an inflection point is observed near pH 10. Thus a solution of $H_7P_5O_{16}$ can be titrated as a strong pentabasic acid, and then less accurately as a weak dibasic acid. When the chain is relatively short such as in tri- and diphosphoric acids, the neutralization of the first hydrogen on one terminal phosphate group results in the weakening of the hydrogen on the other end. In triphosphoric acid there is a difference between pK_4 and pK_5 of about 3, sufficient to cause another inflection point in the titration curve near pH 7 (341).

The third hydrogen of orthophosphoric acid is too weak to be titrated directly (p$K_a = 13$). However the addition of a precipitating agent such as Ag^+ results in the release of the hydrogen ion (93).

The titration method is useful in the analysis of soluble condensed phosphates. The titration between pH 4.5 and pH 9 gives the number of terminal groups and orthophosphates. The titration of the third hydrogen on the orthophosphates gives the number of orthophosphates present. The total number of phosphorus atoms can be determined by the total hydrolysis of the mixture using heat in the presence of a strong mineral acid, followed by titration with a strong base (75,328). The use of a volatile acid such as HCl is preferred. Removal of HCl by evaporation is recommended only in the presence of NaCl or KCl. In the absence of these salts, H_3PO_4, as a result of hydrolysis, re-condenses as the solution approaches dryness. If the alkali chlorides are present they react with H_3PO_4 to form volatile HCl leaving Na_3PO_4

residue (103). The number of middle-group phosphates could then be determined.

As shown, the titration method is capable of determining the number of ortho-, middle-, and end-group phosphates in a sample. However, like any functional group method of analysis, it does not have the capacity of analyzing for individual molecular species. It should be noted that the average chain length, $\bar{n}$, of linear phosphates can be determined in a mixture that does not contain ring phosphates or their amount can be corrected for (328).

b. Titration in Molten Salts. Acid–base titration in fused salts has recently attracted attention. According to the Lux concept (171), any oxyanion such as $Cr_2O_7^{2-}$, VO_3^-, or middle-group phosphate PO_3^- in molten salts acts as a Lewis acid which takes its primary oxide, O^{2-}, from a base, e.g., CO_3^{2-}, NO_3^{2-}, or Na_2O_2. The concept is illustrated in the following reactions:

$$2\,PO_3^- + O^{2-} \rightarrow P_2O_7^{4-} \tag{1}$$

$$P_2O_7^{4-} + O^{2-} \rightarrow 2\,PO_4^{3-} \tag{2}$$

$$VO_3^- + O^{2-} \rightarrow VO_4^{3-} \tag{3}$$

The middle-group phosphates in $Na_3P_3O_9$ react with sodium chromate, silicate, vandate, molybdate, and tungstate (128), orthophosphate (224,225), nitrite (169), carbonate (168), and $NaX + O_2$ (X = F, Cl, Br, or I) (127), under anhydrous conditions in molten salts, to give sodium pyrophosphate. In these reactions, side products may be formed. Both the reaction rate and the formation of the side products are temperature-dependent. The reaction rate is limited by the rate of diffusion of the reacting substance through the layer of the primary compound: the higher the temperature the faster the reaction (168,169). The side reactions occur when $P_2O_7^{4-}$ reacts further with PO_3^-, the Lewis base, or its decomposition products.

The analytical applications of the titrations in molten salts for the determination of middle phosphate groups has been explored. Van Norman and Osteryoung (319) determined both VO_3^- and PO_3^- from the number of moles of carbon dioxide evolved when excess sodium carbonate is added at 400° to a LiCl–KCl eutectic containing one of these oxyanions. The acid–base reaction was formulated as

$$PO_3^- + CO_3^{2-} \rightarrow PO_4^{3-} + CO_2 \tag{4}$$

The % error ranged between -0.7 and 3.2.

The first equation indicates that middle-group phosphates are transferred into end groups; the second equation indicates the conversion of end groups into orthophosphates. Both reactions (350–450°)

were studied potentiometrically with an oxygen indicating electrode (Pt/O) and a Ag/Ag^+ reference electrode (265,266). The solvents arranged in order of their decreasing basicity are KNO_3, $NaNO_3$, and LiCl–KCl eutectic. Both reactions are complete in KNO_3. A linear relationship was found between the amount of titrant (Na_2O_2) and the amount of PO_3^- present in the melt (266). In $NaNO_3$, the first end-point was less than theoretical owing to the acidic nature of the solvent (47). In the chloride melt, the most acidic, only one break corresponding to the first endpoint was observed (266).

In order to define the acidity or basicity of a certain oxyanion in molten salts, Shams El Din et al. (265) developed two quantitative procedures. The first involved the measurement of the potential difference between a nitrate melt which is $10^{-2}M$ with respect to the oxyanion and another solution containing no oxyanion. The "acidity number" of a particular oxyanion was obtained by dividing the mentioned quantity ($E_{nitrate} - E_{oxyanion}$) by the value $2.3RT/2F$ at the working temperature. This procedure assumes a reversible reaction. The second procedure was based on the calculation of the equilibrium constant of the acid–base reaction from the titration curve. With the exception of VO_3^-, PO_3^- was the strongest acid (higher negative acidity number) in the group consisting of AsO_3^-, $P_2O_7^{4-}$, $As_2O_7^{4-}$, $V_2O_7^{4-}$, $H_2PO_4^-$, H_2AsO_4, HPO_4^{2-}, and $HAsO_4^{2-}$. The equilibrium constants of the following reactions were calculated as:

$$2\,PO_3^- + O^{2-} \rightarrow P_2O_7^{4-} \qquad K = 3.1 \times 10^{15}$$
$$2\,H_2PO_4^- + O^{2-} \rightarrow 2\,HPO_4^{2-} + H_2O \qquad K = 3.9 \times 10^{13}$$
$$2\,HPO_4^{2-} + O^{2-} \rightarrow 2\,PO_4^{3-} + H_2O \qquad K = 2.8 \times 10^5$$
$$V_2O_7^{4-} + O^{2-} \rightarrow 2\,VO_4^{3-} \qquad K = 4.5 \times 10^6$$
$$P_2O_7^{4-} + O^{2-} \rightarrow 2\,PO_4^{3-} \qquad K = 6.6 \times 10^5$$
$$2\,HAsO_4^{2-} + O^{2-} \rightarrow 2\,AsO_4^{3-} + H_2O \qquad K = 2.6 \times 10^2$$

It is interesting to compare titrations in molten salts to those in aqueous solution (pH). In molten salts, both end and middle groups can be determined directly from one titration; this is not possible in a pH titration. pH titrations are easier to perform and the reagents are easier to handle in comparison to titrations in molten salts. Hydrolysis in aqueous solution is a source of error in the pH titration. In molten salts, the instability of some phosphate species such as $H_2PO_4^-$, HPO_4^{2-}, and protonated end groups are expected to be a source of error. The condensation polymerization of such species has previously been mentioned (cyclophosphate synthesis). The magnitude of the error is expected to be a function of the rate constants of a particular reaction, concentration of reactants, and the titration time.

PO_4^{3-} is a very weak acid to be titrated in the known molten salts. The release of the weak hydrogen (HPO_4^-) by a precipitating cation, e.g., Ag^+, allows the determination of PO_4^{3-} in a mixture by pH titration. The presence of HPO_4^{2-} or $H_2PO_4^-$ would interfere with the titration of middle and end groups in molten salts. Theoretically, both $H_2PO_4^-$ and PO_3^- would simultaneously be titrated (the K ratio $\simeq$ 100). HPO_4^{2-} and end groups would also be titrated together because of their similar K values. However this interference could be used to determine the four groups. When a sample containing this mixture is titrated, two endpoints are observed. $H_2PO_4^-$ and PO_3^- are titrated at first, while HPO_4^{2-} and middle group phosphates are titrated at the second endpoint. If another sample in which all the protons are exchanged with sodium is titrated, the first and second endpoints correspond to PO_3^- and middle group phosphates, respectively. The difference in the number of equivalents in the corresponding endpoints of both titrations would give the number of equivalents of $H_2PO_4^-$ and HPO_4^{2-}, respectively.

2. SPECTROSCOPY

a. NMR. The ³¹P NMR spectrum studies of condensed phosphates show well-resolved chemical shifts for isolated, end, and middle phosphate groups (51,52,89,107,200,231,325,326). The fine structure due to phosphorus–phosphorus spin coupling was also reported. The subject has recently been thoroughly reviewed (165,175,323,331).

b. Ultraviolet. Among cyclophosphates, only the ultraviolet spectra of $Na_3P_3O_9$ in solvents containing water or deuterium oxide have been investigated (121). The spectrum of $Na_3P_3O_9$ in aqueous solution obeyed the Beers-Lambert law. The temperature coefficient of the absorption edge was evaluated. The spectrum is comprised of at least three bands. Weak bands above 200 nm were assigned to internal electronic transition, while the steep absorption edges below 200 nm were tentatively assigned to a "charge transfer to solvent" (CTTS) type of transition.

c. X-Ray Diffraction. X-ray diffraction and X-ray powder diagram studies have been made on several cyclophosphates to elucidate their structure (3,4,35,46,50,58,60,135,136,139,172,173,203–207,215,241,242). Crystal and molecular data for some cyclophosphates have recently been tabulated by Haiduc (119). A bibliography on this type of work has also been given by Van Wazer (321).

d. Vibrational Spectrum. High-resolution vibrational studies by both IR absorption and Raman scattering on both the cyclic trimer and tetramer have been done (see section on structure). Spectra correlation has been made and recent surveys on the subject were published (43).

B. Methods Requiring Separation

1. PRECIPITATION

a. Inorganic and Organic Cations. Fractional precipitation of phosphates has been carried out by the addition of cations or the addition of organic solvents which are miscible with water. The former method was popular among the classical methods for phosphate analysis, particularly for cyclotriphosphate. It is based on the difference in solubilities of phosphate salts of different cations as a function of the pH of the solution. Ba^{2+}, Pb^{2+}, Ag^{2+}, Mn^{2+}, and different cobaltic complexes are examples of these inorganic cations, while quaternary amines, protonated amines, and proteins are examples of organic cations used to separate phosphate mixtures.

Mellor (182) pointed out that cyclotriphosphate was not precipitated by any metallic cation, including silver and barium. Jones (138) confirmed this and found that in the presence of alcohol and silver ion most of the cyclic trimer remained in the filtrate. This property was used by Jones to separate and estimate the cyclotriphosphate in a mixture that included ortho-, pyro-, and long-chain phosphates. $BaCl_2$ was added to the acidified mixture and then the solution was made alkaline (pH 9). All phosphates precipitated with the exception of cyclotriphosphate. The filtrate was then hydrolyzed and the P_2O_5 content of cyclotriphosphate was volumetrically determined. Other procedures were devised (25,73,353).

The precipitation technique is subject to different sources of error which limit its analytical value. One stems from the hydrolysis of the ring phosphate. The hydrolysis rate is a function of temperature, pH, and the presence of foreign cations. Another is caused by coprecipitation of the cyclotriphosphates with other linear phosphates. A third may be due to incomplete precipitation by barium ion of the linear phosphates that are present in solution (124,247). Dewald and Schmidt (57) demonstrated that determination of cyclotriphosphate was erratic under these conditions.

McCune and Arquett (179) studied the use of tris(ethylenediamine), cobalt(III) chloride, and hexamine cobalt(III) chloride as precipitating agents for condensed phosphates. At pH 12.5 and 7.5 all chain phosphates ($n = 1$–11) along with cyclotetraphosphate are precipitated. Only cyclotriphosphate remains in solution. However the method has not yet been developed for assaying for cyclotriphosphate.

Several organic amines, proteins, and cationic dyes (56,67–69,197,198) have also been tested for their ability as fractionating agents for condensed phosphates. Few proved promising in fractionating ring from chain phosphates. However, as in the previous case, no quantitative data for assaying ring phosphates have been given.

Although the use of the precipitation technique can at best be used for assaying cyclotriphosphate, it is useful as a fractionation technique. The use of cations, organic or inorganic, leads to fractions that are richer in cyclic structures from original mixtures that also contained linear condensed phosphates. Repeating the precipitation process may lead to pure fractions of the desired cyclic structure. The fractional precipitation by both acetone and hexaminocobalt(III) chloride was used by Thilo and Schülke (304) to isolate both cyclic pentamer and hexamer phosphates from Graham's salt.

b. Organic Solvents. The solubility fractionation procedure consists of successive additions of a poorly miscible solvent to a solution containing a high-molecular-weight polymer. This leads to the precipitation of different fractions of the solute, each containing a mixture of molecules less complex than the original mixture. The high-molecular-weight fractions tend to concentrate in the precipitate while those of low molecular weight prefer the supernatant liquid. The technique, which was originally used in the fractionation of organic polymers, was employed by Van Wazer (320) to fractionate inorganic phosphate polymers. Although methyl, ethyl, and isopropyl alcohol are adequate in causing solubility fractionation, acetone was found superior to them. A sample of Graham's salt was fractionated into eight successive fractions which had a mean degree of polymerization ($\bar{n}$) of 550, 382, 398, 322, 214, 112, 35, and a residue of 112. Cyclotriphosphate and presumably other ring phosphates remained in the last fraction. The pH of the solution which ranged between 4 and 9 had no effect on the final results. This was explained by noting that equivalent amounts of anions and cations were precipitated out on the addition of organic solvents. The fractionation was carried out as fast as possible at room temperature to avoid hydrolysis. This technique was later improved by Martens and Rieman (177).

Solubility fractionation with an organic solvent has one great advantage over the use of cations: the various fractions are not contaminated with salts other than phosphates. This facilitates

further handling of these fractions.   Furthermore, it is possible to carry
out this process on a relatively large scale.   However the disadvantage
of this method is that organic solvents catalyze certain degradations of
long-chain phosphates (177).

2. CHROMATOGRAPHY

a. *Paper Chromatography*.   Paper chromatography of phosphates
has been the subject of intense study by many investigators since it was
introduced by Ebel et al. (61,62,64,70) in France, Westman and co-
workers (346,347) in Canada, and Ando et al. (1) in Japan.   Much of the
progress in establishing the identity of condensed phosphates, the
identification of reaction products, and analysis of technical products
has been due to this technique.   The literature on the subject is
voluminous.   The recent review by Ebel (66) contains an extensive
bibliography.

Papers of different grades made primarily by Whatman and Schleicher
& Schull have been used in these studies.   Some of the papers received
preliminary treatment.   Different shapes and configurations including
rectangular, circular, tongue-shaped, and others have been employed
(126).

Three development techniques have been reported.   Ebel (63) used
the ascending technique which appears to give the best results.   Westman et al. (347) and Crowther (49) used the descending method which
requires between 3 and 16 hr for separation.   Koberlein and Mair-
Waldburg (151) and Schormuller and Wurdig (258) successfully used
circular chromatography.   Uhlik (318) adapted this technique to a
centrifuge tube and was able to shorten the development time to 45 min.

The developers are divided into acidic and basic solvents.   Acidic
solvents are miscible in both water and organic solvents.   Another
feature is that all contain an organic acid, e.g., CCl_3COOH and a small
quantity of ammonia.   The basic solvents are a mixture of alcohol,
water, and ammonia.   Acidic solvents are suitable for the separation
of chain phosphates while basic solvents are suitable for the separation
of cyclophosphates.   A list of both solvents along with the respective
R_f values of condensed phosphates has recently been compiled by Ebel
(66).

The influence of water content in both the acidic and basic solvents
and the influence of a small quantity of ammonia water in the acidic
solvent on the R_f values of condensed phosphates was studied by Ebel
(65).   The presence of ammonia was found necessary to improve the
separation since the different phosphate ammonium salts have different

solubilities. Thilo and Feldmann (296) described the previous influences and pointed out that the presence of an acid is necessary since the phosphate salts have difficulty migrating. Water content, quantity of ammonia water, quantity of trichloroacetic acid, and quantity of acetic acid present in the acidic solvent have recently been studied by Iida and Yamabe (130) to determine their influence on R_f values, the ratio of thickness of the mobile phase and the stationary phase, and the free energy necessary to transport the structural units of condensed phosphates from the stationary phase to the mobile phase. A similar study was performed on the basic solvent with the changes in water and ammonia water content. Water content and pH value had great influence on the four variables mentioned above. The influence of the other variables was found to be minor.

The paper can be used for either a uni- or two-dimensional chromatogram (62,70). Unidimensional chromatography in either solvent does not permit the total separation of all phosphate compounds. The advantage of two-dimensional chromatography is the unequivocal distinction between the ring family and the chain family phosphates. In this technique, the papers are eluted with one solvent (basic); after it has been dried, the paper is eluted with the other solvent (acidic) perpendicular to the direction of the first elution. In this way the chain phosphates are positioned differently than the ring phosphates. Each phosphate member in each family occupies a certain place within the family according to its molecular weight. Thus from the chromatographs important structural information and identification of a substance becomes possible.

Visualization of the spots (detection) on the chromatographs has been accomplished by the reaction of molybdate anion with PO_4^{3-} (122), which results from the hydrolysis of condensed phosphate by a strong acid. This is followed by the treatment of the phosphomolybdic complex with a reducing agent. The reducing agent may be stannous chloride (143,347), H_2S (85), UV light (251), or 1-amino-2-naphthol-4-sulfonic acid (154,219), hydrazine hydrochloride (49), or ascorbic acid (90). The use of labeled phosphorus, ^{32}P, can accomplish the same purpose of detecting the spots (256).

Three methods were found suitable for quantitative evaluation of the condensed phosphate spots. One method depends on cutting out the spots after development, hydrolyzing the condensed phosphates, and colorimetrically determining them. There are different techniques associated with this method. One depends on the extraction of the phosphomolybdic complex with ammonia, H_2SO_4, or alkylacetate (9). Another method uses 60% perchloric acid and heat (550° for 3 hr) for the calcination of the condensed phosphates. A second quantitative method consists of direct photometric determination of the spots by a densitometer. The color is developed as a result of reducing the phosphomolybdic complex by $SnCl_2$, UV light, or an ammoniacal solution of 1-amino-2-naphthol-4-sulfonic acid. The third method is based on the activity of ^{32}P in a condensed phosphate spot. This technique is more sensitive than the previous methods. The detection limit is of the order of 10^{-9} g. Details of these methods have been discussed by Hettler (126) and Ebel (66).

The quantitative determination of phosphates by paper chromatography is subject to at least two sources of error. The first is inherent in the technique itself, that is the spotting of a small volume on paper. The second source of error is due to the hydrolysis of the phosphates during development. In general, both errors do not exceed 3% (66,126).

b. Column Chromatography. Rieman et al. (24,218,247,359) studied the theory and technique of elution of different condensed phosphates on Dowex 1-X10 as a function of pH and elutant concentration. The technique has been utilized to separate the constituents of a mixture made up of cyclic trimer, tetramer, probably pentamer, and linear phosphates ranging from ortho- to tridecaphosphate using an Amberlite XE-119 resin column (247). Using an Amberlite IRA-400 resin, Ohashi et al. (202) demonstrated that the cyclic phosphates mentioned above can be separated from each other. However higher cyclophosphates are eluted together. The cyclophosphates are eluted in decreasing order of their molecular weights. Different resins and eluting mixtures have been employed by different investigators (8,69,99,150, 268,282).

The eluted fractions can be hydrolyzed and assayed colorimetrically. Wiecker (348) has shown that the quantitative results obtained by this method are superior to those obtained by paper chromatography. The ion exchange procedure has also been subjected to automation which makes it relatively easy to analyze phosphate mixtures on a routine basis (8,21,170,222,223).

Earlier attempts by Ebel and Bush (68) to reproduce on a column the conditions which were used in paper chromatography to separate phosphate mixtures had limited success. A column of cellulose powder was used and an acidic solvent served as the elutant. The order of separation was the same as in paper chromatography. Hydrolysis of higher-chain phosphates ($n > 3$) occurred on the column; this could not be eliminated even at low temperature. However satisfactory

separation (up to $n = 13$) has been achieved using Sephadex G100 and
0.05M NaCl as the elutant (84). The molecular sieving effect causes
the separation of the cyclic phosphates.

A combination of both gel and anion exchange resin to separate
condensed phosphates has been recently used by Ohashi and Kura
(156). A QAE-Sephadex A-25 gel column was used in this investigation. The distribution ratios of cyclic phosphates from the trimer to
the octamer were measured as a function of the eluting agent (0.25M
KCl). The authors were able to separate cyclotri-, cyclotetra-, cyclo-
penta-, and cycloheptaphosphate from each other and from cyclohexa-
and cyclooctaphosphate using 0.30M KCl. The latter two were eluted
together. Their separation was achieved by elution with 0.25M KCl.
This technique is useful in the preparation of both cyclopenta- and
cycloheptaphosphate from Graham's salt on a macroscale. A number
of other types of gels were also used (195,216).

   c. *Thin Layer Chromatography.* Although the literature on the
applications of paper chromatography to the separation of condensed
phosphates is extensive, relatively few papers on the thin layer chromatography (TLC) of the condensed phosphates have appeared. Most
of the separations by the TLC technique were achieved by partition
chromatography, using cellulose (7,12,42,130), starch layers (36),
and silica gel (26,245,246,278,311,336,354). The presence of calcium
sulfate in silica gel layers is not desirable since it reacts with phosphates;
other binders had to be used (354). Pollard et al. (221) observed that
silica gel was slower running than cellulose. Frontal analysis was
found with silica gel, while cellulose gave tailing spots. An equal
mixture of both cellulose and silica gel reduced the tailing shown by
cellulose and prevented the frontal analysis exhibited by silica gel. As
in paper chromatography, acidic and basic solvents were used as
eluants in connection with uni- or two-dimensional development
techniques. The development time ranged between 1 and 3 hr.
Better separation was later achieved and the development time was
shortened by the use of prepared (354) or commercially available plates
(311).

   Recently anion exchange TLC has been used to separate condensed
phosphates. Berger et al. (22) reported successful separation of chain
phosphates on thin layers of Biorex 5 ion exchanger. Tanzer et al.
(280) achieved good separation of several linear and cyclic phosphates
by one-dimensional development using anion exchange TLC on
polyethyleneimine (PEI) impregnated Avicel coated thin layers. Iida
and Yamabe (360) used PEI-cellulose F precoated (Merck A.G.

Darmstadt, G.F.R., 0.10 mm) plates to study the relationship between R_m and the valency of condensed phosphates in aqueous solution. In $1N$ NaCl the apparent valencies for cyclotri-, cyclotetra-, and cyclohexaphosphates were 3.26, 4.08, and 5.56, respectively.

TLC has several advantages over paper chromatography. Among these are the facts that the composition of the substrates can be varied to give a better separation, the separated zones are less diffuse and the background is more opaque, a shorter development time is needed (does not exceed 3 hr), and thin layers can withstand more severe conditions of spraying and heating than can paper chromatograms.

The main advantage of paper chromatography over TLC is that separation of species with close R_f values can be better achieved with paper chromatography by the descending technique. Also, only with paper chromatography can the length of time of development be extended to as long as desired. Quantitative TLC of condensed phosphates has been reported to be equivalent to paper chromatography (336).

d. *Electrophoresis.* The differential analysis by electrophoresis, known variously as electrochromatography, iongraphy, zone electrophoresis, etc., has scored some success in the fractionation, separation, and identification of micro amounts of condensed phosphates. Most studies have been carried out using filter paper as a buffer stabilizer. Both high (50–100 v/cm strip length) and low (2–10 v/cm) potential gradients have been applied. An example of each will be cited here. Wade and Morgan (335) used a low potential gradient to separate a mixture of ortho-, di-, tri-, cyclotri-, and cyclotetraphosphates on sheets of paper impregnated with an aqueous solution of 9.2% (v/v) n-butyric acid and 0.1% (w/v) NaOH at an ambient temperature of 20°. Also, a two-dimensional fractionation was carried out by a combination of electrophoresis and chromatography. Sato (255) succeeded in separating cyclotri-, cyclotetra-, tri-, and diphosphoric acids along with hypophosphorus, phosphorus, phosphoric, and condensed phosphoric acids by a single migration in acetic acid in a high potential gradient (10 to 15 v/cm). The acids are arranged in the order of decreasing mobility. Other papers on both methods have appeared in the literature (164,252,253,338,355).

Although electrophoresis is a sensitive and effective technique for the examination of oxyacids of phosphorus, particularly those of long-chain phosphates, the reproducibility for condensed phosphates of $n < 10$ is poor in comparison to paper chromatography.

VI. Thermochemistry

The most widely used means for the determination of the enthalpy of the cyclotri- and cyclotetraphosphate systems is solution calorimetry. The sodium salt is hydrolyzed by means of a strong mineral acid (HCl or $HClO_4$) to the orthophosphate and the heats of solution are measured. Also, in the case of $Na_3P_3O_9$ trimetaphosphatase enzyme has been used in the hydrolysis at pH 7 and a temperature of 33° (183).

The heat of formation of $Na_3P_3O_9$ was reported to be -878.7 and -879.0 kcal/mole by Thomsen (310) and Giran (94), respectively. These values were found to be in good agreement with the recent work of Irving and McKerrell (134) who reported -878.6 kcal/mole. Meadowcroft and Richardson (181) reported a value about 2 kcal/mole higher than these values. Compared to the value obtained by Giran from acid hydrolysis, the results obtained by enzyme hydrolysis of $Na_3P_3O_9$ were higher by about 9 kcal/mole. The ΔH of hydrolysis by the enzyme method was found to be -18.6 kcal/mole at pH 7 and 33°. On the other hand, a ΔH of -288.6 for the $NaPO_3$ unit is reported by Van Wazer (321) (taken from Circular 500 of the National Bureau of Standards). The latter value would be lower by approximately 13 kcal/mole for $Na_3P_3O_9$ (c) when compared to either of the early values or that of Irving and McKerrell.

The only available value for the heat of formation of $Na_4P_4O_{12}$ (c) is that of Irving and McKerrell (134) who reported -1169.4 kcal/mole. This gives ΔH_f of the $NaPO_3$ (c) unit a value of -292.3 kcal/mole compared to -292.9 in the trimer indicating little difference in ring strain. If one considers -293 kcal/mole an average ΔH_f for $NaPO_3$ (c), one would expect ΔH_f of the higher members in the series to be $-n \times 293$ kcal/mole.

The enthalpies of HPO_3 (c) and HPO_3 (aq) have been compiled by Latimer (160); they are in accord with values in Circular 500 of the National Bureau of Standards. The values are -228.2 and -234.8 kcal/mole respectively. ΔF (s) for the latter is listed as -215.8 kcal/mole, while its standard entropy is estimated as 36 cal/deg.

Recently the entropy of $Na_3P_3O_9$ (c) from 10 to 320°K has been calculated by Andon et al. (2) from the measurements of low-temperature heat capacity. The value S_{298} was found to be 68.47 cal/(deg)-(mole). The authors compared the entropies calculated by summing S_{298} for Na_2O [18.2 cal/(deg)(mole)] and P_2O_5 [27.35 cal/(deg)(mole)] present in Na_3PO_4, $Na_4P_2O_7$, $Na_3P_3O_9$, and $Na_5P_3O_{10}$ with the experimentally determined values. The agreement between them was good.

For example, $Na_3P_3O_9 = \frac{3}{2} (Na_2O + P_2O_5) = \frac{3}{2} (18.2 + 27.3) = 68.3$ cal/(deg)(mole) compared to the experimentally determined value of 68.5. One may use this calculation to predict the entropy for $Na_nP_nO_{3n}$ as $n/2$ $(S_{NA_2O} + S_{P_2O_5})$ cal/(deg)(mole). Furthermore, the authors extended this method of estimation to acidic phosphates by a similar calculation of the contribution of H_2O $[S_{298} = 7.000$ cal/(deg)(mole)]. In the systems NaH_2PO_4, Na_2HPO_4, and $Na_2H_2P_2O_7$, the calculated S value differed from the experimental by about 1%. This may lead to the estimation of the entropy of the cyclophosphate acids by computing the $n/2$ $(S_{H_2O} + S_{P_2O_5})$ value.

Recently, Beglov (15) calculated the standard entropy of formation of potassium phosphate salts. This was based on a comparison with the experimental data which are available for the sodium compounds. The values for $K_3P_3O_9$ and KPO_3 were 77.47 and 25.83 cal/(deg)(mole), respectively. An identical value, determined from lower heat capacity measurements, was reported earlier (71). It does not appear that an exact value for $S°$ for K_2O (c) has yet been measured. However an estimated value of 20.8 cal/(deg)(mole) was given by Latimer (160). Using the estimation method, one obtains a value of 24 and 72 cal/(deg)(mole) for KPO_3 (c) and $K_3P_3O_9$ (c), respectively. The latter values are lower than the previous by about 8%. The comparison will be more meaningful when an exact value for $S°$ of K_2O (c) is available.

VII. Physical and Chemical Properties

A. Physical Properties

The polymorphism of hydrous and anhydrous $Na_3P_3O_9$ and $Na_4P_4O_{12}$ and the methods of their preparation were reviewed (321). The same reference discussed the solubility of the two salts in water as a function of temperature along with their thermal behavior. The solubilities of the known alkali-metal ring phosphates are generally higher than those of chain phosphates with the same n. A saturated solution of $Na_3P_3O_9$ contained 22.3% (by weight) of the salt at 30°, while $Na_4P_4O_{12}$ was 13% of its saturated solution at the same temperature (107). The aqueous phase diagrams of $Na_4P_4O_{12}$–$Na_3P_3O_9$–H_2O and $Na_3P_3O_9$–$Na_4P_2O_7$–H_2O systems were also determined (107). The solubility of the alkaline earths, heavy metal ions, and organic bases of $P_3O_9^{3-}$ and $P_4O_{12}^{4-}$ salts were mentioned (see section on the analysis by precipitation).

A linear relationship between the density of aqueous solutions of $Na_3P_3O_9$ at 25° and its concentration was found. $Na_4P_4O_{12}$ behaved similarly (104).

The apparent molar refractions of aqueous $P_3O_9^{3-}$ and $P_4O_{12}^{4-}$ were found to be 34.2 and 45.6, respectively, yielding $R_{apparent}$ for PO_3^- of 11.4 (32). An earlier study reported 19.6 for the same value (104). The latter value was higher probably due to an error in the calculations (32). Since the ionic refraction of condensed phosphate anions is an additive function of the number of end and middle groups present, it is expected that the apparent molar refraction at a fixed wavelength and 25° would be 11.4 × n for cyclophosphates.

B. Chemical Properties and Reactions

The reactions of cyclotriphosphates can be classified into four categories (119): (1) reactions involving the cleavage of the ring with the formation of linear phosphates; (2) polymerization to high-molecular-weight chain phosphates; (3) reactions with ring rearrangement (ring contraction); and (4) reactions with ring preservation (metathetic exchange of cations and complex or ion-pair formation).

1. REACTIONS INVOLVING THE CLEAVAGE OF THE RING WITH THE FORMATION OF LINEAR PHOSPHATES

The cyclotriphosphate ring is cleaved by ammonia, amines, and phenols in alkaline media to give the linear triphosphate additive anion. Similarly, water alkali metal fluorides and hydrogen peroxides can bring about the ring cleavage (for review see Refs. 79,119,292,293,295). Some recent reactions of interest will be discussed here. The phosphorylation of alcohols and carbohydrates by cyclotriphosphates in alkaline solution (80) takes place according to:

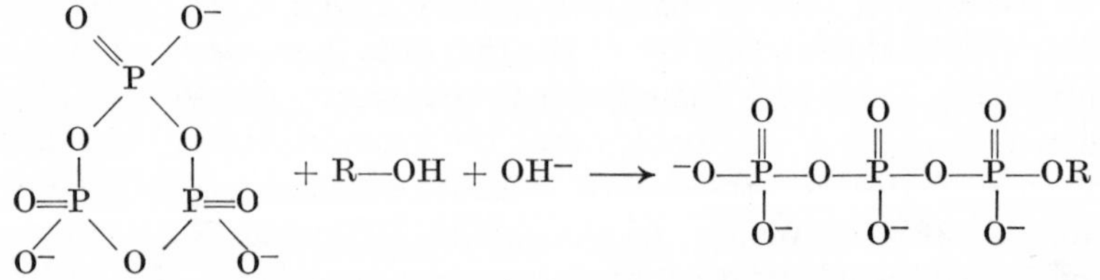

Under similar conditions, adenosine is phosphorylated to yield exclusively the 2′- and 3′-monophosphates (261).

Aminoacids such as β-alanine or glycine in alkaline solution react to form open-chain phosphoraminoacid derivatives which revert to amino acids and cyclotriphosphate on acidification (81). In the presence of excess glycine or DL-alanine, condensation of amino acids took place. When NH_3 was present, glycinamide was also formed (81). The mechanism of the cyclotriphosphate-induced peptide synthesis from glycine was investigated. Glycine reacts with cyclotriphosphate to form an open-chain phosphoramidate which forms a cyclic acylphosphoamidate and pyrophosphate. The cyclic compound then reacts either with water or a second free amine group of glycine or diglycine to give diglycine-N-phosphate or triglycine-N-phosphate (40).

Cleavage of the cyclotriphosphate ring can take place during grinding. Degradation to ortho- and pyrophosphate or polymerization to higher-chain phosphates can occur depending on the grinding conditions. Grinding under N_2, CO_2, low pressure, dry or humid air were investigated (191–193).

Solid-state reactions in which the cyclophosphate ring is cleaved have already been mentioned (titration in molten salts). Ring cleavage as a result of electrochemical reduction in molten salts is suspected to occur. The reaction was investigated by voltammetry, chronopotentiometry, and coulometry in molten LiCl–KCl over the temperature range 450–700°. A two-electron reduction was found. The electrochemical and chemical reaction was proposed as follows (159).

$$P_3O_9^{3-} + Cl^- \xrightarrow{450°} P_3O_9Cl^{4-} + 2e^- \longrightarrow P_3O_9Cl^{6-} \longrightarrow P_2O_7^{4-} + PO_2Cl^-$$

$$2\,Cl^- \downarrow {}^{>450°} \qquad\qquad\qquad\qquad\qquad \downarrow$$

$$P_3O_9Cl_3^{6-} \qquad\qquad\qquad\qquad\qquad\qquad P + other$$

$$\downarrow$$

$$3\,PO_3Cl^{2-} + 3e^- \longrightarrow [3\,PO_3Cl^{3-}] \longrightarrow Products$$

2. POLYMERIZATION TO HIGH MOLECULAR WEIGHT AND

3. REACTIONS WITH RING REARRANGEMENT

Both reactions were recently reviewed (119).

4. REACTIONS WITH RING PRESERVATION (METATHETIC EXCHANGE OF CATIONS AND COMPLEX OR ION PAIR FORMATION)

The metathetic exchange of cations by precipitation or cation ion exchange was mentioned (syntheses). Although interest in complexes of condensed linear phosphates arose earlier due to their ability to soften water (120), cyclophosphate complexes did not draw as much attention because of their relative weakness. Cyclophosphoric acids behave as strong acids when titrated with a base (106,248,282,313,324, 337). The little deviation of the NMR of ^{31}P chemical shift at a pH of 2.0 or lower for $P_4O_{12}^{4-}$ and hardly any for $P_3O_9^{3-}$ (51) demonstrate the acid strength. The near neutrality of the alkali metal salt solutions also supports this view.

The first quantitative information about the dissociation constants of $H_3P_3O_9$ and $H_4P_4O_{12}$ was obtained by Davies and Monk (54) by considering the deviation from Onsager's formula, which uses the slope of a plot of conductance versus the square root of the concentration of the particular acid. All other studies were performed by measuring the pH change of a strong acid in the presence of a ring phosphate salt, as indicated by a glass electrode. The cations present in the solution should form a very weak complex with the ring, e.g., tetraalkyl-ammonium ions. Published data indicate $P_4O_{12}^{4-}$ is appreciably more basic than $P_3O_9^{3-}$. In both cases, one hydrogen ion associates with the ring. The exception that two hydrogen ions associate with the $P_3O_9^{3-}$ ring was made by Elesin et al. (72). The pK_a values of $H_3P_3O_9$ and $H_4P_4O_{12}$ are given in Table I. It is interesting to note that the pK_3 reported by Wells and Salam (345) for $H_3P_3O_9$ of 0.92 is based on concentration which, corrected for H^+ activity, becomes 0.78, in close agreement with the value of 0.68 reported recently by Watters and co-workers (339).

Physical evidence about the interaction between different metal ions and cyclophosphates have been obtained by different methods. Botre and co-workers (30) used an ion-selective electrode prepared by incorporating polystyrenesulfonic acid in a collodion matrix about 0.1 mm thick. They studied the suppressing action of cyclotri- and cyclotetraphosphates toward sodium and calcium. However they did not reach a definite conclusion about the stiochiometric composition of any complex, but rather they interpreted this interaction as due to coulombic effect.

Refractometry was used by Nelson (196) to obtain evidence of complex formation in aqueous solutions between cyclotri- and cyclo-tetraphosphates and different metal ions. The method depends on the fact that light travels more slowly through a solution than it does through pure water. Thus the refractive index of a solution is greater than that of water. The refractive index of a medium is approximately linear with respect to the polarizability of the molecules or ions com-posing the media. Thus the refractive index depends on the concen-tration and the nature of the solutes. Nelson measured the refractive index in connection with Job's method of continuous variations and concluded that there is complex ion formation between Zn^{2+} and Cd^{2+} with cyclotetraphosphate and between Al^{3+} and cyclotriphosphate.

Another study (52) used high-resolution NMR spectra of ^{31}P to study the complexes of lithium, calcium, and magnesium with condensed phosphate anions. Complex formation between the added cations and the tetramethylammonium salts of cyclotri- and cyclotetraphosphates produced observable changes in the phosphorus chemical shifts and spin coupling constants. Both Li^+ and Ca^{2+} had little effect on the ^{31}P chemical shift, but Mg^{2+} produced a definite shift to a higher field which corresponds to increased shielding of ^{31}P. This was interpreted as evidence that the metal is electrostatically attracted to the ring and occupies positions centered above and below the plane of the ring where a minimum of electrostatic potential wells are available. The specific site binding of cations by electrostatic attraction is energetically possible only if sufficient water of hydration is lost to permit close approach of the anion and cation where the effective dielectric constant might be considerably less than the macroscopic dielectric constant of water solvent.

The large degree of dissociation of both cyclophosphoric acids eliminated in most cases the possibility of investigating metal complex formation by means of pH titrations which have been extensively and successfully applied to the study of the complexes of linear phosphates. The interactions of $P_3O_9^{3-}$ and $P_4O_{12}^{4-}$ with different cations have been investigated by Monk and co-workers (54,137) by conductometric and solubility procedures. In the solubility procedure the stability constants of calcium and strontium complexes of cyclotri- and cyclotetraphosphates were evaluated from the solubility of respective metal iodates in phosphate solutions and in pure water. The free metal ion concentration was calculated from the solubility of the iodate in pure water; by assuming formulas for the complex, the dissociation constants were calculated. Association of ions other than the considered complex were corrected for and the theoretical activity coefficients were used. The conductivity method was also applied to study the association of [V-phenenyltris(oxyethylene)]tris[triethylammonium] ion and $P_3O_9^{3-}$ (238).

Other studies were performed by the equilibrium technique of ion exchange. Radioactive elements were used as tracers to measure the distribution ratio of the particular element between the cation exchange resin and the solution. A linear relationship between the molar concentrations of cyclophosphates and the reciprocal of the distribution coefficient of the element between the resin and the solution was found and was cited as evidence of 1:1 complexes. Calcium and strontium complexes with cyclotri- and cyclotetraphosphates were investigated by Gosselin and Coglan (97), and Schwarzenbach and co-workers (96), respectively. Americium, curium, and promethium complexes with cyclotriphosphate were studied by Elesin and co-workers (72).

Roppongi and Kato (244) performed another study on the complexes of alkali earth metals and cyclotri- and cyclotetraphosphate. They used the phosphate salts to displace phthaleine from its calcium, strontium, and barium complexes and Eriochrome Black T from its magnesium complex. Both the organic dye and their metal chelate concentrations were evaluated spectrophotometrically. Using this technique, they were able to evaluate the displacement constant in each system; consequently they obtained formation constants of the alkali earths with cyclotri- and cyclotetraphosphate from the knowledge of the formation constant between the metal and the organic dye.

A kinetic approach has been taken by Wells and Salam (345) to study the nature of the complexes of ferrous ions with cyclotriphosphate. This involves studying the rate constant of oxidation of Fe(II) to Fe(III) in the presence and absence of cyclotriphosphate. They reported the presence of $FeP_3O_9^-$ and $FeHP_3O_9$.

Potentiometric methods employing metal electrodes and the more recently developed specific ion electrodes are capable of exceptional reliability and accuracy in indicating the free metal ion concentration provided the electrode reactions are reversible in the Nernstian sense. Of the metal electrodes, only the copper amalgam electrode has been applied in studying cyclophosphate systems. Indelli (133) and Gross and Gryder (111) used the two phase quiet copper amalgam electrode to study copper complexes of $P_3O_9^{3-}$ and $P_4O_{12}^{4-}$ respectively. The dropping copper amalgam electrode, 0.0005% copper by weight, was used to investigate copper cyclotri- and cyclotetraphosphates by Watters et al. (339). The anodic–cathodic polarographic waves were essentially reversible. They were perfectly continuous and have slopes of about 0.032. The stabilities of the complexes have been determined from the shift in potential due to complex formation. The potentiometric data indicated that two of the $P_3O_9^{3-}$ or $P_4O_{12}^{4-}$ ions can associate with one Cu(II) ion in solution. The dropping copper amalgam electrode was also used as an indicating electrode to study Ca^{2+} and Zn^{2+} complexes with cyclotetraphosphate (340).

The cation ion selective electrodes have been recently applied to study the complexes of cyclotri- and cyclotetraphosphates. Their application in studying the ionic equilibrium is due to the ability of these electrodes to determine the concentration of an uncomplexed cation in solution by direct potentiometry. They have the advantage of direct measurements for a number of metal ion activities for which classical electrodes are not feasible. Furthermore they have the advantage of being fairly specific, and thus make it possible to determine activity measurements in the presence of other cations and anions. However the interferences of these cations and anions must be determined prior to the study.

Gardner and Nancollas (91) and Kalliney (140) used a sodium glass electrode to investigate the cyclotri- and cyclotetraphosphates with sodium ion. Similarly, a univalent glass electrode was employed to investigate the complexes of K^+ and NH_4^+ (140,141).

Selective ion electrodes have also been applied to study the divalent metal ion complexes. The calcium selective liquid ion exchange electrode was applied to determine the stability constants with cyclotri- and cyclotetraphosphate ions (340). Similarly, the complexes of Mg^{2+}, Co^{2+}, Ni^{2+}, Fe^{2+}, Zn^{2+}, and Cd^{2+} ions were studied by the divalent metal ion electrode. These cations formed a 1:1 complex with $P_3O_9^{3-}$ and $P_4O_{12}^{4-}$. In addition, two cations associated with one $P_4O_{12}^{4-}$ when the cation concentration exceeded that of $P_4O_{12}^{4-}$. Sr^{2+} formed a 1:1 complex with $P_3O_9^{3-}$. All solutions contained $(CH_3)_4N^+$ at 25° (140).

The cadmium and lead ion electrodes were used to study the Cd^{2+} and Pb^{2+} complexes with $P_4O_{12}^{4-}$ under the above-mentioned conditions. Both electrodes stop functioning after they have been used for a period of time. It appears that the two cyclophosphates cause surface problems for both electrodes.

The method for the use of specific ion electrodes depends on constructing a calibration curve in which the electrode potential is plotted as a function of known concentrations of a certain metal ion (at constant ionic strength) in the absence of a complexing agent. The electrode potential is then measured when it is placed in a solution containing known concentrations of a ligand and a metal ion. The free metal ion is evaluated from the calibration curve. In the concentration range where one metal ion can associate with one or more anions, the formation constants are obtained using an expression equivalent to Leden's equation (340). In contrast, in the region where the metal ion concentration exceeds that of the $P_4O_{12}^{4-}$ anion, two cations can associate with one anion. An expression equivalent to $\bar{n}$, $\bar{m}$, was defined as the average metal ion per ligand.

$$\bar{m} = \frac{[M]_t - [M]}{[L]_t} = \frac{M_1L + 2\,M_2L + 3\,M_3L + \ldots n\,M_nL}{L + ML + M_2L + \ldots M_nL} \tag{5}$$

In the above equation, $[M]_t$ and $[M]$ are concentrations of total and free metal ion respectively, and $[L]_t$ and $[L]$ are concentrations of total and free ligand (cyclophosphate) respectively. Charges on the complexes are omitted for simplicity.

Substitution of the concentrations of the species in terms of β and rearranging leads to

$$\bar{m} = (1 - \bar{m})[M]\beta_1 + (2 - \bar{m})[M]^2\beta_{12} + \ldots (n - \bar{m})[M]^n\beta_{1n} \quad (6)$$

If only three species (M, M_1L, and M_2L) are present in significant concentrations, this equation can be arranged to

$$\frac{\bar{m}}{(2 - \bar{m})[M]^2} = \frac{(1 - \bar{m})}{(2 - \bar{m})[M]}\beta_1 + \beta_{12} \quad (7)$$

The terms β_1 and β_{12} can be evaluated graphically.

The relation between $\bar{m}$ and pM for systems where only a 1:1 metal–ligand complex is formed is derived from Eq. (5). By substitution in terms of β and arranging

$$pM = \log\frac{1 - \bar{m}}{\bar{m}} + \log K \quad (8)$$

Assuming different values for $\bar{m}$ yields the corresponding pM.

If the system has three species M, ML, and M_2L Eq. (5) can be arranged to:

$$[1 - \bar{m}][M]K + (2 - \bar{m})[M]^2\beta_{12} - \bar{m} = 0$$

$$[M] = \frac{K(\bar{m} - 1) \pm \sqrt{(\bar{m} - 1)^2K^2 + 4\bar{m}(2 - \bar{m})\beta_{12}}}{2(2 - \bar{m})\beta_{12}} \quad (9)$$

By assuming a certain value for $\bar{m}$, a corresponding value for $[M]$ is obtained.

The different fractions of metal complexes in the cyclotri- and cyclotetraphosphate systems were calculated as a function of total cyclophosphate present. When only $[M]$ and $[ML]$ are present

$$K = \frac{[ML]}{[M][L]} = \frac{[M]_T - f[M]_T}{f[M]_T[L]}$$

where f is the fraction of free metal ion.

$$[L] = \frac{1 - f}{fk} \quad \text{and} \quad [L]_T = (1 - f)[M]_T + \frac{1 - f}{fk} \quad (10)$$

If there are three species [M], [ML], and [M_2L], a metal conservation equation is written and [L] is evaluated from it.

$$[M]_T = [M] + [M][L]K + 2[M]^2[L]\beta_{12} \tag{11}$$

$$[L] = \frac{[M]_T - [M]}{[M]K + 2[M]^2\beta_{12}}$$

But from the mass conservation of ligand

$$[L]_T = [L] + [ML] + [M_2L]$$
$$= [L](1 + [M]K + [M]^2\beta_{12}) \tag{12}$$

Again by assuming a certain value for [M], [L] is obtained and [L]$_T$ can be calculated.

A series of data and results for different cations with both cyclotri- and cyclotetraphosphate have been obtained (140). The formation constants are given in Table I.

From the data presented in Table I, it appears that the thermodynamic constants reported by Monk and co-workers (54,137) are larger than the values reported by other workers. In fact, Monk's work has been the subject of criticism by others (111,324). The difference between $\log K_{\text{thermodynamic}}$ and $\log K_{\text{apparent}}$ at ionic strength of 0.1 can be obtained by calculating the values of the activity coefficients of the different species by Davies' (53) extended Debye-Huckel equation:

$$-\log f_i = Az_i{}^2 \left[\frac{\sqrt{I}}{1 + \sqrt{I}} - 0.30I \right]$$

where f_i is the activity coefficient, z_i is the valency of the ion, A is a constant calculated to be 0.50, and I is the ionic strength. Using the above equation and $I = 0.1, f_{\text{M}^\text{I}} = f_{\text{M}^\text{I}\text{P}_3\text{O}_9} = 0.785; f_{\text{M}^\text{II}} = f_{\text{M}^\text{II}\text{P}_4\text{O}_{12}} = f_{\text{M}^\text{I}\text{P}_3\text{O}_9} = 0.379; f_{\text{P}_3\text{O}_9} = f_{\text{M}^\text{I}\text{P}_4\text{O}_{12}} = 0.114;$ and $f_{\text{P}_4\text{O}_{12}} = 0.0209$. Substitution of f values into the expression

$$\log K_{\text{apparent}} = \log K_{\text{thermodynamic}} - \log \frac{f_{\text{MP}_3\text{O}_9}}{f_{\text{P}_3\text{O}_9} f_{\text{M}}}$$

In $\text{M}^\text{I}\text{P}_3\text{O}_9$

$$\log K_{\text{apparent}} = \log K_{\text{thermodynamic}} - 0.63$$

In the $\text{M}^\text{II}\text{P}_3\text{O}_9$ system,

$$\log K_{\text{apparent}} = \log K_{\text{thermodynamic}} - 1.26$$

In the $\text{M}^\text{I}\text{P}_4\text{O}_{12}$ system,

$$\log K_{\text{apparent}} = \log K_{\text{thermodynamic}} - 0.84$$

And in the $\text{M}^\text{II}\text{P}_4\text{O}_{12}$ system,

$$\log K_{\text{apparent}} = \log K_{\text{thermodynamic}} - 1.68$$

TABLE I

Stability Constants

Metal	Method	Temp. (°C)	Medium	$P_3O_9^{3-}$ log K	$P_4O_{12}^{4-}$ log K, log K_{12}	$P_6O_{18}^{6-}$ log K	Ref. $P_3O_9^{3-}$	Ref. $P_4O_{12}^{4-}$	Ref. $P_6O_{18}^{6-}$
H^+	pH	25	0 corr.	no ev cpx			281		
	con	20	0.1 KCl	2.05	2.74		54	54	
	H	25	0 corr.	1.35			357		
	gl	25	0 corr.	no ev cpx	2.78		24	24	
	gl	25	1.0 $NaClO_4$	0.92			345		
	gl	25	0.2 NH_4ClO_4	1.64, K_2 2.07			72		
	Kin	50–70		ev. HL^{2-}			124		
	gl	25	1.0 TMA·NO_3	0.65	1.53		339	339	
	con	25	0	1.17	2.05		54	54	
Na^+	CuHg	25?	0.56? (Me_4NNO_3)	−0.1, β_{12} 0.0			133		
	MHg	30	1 (Me_4NNO_3)	1.40	0.81		91	91	
	SIE	25	0 Corr.	0.88	2.12		140	140	
	SIE	25	0.1 $Me_4N\cdot NO_3$	0.97	1.42, 1.13		141	141	
NH_4^+	SIE	25	0.1 $Me_4N\cdot NO_3$	no ev cpx	1.62, 1.29		140	140	
K^+	SIE	25	0.1 $Me_4N\cdot NO_3$		1.26, 0.92			361	
Mg^{2+}	con	25	0	3.31	5.17		137	361	
	H	20	0.1 KCl	1.11	4.52		357	140	
Ca^{2+}	sp, gl	?	0.1 NH_4Cl	2.74	3.47, 2.00		244	54	
	SIE	25	0.1 $Me_4N\cdot Cl$	1.80	4.89, 2.66		140	361	
	sol	25	0 corr.	3.48	5.32, 2.60		54	361	
	sol	25	0 corr.	3.45	5.42, 2.70		54	97	
	con	25	0	2.50	3.36		137		
	oix	37	0.15 NaCl	1.68	3.77		97		
	sp, gl	?	0.1 NH_4Cl	1.64			244		
	SIE	25	1.0 $Me_4N\cdot NO_3$		3.04		140		

Ion	Method	t (°C)	Medium				Ref	Ref
Sr^{2+}	con	25	0	3.35	5.15		184	184
	sol	25	0 Corr	3.35	5.08	2.46	184	184
	cix	20	0.15 NaCl	1.95	2.80		96	96
	sp, gl	?	$0.1\ NH_4Cl$	0.62	1.46		244	244
	SIE	25	$0.1\ Me_4N \cdot Cl$	1.91			140	
Ba^{2+}	con	25	0	3.35	4.99		137	361
	sp, gl	?	$0.1\ NH\ Cl$	0.08	1.00		244	244
La^{3+}	con	25	0	5.70	6.66		185	185
Mn^{2+}	con	25	0	3.57	5.74		137	361
Co^{2+}	SIE	25	$0.1\ Me_4N \cdot NO_3$	1.79	3.56	1.16	140	140
$Co(NH_3)_6{}^{3+}$	con	25	0	4.44	5.74		185	185
Ni^{2+}	con	25	0	3.22	4.95		137	361
	MHg	30	$1\ (Me_4NNO_3)$	1.82	$2.63,\ K_2\ 0.85$		111	111
	SIE	25	$0.1\ Me_4N\ NO_3$		$3.38,\ 1.35$		140	140
Cu^{2+}	MHg	25	$0.56?\ (Me_4NNO_3)$	$1.55\ \text{ev}\ Cu_2L^+$			133	
	MHg	30	$1\ (Me_4NNO_3)$	$1.58,\ K_2\ 0.60$	$3.18,\ K_2\ 1.46$		339	111
	MHg	25	$1.0\ Me_4NNO_3$		$3.05,\ K_2\ 1.24$		339	339
Fe^{2+}	kin	25	$0.1\ NaClO_4$	$1.13,\ K_{11}\ 0.97$			345	
	SIE	25	$0.1\ Me_4N \cdot NO_3$	1.61	3.39	2.2	140	140
Zn^{2+}	oth	25	var	2.00	$\text{ev}\ ZnL^{2-}$		196	196
	SIE	25	$0.1\ Me_4N \cdot NO_3$		3.63		140	140
	MHg	25	$1.0\ Me_4N \cdot NO_3$		3.01		140	140

(continued overleaf)

TABLE I (continued)

Metal	Method	Temp. (°C)	Medium	$P_3O_9^{3-}$ log K	$P_4O_{12}^{4-}$ log K, log K_{12}	$P_6O_{18}^{6-}$ log K	References $P_3O_9^{3-}$	References $P_4O_{12}^{4-}$	References $P_6O_{18}^{6-}$
Cd^{2+}	oth	25	var		ev CdL^{2-}			196	
	SIE	25	0.1 $Me_4N \cdot NO_3$	2.21	3.55 3.25			140	
Pb^{2+}	SIE	25	0.1 $Et_4N \cdot ClO_4$		4.31			140	
Syn^{3+}	con	25	0 Corr	3.21			238		
Al^{3+}	oth	25	var	ev AlL	ev AlL⁻		196	196	
Am^{3+}	cix	25	0.2 NH_4ClO_4	3.48			72		
	cix	25	0 Corr	5.94			72		
Cm^{3+}	cix	25	0.2 NH_4ClO_4	3.64			72		
	cix	25	0 Corr	6.08			72		
Pm^{3+}	cix	25	0.2 NH_4ClO_4	3.80			72		
	cix	25	0 Corr	6.25			72		
Hb	Eq. dialysis	25	0.05M cacodylic buffer		5.00, K_2 3.70	5.40, K_2 4.15 K_3 3.00		37	37
HbO_2	Eq. dialysis	25	0.05M cacodylic buffer		5.18, K_2 3.23	5.18, K_2 3.70 K_3 3.00		37	37
MetHB	Eq. dialysis	25	0.05M cacodylic buffer		5.00, K_2 3.81	5.00, K_2 3.59		37	37

NOTE: All symbols have their usual meaning as given in Sillen, L. G., and A. E. Martell, *Stability Constants of Metal Ion Complexes*, The Chemical Society, London, 1964. The exceptions are: SIE = specific ion electrode; Hb, HbO_2, MetHB are human deoxy-, oxy-, and methemoglobin, respectively.

Application of these corrections to Gardner and Nancollas (91) and Monk's (54) values listed in Table I gives $\log K_{apparent}$ at $I = 0.1$ of 0.77 and 0.73 for $NaP_3O_9^{2-}$. These values may be compared to 0.88 obtained by Kalliney (140) under similar conditions. Similarly for $NaP_4O_{12}^{3-}$ $\log K_{apparent}$ of 1.28 and 1.22 obtained by Gardner and Nancollas (91) and Monk (54), respectively, can be compared to 1.42 obtained by this author.

These corrections can be applied to $\log K_{thermodynamic}$ values for $CaP_3O_9^{1-}$ and $CaP_4O_{12}^{2-}$ obtained by Monk et al. (54,137) to give $\log K_{apparent}$ at $I = 0.1$ of 2.19 and 3.74, respectively. These values agree favorably with those obtained by Roppongi and Kato (244) and Gosselin and Coghlan (97) who reported values of 3.77 and 3.36 for $CaP_4O_{12}^{2-}$ and 2.50 for $CaP_3O_9^{1-}$ at the conditions mentioned in Table I. Watters et al. (340) reported $\log K$ of 3.04 and 1.64 for $CaP_4O_{12}^{2-}$ and $CaP_3O_9^{1-}$ respectively at $I = 1$ and 25°. These values, which are relatively low, can be attributed to the ionic strength. Experimentally, decreasing I from 1.0 to 0.1 in the case of the $ZnP_4O_{12}^{2-}$ complex increased $\log K$ by about 0.6 units. The estimated values for $\log K$'s at $I = 0.1$ in the Watters study would be 3.7 for $CaP_4O_{12}^{2-}$ and 2.2 for $CaP_3O_9^{1-}$. By the same argument, the difference of 0.6 units in the $\log K$ value for $NaP_4O_{12}^{3-}$ reported by Gross and Gryder (111) and the apparent value of 1.4 at $I = 0.1$ reported by the rest of the studies, can be explained by the change in ionic strength from 1.0 to 0.1.

Complexes of chain phosphates with transition metal ions are more stable than the analog cyclotri- and cyclotetraphosphates. Both of these cyclophosphates are not flexible enough to take different configurations around the transition metal ions which would allow d orbitals to participate in bonding. This is in contrast to chain phosphates which are flexible and can fit around a transition metal ion in a square of octahedral configuration, an arrangement favorable to d orbital participation in bonding. This argument is further strengthened by the fact that calcium and copper(II) which have nearly identical Goldschmidt ionic radii of 0.94 and 0.92 Å, respectively, also have nearly identical formation constants for their complexes with $P_3O_9^{3-}$ and $P_4O_{12}^{4-}$. This is in sharp constrast with the constants for the chain pyro-, tri-, and tetraphosphates of the two metals. The first formation constants for the complexes of copper with chain phosphates (343) are larger than the corresponding calcium complexes (342) by the factor $10^{3.3}$–$10^{4.1}$. It is probable that the difference is related to the availability of d orbitals for participation in bonding by copper(II) but not by calcium. It is likely that the bonds between the cations and the two cyclophosphates are mainly ionic. If there is any covalent character, it must be due to a weaker sp^3 tetrahedral or an equivalent bonding based on molecular orbital arguments. This also may explain the increase in the formation constants between the studied cations and the two cyclophosphates with the increase in their charge. The log K's for the divalent cations with either one of the cyclophosphates is 2.5 times greater than the respective value for the monovalent complex. As expected, this ratio is even greater for the trivalent cations.

$P_4O_{12}^{4-}$ is more basic and its complexes with different cations are more stable than those of $P_3O_9^{3-}$. The $P_4O_{12}^{4-}$ ring is larger than that of $P_3O_9^{3-}$ and as a consequence is more flexible. The flexibility of the $P_4O_{12}^{4-}$ ring is evident from its different conformation in the presence of different cations (see the section on structure). It is feasible to bring two or more terminal oxygens of $P_4O_{12}^{4-}$ into proximity with a cation. In contrast, while more than one terminal oxygen atom of $P_3O_9^{3-}$ can be made to approach the cation, its structure seems to be too stiff to permit optimum fitting of the bonding atoms. In comparison to the $P_3O_9^{3-}$ anion, the larger size of $P_4O_{12}^{4-}$ allows more available positions where a cation can approach a cluster of two to four terminal oxygens. The addition of a second cation to $P_4O_{12}^{4-}$ is evident with many of the cations that have been studied. This was not found in the $P_3O_9^{3-}$ systems. The previous argument about the ring size of $P_4O_{12}^{4-}$ and the number of available oxygen atoms in comparison to $P_3O_9^{3-}$ can also be used to explain this.

The interaction of cyclophosphates with hydrated cations is not by any means a limiting case. Cyclotetraphosphates and cyclohexaphosphates were found to interact also with some charged complexes. These complexes were human deoxy-, oxy-, and methemoglobin. The interaction was studied by equilibrium dialysis in a cacodylate buffer of pH 6.5 (37). This interaction might be of interest since the hemoglobin affinity to oxygen decreased the more strongly it formed complexes with phosphates (38). This is probably due to the competition with molecular oxygen to form complex with hemin. Cyclohexaphosphate forms a stronger complex and has more binding sites with hemoglobin than cyclotetraphosphate which is consistent with the effect of ring size and flexibility on complex formation. The complex formation of the two cyclophosphates with hemoglobin is similar to the complex formation between the cyclic trimer and tetramer with hydrated Fe^{2+} and Fe^{3+} since the difference between them is that the porphin ring in hemoglobin occupies positions previously occupied by the water of hydration. The K values are given in Table I. It might be of interest to examine chlorophyll for possible interaction with organic and inorganic phosphates.

With the exception of cyclohexaphosphate formation constants with hemoglobin, no data on the basicity or formation constants between cyclophosphates with $n > 4$ and different cations are available. It is expected that the increase in the ring size would increase the complex stability. Also, as n increases, the behavior of cyclophosphates toward cations would approach those of linear phosphates with a similar degree of polymerization.

VIII. Applications

During the last several years there has been a rapidly increasing interest in the application of cyclophosphate salts. Use as a cleanser in the soap industry, as a polishing agent, in cheese processing, in tanning leather, and in fluorescent and mercury lamps has been suggested.

The use of sodium triphosphate ($Na_5P_3O_{10}$) in the detergent industry is well known. This compound can be obtained upon the hydrolysis of $Na_3P_3O_9$ with NaOH or Na_2CO_3. The use of this reaction has been suggested as a replacement for using the compound $Na_5P_3O_{10}$. It is claimed that the mixture of $Na_3P_3O_9$ and either alkali is easier to store and is not affected by moisture as is $Na_5P_3O_{10}$. The mixture is also claimed to be more reproducible and has a lower slurry viscosity (201,269,344).

Polyvalent metal cyclophosphates may be used advantageously as a polishing or a tartar-removing tooth cleaning agent in a water-soluble diglycolate toothpaste (220). When alkali metal cyclophosphates, for example, cyclotri- and cyclotetraphosphate, are treated with various oxides such as Al_2O_3, ZnO, Fe_2O_3, B_2O_3, asbestos, fullers earth, or china clay, they form molded polymeric products. These products with appropriate fillers can be used for abrasives or surfaces of high friction (92).

The strontium salt of cyclotriphosphate activated by Cu^+ has been described for use in fluorescent and mercury lamps. The salt has a deep red emission and is relatively temperature independent. The emission maximum is 6400 nm while the excitation maximum is 3000 nm (243).

Linear phosphates have been used in tanning chrome leather. Two difficulties were encountered with this process: formation of a precipitate and the inability to dye the leather with acid dyes to obtain full shades. To overcome these difficulties, $Na_3P_3O_9$ and $Na_4P_4O_{12}$ have been suggested as replacements in this process (20,74,237).

The sodium salt of cyclotetraphosphate has found application in cheese processing. A homogenous cheese was obtained when $Na_4P_4O_{12}$ was used as a melting salt, but not with $Na_3P_3O_9$. The action was explained on the basis of the calcium-binding ability of $P_4O_{12}{}^{4-}$ which is within the range of the usual melting salt (199).

Some cyclophosphate salts might find application as cariostatic agents, at least in animals. Recent experiments on the dental development in rats showed cyclohexa- and cyclotriphosphate sodium salts were more effective as cariostatic agents than the sodium salt of ortho-, pyro-, and triphosphate. $Na_3P_3O_9$ had the highest activity among the tested salts (123).

Acknowledgments

The author wishes to express appreciation to Dr. P. Kabasakalian for advice and encouragement, to the Schering Corporation, for permission to publish, and thanks to my wife, Elizabeth, for her help in preparing the manuscript.

REFERENCES

1. Ando, T., J. Ito, S. H. Ishii, and T. Soda, *Bull. Chem. Soc. Japan*, **25**, 78 (1952).
2. Andon, R. J. L., J. F. Counsell, J. F. Martin, and C. J. Mash, *J. Appl. Chem.*, **17**, 65 (1967).
3. Andress, K. R., and K. Fisher, *Acta Cryst.*, **3**, 399 (1950).
4. Andress, K. R., W. Gehring, and K. Fisher, *Z. Anorg. Allgem. Chem.*, **260**, 331 (1949).
5. Anghileri, L. J., *Agr. Rep. Com. Nacl. Energia At. Inform.*, No. 117 (1964); *Chem. Abstr.*, **62**, 2227d (1965).
6. Aurenge, J., *Bull. Soc. Chim. Fr.*, **1963**, 1525.
7. Aurenge, J., M. Degeorges, and J. Normand, *Bull. Soc. Chim. Fr.*, **1964**, 508.
8. *Automat. Anal. Chem. Technicon Symp.*, 3rd, **2**, 247, Mediad, Inc., White Plains, New York, 1968.
9. Babko, A. K., Yu. F. Shkaravskii, and E. M. Ivashkovich, *Zh. Anal. Khim.*, **26**, 854 (1971); *Chem. Abstr.*, **75**, 58380p (1971).
10. Bammann, E., and M. Meisenheimer., *Chem. Ber.*, **71**, 2086 (1938).
11. Barney, D. L., and J. W. Gryder, *J. Amer. Chem. Soc.*, **77**, 3195 (1955).
12. Baudler, M., and M. Mengel, *Z. Anal. Chem.*, **206**, 8 (1964).
13. Baudler, M., and F. Stuhlmann, *Naturwissenschaften*, **51**, 57 (1964).
14. Beans, H. T., and S. J. Kiehl, *J. Amer. Chem. Soc.*, **49**, 1878 (1927).
15. Beglov, B. M., *Akad. Nauk. Uzb. SSSR*, **27**, 30 (1970).

CYCLOPHOSPHATES 301

16. Bekturov, A. B., E. V. Poletaev, Yu. A. Kushnikov, *Khim. Technol. Kondens. Fosfatov, Tr. Vses. Soveshch.*, 2nd, 1968 (Pub. 1970) p. 146. Edited by Serazetdinov, D. Z. "Nauka" KaZ. SSR: Alma-Ata KaZ. SSR.; *Chem. Abstr.*, **75**, 70792g (1971).
17. Bell, R. N., *Ind. Eng. Chem.*, **39**, 136 (1947).
18. Bell, R. N., *Inorg. Syn.*, **3**, 103 (1950).
19. Bell, R. N., L. F. Audrieth, and O. F. Hill, *Ind. Eng. Chem.*, **44**, 568 (1952).
20. Benckiser, G. M. B. H., Ger. Pat. 1,256,834 (Dec. 21, 1967), Appl. Nov. 6, 1965; *Chem. Abstr.*, **68**, 41000x (1968).
21. Benz, C., and L. M. Paixao, *Chim. Anal. (Paris)*, **50**, 247 (1968).
22. Berger, J. A., G. Meyniel, and J. Petit, *J. Chromatog.*, **29**, 190 (1967).
23. Berzelius, J. J., *Ann. Physik.*, **54**, 31 (1816).
24. Beukenkamp, J., W. Rieman III, and S. Lindenbaum, *Anal. Chem.*, **26**, 505 (1954).
25. Bhargava, H. N., and D. C. Srivastava, *Anal. Chim. Acta*, **37**, 269 (1967).
26. Bonnemann, P., *Compt. Rend.*, **204**, 433 (1937).
27. Bonnemann, P., *Compt. Rend.*, **204**, 865 (1937).
28. Bonnemann-Bemia, P., *Ann. Chim. (Rome)*, **16**, 395 (1941).
29. Bonnemann-Bemia, P., *Ann. Chim. (Rome)*, **16**, 451 (1941).
30. Botre, C., V. Crescenzi, and A. Mele, *J. Inorg. Nucl. Chem.*, **8**, 369 (1958).
31. Boulle, A., *Compt. Rend.*, **206**, 1732 (1938).
32. Brasted, R. C., and A. K. Nelson, *J. Phys. Chem.*, **66**, 377 (1962).
33. Bronnikov, A. Kh., *J. Appl. Chem. (USSR)*, **12**, 1287 (1939); *Chem. Abstr.*, **34**, 3609 (1940).
34. Brovkina, I. A., *Zh. Obshch. Khim.*, **22**, 1917 (1952); *Chem. Abstr.*, **47**, 5289b (1953).
35. Cagliotti, V., G. Giacomello, and E. Bianchi, *Atti R. Accad. Italia, Rend. Classe Sc. Fis. Natl. Natur.*, **3**, 761 (1942).
36. Canic, V. D., M. N. Turcic, S. M. Petrovic, and S. E. Petrovic, *Anal. Chem.*, **37**, 1576 (1965).
37. Chanutin, A., and E. Hermann, *Arch. Biochem. Biophys.*, **131**, 180 (1969).
38. Chanutin, A., and R. R. Curnish, *Arch. Biochem. Biophys.*, **121**, 96 (1967).
39. Chemische Werke Albert, Neth. Pat., Appl. 6,509,168; *Chem. Abstr.*, **67**, 34544d (1967).
40. Chung, N. M., R. Lohrmann, L. E. Orgel, and J. Rabinowitz, *Tetrahedron*, **1971**, 1205, and references cited.
41. Clark, T., *Edinburgh J. Sci.*, **7**, 298 (1827).
42. Cleserci, N. L., and G. F. Lee, *Anal. Chem.*, **36**, 2207 (1964).
43. Corbridge, D. E. C. in *Topics in Phosphorus Chemistry*, Vol. 3, E. J. Griffith and M. Grayson, Eds., Interscience, New York, 1966.
44. Corbridge, D. E. C. in *Topics in Phosphorus Chemistry*, Vol. 6, E. J. Griffith and M. Grayson, Eds., Interscience, New York, 1969.
45. Corbridge, D. E. C., and E. J. Lowe, *J. Chem. Soc.*, **1954**, 493, 4555.
46. Corbridge, D. E. C., and F. R. Thomas, *Anal. Chem.*, **30**, 1101 (1958).
47. Coumert, N., M. Porthault, J. Merlin, *Bull. Soc. Chem. France*, **1965**, 910.
48. Crowter, J. P., and A. E. R. Westman, *Can. J. Chem.*, **34**, 967 (1956).
49. Crowther, J., *Anal. Chem.*, **26**, 1383 (1954).
50. Cruickshank, D. W. J., *Acta Cryst.*, **17**, 674 (1964).
51. Crutchfield, M. M., C. F. Callis, R. R. Irani, and G. C. Roth, *Inorg. Chem.*, **1**, 813 (1962).

52. Crutchfield, M. M., and R. R. Irani, *J. Amer. Chem. Soc.*, **87**, 2815 (1965).
53. Davies, C. W., *Ion Association*, Butterworths, Washington, 1962.
54. Davies, C. W., and C. B. Monk, *J. Chem. Soc.*, **1949**, 413.
55. DeSallier Dupin, A., B. Hogan, and A. Boulle, *C.R. Acad. Sci.*, *Ser. C*, **272**, 1491 (1971); *Chem. Abstr.*, **75**, 29419c (1971).
56. Dewald, W., and H. Schmidt, *Z. Anal. Chem.*, **136**, 420 (1952).
57. Dewald, W., and H. Schmidt, *Z. Anal. Chem.*, **137**, 178 (1952).
58. Dornberger-Schiff, K., *Acta Cryst.*, **17**, 482 (1964).
59. Durif, A., C. Martin, J. Jordjmann, and D. Tranqui, *Bull. Soc. Fr. Mineral.*, **89**, 439 (1966).
60. Eanes, E. D., and H. M. Ondik, *Acta Cryst.*, **15**, 1280 (1962).
61. Ebel, J. P., *Compt. Rend.*, **234**, 621 (1952).
62. Ebel, J. P., *Bull. Soc. Chim. France*, **1953**, 991.
63. Ebel, J. P., *Bull. Soc. Chim. France*, **1953**, 998.
64. Ebel, J. P., *Bull. Soc. Chim. France*, **1953**, 1089, 1096.
65. Ebel, J. P., *Mikrochim. Acta*, **1954**, 679.
66. Ebel, J. P., *Bull. Soc. Chim. France*, **1968**, 1663.
67. Ebel, J. P., J. Collas, and N. Busch, *Bull. Soc. Chim. France*, **1955**, 1087.
68. Ebel, J. P., and N. Busch, *Bull. Soc. Chim. France*, **1956**, 758.
69. Ebel, J. P., and N. Busch, *Compt. Rend.*, **242**, 647 (1956).
70. Ebel, J. P., and Y. Volmar, *Compt. Rend.*, **233**, 415 (1951).
71. Egan, E. P., Jr. and Z. T. Wakefield, *J. Phys. Chem.*, **64**, 1955 (1960).
72. Elesin, A. A., I. A. Lebedev, E. M. Piskunov, and G. N. Yakovlev, *Sov. Radiochem.*, **9**, 159 (1967).
73. Ender, G., *Z. Anal. Chem.*, **138**, 401 (1953).
74. Erasmus, A. (Benckiser Knapsack G.M.B.H., Ludwigshafen. Ger.) AQEIC (Association Quim. Espan. Ind. Cuero), Bol. Tech., 1970, **20** (2) 44–50 (Span.); *Chem. Abstr.*, **73**, 26642t (1970).
75. Etienne, H., *Ind. Chim. Belge*, **18**, 1340 (1953); *Chem. Abstr.*, **47**, 8583 (1953).
76. Faber, W., Ger. Pat. 734,511 (1943).
77. Feldmann, W., *Z. Chem.*, **5**, 26 (1965).
78. Feldmann, W., *Z. Anorg. Allgem. Chem.*, **338**, 235 (1965).
79. Feldmann, W., *Chem. Ber.*, **99**, 3251 (1966).
80. Feldmann, W., *Chem. Ber.*, **100**, 3850 (1967).
81. Feldmann, W., *Z. Chem.*, **9**, 154 (1969); *Chem. Abstr.*, **71**, 3643 (1969).
82. Feldmann, W., and E. Thilo, *Z. Anorg. Allgem. Chem.*, **327**, 159 (1964).
83. Feldmann, W., and E. Thilo, *Z. Anorg. Allgem. Chem.*, **328**, 113 (1964).
84. Felter, S., G. Dirheimer, and J. P. Ebel, *J. Chromatog.*, **35**, 207 (1968).
85. Fleckenstein, J., *Naturwissenschaften*, **40**, 462 (1953).
86. Fleitmann, T., *Poggendorffs Ann.*, **78**, 233 (1849).
87. Fleitmann, T., *Poggendorffs Ann.*, **78**, 338 (1849).
88. Fleitmann, T., and R. Henneberg, *Ann. Chem.*, **65**, 304 (1848).
89. Fluck, E., *Z. Naturforsch.*, **20b** 505 (1965).
90. Frank, W. H., *Angew. Chem.*, **68**, 586 (1956).
91. Gardner, G. L., and G. H. Nancollas, *Anal. Chem.*, **41**, 202 (1969).
92. Gebbett, D., A. Hourd, and P. R. Bloomfield, Brit. Pat. 1,018,401; *Chem. Abstr.*, **64**, 9407b (1966).
93. Gerber, A. B., and F. T. Miles, *Ind. Eng. Chem., Anal. Ed.*, **13**, 406 (1941).
94. Giran, H., *Ann. Chim. Phys.*, **30**, 203 (1903).

95. Glatzel, A., *Inaugural Dissertation*, Wurtzburg, 1880.
96. Gnepf, H., O. Gubeli, and G. Schwarzenbach, *Helv. Chim. Acta*, **45**, 1171 (1962).
97. Gosselin, R. E., and E. R. Coghlan, *Arch. Biochem. Biophys.*, **45**, 301 (1953).
98. Graham, T., *Phil. Trans. Roy. Soc. London, Ser. A.*, **123**, 253 (1833).
99. Grande, J. A., and J. Boukenkamp, *Anal. Chem.* **28**, 1497 (1956).
100. Grenier, J., and R. Masse, *Bull. Soc. Fr. Mineral*, **91**, 428 (1968).
101. Griffith, E. J., *J. Amer. Chem. Soc.*, **76**, 5892 (1954).
102. Griffith, E. J., *J. Amer. Chem. Soc.*, **78**, 3867 (1956).
103. Griffith, E. J., *Anal. Chem.*, **28**, 525 (1956).
104. Griffith, E. J., *J. Amer. Chem. Soc.*, **79**, 509 (1957).
105. Griffith, E. J., *Inorg. Chem.*, **1**, 962 (1962).
106. Griffith, E. J., and R. L. Buxton, *Inorg. Chem.*, **4**, 549 (1965).
107. Griffith, E. J., and R. L. Buxton, *J. Chem. Eng. Data*, **13**, 145 (1968).
108. Griffith, E. J., and J. R. Van Wazer, *J. Amer. Chem. Soc.*, **77**, 4222 (1955).
109. Griffith, W. P., *J. Chem. Soc. (A)*, **1967**, 905.
110. Griffith, W. P., and K. J. Rutt, *J. Chem. Soc. (A)*, **1968**, 2331.
111. Gross, R. J., and J. W. Gryder, *J. Amer. Chem. Soc.*, **77**, 3695 (1955).
112. Grunze, H., and E. Thilo, *Z. Anorg. Allgem. Chem.*, **281**, 284 (1955).
113. Grunze, H., and E. Thilo, *Monatsber. Deut. Akad. Wiss. Berlin*, **1**, 510 (1959).
114. Grunze, I., K. Dostal, and E. Thilo, *Z. Anorg. Allgem. Chem.*, **302**, 221 (1959).
115. Grunze, I., E. Thilo, and H. Grunze, *Chem. Ber.*, **93**, 2631 (1960).
116. Gutsov, I., and I. Konstantinov, *Dokl. Bolg. Akad. Nauk.*, **23**, 1107 (1970); *Chem. Abstr.*, **74**, 35415u (1971).
117. Gutzow, I., *Z. Anorg. Allgem. Chem.*, **302**, 259 (1960).
118. Gutzow, I., Z. Toschev, M. Marinov, and E. Popov, *Proc. Conf. Silicate Ind.*, 1967 (Pub. 1968), **9**, 65; *Chem. Abstr.*, **70**, 70763v (1969).
119. Haiduc, I., *The Chemistry of Inorganic Ring Systems*, Vols. 1 and 2, Interscience, New York, 1970.
120. Hall, R. E. (to Hall Laboratories, Inc.). U.S. Pat. 1,956,515 (April 24, 1934); Reissue (1935).
121. Halmann, H., and I. Platzner, *J. Chem. Soc.*, **1965**, 1440.
122. Hanes, C. S., and F. A. Isherwood, *Nature*, **164**, 1107 (1949).
123. Harris, R., A. E. Nizel, and N. B. Walsh, *J. Dent. Res.*, **46**, 290 (1967); *Chem. Abstr.*, **66**, 93320e (1967).
124. Healy, R. M., and M. L. Kilpatrick, *J. Amer. Chem. Soc.*, **77**, 5258 (1955).
125. Hecht, H., *Z. Anal. Chem.*, **143**, 93 (1954).
126. Hettler, H., *J. Chromatog.*, **1**, 389 (1958); *Chromatog. Rev.*, **1**, 225 (1959).
127. Hisar, R. S., *Bull. Soc. Chim. France*, **1952**, 308.
128. Hisar, R. S., *Bull. Soc. Chim. France*, **1959**, 158, 162.
129. Hofmann, H. J., and K. R. Andress, *Naturwissenschaften*, **41**, 94 (1954).
130. Iida, T., and T. Yamabe, *J. Chromatog.*, **54**, 413 (1971).
131. Indelli, A., *Ann. Chim. (Rome)*, **46**, 717 (1956).
132. Indelli, A., *Ann. Chim. (Rome)*, **47**, 586 (1957).
133. Indelli, A., *Ann. Chim. (Rome)*, **48**, 345 (1958).
134. Irving, R. J., and H. McKerrell, *Trans. Faraday Soc.*, **64**, 879 (1968).
135. Jarchow, O. H., and K. Dornberger-Schiff, *Acta Cryst.*, **13**, 1020 (1960).
136. Jarchow, O. H., *Acta Cryst.*, **17**, 1253 (1964).

137. Jones, H. W., C. B. Monk, and C. W. Davies, *J. Chem. Soc.*, **1949**, 2693.
138. Jones, L. T., *Ind. Eng. Chem., Anal. Ed.*, **14**, 536 (1942).
139. Jost, K. H., *Acta Cryst.*, **19**, 555 (1965).
140. Kalliney, S., Ph.D. Thesis, The Ohio State University, 1970.
141. Kalliney, S., unpublished results.
142. Karbe, K., and G. Jander, *Kolloid Beihefte*, **54**, 1 (1942).
143. Karl-Kroupa, E., *Anal. Chem.*, **28**, 1091 (1956).
144. Karl-Kroupa, E., J. R. Van Wazer, and C. H. Russell, *Scott's Standard Methods of Chemical Analysis*, N. H. Furman, Ed., 6th ed., Vol. 1, Chapter 35, Van Nostrand, New York, 1962.
145. Khodakov, Yu. V., *Tr. Mosk. Aviats Inst. Sb. Statei*, **52**, 20 (1955); *Ref. Zh. Khim.*, **1956**, 57303.
146. Khodakov, Yu. V., *Zh. Neorgan Khim.*, **1**, 368 (1956).
147. Klement, R., *Handbuch der Analytischen Chemie*, W. Fresenius and G. Jander, Eds., Springer, Berlin, Sec. 3, Vol. Vaβ, 1953, pp. 305–322.
148. Klement, R., and R. Popp, *Chem. Ber.*, **93**, 156 (1960).
149. Knorre, G., *Z. Anorg. Allgem. Chem.*, **24**, 369 (1900).
150. Kobayashi, E., *Nippon Kagaku Zasshi*, **85**, 317 (1964).
151. Koberlein, W., and H. Mair-Waldburg, *Z. Lebensm-Untersuch. U.-Forsch*, **102**, 231 (1955).
152. Kolitowska, H. J., *Bull. Acad. Polon. Sci., Ser. Sci. Chim.*, **7**, 369 (1959).
153. Kolitowska, H. J., and M. Maczynski, *Bull. Acad. Polon. Sci., Ser. Sci. Chim.*, **8**, 449 (1960).
154. Kolloff, R. H., *Anal. Chem.*, **33**, 373 (1961).
155. Kunovits, G., *Seifen-Oele-Fette-Wachse*, **92**, 591 (1966).
156. Kura, G., and S. Ohashi, *J. Chromatog.*, **56**, 111 (1971).
157. Kushnikov, Yu. A., E. V. Poletoaev, and S. M. Divnenko, *Zh. Neog. Khim.*, **12**, 2355 (1967); *Chem. Abstr.*, **68**, 56029n (1968).
158. Kuzmichev, S. I., *Tr. Mosk. Aviats. Inst.*, **53**, 36 (1955).
159. Laitnen, H. A., and K. R. Lucas, *J. Electroanal. Chem.*, **12**, 553 (1966).
160. Latimer, W. M., *Oxidation Potentials*, 2nd Ed., Prentice-Hall, New York, 1952.
161. Lederle, P., *Z. Anal. Chem.*, **121**, 403 (1941).
162. Lee, J. D., and A. H. Bond, *J. Appl. Chem.*, **18**, 345 (1968).
163. Lee, J. D., and L. S. Browne, *J. Chem. Soc. (A)*, **1968**, 559.
164. Lenzi, M., and E. Mariani, *Rass. Chim.*, **3**, 11 (1959); *Chem. Abstr.* **54** 4238d (1960).
165. Letcher, J. H., and J. R. Van Wazer in *Topics in Phosphorus Chemistry*, Vol. 5, E. J. Griffith and M. Grayson, Eds., Interscience, New York, 1967.
166. Liddell, R. W., *J. Amer. Chem. Soc.*, **71**, 207 (1949).
167. Lindboom, C. G., *Chem. Ber.*, **8**, 122 (1875).
168. Liteanu, C., I. Lukacs, and C. Strusievici, *Studii Cercetari Chim. (Cluj)*, **9**, 101 (1958); *Chem. Abstr.*, **53**, 18717 (1959).
169. Liteanu, C., I. Lukacs, and C. Strusievici, *Studii Cercetari Chim. (Cluj)*, **10**, 119 (1959); *Chem. Abstr.*, **54**, 9588h (1960).
170. Lundgren, D. P., and N. P. Leob, *Anal. Chem.* **33**, 367 (1961).
171. Lux, H., *Z. Electrochem.*, **45**, 303 (1939).
172. MacGillavry, C. H., H. C. I. DeDecker, and L. M. Nijland, *Nature*, **164**, 448 (1949).
173. MacGillavry, C. H., and C. Romers, *Nature*, **164**, 960 (1949).

174. Maddrell, R., *Ann. Chem.*, **61**, 53 (1847).
175. Mark, V., C. H. Dungan, M. M. Crutchfield, and J. R. Van Wazer, in *Topics in Phosphorus Chemistry*, Vol. 5, E. J. Griffith and M. Grayson, Eds., Interscience, New York, 1967.
176. Markovitz, M. M., *J. Chem. Educ.*, **33**, 36 (1956).
177. Martens, W. S., and W. Rieman, *J. Polymer. Sci.*, **54**, 603 (1961).
178. Masse, R., J. C. Grenier, M. T. Overbuch-Pouchot, T. Q. Due, and A. Durif, *Bull. Soc. Fr. Mineral Cristollogr.*, **90**, 158 (1967).
179. McCune, H. W., and G. H. Arquette, *Anal. Chem.*, **27**, 401 (1955).
180. McGilvery, J. D., and A. E. Scott, *Can. J. Chem.*, **32**, 1100 (1954).
181. Meadowcroft, T. R., and F. D. Richardson, *Trans. Faraday Soc.*, **59**, 1564 (1963).
182. Mellor, J. W., *Comprehensive Treatise on Inorganic and Theoretical Chemistry*, Vol. 8, Longmans, Green, New York, 1928, p. 987.
183. Meyerhof, O., R. Shatas, and A. Kaplan, *Biochem. Biophys. Acta*, **12**, 121 (1953).
184. Monk, C. B., *J. Chem. Soc.*, **1952**, 1314.
185. Monk, C. B., *J. Chem. Soc.*, **1952**, 1317.
186. Moore, E. L., and C. Y. Shen., U.S. Pat. 3,367,737 (Cl-23-106) (Feb. 6, 1968); *Chem. Abstr.*, **68**, 80044 (1968).
187. Morey, G. W., *J. Amer. Chem. Soc.*, **74**, 5783 (1952).
188. Morey, G. W., *J. Amer. Chem. Soc.*, **76**, 4724 (1954).
189. Morey, G. W., F. R. Boyd, J. L. England, and W. T. Chen, *J. Amer. Chem. Soc.*, **77**, 5003 (1955).
190. Morey, G. W., and E. Ingerson, *Amer. J. Sci.*, **242**, 1 (1944).
191. Motooka, I., G. Hashizum, and M. Kobayashi, *Bull. Chem. Soc. Japan*, **40**, 2095 (1967); *Chem. Abstr.*, **69**, 53145g (1968).
192. Motooka, I., G. Hashizum, and M. Kobayashi, *Bull. Chem. Soc. Japan*, **41**, 2040 (1968); *Chem. Abstr.*, **69**, 97945x (1968).
193. Motooka, I., G. Hashizum, and M. Kobayashi, *Kogyo Kagaku Zasshi*, **71**, 1412 (1968); *Chem. Abstr.*, **70**, 59229e (1969).
194. Narath, A., F. H. Lohman, and O. T. Quimby, *J. Amer. Chem. Soc.*, **78**, 4493 (1956).
195. Neddermeyer, P. A., and L. B. Rogers, *Anal. Chem.*, **41**, 95 (1969).
196. Nelson, A. K., Ph.D. Thesis, University of Minnesota, 1959.
197. Neu, R., *Anal. Chem.*, **131**, 102 (1950).
198. Neu, R., *Fette U. Seifen*, **53**, 148 (1951).
199. Ney, K. H., and O. P. Garg (Unilever Farschungslab., Hamburg, Ger.), *Fette Seifen. Anstrichm*, **72**, 279 (1970); *Chem. Abstr.*, **73**, 23966r (1970).
200. Nielson, M. L., and J. V. Pustinger, *J. Phys. Chem.*, **68**, 152 (1964).
201. Novak, D. A., Jr., U.S. Pat. 3,325,413; *Chem. Abstr.*, **67**, 65783k (1967).
202. Ohashi, S., G. Kura, and M. Kamo, *Mem. Fac. Sci. Kyushu Univ.*, Ser. C, **7**, 43 (1970) quoted in Ref. 156.
203. Ondik, H. M., *Acta Cryst.*, **16**, 31A (1963).
204. Ondik, H. M., *Acta Cryst.*, **17**, 1139 (1964).
205. Ondik, H. M., *Acta Cryst.*, **18**, 226 (1965).
206. Ondik, H. M., S. Block, and C. H. MacGillavry, *Acta Cryst.*, **14**, 555 (1961).
207. Ondik, H. M., and J. W. Gryder, *J. Inorg. Nucl. Chem.*, **14**, 240 (1960).
208. Partridge, E. P., *Chem. Eng. News*, **27**, 214 (1949).
209. Pascal, P., *Bull. Soc. Chim.*, **33**, 1611 (1923).

210. Pascal, P., *Compt. Rend.*, **176**, 1398 (1923).
211. Pascal, P., *Compt. Rend.*, **178**, 211 (1924).
212. Pascal, P., *Compt. Rend.*, **178**, 1906 (1924).
213. Pascal, P., *Bull. Soc. Chim. France*, **35**, 1119 (1924).
214. Pascal, P., and M. Rechid, *Compt. Rend.*, **196**, 860 (1933).
215. Pauling, L., and J. Sherman, *Z. Krist.*, **96**, 481 (1937).
216. Pechkovskii, V. V., A. S. Shulman, M. I. Kuzmenkov, G. Kh. Cherchers, *Izv. Akad. Nauk. SSR, Neorg. Mater.*, **6** (11), 1984 (1970); *Chem. Abstr.*, **74**, 35 5 w (1971).
217. Pepinsky, R., C. M. McCarty, E. Zemyan, and K. Drenck, *Phys. Rev.*, **86**, 793 (1952).
218. Peters, T. V., Jr., and W. Reiman III, *Anal. Chim. Acta*, **14**, 131 (1956).
219. Pfrengle, O., *Z. Anal. Chem.*, **158**, 81 (1957).
220. Pfrengle, O., and C. Pietruk (Chemische Fabrick Budenheim Rudolf A. Oetkerk-G), Ger. Pat. 1,251,468, *Chem. Abstr.*, **67**, 120198a (1967).
221. Pollard, F. H., G. Nickless, K. Burton, and J. Hubbard, *Microchem. J.*, **10**, 131 (1966).
222. Pollard, F. H., G. Nickless, and J. D. Murray, *J. Chromatog.*, **22**, 139 (1966).
223. Pollard, F. H., G. Nickless, D. E. Rogers, and M. T. Rothell, *J. Chromatog.*, **17**, 157 (1965).
224. Porthault, M., *Rev. Chim. Minerale*, **1**, 125 (1964); *Chem. Abstr.*, **67**, 49946k (1967).
225. Porthault, M., and J. C. Merlin, *Compt. Rend*, **246**, 2763 (1958).
226. Porthault, M., and J. C. Merlin, *Compt. Rend.*, **250**, 3332 (1960).
227. Proust, J. L., *Ann. Chim. Phys.*, **14**, 281 (1820).
228. Quimby, O. T., *Chem. Rev.*, **40**, 141 (1947).
229. Quimby, O. T., *J. Phys. Chem.*, **58**, 603 (1954).
230. Quimby, O. T., and T. J. Flautt, *Z. Anorg. Allgem. Chem.*, **296**, 220 (1958).
231. Quimby, O. T., and T. J. Flautt, *Z. Anorg. Allgem. Chem.*, **296**, 224 (1958).
232. Quimby, O. T., and F. P. Krause, *Inorg. Syn.*, **5**, 98 (1957).
233. Quimby, O. T., A. Narath, and F. H. Lohman, *J. Amer. Chem. Soc.*, **82**, 1099 (1960).
234. Rabinowitz, J., *Helv. Chem. Acta*, **52**, 2663 (1969).
235. Raistrick, B., *Disc. Faraday Soc.*, **5**, 234 (1949).
236. Raistrick, B., *J. Roy. Coll. Sci.*, **19**, 9 (1949).
237. Riess, C., Fr. Pat. 1,498,220, (Cl. C14C), (Oct. 13, 1967), Applied Nov. 6, 1965; *Chem. Abstr.*, **69**, 68269z (1969).
238. Rizzardi, G., and A. Indelli, *Electrochima Acta*, **14**, 845 (1969).
239. Rodionova, N. I., and Yu. V. Khodakov, *Zh. Obshch. Khim.*, **20**, 1347 (1950).
240. Rohlfs, H. A. (Chemische Werke Albert), Ger. Pat. 1,228,593; *Chem. Abstr.*, **66**, 30594b (1967).
241. Romers, C., J. A. A. Ketelaar, and C. H. MacGillavry, *Nature*, **164**, 960 (1949).
242. Romers, C., J. A. A. Ketelaar, and C. H. MacGillavry, *Acta Cryst.*, **4**, 114 (1951).
243. Ropp, R. C., and H. D. Layman (Sylvania Elect. Production, Inc.), U.S. Pat. 3,318,818 (1967); *Chem. Abstr.*, **67**, 77381r (1967).
244. Roppongi, A., and T. Kato, *Bull. Chem. Soc. Japan*, **35**, 1086 (1967).
245. Rossel, T., *Z. Anal. Chem.*, **197**, 333 (1963).

246. Rossel, T., and H. Kiesslich, *Z. Anal. Chem.*, **229**, 96 (1967).
247. Rothbart, H. L., H. W. Weymouth, and W. Riemann III, *Talanta*, **11**, 33 (1964).
248. Rudy, H., and H. Schloesser, *Chem. Ber.*, **73**, 484 (1940).
249. Naray-Szabo, St., *Z. Krist*, **75**, 387 (1930).
250. Samuelson, O., *Svensk. Kem. Tidskr.*, **56**, 343 (1944).
251. Sansoni, B., *Angew. Chem.*, **65**, 423 (1953).
252. Sansoni, B., *Angew. Chem.*, **67**, 327 (1955).
253. Sansoni, B., and L. Baumgartner, *Z. Anal. Chem.*, **158**, 241 (1957).
254. Sansoni, B., and R. Klement, *Angew. Chem.*, **65**, 423 (1953).
255. Sato, T. R., *Anal. Chem.*, **31**, 841 (1959).
256. Sato, T. R., W. E. Kisielski, W. P. Norris, and H. H. Strain, *Anal. Chem.*, **25**, 438 (1953).
257. Schaerer, T., *J. Prakt. Chem.*, **75**, 113 (1858).
258. Schormuller, J., and G. Wurdig, *Z. Lebens Untersuch. U. Forsch.*, **107**, 415 (1958).
259. Schülke, U., *Angew. Chem., Intern. Ed. (Engl.)*, **7**, 71 (1968).
260. Schülke, U., *Z. Anorg. Allgem. Chem.*, **360**, 231 (1968).
261. Schwartz, A. W., *Chem. Commun.*, **1969**, 1393.
262. Seiler, H., *Helv. Chim. Acta*, **44**, 1753 (1961).
263. Serra, O. A., and E. Giesbrecht, *J. Inorg. Nucl. Chem.*, **30**, 793 (1968).
264. Shaffery, R. J., J. Strumpf, and B. P. Leber, U.S. Pat. 3,382,037 (1968); *Chem. Abstr.*, **69**, 11640j (1969).
265. Shams El Din, A. M., and A. A. A. Gerges, *Electrochim. Acta*, **9**, 123 (1964).
266. Shams El Din, A. M., A. A. A. Gerges, and A. A. El Hosary, *J. Electroanal. Chem.*, **6**, 131 (1963).
267. Shaw, R. A., B. W. Fitzsimmons, and B. C. Smith, *Chem. Rev.*, **62**, 247 (1962).
268. Shiraishi, N., and T. Iba, *Bunse Kagoku*, **13** (9), 883 (1964); *Chem. Abstr.*, **61**, 12603g (1964).
269. Silvis, S. J., and M. Ballestra, *Detergent Age*, **4**, 22 (1967); *Chem. Abstr.*, **67**, 101291j (1967).
270. Simon, A., and E. Steger, *Z. Anorg. Allgem. Chem.*, **277**, 209 (1954).
271. Soklakov, A. I., N. L. Portnova, V. V. Nechaeva, *Khim. Tekhol Kondens. Fosfatov, Tr. Vses. Soveshch.*, 2nd, 1968 (Pub. 1970), p. 17; *Chem. Abstr.*, **75**, 68510v (1971).
272. Steger, E., *Angew. Chem.*, **70**, 376 (1958).
273. Steger, E., *Z. Anorg. Allgem. Chem.*, **296**, 305 (1958).
274. Steger, E., and A. Simon, *Silicuim-Schwefel Phosphate*, IUPAC Coloquium, Munster/Westf., 1954.
275. Steger, E., and A. Simon, *Z. Anorg. Allgem. Chem.*, **291**, 76 (1957).
276. Steger, E., and A. Simon, *Z. Anorg. Allgem. Chem.*, **294**, 1 (1958).
277. Stranski, I. N., and R. Kaischew, *Physik Z.*, **36**, 293 (1935).
278. Takai, N., T. Iida, and T. Yamabe, *Sesian-Kenkyu*, **19**, 87 (1967); *Chem. Abstr.*, **69**, 24166j (1969).
279. Tammann, G., *J. Prakt. Chem.*, **45**, 417 (1892).
280. Tanzer, J. M., M. I. Krichevsky, and B. Chassy, *J. Chromatog.*, **38**, 526 (1968).
281. Teichert, W., *Acta Chem. Scand.*, **2**, 414 (1948).
282. Teichert, W., and K. Rinman, *Acta Chem. Scand.*, **2**, 225 (1948).
283. Thilo, E., *Chem. Tech. (Berlin)*, **4**, 345 (1952).

284. Thilo, E., *Angew. Chem.*, **67**, 141 (1955).
285. Thilo, E., *Zh. Prikl. Khim.*, **29**, 1621 (1956).
286. Thilo, E., *Acta Chim. Acad. Sci. Hung.*, **12**, 221 (1957).
287. Thilo, E., *Kondensierte Phosphate in Lebensmuteln*, Symposium, Mainz, 1957, 5 (Publ. 1958).
288. Thilo, E., *Österr. Chem. Ztg.*, **59**, 1 (1958).
289. Thilo, E., *Chem. Tech. (Berlin)*, **10**, 70 (1958).
290. Thilo, E., *Naturwissenschaften*, **46**, 367 (1959).
291. Thilo, E., *Chem. Soc. Spec. Publ.*, **15**, 33 (1961).
292. Thilo, E., *Advan. Inorg. Chem. Radiochem.*, **4**, 1 (1962).
293. Thilo, E., *Angew. Chem.*, **77**, 1056 (1965); *Angew. Chem. Intern. Ed. (Engl.)*, **4**, 1039 (1965).
294. Thilo, E., *Pure Appl. Chem.*, **12**, 463 (1966).
295. Thilo, E., *Bull. Soc. Chem. Fr.*, **1968**, 1725.
296. Thilo, E., and W. Feldmann, *Z. Anorg. Allgem. Chem.*, **298**, 316 (1959).
297. Thilo, E., and I. Grunze, *Z. Anorg. Allgem. Chem.*, **281**, 262 (1955).
298. Thilo, E., and I. Grunze, *Z. Anorg. Allgem, Chem.*, **290**, 209 (1957).
299. Thilo, E., and I. Grunze, *Z. Anorg. Allgem. Chem.*, **290**, 223 (1957).
300. Thilo, E., and I. Grunze, Ger. Pat. 49,837 (1966); *Chem. Abstr.*, **66**, 12590t (1967).
301. Thilo, E., I. Grunze, and H. Grunze, *Monatsber. Deuts. Akad. Wiss. Berlin*, **1**, 40 (1959).
302. Thilo, E., and R. Ratz, *Z. Anorg. Chem.*, **258**, 33 (1949).
303. Thilo, E., and R. Ratz, *Z. Anorg. Allgem. Chem*, **260**, 255 (1949).
304. Thilo, E., and U. Schülke, *Angew. Chem.*, **75**, 1175 (1963).
305. Thilo, E., and U. Schülke, *Internat. Symposium on Macromol. Chem.*, Prague, Sept. 1965, Preprint No. 252.
306. Thilo, E., and D. Schülke, *Z. Anorg. Allgem. Chem.*, **341**, 293 (1965).
307. Thilo, E., and D. Schülte, *Chem. Ber.*, **93**, 2430 (1960).
308. Thilo, E., and W. Wicker, *Z. Anorg. Allgem. Chem.*, **277**, 27 (1954).
309. Thilo, E., and W. Wieker, *Z. Anorg. Allgem. Chem.*, **291**, 164 (1957).
310. Thomsen, J., *Thermochemische Untersuchungen*, Barth, Leipzig, 1882–1886.
311. Thorburn Burns, D., and J. D. Lee, *Mikrochim. Acta*, **1969**, 202.
312. Thorburn Burns, D., and J. D. Lee, *Mikrochim. Acta*, **1969**, 206.
313. Topley, B., *Quart. Rev. (London)*, **3**, 345 (1949).
314. Travers, A., and Y. K. Chu, *Compt. Rend.*, **198**, 2100 (1934).
315. Travers, A., and Y. K. Chu, *Compt. Rend.*, **198**, 2169 (1934).
316. Treadwell, W. D., and F. Leutwyler, *Helv. Chim. Acta*, **20**, 931 (1937).
317. Treadwell, W. D., and F. Leutwyler, *Helv. Chim. Acta*, **21**, 1450 (1938).
318. Uhlik, Z., *Coll. Czech. Chem. Comm.*, **30**, 3958 (1965).
319. Van Norman, J. D., and R. A. Osteryoung, *Anal. Chem.*, **32**, 398 (1960).
320. Van Wazer, J. R., *J. Amer. Chem. Soc.*, **72**, 647 (1950).
321. Van Wazer, J. R., *Phosphorus and its Compounds*, Vol. 1, Interscience, New York, 1958.
322. Van Wazer, J. R., *Record Chem. Progr. (Kresge-Hooker Sci. Lib.)*, **19**, 113 (1958).
323. Van Wazer, J. R., *Bull. Soc. Chim. Fr.*, **1968**, 1732.
324. Van Wazer, J. R., and C. F. Callis, *Chem. Rev.*, **58**, 1011 (1958).
325. Van Wazer, J. R., C. F. Callis, and J. N. Shoolery, *J. Amer. Chem. Soc.*, **77**, 4945 (1955).

326. Van Wazer, J. R., F. F. Callis, J. N. Shoolery, and R. C. Jones, *J. Amer. Chem. Soc.*, **78**, 5715 (1957).

327. Van Wazer, J. R., and E. J. Griffith, *J. Amer. Chem. Soc.*, **77**, 6140 (1955).

328. Van Wazer, J. R., E. J. Griffith, and J. F. McCullough, *Anal. Chem.*, **26**, 1755 (1954).

329. Van Wazer, J. R., and K. A. Holst, *J. Amer. Chem. Soc.*, **72**, 639 (1950).

330. Van Wazer, J. R., and E. Karl-Kroupa, *J. Amer. Chem. Soc.*, **78**, 1772 (1956).

331. Van Wazer, J. R., and J. H. Letcher in *Topics in Phosphorus Chemistry*, Vol. 5, E. J. Griffith and M. Grayson, Eds., Interscience, New York, 1967.

332. Verdier, J. M., and A. Boulle, Fr. Pat. 1,519,730 (Cl. Co. 1b) (Apr. 5, 1968); Applied 20 Feb. 1967; *Chem. Abstr.*, **70**, 116727s (1969).

333. Vogel, R. C., and H. Poddal, *J. Amer. Chem. Soc.* **72**, 1420 (1950).

334. Volmar, Y., J. P. Ebel, and F. Bassili, *Bull. Soc. Chim. France*, **1953**, 1085.

335. Wade, H. E., and D. M. Morgan, *Biochem. J.*, **60**, 264 (1955).

336. Wagner, E. F., *Seifen-Oele-Fette-Wachse*, **94**, 443 (1968); *Chem. Abstr.*, **69**, 73743d (1968).

337. Warschauer, F., *Z. Anorg. Chem.*, **36**, 137 (1903).

338. Watanabe, M., T. Takahara, and T. Yamada, *Chuba Kogyo Daigaku Kiyo*, **1967** (3), 160; *Chem. Abstr.*, **70**, 16752 (1969).

339. Watters, J. I., S. Kalliney, and R. Machen, *J. Inorg. Nucl. Chem.*, **31**, 3823 (1969).

340. Watters, J. I., S. Kalliney, and R. Machen, *J. Inorg. Nucl. Chem.*, **31**, 3829 (1969).

341. Watters, J. I., E. D. Loughran, and S. M. Lambert, *J. Amer. Chem. Soc.*, **78**, 4855 (1956).

342. Watters, J. I., and R. Machen, *J. Inorg. Nucl. Chem*, **30**, 2163 (1968).

343. Watters, J. I., and S. Matsumato, *Inorg. Chem.* **5**, 361 (1966).

344. Weldes, H. H., and R. N. Horikawo, *Soap. Chem. Spec.*, **43**, 51 (1967); *Chem. Abstr.*, **67**, 83237q (1967).

345. Wells, C. F., and M. A. Salam, *J. Chem. Soc. (A)*, **1968**, 308.

346. Westman, A. E. R., and A. E. Scott, *Nature*, **168**, 740 (1951).

347. Westman, A. E. R., A. E. Scott, and J. T. Pedley, *Chem. Can.*, **4**, 35 (1952).

348. Wiecker, W., *Z. Chem.*, **1**, 19 (1960).

349. Wieker, W., *Z. Electrochem.*, **64**, 1047 (1960).

350. Wieker, W., *Z. Anorg. Allgem. Chem.*, **313**, 309 (1962).

351. Wiesler, A., *Z. Anorg. Allgem. Chem.*, **28**, 177 (1901).

352. Winkler, A., and E. Thilo, *Z. Anorg. Allgem. Chem.*, **298**, 302 (1959).

353. Wurzschmitt, B., and W. Schuhknecht, *Angew. Chem.*, **52**, 711 (1939).

354. Yamabe, T., T. Iida, and N. Takai, *Bull. Chem. Soc. Japan*, **41**, 1959 (1968).

355. Yamada, T., and Y. Ueda, *Nagoya Kogyo Daigaku Gakuho*, **9**, 163 (1957); *Chem. Abstr.*, **52**, 6893c (1958).

356. Yost, D. M., and H. Russell, Jr., *Systematic Inorganic Chemistry*. Prentice-Hall, New York, 1944, pp. 210–224.

357. Zurc, J., Dissertation, University of Zurich, 1949.

358. Liebig, J., *J. Ann. Pharm.*, **26**, 1 (1838), quoted in Ref. 292.

359. Lindenbaum, S., T. U. Peters, and W. Rieman, III, *Anal. Chim. Acta*, **11**, 530 (1954).

360. Iida, T., and T. Yamabe, *J. Chromatog.*, **56**, 373 (1971).

361. Jones, H. W., and C. B. Monk, *J. Chem. Soc.*, **1950**, 3475.

Compilation of Physical Data of Phosphazo Trihalides

M. BERMANN

Department of Chemistry, Vanderbilt University, Nashville, Tennessee

CONTENTS

I. Introduction

A wide variety of compounds which may formally be considered to contain the $-N=PX_3$ group (X = halogen) is known. The chemistry of these phosphazo compounds has thoroughly been reviewed recently (27). Note that in addition to the monomeric structure, $RNPX_3$ (where R is any substituent), in which the phosphorus atom is bonded to four nearest-neighbor atoms, one must also consider polymers such

$$X_3$$
$$P$$

as the well-known dimers, $RN \diagdown^{\diagup} NR$, in which the phosphorus has

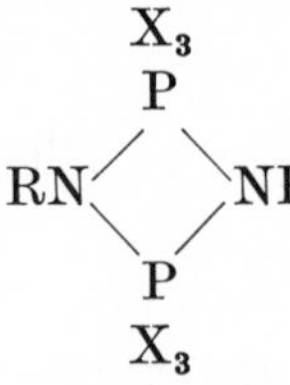

$$P$$
$$X_3$$

five nearest neighbor atoms.   Because of the widespread scatter in the literature of physical data for a given phosphazo trihalide, it was felt desirable to collect these data in the form of this chapter.   Moreover, about 50 new papers on this subject are being published each year, covering the range from purely physical investigations to preparative work, so that abstract searching for physical data for this class of compounds may become rather painful and can easily be incomplete. Some physical data for these phosphazo trihalides have been mentioned in a few reviews (59,81,224).

The compounds have been classified into six main sections in this chapter: Section III deals with ionic compounds containing the —N=PX$_3$ group (for example $[Cl_3P$≡N≡$PCl_3]^+PCl_6^-$); Section IV comprises the N-phosphorylated phosphazo trihalides (e.g., $Cl_2P(O)N$= PCl_3); Section V treats the sulfonylphosphazo trihalides, RSO_2N=PX_3; Section VI are the carbonylphosphazo trichlorides, $RCON$=PCl_3; Section VII deals with aryl- and alkylphosphazo trihalides, $(RNPX_3)_n$ ($n = 1, 2$); and Section VIII reports miscellaneous compounds, belonging in neither of the previous sections (e.g., diazaphosphetidinones $\underline{RN$—CO—$NR}$—PCl_3).

Some of the compounds mentioned above are dimeric (cyclic phosphorus–nitrogen compounds having four atoms per ring) and these are included in the respective sections as if they were just composed of two monomeric units.   Thus, for instance, the four-membered ring

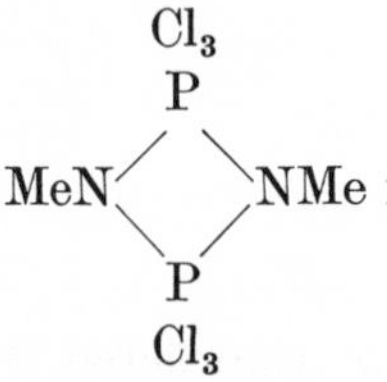

$$Cl_3$$
$$P$$

$MeN \diagdown^{\diagup} NMe$ is reported as $(MeN$=$PCl_3)_2$ and not as a bistrichloro-

$$P$$
$$Cl_3$$

phosphazo derivative.

The subclassification in each section follows the general pattern of showing the less complicated structures first.   As an example, the ionic structures (Section III) are subdivided as (III.A) compounds with symmetrical cation containing 2 phosphorus atoms, (III.B) higher ionic P–N compounds with more than two phosphorus atoms, (III.C) ionic compounds with [P═N═C(R)═N═P] units, and (III.D) unsymmetrical ionic P–N compounds.

Within a given subgroup compounds with one —N=PX$_3$ group appear before those with more than one. The further arrangement is in the order: alkyl-, aryl-, other substituents; and, where appropriate, fluorine comes before chlorine, which comes before bromine.

The first column of the tables making up Sections III through VIII gives the formula. The second column reports all available melting and/or boiling points of these compounds; in general, those melting points for a given substance which differ by not more than $\pm 1°$ are only mentioned once. Most of the melting points are taken from the papers dealing with the original syntheses. When there is appreciable disagreement between the reported melting points, all values are mentioned in the table, since an objective judgment as to which is the correct value cannot properly be made by simply reviewing the literature; moreover, differences (especially in the boiling points) may arise from different syntheses and different methods of measurement by the various authors.

All available ^{31}P NMR chemical shifts (vs. 85% H$_3$PO$_4$, external) are presented in the third column; the solvents used for their determination are not listed and, indeed, they are sometimes not even mentioned in the original work. The fourth column lists NMR coupling constants, densities and refractive indices, as well as references to ^{1}H, ^{11}B, and ^{19}F NMR work, infrared spectra, NQR work, electron diffraction, and crystallographic studies, etc., when reported in the literature.

The individual literature citations are presented alphabetically in the References section, with the literature being thoroughly covered through January 1972.

The abbreviations used below are: Me = methyl, Et = ethyl, Pr = n-propyl, Bu = n-butyl, Am = n-amyl, Ph = phenyl, R = alkyl, Ar = aryl.

II. Some Generalizations

From the data making up Sections III through VIII, some correlations may be tentatively obtained. These mainly involve the change (with respect to the influence of a substituent) of the ^{31}P NMR chemical shift (δ_P) in a series of compounds.

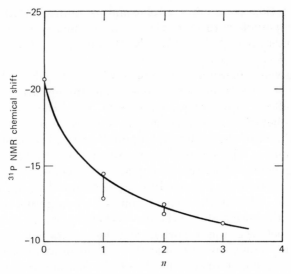

Fig. 1.   Variation of δ_{P_A} of $[Cl_3P_A{=\!=\!=}N(P_BCl_2N)_n{=\!=\!=}P_ACl_3]^+X^-$ with n.

Generally, for the ionic compounds (Sections III.A through III.D) it may be noted that the chemical shifts of the phosphorus atoms in the cation are, as expected, independent of the anion.

The chemical shift, δ_{P_A}, of the trichlorophosphazo terminal groups in the series of compounds $[Cl_3P_A{=\!=\!=}N(P_BCl_2)_n{=\!=\!=}P_ACl_3]^+X^-$ ($n = 0, 1, 2,$ $3 \ldots$) (Sections III.A and III.B) approaches a limiting value of $\delta_{P_A} =$ ca. -11 ppm for higher values of n (Fig. 1); whereas, the chemical shift δ_{P_B} of the bridging —Cl_2P_BN unit remains fairly constant at ca. $+13$ to $+14$ ppm.   Although no structure determination [except for IR studies on the first member ($n = 0$) (6)] on these types of compounds has been carried out, these observations may be sensibly treated on the basis of the usual theoretical interpretation in terms of the "neighbor anisotropy effect."*

The great influence on the chemical shift δ_{P_A} of the substituent R in the series of compounds $[Cl_3P_A{=\!=\!=}N{=\!=\!=}C(R){=\!=\!=}N{=\!=}P_ACl_3]^+X^-$ (Section III.C) should be noted.   Figure 2 shows a plot of δ_{P_A} versus the Hammett σ_p values of the substituents R.   This correlation is made on the assumption that the bond distances, bond angles, etc., are not much affected by changing R; deviations of the points from the smooth curve indicate that this assumption is a simplification.   Plotting δ_{P_A} versus σ_m gives only a shot-gun pattern.

* J. R. Van Wazer and D. Grant, *J. Amer. Chem. Soc.*, **88**, 1450 (1964) and loc. cit.

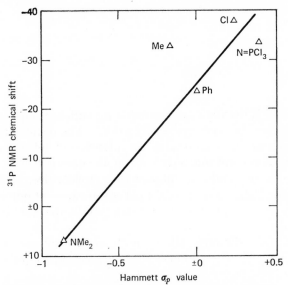

Fig. 2.   Plot of δ_{P_A} versus σ_p of R for compounds of the type

$$[(Cl_3P_A\!=\!\!=\!\!N\!=\!\!=)_2C(R)]^+X^-.$$

σ_p values are taken from G. B. Barlin and D. D. Perrin, *Quart. Revs.*, **20,** 75 (1966), and from Ref. 256.

More conclusions may be drawn from the N-phosphorylated phosphazo trihalides, due to the fact that more data are available. Thus there is relatively little change in the chemical shifts, δ_{P_A}, of the —$N\!=\!PCl_3$ group in the series $F_nCl_{2-n}(X)P_BN\!=\!P_ACl_3$ ($n = 0$, 1, 2) when passing from $X = O$ to the thio analog with $X = S$; the mean value of $\Delta\delta_{P_A}$, which is surely influenced by measurement variations between different authors, ranges from 2 to 3.5 ppm. This means that for identification purposes the measurement of the chemical shift δ_{P_A} is not at all suitable, but the compounds may be distinguished by observation of δ_{P_B}. The small change of δ_{P_A} in one such series indicates moreover that the environment of P_A is not notably affected by the substitution of chlorine by fluorine on P_B. The same close range for δ_{P_A} can be found in the series $F_nCl_{2-n}(S)P_BN\!=\!P_AF_3$. Of course, substitution on P_A has, as expected, a great influence.

Similarly, δ_{P_A} of the —$N\!=\!P_ACl_3$ group in derivatives of the cyclophosphazenes $(PNCl_2)_n$ ($n = 3$, 4) is very much influenced by the other substituent attached to the phosphorus atom P_B of the ring. Introduction of a second —$N\!=\!PCl_3$ group on the same phosphorus atom, P_B, shifts P_A regularly to higher field.

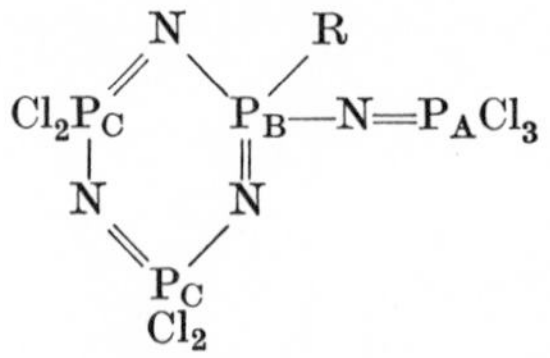

$$\mathrm{Cl_2P_C} \overset{\displaystyle N}{\underset{\displaystyle N}{\diagdown}} \overset{R}{\underset{\displaystyle N}{P_B}} - N = P_A Cl_3$$

$$\underset{\mathrm{Cl_2}}{\overset{\mathrm{P_C}}{}}$$

Not enough ^{31}P NMR data are available for sulfonyl- and carbonyl-phosphazo trihalides to allow systematization. The chemical shifts of these compounds are in accord with the presence of tetra coordinate phosphorus and hence are indicative of the monomeric form.

Monomeric alkyl- and arylphosphazo trichlorides, $RN{=}PCl_3$, exhibit chemical shifts from $\simeq -10$ ppm up to ca. $+50$ ppm, a value which seems to be the upper limit for tetra coordinated phosphorus in such compounds.

In contrast, all dimeric alkyl- and arylphosphazo trihalides $(RN{=}PX_3)_2$ ($X = Cl$, F) exhibit ^{31}P resonances in a quite narrow region, namely between ca. $+75$ to $+81$ ppm.

The existence of the alkyl- or arylphosphazo trihalides, $(RN{=}PCl_3)_n$ ($n = 1, 2$) in either the monomeric form or the dimeric form (diazadiphosphetidine) has often been correlated to the base strength K_B (in aqueous solution) of the parent amine; so that weak bases with $K_B = 10^{-11}$ to 10^{-14} give only monomers, bases with K_B from 10^{-9} to 10^{-11} give either monomeric or dimeric products, and the stronger amines with $K_B > 10^{-9}$, only dimers. This general pattern, however, holds only when no steric effects are encountered (as in n-alkyl amines). For α-isoalkyl amines, the branching forces the structure to be monomeric even though the base strength of the amine would normally lead to dimerization (28,79). Further, β branching may give either monomeric or dimeric species for strong amines, while γ and more distant branching has no effect. This behavior has so far only been verified for phosphazo trihalides; substitution of a chlorine or a fluorine atom with other substituents (e.g., phenyl) does not generally follow this scheme.

Another correlation of physical properties is to be found in a plot of the melting points of the hexachlorodi-n-alkyldiazadiphosphetidines $(RN{=}PCl_3)_2$ ($R = $ Me to n-decyl) with respect to their chain length (79); a regular decrease in the melting point with growing chain length is observed until the n-octyl compound. From there on, the melting point is observed to rise.

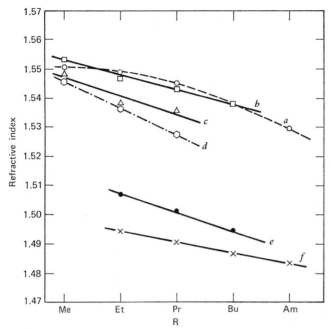

Fig. 3.   Plot of n_D^{20} versus the chain length R in the series $RN{=}PCl_3$.   Curve a: $RCCl_2CCl_2N{=}PCl_3$; curve b: $R(Ph)C(CN)N{=}PCl_3$; curve c: $RCCl(CN)CCl_2N{=}PCl_3$; curve d: $RCCl(COCl)CCl_2N{=}PCl_3$; curve e: $R_2NSO_2N{=}PCl_3$; curve f: $R_2C(CN)CON{=}PCl_3$.

Figure 3 shows a plot of the refractive index, n_D^{20}, versus the chain length, R, for several series of n-alkyl phosphazo trichlorides.   As can be seen, the index of refraction decreases linearly with increasing chain length.*

The same general trend as was found for the refractive indices is observed with the densities, d_{20}, for the same series of compounds, so that, with increasing chain length, R, the density decreases (Fig. 4).†

The only other case where sufficient data are available for correlation, namely $(RN{=}PF_3)_2$ shows, in contrast to the aforementioned results, that the density decreases, the larger the chain length while the refractive index increases.   This difference in behavior might be attributed to the fact that these compounds are dimeric, i.e., composed of a four-membered ring, so that the variation in chain length does not have a decisive influence.

* The obviously erroneous report (244) of n_D^{20} for the compound $BuCCl_2CCl_2N{=}PCl_3$ (1.5003) has been deleted from Figure 3.

† Here again, the reported value (244) of $d_{20} = 1.6121$ for $BuCCl_2CCl_2N{=}PCl_3$ is questionable.

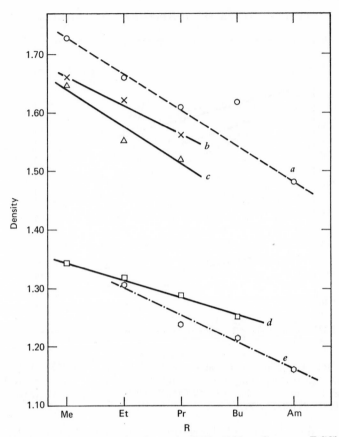

Fig. 4. Plot of d_{20} versus R in the series $RN{=}PCl_3$. Curve a: $RCCl_2CCl_2N{=}$ PCl_3; curve b: $RCCl(COCl)CCl_2N{=}PCl_3$; curve c: $RCCl(CN)CCl_2N{=}PCl_3$; curve d: $R(Ph)C(CN)N{=}PCl_3$; curve e: $R_2C(CN)CON{=}PCl_3$.

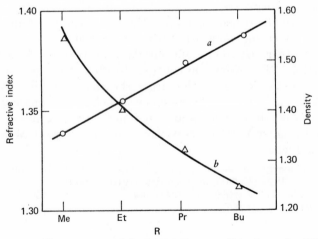

Fig. 5. Plot of n_D^{20} (curve a) and d_{20} (curve b) versus R in the series $(RN{=}PF_3)_2$.

III. Ionic Compounds

A. Ionic Compounds with Symmetrical Cation Containing Two Phosphorus Atoms

TABLE I

Compound	Fp or bp/Torr (°C)	^{31}P NMR (85% H_3PO_4) δ_P (ppm)	Remarks
$[Cl_3P=N=PCl_3]^+Cl^-$	subl. 90–100/0.01 (15)	−21.4 (71)	IR (6); el. cond. (20)
$[Cl_3P=N=PCl_3]^+Cl^-\cdot$ sym $C_2H_2Cl_4$	165 (dec) (15)		loses solvent at 100/0.05 (11)
$[Cl_3P=N=PCl_3]^+AlCl_4^-$	27		IR (169)
	~ 240/0.1 (169,170)		
$[Cl_3P=N=PCl_3]^+BCl_4^-$	174–175 (259a)	−19.8 ± 0.2 (259a)	$\delta_B = \sim -12$ ppm (171)
	215 (33)	−21.8 (171)	
	195 (170a)	−21.6 (170a)	$\delta_B = -7.1$ (33)
		−21.4 (33)	
$[Cl_3P=N=PCl_3]^+FeCl_4^-$	no details (259a)	$P_A = -21.4$	IR (6,38a); el. cond. (14);
$[Cl_3P_A=N=P_ACl_3]^+P_BCl_6^-$	310–315	$P_B = +300.5$ (7,14,69)	NQR (86,97); d-orb. exp. calc. (38b)
$[Cl_3P=N=PCl_3]^+SbCl_6^-$	subl. 150/0.01 (14)	−21.7 (211)	IR (210)
$[Br_3P=N=PBr_3]^+Br^-$	>420 (210)	NMR (94)	IR, debyeogram (94)
$[Br_3P=N=PBr_3]^+Br_3^-$		NMR (94)	IR, debyeogram (94)
$[F_3P=NPF_2=N]^-[AsPh_4]^+$			identified by IR (188)

B. *Higher Ionic Compounds with More Than Two Phosphorus Atoms*

TABLE II

Compound	Fp or bp/Torr (°C)	^{31}P NMR (85% H_3PO_4) δ_P (ppm)	Remarks
$[Cl_3P=NPCl_2=NPCl_3]^+Cl^-$	subl. 170/0.05 (15)		
$[F_3P=(NPF_2)_2=N]^+[AsPh_4]^+$	(188)		
$[F_3P=(NPF_2)_2=N]^-Cs^+$	(61)		
$[Cl_3P_A=NP_BCl_2N=P_ACl_3]^+BCl_4^-$		$P_A = -15.1$ (170a) $P_B = +9.8$	$J_{P_AP_B} = 42.0 \pm 2.4$ Hz (170a)
$[Cl_3P_A=NP_BCl_2=NP_ACl_3]^+[P_CCl_6]^-$	228 (dec) (15)	$P_B = +9.8$ (166) $+14.0 \pm 1$ (10,69) $P_A = -14.4$ (166) -12.5 ± 1 (10,69) $P_C = +305 \pm 5$ (10,69)	$J_{P_AP_B} = 45.3 \pm 1.1$ Hz (69); IR (38a); NQR (97) el. cond. (10)
$[Cl_3P=NPCl_2=NPCl_3]^+SbCl_6^-$	235–250 (169)		IR (169,170)
$[Cl_3P=NPCl_2NPCl_2=NPCl_3]^+AlCl_4^-$	oil (166) 420–422/1.1 (169,170)		
$[Cl_3P_A=NP_BCl_2NP_BCl_2=NP_ACl_3]^+BCl_4^-$	oil (166)	$P_A = -12.5$ (170a) $P_B = +11.5$	$J_{P_AP_B} = 33.0 \pm 1$ Hz; $J_{P_AP_BP_B} = 7.8 \pm 0.3$ Hz (170a)
$[Cl_3P=NPCl_2NPCl_2=NPCl_3]^+M^mCl_{m+1}$	oils (167)		$M^m = $ Nb^V, Mo^V, Ta^V, Pt^{IV}, W^{VI}, Ru^{IV} (167)

Compound	State / m.p.	^{31}P NMR	Notes
$[Cl_3P_A=NP_BCl_2NP_BCl_2=NP_ACl_3]^+[P_CCl_6]^-$	63 (16), 95–96 (166)	$P_B = +13.5$, $P_A = -11.3$ (70), $P_C \sim +300$; $P_B = +12.2$, $P_A = -12.2$ (166), $P_C = +297$	
$[Cl_3P=(NPCl_2)_3N=PCl_3]^+M^mCl_{m+1}$	oils (167)		$M^m = Pt^{IV}, Ru^{IV}, Nb^V, Mo^V, Ta^V, W^{VI}$ (167)
$[Cl_3P_A=NP_BCl_2{-}NP_BCl_2{-}NP_BCl_2=NP_ACl_3]^+[P_CCl_6]^-$	98–100 (166)	$P_B = +13.4$, $P_A = -11.5$ (166), $P_C = +297$	
$[Cl_3P=(NPCl_2)_nN=PCl_3]^+X^-$ ($n = 3, 4$)	oils (186)		X = OPh, OEt, NMe$_2$ (186)
$[Cl_3P_A=(NP_BCl_2)_nN=P_ACl_3]^+ZnCl_3^-$ ($n = 1{-}10$)	solid (170)	$P_A = -11$ (170), $P_B = +15$	
$[Cl_3P(=NPCl_2)_nN=PCl_3]^+Cl^-$ ($n = 10{-}15$)	oils (32)		
$[P_B(N=P_ACl_3)_4]^+Cl^-$	no details (215)	$P_B = +38.5$, $P_A = +3.4$ (215)	$J_{PNP} = 29.9$ Hz (215)
$[P_B(N=P_ACl_3)_4]^+Cl_2I$	no details (215)		
$[P_B(N=P_ACl_3)_4]^+HgI_3$	no details (215)	$P_B = +39.3$, $P_A = +4.1$ (215)	$J_{PNP} = 28.7$ Hz (215)

C. Ionic Compounds with (P=N=C(R)=N=P) Units

TABLE III

Compound	Fp or bp/Torr (°C)	^{31}P NMR (85% H_3PO_4) δ_P (ppm)	Remarks
$[Me_2NPCl_3]^+BCl_4^-$	245 (35)	−61.0 (35)	δ_B = −6.6 (35)
$[Me_2NPCl_3]^+SbCl_6^-$	310 (210)		
$[Cl_3P_A=N-C{=\!=}N-P_ACl_3]^+P_BCl_6^-$ (Cl)	167–168 (13)	P_A = −38.5 P_B = +297.5 (8,13,149)	
$[Cl_3P=N-C{=\!=}N{=\!=}PCl_3]^+SbCl_6^-$ (Me)	179–180 (211)	−32.7 (211)	IR, ^{1}H NMR; J_{HP} = 2.55 Hz (211)
$[Cl_3P=N-C{=\!=}N{=\!=}PCl_3]^+SbCl_6^-$ (Ph)	127–128 (211)	−23.4 (211)	IR (211)
$[Cl_3P=N-C{=\!=}N{=\!=}PCl_3]^+Cl^-$ (CCl$_3$)	no details (211)		
$[Cl_3P=N-C{=\!=}N{=\!=}PCl_3]^+Cl^-$ (CCl$_3$)	203–205 (211)	+6.3 (211)	IR; ^{1}H NMR; J_{PH} = 2.5 Hz (211) cis-trans-isomerism
$[Ph_2PCl=N-C{=\!=}NPCl_3]^+Cl^-$ (NMe$_2$)	63–67 (212)		
$[Ph_2PCl=N-C{=\!=}NPCl_3]^+SbCl_6^-$ (NMe$_2$)	128–130 (212)		
$\left[\begin{smallmatrix}Cl_3P=N\\ \quad\ \ C=NPCl_3\\ Cl_3P=N\end{smallmatrix}\right]^+ Cl^-$	155–157 (13)	−33.5 (13,148)	2 peaks (?) in ^{31}P NMR (218)
$\left[\begin{smallmatrix}Cl_3P=N\\ \quad\ \ C=NPCl_3\\ Cl_3P=N\end{smallmatrix}\right]^+ SbCl_6^-$	175 (211)	−32.6 (211)	IR (211)

322

D. Unsymmetrical Ionic Compounds

TABLE IV

Compound	Fp or bp/Torr (°C)	^{31}P NMR (85% H_3PO_4) δ_P (ppm)	Remarks
[BuOCCl=NPCl$_3$]$^+$PCl$_6^-$	105–108 (141a)		
[i-BuOCCl=NPCl$_3$]$^+$PCl$_6^-$	oil (141a)		
[PhOCCl=NPCl$_3$]$^+$PCl$_6^-$	114–116 (239)		
[p-ClC$_6$H$_4$OCCl=NPCl$_3$]$^+$PCl$_6^-$	140–142 (239)		
[o-BrC$_6$H$_4$OCCl=NPCl$_3$]$^+$PCl$_6^-$	109–112 (239)		
[p-BrC$_6$H$_4$OCCl=NPCl$_3$]$^+$PCl$_6^-$	133–135 (239)		
[o-MeC$_6$H$_4$OCCl=NPCl$_3$]$^+$PCl$_6^-$	102–104 (239)		
[m-MeC$_6$H$_4$OCCl=NPCl$_3$]$^+$PCl$_6^-$	138–139 (239)		
[p-MeC$_6$H$_4$OCCl=NPCl$_3$]$^+$PCl$_6^-$	113–116 (239)		
[3,5-Me$_2$C$_6$H$_3$OCCl=NPCl$_3$]$^+$PCl$_6^-$	185–187 (239)		
[1-C$_{10}$H$_7$OCCl=NPCl$_3$]$^+$PCl$_6^-$	126–128 (239)		
[Me$_2$NCCl=NPCl$_3$]$^+$PCl$_6^-$	193–194 (135a)		
[Et$_2$NCCl=NPCl$_3$]$^+$PCl$_6^-$	181–183 (135a)		
[Bu$_2$NCCl=NPCl$_3$]$^+$PCl$_6^-$	79–80 (135a)		
[MeSCCl=NPCl$_3$]$^+$PCl$_6^-$	100–104 (237)		
[EtSCCl=NPCl$_3$]$^+$PCl$_6^-$	94–96 (237)		
[PrSCCl=NPCl$_3$]$^+$PCl$_6^-$	81–84 (237)		
[i-PrSCCl=NPCl$_3$]$^+$PCl$_6^-$	102–104 (237)		
[BuSCCl=NPCl$_3$]$^+$PCl$_6^-$	84–88 (237)		
[i-BuSCCl=NPCl$_3$]$^+$PCl$_6^-$	88–90 (237)		
[AmSCCl=NPCl$_3$]$^+$PCl$_6^-$	77–79 (237)		
[i-AmSCCl=NPCl$_3$]$^+$PCl$_6^-$	92–95 (237)		
[PhSCCl=NPCl$_3$]$^+$PCl$_6^-$	84–88 (238)		
[p-MeC$_6$H$_4$SCCl=NPCl$_3$]$^+$PCl$_6^-$	69–74 (238)		
[p-O$_2$NC$_6$H$_4$SCCl=NPCl$_3$]$^+$PCl$_6^-$	122–124 (238)		
[Ph$_2$P$_B$Cl=NP$_A$Cl$_3$]$^+$P$_C$Cl$_6^-$	163–166 (12)	$P_B = -42.3$ $P_A = -14.3$ (12,148) $P_C = +305$	
[Ph$_3$P=N=PCl$_3$]$^+$Br$^-$	no details (4)		
[Ph$_3$P=N=PBr$_3$]$^+$Br$^-$	no details (4)		

323

(continued overleaf)

TABLE IV (continued)

Compound	Fp or bp/Torr (°C)	³¹P NMR (85% H₃PO₄) δ_P (ppm)	Remarks
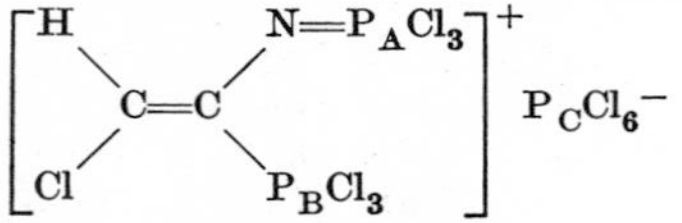 $\left[\begin{matrix} H \\ Cl \end{matrix} C{=}C \begin{matrix} P_BCl_3 \\ N{=}P_ACl_3 \end{matrix}\right]^+ P_CCl_6^-$		$P_A = -15.9$ $P_B = -83.1;$ $\quad -80.7\ (8,149)$ $P_C = +297$	cis-trans-isomerism
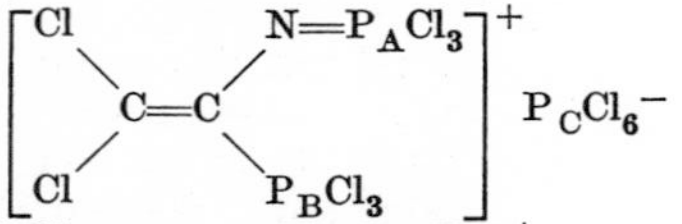 $\left[\begin{matrix} Cl \\ Cl \end{matrix} C{=}C \begin{matrix} P_BCl_3 \\ N{=}P_ACl_3 \end{matrix}\right]^+ P_CCl_6^-$		$P_A = -14.2$ $P_B = -85.0\ (8,149)$ $P_C = +296$	
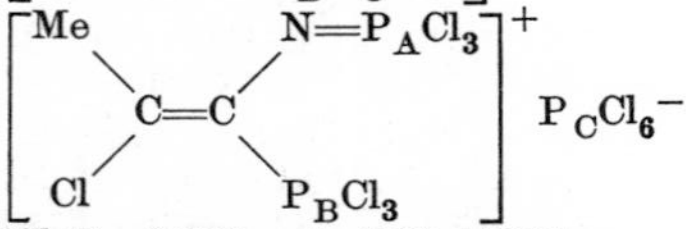 $\left[\begin{matrix} Me \\ Cl \end{matrix} C{=}C \begin{matrix} P_BCl_3 \\ N{=}P_ACl_3 \end{matrix}\right]^+ P_CCl_6^-$		$P_A = -7.8$ $P_B = -84.0;$ $\quad -86.0\ (149)$ $P_C = +292$	cis-trans-isomerism
$[Cl_3P{=}NCCl{=}C{-}PCl_3]^+PCl_6^-$ $\qquad\qquad\qquad\ \ \overset{\displaystyle R}{\mid}$	no details (244)		R = Me, ClCH₂, Et, Pr, Bu, i-Bu, Am (244)
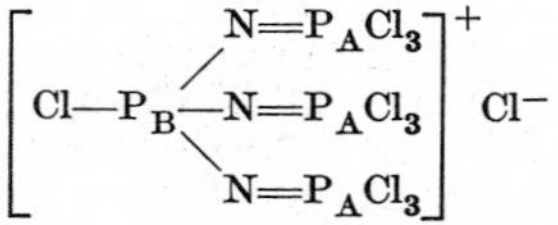 $\left[Cl{-}P_B \begin{matrix} N{=}P_ACl_3 \\ N{=}P_ACl_3 \end{matrix}\right]^+ Cl^-$	200–202 (19)	$P_B = +26.8$ $P_A = -6.5\ (146)$	$J_{P_A P_B} = 29 \pm 1$ Hz; el. cond. (146)
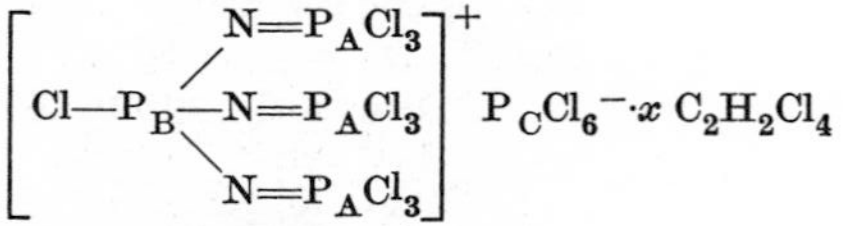 $\left[Cl{-}P_B \begin{matrix} N{=}P_ACl_3 \\ N{=}P_ACl_3 \end{matrix}\right]^+ P_CCl_6^-{\cdot}x\ C_2H_2Cl_4$	no details (146)	$P_B = +26.8$ $P_A = -6.5\ (146)$ $P_C = +300$	el. cond. (146)
$[PhC({=}NPCl_3)OEt]^+Cl^-$	90–92 (57)		el. cond. (57)
$[p\text{-}BrC_6H_4C({=}NPCl_3)OEt]^+Cl^-$	131–132 (57)		el. cond. (57)
$[(MeO)_2C{=}NPCl_3]^+Cl^-$	119–120 (57)		
$[(EtO)_2C{=}NPCl_3]^+Cl^-$	46–48 (57)	-20.4 (249)	

324

IV. N-Phosphorylated Phosphazo Trihalides

TABLE V

Compound	Fp or bp/Torr (°C)	^{31}P NMR (85% H$_3$PO$_4$) δ_P (ppm)	Remarks
$F_Y{}^2P_BN=P_AF_X{}^3$	-87.2 ± 0.7; 12.5/760 (extr.) (77a)	$P_A = +43.6$; $P_B = -129$ (77a)	IR, ^{19}F NMR, $J_{P_AF_X} = 1031$ Hz, $J_{P_BF_Y} = 1279$ Hz, $J_{P_AF_Y} = 24.1$ Hz; mass sp., vapor pressure (77a)
$(C_3F_7)_2PN=PCl_3$ $F_2(O)PN=PF_3$	88–90/0.02 (182); 5.2 ± 0.9 (158); 30/17 (199); 26/~39 (extr.) (158)		mass sp. (158,199); IR (158)
$F_X{}^2(O)P_BN=P_ACl_3$	12–14 (131); 17 ± 1 (75); 38/0.5 (204); 44/1.2 (131); 50–51/1.3 (75)	$P_A =$ -6.4 (204); -8 (77); -6.5 (75); $P_B =$ $+25.1$ (204); $+26$ (77); $+25$ (75)	^{19}F NMR (75,77,204); $J_{P_BF_X} = \pm 974$ Hz (204), ± 973.5 Hz (75); $J_{P_AF_X} = \pm 21.5$ Hz (75,204); $J_{P_AP_B} = 70$ Hz (75,204); $d_{27} = 1.7873$ (131); IR (75,131,204)
$F_XCl(O)P_BN=P_ACl_3$	6 ± 1 (75); 47–48/0.03 (75); 65/0.5 (204)	$P_A = -3.9$; $P_B = +14.5$ (75)	^{19}F NMR (75,204); $J_{P_AF_X} = 22.5$ Hz (204), 23 Hz (75); $J_{P_BF_X} = 1041$ Hz (75); $J_{P_AP_B} = 46.5$ Hz (75)

(continued overleaf)

TABLE V (continued)

Compound	Fp or bp/Torr (°C)	³¹P NMR (85% H₃PO₄) δ_P (ppm)	Remarks
$Cl_2(O)P_B N=P_A Cl_3$	32 (64b) 35.5 (9,19) 55–58/0.005 (19) 55/0.1, 92/1, 138/10, 198/100, 270–280/760 (96) 85/0.2 (217) 90/0.3 (9) 94/0.5 (74) 105/1 (9) 110–115/0.1 (?) (64b)	$P_A = +0.1 \pm 0.5$ $P_B = +14.2 \pm 0.5$ } (68) $P_A = +1.1$ $P_B = +12.7$ } (1)	IR (9,21,38a,75); Raman (129); NQR (97); $J_{P_A P_B} = 15.4 \pm 0.3$ Hz (68), 19.5 Hz (1); $n_D^{25} = 1.5305$ (9); $n_D^{38} = 1.5260$; $d_{38} = 1.796$ (96); $M_D^{38} = 46.02$ (96)
$Cl_2(O)PN=PCl_3 \cdot BF_3$			³¹P, ¹⁹F, ¹¹B NMR (33)
$Cl_2(O)PN=PCl_3 \cdot BCl_3$			³¹P, ¹⁹F, ¹¹B NMR (33)
$Cl_2(O)PN=PCl_3 \cdot PF_5$			³¹P, ¹⁹F, ¹¹B NMR (33)
$2Cl_2(O)P_B N=P_A Cl_3 \cdot TiCl_4$	123–124 (21)	$P_A = -4$ (21) $P_B = +4$	IR (21)
$(PhO)_2 P_B(O)N=P_A Cl_3$	oil (127)	$P_A = P_B = +11.2$ (173)	³¹P NMR: 1 peak, half bandwidth: 3 Hz (172) compound is rather the isomer $MePCl_2=NPOCl_2$ (248)
$Me(O)PClN=PCl_3$ (?)	71–74 128–130/0.03 (247)		
$Me(O)P(OPh)N=PCl_3$ (?)	160–162/0.02 (247)		$d_{20} = 1.4811$, $n_D^{20} = 1.5532$ (247); see (248)
$Me(p\text{-}ClC_6H_4O)P(O)N=PCl_3$ (?)	190–191/0.45 (247)		$d_{20} = 1.5460$, $n_D^{20} = 1.5612$ (247), see (248)
$Me(p\text{-}MeC_6H_4O)P(O)N=PCl_3$ (?)	163–165/0.05 (247)		$d_{20} = 1.4283$, $n_D^{20} = 1.5464$ (247), see (248)
$CH_2ClP(O)ClN=PCl_3$ (?)	119–120/0.06 (247)		$d_{20} = 1.7593$, $n_D^{20} = 1.5449$ (247), see (248)

Compound	b.p./mm (m.p.) (ref.)	³¹P shifts	Other data
CH₂Cl(PhO)P(O)N=PCl₃ (?)	166–167/0.02 (247)		$d_{20} = 1.5554$ $n_D^{20} = 1.5639$ (247), see (248)
CH₂Cl(p-MeC₆H₄O)P(O)N=PCl₃ (?)	181–183/0.1 (247)		$d_{20} = 1.4818$, $n_D^{20} = 1.5463$ (247), see (248)
CH₂Cl(p-O₂NC₆H₄O)P(O)N=PCl₃ (?)	173–174/0.5 (247)		$d_{20} = 1.5866$, $n_D^{20} = 1.5589$ (247), see (248)
Et₂P(O)N=PCl₃			is rather Et₂PCl=NPOCl₂ (266) ¹⁹F NMR, (158,194), see also (73)
$F_X^2(S)P_BN=P_AF_Y^3$	−49.5 ± 1.5 (158) 17/94.2 (extr.) (158) 31/242 (196) 32/214 (?) (194)	$P_A = +36.4$ $P_B = -42.0$ (73)	$J_{P_AP_B} = 137$ Hz; $J_{P_AF_Y} = 1020$ Hz; $J_{P_BF_X} = 1080$ Hz; $J_{P_AF_X} = 30$ Hz; $J_{P_BF_Y} = 3$ Hz; (73) mass sp. (158,194); IR(158,194,196); vapor pressure (158)
$F_XCl(S)P_BN=P_AF_Y^3$	31/58 (194) 34/58 (196)	$P_A = +39.0$ $P_B = -44.3$ (73)	IR (194,196); ¹⁹F NMR (194,196), see also (73); mass sp. (194)
$Cl_2(S)P_BN=P_AF_3$	41/25 (194)	$P_A = +40.8$ $P_B = -29.7$ (73)	IR, mass sp. (194); ¹⁹F NMR (194), see also (73);
$F_X^2(S)PN=PF_Y^2Cl$	31/37 (194)		$J_{P_AP_B} = 75$ Hz; $J_{P_AF} = 1040$ Hz; $J_{P_BF} = 2$ Hz (73) IR, ¹⁹F NMR, mass sp. (194)
$F_XCl(S)PN=PF_Y^2Cl$	35/14 (194)		IR, ¹⁹F NMR, mass sp. (194)

327

(continued overleaf)

TABLE V (continued)

Compound	Fp or bp/Torr (°C)	^{31}P NMR (85% H_3PO_4) δ_P (ppm)	Remarks
$Cl_2(S)P_BN=P_AF_2Cl$	40/0.05–0.01 (194)	$P_A = +16.7$ (73) $P_B = -29.4$	IR, mass sp. (194); ^{19}F NMR (194), see also (73); $J_{P_AP_B} = 60$ Hz; $J_{P_AF} = 1091$ Hz; $J_{P_BF} = 5$ Hz (73)
$F_2(S)PN=PF_2Br$	30/15 (195)		IR, ^{19}F NMR, mass sp. (195)
$FCl(S)PN=PF_2Br$	25/0.01 (195)		IR, ^{19}F NMR, mass sp. (195)
$Cl_2(S)PN=PFCl_2$	60/0.01 (195)		IR, ^{19}F NMR, mass sp. (195)
$F_2(S)P_BN=P_ACl_3$	-29—28 (76) 43/0.15 (76) 35–36/0.05 (189)	$P_A = -4.4$ $P_B = -40.7$ (189)	IR, mass sp., ^{19}F NMR (189) $d_{20} = 1.7332$; $n_D^{20} = 1.49977$; $J_{P_AP_B} = 70$ Hz; $J_{P_AF} = \pm 22$ Hz; $J_{P_BF} = \pm 1085$ Hz (189)
$FCl(S)P_BN=P_ACl_3$	50–51/0.05 (76,189)	$P_A = -0.8$ $P_B = -42.3$ (189)	IR, mass sp, ^{19}F NMR (189); $d_{20} = 1.7506$; $n_D^{20} = 1.5433$ (189); $J_{P_AP_B} = 40$ Hz; $J_{P_AF} = \pm 21.5$ Hz; $J_{P_BF} = \pm 1115$ Hz (189)

328

$Cl_2(S)P_BN{=}P_ACl_3$	$\left.\begin{array}{l}35 \\ 68/0.01 \\ 103{-}105/0.5\end{array}\right\}$ (15) $93{-}95/0.5$ (33)	$\begin{array}{l}P_A = +3.4 \pm 1 \text{ (70)} \\ +2.9 \text{ (1)} \\ P_B = -28.4 \pm 1 \text{ (1,70)} \\ P_A = +3.4 \\ P_B = -28.4 \text{ (33)}\end{array}$	NQR (97); IR (38a) $J_{P_AP_B} = 4$ Hz (1)
$Et(S)FP_BN{=}P_ACl_3$	oil (199)		IR, ^1H, ^{19}F NMR (199); $\lvert J_{P_AF} + J_{P_BF}\rvert$ $= 1052$ Hz (199)
$ClSO_2NP_BCl_2N{=}P_ACl_3$		$\begin{array}{l}P_B = +7.0 \\ P_A = -9.3 \text{ (148)}\end{array}$	IR (38a)
$(PhO)_2P(S)N{=}PCl_3$ $Cl_2(O)P_CN{=}P_BCl_2N{=}P_ACl_3$	oil (127) 34 (15) 118–124/0.05 (15) oil (260)	$\left.\begin{array}{l}P_B = +20.0 \pm 0.5 \\ P_C = +13.4 \pm 0.5 \\ P_A = -7.1 \pm 0.5\end{array}\right\}$ (69)	$J_{P_AP_B} = 29.5 \pm 1$ Hz; $J_{P_BP_C} = 26.7 \pm 1$ Hz (69)
$Cl_2(S)P_CN{=}P_BF_2N{=}P_AF_3$	30/0.01 (196a)	$\left.\begin{array}{l}P_A = +40.3 \\ P_B = +29.0 \\ P_C = -37.2\end{array}\right\}$ (196a)	$J_{P_AP_B} = 175$ Hz; $J_{P_BP_C} = 74.6$ Hz; $J_{P_AF} = 1030$ Hz; $J_{P_BF} = 925$ Hz; IR, ^{19}F NMR, mass sp. (196a)
$F_2(S)P_CN{=}P_BF_2N{=}P_AF_2Cl$	30/0.05 (196a)	$\left.\begin{array}{l}P_A = +12.3 \\ P_B = +23.2 \\ P_C = -46.8\end{array}\right\}$ (196a)	$J_{P_AP_B} = 148$ Hz; $J_{P_BP_C} = 122$ Hz; $J_{P_AF} = 1095$ Hz; $J_{P_BF} = 924$ Hz; $J_{P_CF} = 1062$ Hz; IR, ^{19}F NMR (196a)
$F_2(S)P_CN{=}P_BF_2N{=}P_AFCl_2$	31/0.01 (196a)	$\left.\begin{array}{l}P_A = -4.3 \\ P_B = +24.3 \\ P_C = -46.2\end{array}\right\}$ (196a)	$J_{P_AP_B} = 116$ Hz; $J_{P_BP_C} = 118$ Hz; $J_{P_AF} = 1130$ Hz; $J_{P_BF} = 936$ Hz; $J_{P_CF} = 1065$ Hz; IR, ^{19}F NMR (196a)
$F_2(S)PN{=}PFClN{=}PFCl_2$	no details (196a)		

329

(continued overleaf)

TABLE V (*continued*)

Compound	Fp or bp/Torr (°C)	^{31}P NMR (85% H_3PO_4) δ_P (ppm)	Remarks
$F_2(S)P_CN=P_BF_2N=P_ACl_3$	65/0.01 (193)	$P_A = -6.5$; $P_B = +25.1$; $P_C = -45.8$ (196a)	IR, mass sp. (193); ^{19}F NMR (193,196a)
$FCl(S)P_CN=P_BF_2N=P_ACl_3$	96/0.01 (193)	$P_A = -7.7$; $P_B = +27.6$; $P_C = -47.8$ (196a)	IR, mass sp. (193); ^{19}F NMR, (193,196a)
$Cl_2(S)P_CN=P_BF_2N=P_ACl_3$	130/0.01 (196a)	$P_A = -7.6$; $P_B = +30.9$; $P_C = -34.9$ (196a)	$J_{P_AP_B} = 78.8$ Hz; $J_{P_BP_C} = 55.8$ Hz; $J_{P_BF} = 958.5$ Hz; IR, ^{19}F NMR (196a)
$Cl_2(S)P_CN=P_BFClN=P_ACl_3$	145/0.01 (196a)	$P_A = -4.2$; $P_B = +20.5$; $P_C = -30.9$ (196a)	$J_{P_AP_B} = 54.6$ Hz; $J_{P_BP_C} = 38.0$ Hz; $J_{P_BF} = 1006$ Hz; IR, ^{19}F NMR (196a)
$F_2(S)P_CN=P_BCl_2N=P_ACl_3$	125/0.01 (196a)	$P_A = -3.3$; $P_B = +15.1$; $P_C = -44.6$ (196a)	$J_{P_AP_B} = 32.4$ Hz; $J_{P_BP_C} = 74.0$ Hz; $J_{P_AP_C} = 3.3$ Hz; $J_{P_CF} = 1070.5$ Hz; IR, ^{19}F NMR, mass sp. (196a)
$FCl(S)P_CN=P_BCl_2N=P_ACl_3$	134/0.01 (196a)	$P_A = -3.3$; $P_B = +15.6$; $P_C = -44.7$ (196a)	$J_{P_AP_B} = 29.3$ Hz; $J_{P_BP_C} = 47.6$ Hz; $J_{P_AP_C} = 3.2$ Hz; $J_{P_CF} = 1100.5$ Hz; IR, ^{19}F NMR (196a)

$Cl_2(S)P_C N=P_B Cl_2 N=P_A Cl_3$	23 (15) 150/0.01 (196a)	$P_A = 0.0 \pm 1$ (70), -6.0 (1), -4.3 (196a)	$J_{P_A P_B} = 29.2$ Hz (70), 26.6 Hz (196a), 27.5 Hz (1);
		$P_B = +26.5 \pm 1$ (70), $+20.6$ (1), $+25.9$ (196a)	$J_{P_B P_C} = 12.2$ Hz (70), 12.1 Hz (196a), 10 Hz (1);
		$P_C = -24.0 \pm 1$ (70), -30.2 (1), -25.0 (196a)	$J_{P_A P_C} = 3.5$ Hz (1), 3.3 Hz (196a)
$ClSO_2 NP_B Cl_2 NP_B Cl_2 N=P_A Cl_3$		$P_B = +11.0$ (148) $P_A = -9.3$	
$Cl_2(O)P_D N=P_C Cl_2 N=P_B Cl_2 N=P_A Cl_3$		$\left.\begin{array}{l} P_A = -11 \pm 1 \\ P_B = +14 \pm 1 \\ P_C = +19 \pm 1 \\ P_D = +10.5 \pm 1 \end{array}\right\}$ (64c)	
$Et_2C(Br)CONHP(O)(N=PCl_3)_2$	47 (69)		only the hydrolysis product was isolated (74)
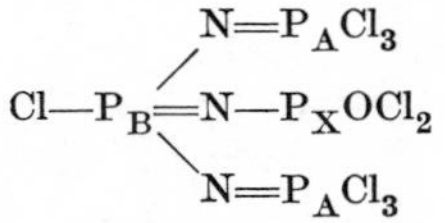 $\begin{array}{c} N=P_A Cl_3 \\ \diagup \\ Cl—P_B=N—P_X OCl_2 \\ \diagdown \\ N=P_A Cl_3 \end{array}$		$\left.\begin{array}{l} P_B = +29.6 \pm 0.5 \\ P_A = +1.9 \pm 0.5 \\ P_X = +13.6 \pm 0.5 \end{array}\right\}$ (69)	$J_{P_A P_B} = 28.7$ Hz, $J_{P_B P_X} = 31.0$ Hz (69)
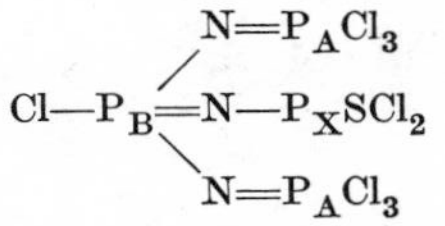 $\begin{array}{c} N=P_A Cl_3 \\ \diagup \\ Cl—P_B=N—P_X SCl_2 \\ \diagdown \\ N=P_A Cl_3 \end{array}$		$\left.\begin{array}{l} P_B = +28.9 \pm 1 \\ P_A = +2.5 \pm 1 \\ P_X = -28.5 \pm 1 \end{array}\right\}$ (70)	$J_{P_A P_B} = 23.8$ Hz (70)
$Cl_2(O)P_B N=P(N=P_A Cl_3)_3$	81–84 (214)	$P_B = +39.1$ (214)	

(continued overleaf)

TABLE V (continued)

Compound	Fp or bp/Torr (°C)	^{31}P NMR (85% H$_3$PO$_4$) δ_P (ppm)	Remarks
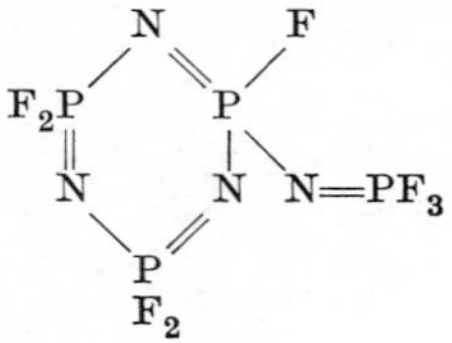 (structure: ring with F, P, N=PF$_3$, F$_2$P, F$_2$)	37/28 (199)		IR. ^{19}F NMR, mass sp. (199)
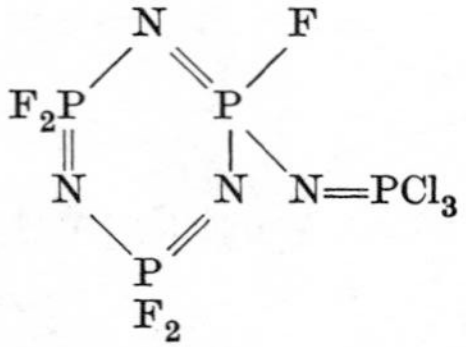 (structure: ring with F, P, N=PCl$_3$, F$_2$P, F$_2$)	63/1 (203)		IR, mass sp. (203); $d_{20} = 1.847$; $n_D^{20} = 1.442$ (203)
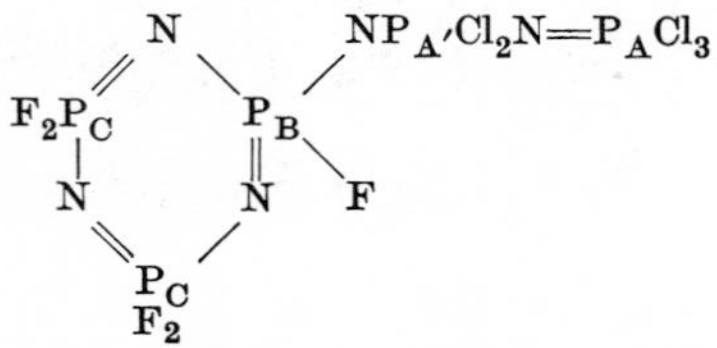 NP$_{A'}$Cl$_2$N=P$_A$Cl$_3$ (ring with P$_B$, F, P$_C$F$_2$, N, F$_2$P$_C$)	130/0.01 (200)	$P_A = -4.5$ $P_{A'} = +18$ $P_B = -3$ (200) $P_C = -9$	IR, ^{19}F NMR, $J_{P_A P_{A'}} = 33$ Hz (200)
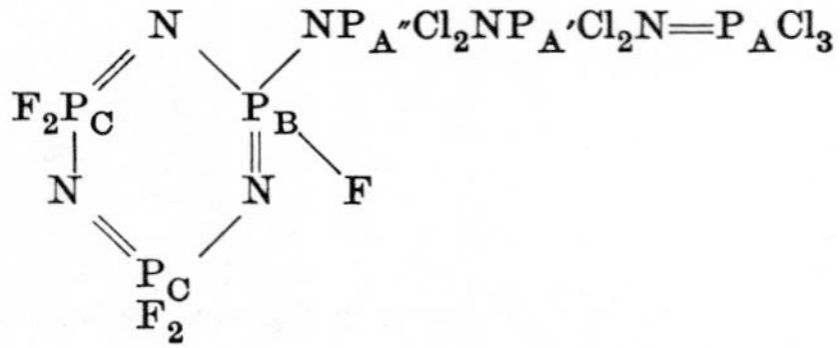 NP$_{A''}$Cl$_2$NP$_{A'}$Cl$_2$N=P$_A$Cl$_3$ (ring with P$_B$, F, P$_C$F$_2$, N, F$_2$P$_C$)	170/0.01 (200)	$P_A = -7.5$ $P_{A'}, P_{A''} = +16-+18$ (200) $P_B = -3$ (200) $P_C = -9$	IR, ^{19}F NMR; $J_{P_A P_{A'}} = 31$ Hz (200)

332

Structure	b.p./m.p.	^{31}P shifts	Coupling / IR
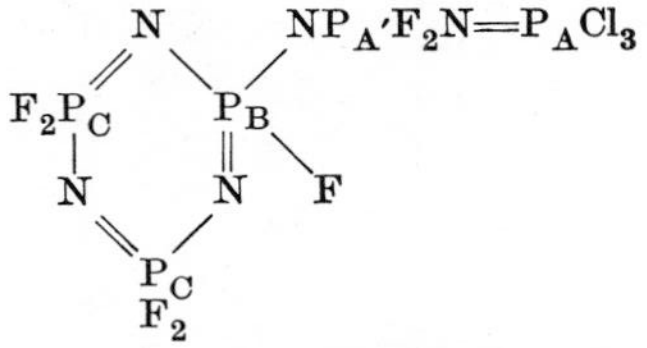 $NP_{A'}F_2N{=}P_ACl_3$ (ring with $P_B{-}F$, F_2P_C)	62/0.01 (200)	$P_A = -8.5$ $P_{A'} = +28$ (200) $P_B = -4$ $P_C = -8.5$	IR, ^{19}F NMR; $J_{P_AP_{A'}} = 78$ Hz (200)
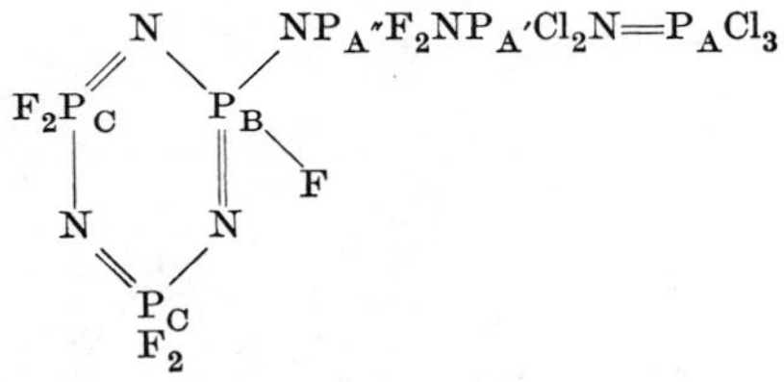 $NP_{A'}F_2NP_{A'}Cl_2N{=}P_ACl_3$ (ring with $P_B{-}F$, F_2P_C)	124/0.01 (200)	$P_A = -6.5$ $P_{A'} = +13$ $P_{A''} = +25$ (200) $P_B = -3$ $P_C = -9$	IR, ^{19}F NMR; $J_{P_AP_{A'}} = 38$ Hz; $J_{P_AP_{A''}} = 78$ Hz (200)
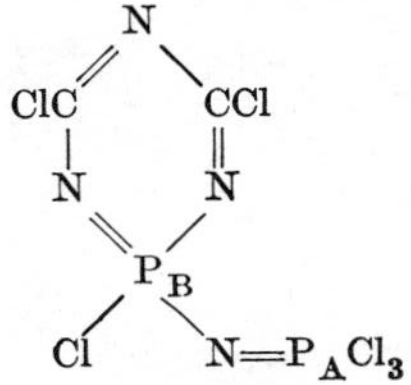 ClC···$N{=}P_ACl_3$ ring (with P_B, Cl, CCl)	108-111 (13) 125-128 (48)	$P_B = -57.0$ (13,148) $P_A = -22.7$	$J_{P_AP_B} \simeq 40$ Hz (13); 40 ± 5 Hz (148)
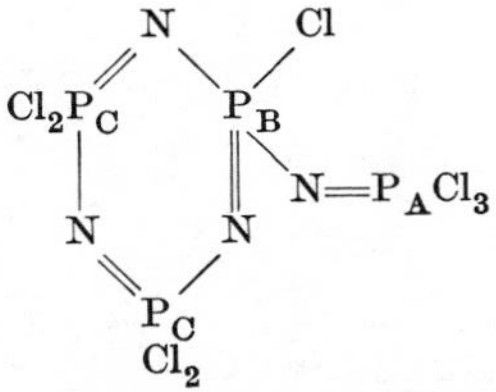 $N{=}P_ACl_3$ ring (with P_B, Cl, Cl_2P_C, Cl_2)	39.5-41 (66)	$P_B = +2.2$ $P_A = +3.3$ (66) $P_C = -20.5$	IR (66); $J_{P_AP_B} = 40.0$ Hz; $J_{P_BP_C} = 58.4$ Hz; $J_{P_AP_C} = 9.5$ Hz (66)

(continued overleaf)

TABLE V (continued)

Compound	Fp or bp/Torr (°C)	^{31}P NMR (85% H$_3$PO$_4$) δ_P (ppm)	Remarks
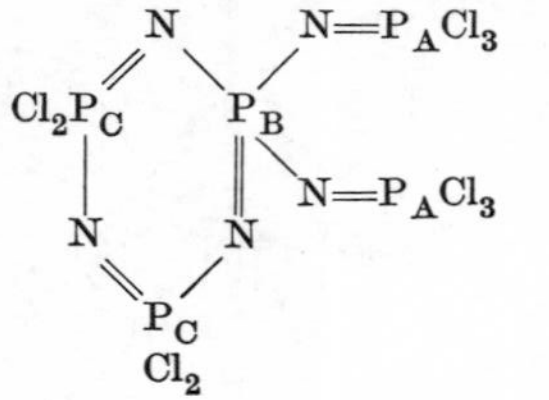 N=P$_A$Cl$_3$ ... P$_B$ ... Cl$_2$P$_C$ ... P$_C^{C}$Cl$_2$ (ring structure)	57.5–59 (66) 57–58 (150) 85/0.01 (150)	P$_B$ = +20.4 P$_A$ = +13.5 ⎫ (66) P$_C$ = −17.5 ⎭ P$_B$ = +20.6 P$_A$ = +12.7 ⎫ (60,150) P$_C$ = −17.7 ⎭	IR (66) $J_{P_AP_B}$ = 34.5 Hz (60,150), 35.5 Hz (66); $J_{P_BP_C}$ = 61.0 Hz (66), 59.0 Hz (60,150); $J_{P_AP_C}$ = 4.2 Hz (66)
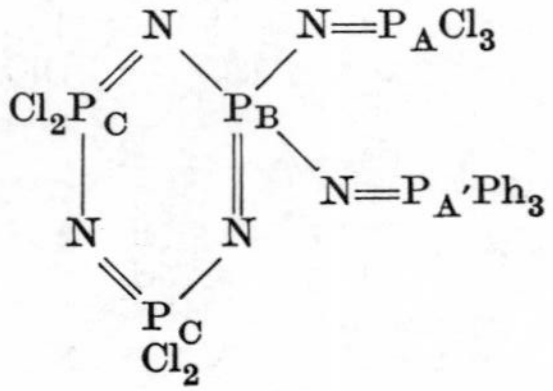 N=P$_A$Cl$_3$ ... P$_B$ ... N=P$_A$·Ph$_3$ (ring structure with Cl$_2$P$_C$, P$_C^{C}$Cl$_2$)	no details (150)	P$_B$ = +20.3 P$_A$ = ∼−19.0 (150) P$_A$' = +11.8 P$_C$ = −18.5	$J_{P_BP_C}$ = 27 ± 2 Hz (60,150)
P$_4$N$_4$F$_7$N=PCl$_3$	42/0.01 (198)		IR (198)
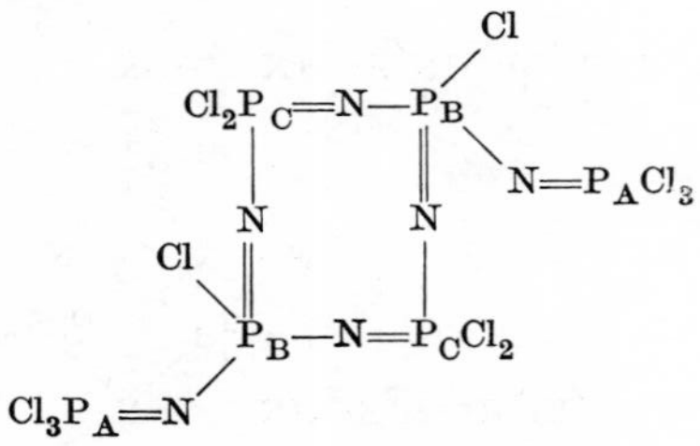 Cl$_2$P$_C$ ... P$_B$—N=P$_A$Cl$_3$ ... P$_C$Cl$_2$ ... P$_B$—N=P$_C$Cl$_2$ ... Cl$_3$P$_A$=N (ring structure)	134 (151)	P$_A$ = +8.0 P$_B$ = +23.5 (151) P$_C$ = +2.1	

334

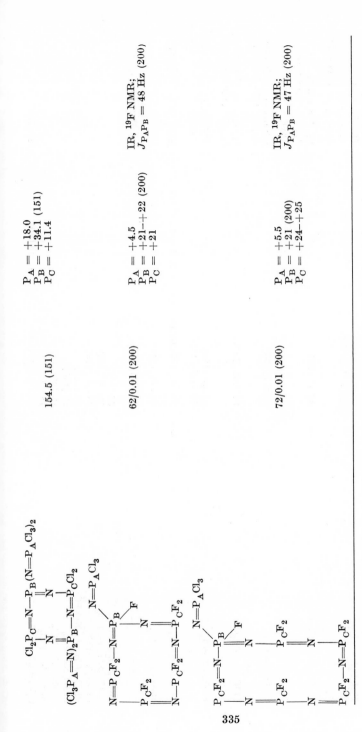

154.5 (151)

$P_A = +18.0$
$P_B = +34.1$ (151)
$P_C = +11.4$

62/0.01 (200)

$P_A = +4.5$
$P_B = +21 - +22$ (200)
$P_C = +21$

IR, ^{19}F NMR;
$J_{P_A P_B} = 48$ Hz (200)

72/0.01 (200)

$P_A = +5.5$
$P_B = +21$ (200)
$P_C = +24 - +25$

IR, ^{19}F NMR;
$J_{P_A P_B} = 47$ Hz (200)

335

V. Sulfonylphosphazo Trihalides

TABLE VI

Compound	Fp or bp/Torr (°C)	^{31}P NMR (85% H$_3$PO$_4$) δ_P (ppm)	Remarks
$F_XSO_2N=PF_Y^3$	-26.0 ± 0.5 24.5/11.5 (158)		IR, ^{19}F NMR; $J_{PNSF} = 16$ Hz; $J_{FPNSF} = 4$ Hz (158)
$F_XSO_2N=PF_Y^2Cl$	46/13 (199)		IR, ^{19}F NMR; $J_{FyP} = 1108$ Hz; $J_{FxP} = 11$ Hz (199)
$ClSO_2N=PF_3$	21/0.1 (197)		IR, mass sp., ^{19}F NMR; $J_{PF} = 1078$ Hz (197)
$ClSO_2N=PF_2Cl$	31/0.01 (197)		IR, mass sp., ^{19}F NMR; $J_{PF} = 1120$ Hz (197)
$FSO_2N=PCl_3$	44-46 (130) 74-75/1 (207)		^{19}F NMR; $J_{PF} = 4$ Hz (207)
$ClSO_2N=PCl_3$	35-36 (100,102,105)	-20.5 ± 0.5 (261)	NQR (87); IR (38a)
$MeSO_2N=PCl_3$	47-50 (113)		
$CF_3SO_2N=PCl_3$	41/0.01 (202)		IR, ^{19}F NMR (202)
$ClCH_2SO_2N=PCl_3$	50 (156)		
$PhCH_2SO_2N=PCl_3$	77-82 (113)		
$EtSO_2N=PCl_3$	18-21 (113)		
$PhCH_2CH_2SO_2N=PCl_3$	57 (156)		
$PrSO_2N=PCl_3$	60 (156)		
$i\text{-}PrSO_2N=PCl_3$	19-22 (113)		
$BuSO_2N=PCl_3$	48-51 (113)		
$C_4F_9SO_2N=PCl_3$	59-60/0.01 (190)	-17.8 (190)	IR (190)

336

Compound	m.p. (°C) (ref.)		Remarks
i-BuSO₂N=PCl₃	-6.5 (156)		
AmSO₂N=PCl₃	35 (156)		
i-AmSO₂N=PCl₃	2 (156)		
C₆H₁₃SO₂N=PCl₃	10 (156)		
cyclo-C₆H₁₁SO₂N=PCl₃	39–41 (113)	-4.0 (88)	
PhSO₂N=PCl₃	53.5 (100,103,106)		
PhSO₂N=PCl₃·DMF	oil (140a)		complex with DMF (140a);
p-FC₆H₄SO₂N=PCl₃	72–73 (264)		
o-ClC₆H₄SO₂N=PCl₃	83–84 (103,155)		
p-ClC₆H₄SO₂N=PCl₃	69–71 (154)		
	71–73 (153)		
	74–75 (175)		
p-BrC₆H₄SO₂N=PCl₃	82–83 (103,155)		
o-MeC₆H₄SO₂N=PCl₃	52.5 (100,103,106)		
p-MeC₆H₄SO₂N=PCl₃	106 (100,103,106)		
	88–90 (254)		
m-CF₃C₆H₄SO₂N=PCl₃	101–103 (263)		IR (263); complex with DMF (140a)
o-O₂NC₆H₄SO₂N=PCl₃	114–115 (175)		
m-O₂NC₆H₄SO₂N=PCl₃	52–54 (264)		
	73–75 (114)		
	72–73 (157)		
p-O₂NC₆H₄SO₂N=PCl₃	82–84 (114)		
p-O₂NC₆H₄SO₂N=PCl₃·DMF	78–80 (154)		
p-MeOC₆H₄SO₂N=PCl₃	118–119 (114)		
	103–104 (140a)		
	72–73 (175)		
p-(PhO)O₂SC₆H₄SO₂N=PCl₃	83–84 (103,155)		
p-Me₂NC₆H₄N=NC₆H₄SO₂N=PCl₃	103–104 (157)		
1-C₁₀H₇SO₂N=PCl₃	no details (115)		
	110–112 (125)		
	117–119 (104)		
2-C₁₀H₇SO₂N=PCl₃	130–132 (125)		
p-ClOCC₆H₄SO₂N=PCl₃	82 (187)		
ClCOC₄H₁₀SO₂N=PCl₃	no details (39)		
Me₂NSO₂N=PCl₃	73–75 (123)		
Et₂NSO₂N=PCl₃	113/0.2 (258)		$n_D^{20} = 1.5072$; $d_{20} = 1.4665$ (123)

337

(continued overleaf)

TABLE VI (continued)

Compound	Fp or bp/Torr (°C)	^{31}P NMR (85% H_3PO_4) δ_P (ppm)	Remarks
$Pr_2NSO_2N{=}PCl_3$	113/0.005 (164,258)		$n_D^{20} = 1.5010$ (164,258)
$Bu_2NSO_2N{=}PCl_3$	140/0.05 (164,258)		$n_D^{20} = 1.4952$ (164,258)
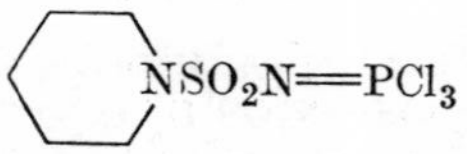 ⟨piperidine⟩$NSO_2N{=}PCl_3$	68–70 (152)		
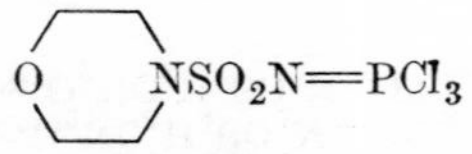 ⟨morpholine⟩$NSO_2N{=}PCl_3$	94 (164,258) 90–91 (152)		
$PhNHSO_2N{=}PCl_3$	61 (261)	-3.5 ± 0.3 (261)	
$p\text{-}ClC_6H_4NHSO_2N{=}PCl_3$	97–98 (261)	-4.2 ± 0.3 (261)	
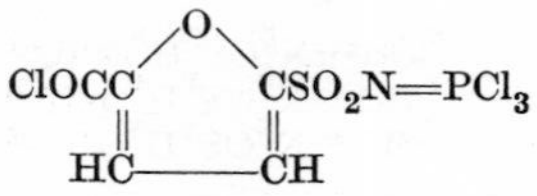 ⟨cyclic ClOCC–CSO₂N=PCl₃ / HC=CH structure⟩	no details (39)		
$FSO_2N{=}PBr_3$	57–58 106–107/0.05 (191)		IR, ^{19}F NMR; $J_{PF} = 2$ Hz (191)
$MeSO_2N{=}PBr_3$	93.5 (205)	$+105.1$ (205)	^{1}H NMR, $J_{PH} = 4$ Hz; IR, mass sp. (205)
$CF_3SO_2N{=}PBr_3$	42 (205)	$+85.1$ (205)	^{1}H, ^{19}F NMR; $J_{PF} = 2.2$ Hz; IR, mass sp. (205)
$PhSO_2N{=}PBr_3$	94–97 (128) 110 (205)	$+101.4$ (205)	^{1}H NMR, IR, mass sp. (205)

338

Compound	m.p. (°C) (ref.)		Spectra
o-MeC₆H₄SO₂N=PBr₃	148–151 (128)		
p-MeC₆H₄SO₂N=PBr₃	138–141 (128)		
p-ClC₆H₄SO₂N=PBr₃	89 (205)	+108.9 (205)	¹H NMR, IR, mass sp. (205)
	118 (205)	+99.5 (205)	¹H NMR, IR, mass sp. (205)
1-C₁₀H₇SO₂N=PBr₃	157–159 (128)		
2-C₁₀H₇SO₂N=PBr₃	150–153 (128)		
ArSO₂N=PBr₃	(160)		
o-Cl₃P=NC₆H₄SO₂N=PCl₃	90–93 (280)		
m-Cl₃P=NC₆H₄SO₂N=PCl₃	149–152 (280)		
p-Cl₃P=NC₆H₄SO₂N=PCl₃	145–146 (175)		
	153–155 (251)		
m-Cl₃P=NCOC₆H₄SO₂N=PCl₃	162–165 (280)		
p-Cl₃P=NCOC₆H₄SO₂N=PCl₃	67–68 (37a)		
SO₂(N=PCl₃)₂	72–75 (183)		
	37.7–41 (107)		
[Cl₃P=N(CH₂)₂SO₂N=PCl₃]₂	40–41 (259)		
m-C₆H₄(SO₂N=PCl₃)₂	178–180 (dec) (278)		
p-C₆H₄(SO₂N=PCl₃)₂	101–104 (116)		
SO₂(MeNSO₂N=PCl₃)₂	153–157 (116)		
MeN(SO₂N=PCl₃)₂	118 (168)		
	103–104 (168)		

Cl₃P=NO₂S — [naphthalene structure with Cl, Cl substituents] — SO₂N=PCl₃

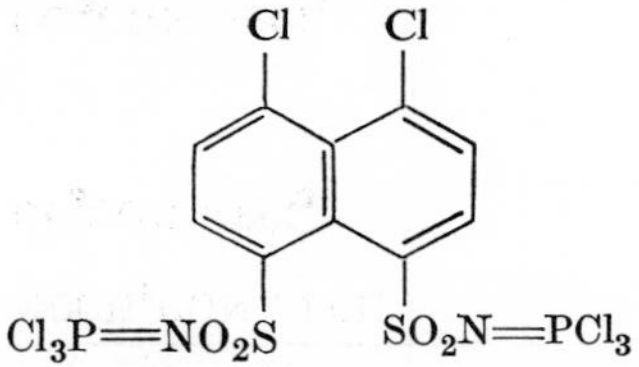

no details (39)

339

VI. Carbonylphosphazo Trichlorides

TABLE VII

Compound	Fp or bp/Torr (°C)	^{31}P NMR (85% H_3PO_4) δ_P (ppm)	Remarks
$NCCH_2CON{=}PCl_3$	96/0.18 (246)		$n_D^{20} = 1.5260$; $d_{20} = 1.700$ (246)
$CF_3CON{=}PCl_3$	31–31.5/1 (95); 146–149/750 (185)		IR (46,95,249); $n_D^{20} = 1.4408$ (95); $n_D^{25.5} = 1.4341$ (185); $d_{20} = 1.7119$ (95)
$CCl_3CON{=}PCl_3$	77–79 (110)	−25.1 (249)	IR (46,64a,249)
$CCl_3CON{=}PCl_3{\cdot}DMA$	79–81 (140a)		
$CCl_3CON{=}PCl_3{\cdot}DMF$	71–72 (140a)		
$NCCCl_2CON{=}PCl_3$	50–51 (246)		
$MeOCCl_2CON{=}PCl_3$	55–57 (120)		
$EtOCCl_2CON{=}PCl_3$	71–72 (101,120)		
$BuOCCl_2CON{=}PCl_3$	−3–+1 (120)		
$i\text{-}BuOCCl_2CON{=}PCl_3$	28–30 (120)		
$cyclo\text{-}C_6H_{11}OCCl_2CON{=}PCl_3$	oil (120)		
$PhOCCl_2CON{=}PCl_3$	49–50 (118,122)		
$p\text{-}ClC_6H_4OCCl_2CON{=}PCl_3$	52–55 (121,122)		
$o\text{-}MeC_6H_4OCCl_2CON{=}PCl_3$	49–51 (122)		
$m\text{-}MeC_6H_4OCCl_2CON{=}PCl_3$	30–34 (121,122)		
$p\text{-}MeC_6H_4OCCl_2CON{=}PCl_3$	74–76 (121,122)		
$1\text{-}C_{10}H_7OCCl_2CON{=}PCl_3$	69–72 (121,122)		
$2\text{-}C_{10}H_7OCCl_2CON{=}PCl_3$	80–83 (121,122)		
$Ph_2ClCICON{=}PCl_3$	60–62 (111)		
$(EtO)_2P(O)CCl_2CON{=}PCl_3$	oil (37)		
$CBr_3CON{=}PCl_3$	115–117 (138)	−8.9 (249)	
$Ph_3CCON{=}PCl_3$	123–125 (111)	−7.3 (249)	
$MeOCON{=}PCl_3$	not isolated (118)		
$EtOCON{=}PCl_3$	oil (101)		$d_{15} = 1.48$ (101); $Et^{18}OC^{18}ON{=}PCl_3$ (2)

340

ROCON=PCl₃	not isolated (118)	R = Pr, i-Pr, Bu, i-Bu
PhOCON=PCl₃	oil (119)	
CF₃CF₂CON=PCl₃	no details (185)	
CHF₂CF₂CON=PCl₃	89/12 (65)	only the hydrolysis product was isolated (65)
CHClFCF₂CON=PCl₃		only the hydrolysis product was isolated (65)
CHCl₂CF₂CON=PCl₃		
MeCCl₂CON=PCl₃	82–85 (53,205a)	
ClCH₂CCl₂CON=PCl₃	64–67 (51)	
MeCHClCON=PCl₃	no details (206a)	
ClCH₂CHClCON=PCl₃	36–39 (51)	
ClCH₂C(Me)ClCON=PCl₃	oil (205a)	
Me₂C(CN)CON=PCl₃	77–79/0.06 (222)	
	48–49 (222)	only the hydrolysis product was isolated (184)
PhCH=CHCON=PCl₃	86–88 (223)	
PhCH=C(CN)CON=PCl₃	159–160 (223)	Ar = 9-anthracenyl-
ArCH=C(CN)CON=PCl₃	162–163 (223)	
p-ClC₆H₄CH=C(CN)CON=PCl₃	108–110 (223)	
o-O₂NC₆H₄CH=C(CN)CON=PCl₃	150–152 (223)	
m-O₂NC₆H₄CH=C(CN)CON=PCl₃	156–158 (223)	
p-O₂NC₆H₄CH=C(CN)CON=PCl₃	121–123 (223)	
p-MeOC₆H₄CH=C(CN)CON=PCl₃	115–117 (223)	
p-MeC₆H₄CH=C(CN)CON=PCl₃	76–78 (223)	
(CH₂)₅C=C(CN)CON=PCl₃	132–133 (223)	
Ph₂C=C(CN)CON=PCl₃	no details (185)	
C₃F₇CON=PCl₃		only the hydrolysis product was isolated (159)
Et₂C(CN)CON=PCl₃	90–91/0.03 (222)	$n_D^{20} = 1.4950$; $d_{20} = 1.3256$ (222)
C₄F₉CON=PCl₃	no details (185)	

341

(continued overleaf)

TABLE VII (continued)

Compound	Fp or bp/Torr (°C)	31P NMR (85% H$_3$PO$_4$) δ_P (ppm)	Remarks
Pr$_2$C(CN)CON=PCl$_3$	102–104/0.03 (222)		n_D^{20} = 1.4900; d_{20} = 1.2260 (222)
Bu$_2$C(CN)CON=PCl$_3$	110–112/0.04 (222)		n_D^{20} = 1.4880; d_{20} = 1.2103 (222)
Am$_2$C(CN)CON=PCl$_3$	128–129/0.04 (222)		n_D^{20} = 1.4855; d_{20} = 1.1659 (222)
(Me$_2$CHCHCH$_2$CH$_2$)$_2$C(CN)CON=PCl$_3$	117–118/0.04 (222)		n_D^{20} = 1.4880; d_{20} = 1.1743 (222)
PhCON=PCl$_3$	60–61 (101,117)	–13.2 (249)	IR (46,249) PhC^{18}ON=PCl$_3$ (142)
o-FC$_6$H$_4$CON=PCl$_3$	49–50 (58)		
m-FC$_6$H$_4$CON=PCl$_3$	11–12 (58)		
p-FC$_6$H$_4$CON=PCl$_3$	50–51 (58)		
o-ClC$_6$H$_4$CON=PCl$_3$	oil (117)	–15.0 (249)	
p-ClC$_6$H$_4$CON=PCl$_3$	62–63 (117)		IR (46,249)
o-BrC$_6$H$_4$CON=PCl$_3$	41–43 (139)		
p-BrC$_6$H$_4$CON=PCl$_3$	64–66 (117)		IR (46)
p-IC$_6$H$_4$CON=PCl$_3$			only the hydrolysis product was isolated (184)
p-MeC$_6$H$_4$CON=PCl$_3$	glassy (112)	–12.2 (249)	
o-O$_2$NC$_6$H$_4$CON=PCl$_3$	oil (117)		
m-O$_2$NC$_6$H$_4$CON=PCl$_3$	103–105 (117)		
p-O$_2$NC$_6$H$_4$CON=PCl$_3$	121–123 (117)	–18.7 (249)	IR (249)
2,4-Cl$_2$C$_6$H$_3$CON=PCl$_3$	oil (117)		
2,4-Cl(O$_2$N)C$_6$H$_3$CON=PCl$_3$	oil (117)		
2,4-(O$_2$N)$_2$C$_6$H$_3$CON=PCl$_3$	62–64 (112)		
3,5-(O$_2$N)$_2$C$_6$H$_3$CON=PCl$_3$	oil (117)		
2,3,6-Cl$_3$C$_6$H$_2$CON=PCl$_3$	125–126 (117)		
	52–56 (250)		

1-C$_{10}$H$_7$CON=PCl$_3$ 66-68 (112)
2-C$_{10}$H$_7$CON=PCl$_3$ —40 (112)
Ph$_2$NCON=PCl$_3$ 112-114 (124)

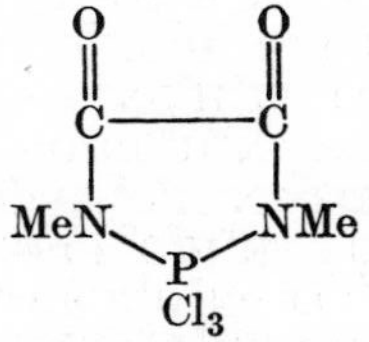

+69 (26)   not isolated

CO(N=PCl$_3$)$_2$ 40-41 (101)
Cl$_2$C(CON=PCl$_3$)$_2$ 165-166 (245)
ClHC(CON=PCl$_3$)$_2$ 118-121 (206)
Br$_2$C(CON=PCl$_3$)$_2$ 96-97 (246)
Br(NO$_2$)(CON=PCl$_3$)$_2$ 165 (dec) (246)
(F$_2$C)$_3$(CON=PCl$_3$)$_2$ 121-123 (246)
(F$_2$C)$_4$(CON=PCl$_3$)$_2$ 44-46 (56,205b)
CCl$_2$(CON=PCl$_3$)$_2$ 76-78 (205b)
CCl$_2$CH$_2$CH$_2$CCl$_2$(CON=PCl$_3$)$_2$ 118-121 (205b)
119-121 (206)
117-119 (205b)
CCl$_2$(CH$_2$)$_5$CCl$_2$(CON=PCl$_3$)$_2$ 119-121 (206,205b)
CCl$_2$(CH$_2$)$_6$CCl$_2$(CON=PCl$_3$)$_2$ 159-161 (206)
F$_2$C(p-C$_6$H$_4$CON=PCl$_3$)$_2$ 82-85 (56,205b)
(F$_2$C)$_2$(p-C$_6$H$_4$CON=PCl$_3$)$_2$ 127-129 (56,205b)
m-C$_6$H$_4$(CON=PCl$_3$)$_2$ 96-99 (112)
p-C$_6$H$_4$(CON=PCl$_3$)$_2$ 118-120 (183)

343

VII. Aryl- and Alkylphosphazo Trihalides
A. Arylphosphazo Trichlorides

TABLE VIII

Compound	Fp or bp/Torr (°C)	^{31}P NMR (85% H_3PO_4) δ_P (ppm)	Remarks
(PhN=PCl$_3$)$_2$	180–182 (266a,266b, 276,277)	+80.2 (145)	monomeric in dioxane
(o-FC$_6$H$_4$N=PCl$_3$)$_2$	128–132 (256)	+79.9 (256)	IR (256)
(m-FC$_6$H$_4$N=PCl$_3$)$_2$	124–125 (256)	+81.1 (256)	IR, ^{19}F NMR (256)
(p-FC$_6$H$_4$N=PCl$_3$)$_2$	129–132 (256)	+80.4 (256)	IR, ^{19}F NMR (256)
(o-ClC$_6$H$_4$N=PCl$_3$)$_2$	127–128 (266a,276,277)		monomeric in dioxane
(m-ClC$_6$H$_4$N=PCl$_3$)$_2$	129–131 (266a,276,277)		monomeric in dioxane
(p-ClC$_6$H$_4$N=PCl$_3$)$_2$	181–183 (266a,276,277)	+11.0 ± 0.5 (261)*	monomeric in dioxane and POCl$_3$
	171–174 (261)		
	180–182 (177)		
(o-BrC$_6$H$_4$N=PCl$_3$)$_2$	121–123 (266a,276)		monomeric in dioxane
(m-BrC$_6$H$_4$N=PCl$_3$)$_2$	125–126 (266a,276)		monomeric in dioxane
(p-BrC$_6$H$_4$N=PCl$_3$)$_2$	184–186 (266a,276)		monomeric in dioxane
(o-MeC$_6$H$_4$N=PCl$_3$)$_2$	124–126 (266a,276,277)		monomeric in dioxane
(m-MeC$_6$H$_4$N=PCl$_3$)$_2$	154–155 (266a,276)		monomeric in dioxane
(p-MeC$_6$H$_4$N=PCl$_3$)$_2$	198–200 (266a,276,277)		monomeric in dioxane
o-CF$_3$C$_6$H$_4$N=PCl$_3$	80–82/0.005 (28)	+49.8 (28)	IR, ^{19}F NMR (28)
(m-CF$_3$C$_6$H$_4$N=PCl$_3$)$_2$	96–99 (28)	+81.0 (28)	IR, ^{19}F NMR (28)
(p-CF$_3$C$_6$H$_4$N=PCl$_3$)$_2$	126–129 (28)	+77.9 (28)	IR, ^{19}F NMR (28)
(o-O$_2$NC$_6$H$_4$N=PCl$_3$)$_2$	109–111 (276)		monomeric in dioxane
(m-O$_2$NC$_6$H$_4$N=PCl$_3$)$_2$	142–143 (276)		monomeric in dioxane
(p-O$_2$NC$_6$H$_4$N=PCl$_3$)$_2$	140–141 (276)		monomeric in dioxane
(p-MeOC$_6$H$_4$N=PCl$_3$)$_2$	196–198 (266a,276)		monomeric in dioxane
(p-EtOC$_6$H$_4$N=PCl$_3$)$_2$	185–187 (276)		monomeric in dioxane
p-ClSO$_2$C$_6$H$_4$N=PCl$_3$	108–111 (30,175)		monomeric in dioxane
o-Me$_2$NSO$_2$C$_6$H$_4$N=PCl$_3$	88–93 (280)		monomeric in dioxane

Compound	m.p. (°C) (Ref.)	^{31}P NMR shift (Ref.)	Notes
(m-Me₂NSO₂C₆H₄N=PCl₃)₂	158–162 (280)		
(p-Me₂NSO₂C₆H₄N=PCl₃)₂	138–141 (280)		
(2,4-F₂C₆H₃N=PCl₃)₂	140 (256)	+78.4 (256)	IR, ¹⁹F NMR (256)
(2,5-F₂C₆H₃N=PCl₃)₂	123 (256)	+78.4 (256)	IR (256)
(2-F,5-CF₃C₆H₃N=PCl₃)₂	92–94 (28)	+40.0; +77.1 (28)	IR, ¹⁹F NMR; equilibrium monomeric-dimeric in CCl₄ (28)
(2,4-Cl₂C₆H₃N=PCl₃)₂	116–118 (46a, 266a, 276,277)		monomeric in dioxane
(3,5-Cl₂C₆H₃N=PCl₃)₂	136–138 (276)		monomeric in dioxane
(2,4-Br₂C₆H₃N=PCl₃)₂	114–115 (276,277)		monomeric in dioxane
(2,4-(O₂N)₂C₆H₃N=PCl₃)₂	75–77 (276)		
(5-Me,2-NOC₆H₃N=PCl₃)₂	114–115 (213)	+38.0 (213)*	monomeric in dioxane
(2,4,6-Cl₃C₆H₂N=PCl₃)₂	oil (276)		monomeric in dioxane
(2,6,4-Cl₂(O₂N)C₆H₂N=PCl₃)₂	71–73 (276)		monomeric in dioxane
(2,4,6-Br₃C₆H₂N=PCl₃)₂	37–40 (276)		monomeric in dioxane
(2,3,4,5-F₄C₆HN=PCl₃)₂	96–97 (256)	+37.8 (256)*	IR, ¹⁹F NMR; monomeric in CCl₄ and CS₂ (256)
(2,3,5,6-F₄C₆HN=PCl₃)₂	94–95 (256)	+31.6 (256)*	IR, ¹⁹F NMR; monomeric in CCl₄ and CS₂ (256)
C₆F₅N=PCl₃	(82)		

* Monomeric form in solution.

B. Alkylphosphazo Trichlorides

TABLE IX

Compound	Fp or bp/Torr (°C)	^{31}P NMR (85% H$_3$PO$_4$) δ_P (ppm)	Remarks
(MeN=PCl$_3$)$_2$	160 (38); 174–176 (266a,271); 180 (174)	+78.2 ± 1 (70); +77.5 (78); +82.2 (92a)	IR (38,252,265); Raman (38,265); ^{1}H NMR (78,92a,252); cryst.struct.(89,90); NQR (262); J_{PH} = 20 Hz (78); mass sp. (255); enthalpy of formation (67)
(OC)$_4$Cr(MeN=PCl$_3$)$_2$	133–135 (278)	+97.4 (93)	^{1}H NMR; J_{PH} = 300(?) Hz (165)
(CH$_2$ClN=PCl$_3$)$_2$	134–135 (62,63)	+80 (165)	
(PhCH$_2$N=PCl$_3$)$_2$	140–141 (165)		
(ClCOCH$_2$N=PCl$_3$)$_2$	167–169 (272)		
(ClCOCH$_2$N=PCl$_3$)$_2$·C$_6$H$_6$	149–151 (267,274,275)		
ClCOCHClN=PCl$_3$	151–155 (dec) (275)		
Ph$_2$CHN=PCl$_3$	84–86/2 (274,275)		n_D^{20} = 1.5408; d_{20} = 1.7167 (274,275)
CCl$_2$(COCl)N=PCl$_3$	110–115/0.05 (279)		n_D^{20} = 1.6000; d_{20} = 1.3240 (279)
CCl$_3$N=PCl$_3$	85–87/2 (275); 53–55/0.05 (237); 56/0.01 (72,80a); 102–103/12 (136); 69–70/2 (137)	+16.3 ± 0.5 (72); +17.4 (99)	n_D^{20} = 1.5445; d_{20} = 1.7610 (275); n_D^{20} = 1.5502; d_{20} = 1.7877 (136); n_D^{20} = 1.5480 (137); IR (98,99,137)
PhC(=NH)N=PCl$_3$	oil (47)		
PhC(=NSO$_2$Ph)N=PCl$_3$	59–61 (44)		
PhC(=NSO$_2$C$_6$H$_4$Cl-p)N=PCl$_3$	oil (44)		
PhC(=NSO$_2$C$_6$H$_4$Me-p)N=PCl$_3$	81–83 (44)		
PhC(=NSO$_2$C$_6$H$_4$NO$_2$-p)N=PCl$_3$	143–145 (44)		
PhC(=NSO$_2$C$_{10}$H$_7$-1)N=PCl$_3$	124–125 (44)		
PhC(=NSO$_2$C$_{10}$H$_7$-2)N=PCl$_3$	97–100 (44)		
PhC(=NP(O)(OPh)$_2$)N=PCl$_3$	oil (44)		
p-O$_2$NC$_6$H$_4$C(=NSO$_2$Ph)N=PCl$_3$	oil (45)		complex with AlCl$_3$, FeCl$_3$, SbCl$_5$ (140b)
PhSO$_2$N=C(Cl)N=PCl$_3$	no details (42)		
p-O$_2$NC$_6$H$_4$SO$_2$N=C(Cl)N=PCl$_3$	no details (42)		
PhCON=C(Cl)N=PCl$_3$	47–49 } (135a); 143–145/0.03		IR (135a)

346

Compound	b.p./mm (ref.)		Physical constants	Spectral / other data
d-ClC$_6$H$_4$CON=C(Cl)N=PCl$_3$	156–158/0.03 (135a)			IR (135a)
(EtN=PCl$_3$)$_2$	119–121 (266a,271)	+78.8 (79)		IR, ^{1}H NMR; J_{PCH_2} = 29 Hz (79); 30 Hz (78)
	122–124 (79)	+79.0 (78)		
(ClCH$_2$CH$_2$N=PCl$_3$)$_2$	145–147 (272)			
(PhCH$_2$CH$_2$N=PCl$_3$)$_2$	173–176 (272)			
(ClSO$_2$(CH$_2$)$_2$N=PCl$_3$)$_2$	172–173 (108)			
NCCCl$_2$CH$_2$N=PCl$_3$	64–65/0.06 (177)		n_D^{20} = 1.5242; d_{20} = 1.5955 (177)	
Me(Ph)C(CN)N=PCl$_3$	92–94/0.045 (179)		n_D^{20} = 1.5517; d_{20} = 1.3436 (179)	
	95–96/0.05 (181)			
[Me(Ph)C(CN)N=PCl$_3$]$_2$	217–218 (179)			
Me(p-ClC$_6$H$_4$)C(CN)N=PCl$_3$	107–109/0.05 (179)		n_D^{20} = 1.5639; d_{20} = 1.4273 (179)	
Me(p-MeC$_6$H$_4$)C(CN)N=PCl$_3$	104–105/0.05 (179)		n_D^{20} = 1.5513; d_{20} = 1.3266 (179)	
Me(2,4-Me$_2$C$_6$H$_3$)C(CN)N=PCl$_3$	114–116/0.055 (179)		n_D^{20} = 1.5557; d_{20} = 1.3041 (179)	
Me(2,5-Me$_2$C$_6$H$_3$)C(CN)N=PCl$_3$	109–110/0.05 (179)		n_D^{20} = 1.5556; d_{20} = 1.3080 (179)	
CF$_3$CCl$_2$N=PCl$_3$	no details (163)	+6.5 (99)		IR (99,249); complex with AlCl$_3$, FeCl$_3$, SbCl$_5$ (140b)
CCl$_3$CCl$_2$N=PCl$_3$	17–18 (275)			IR (98,99,249)
	21–23 (63,109,126,220,221)		n_D^{20} = 1.5649 (134),	
	20–23 (219)		1.5612 (221),	
	66–68/9.2(?) (134)	+10.7 (249)	1.5678, 1.5684 (126),	
	74–76/0.04 (133)		1.5615 (133);	
	94–98/2 (126,221)	+11.1 (148)	n_D^{24} = 1.6010(?) (109);	
	100–105/2 (275)	+13.0 (99)	n_D^{25} = 1.5608 (220);	
	95–96/1 (63)		1.5600 (275);	
	103–104/3 (109,126,219, 220)		d_{20} = 1.821; 1.823 (126);	
	113–114/4 (126)		d_{24} = 1.823 (109,220)	
	117–118/5 (126,220,221)		complex with AlCl$_3$, FeCl$_3$, SbCl$_5$ (140b)	
	273–275/760 (126,220)			
PhCCl$_2$CCl$_2$N=PCl$_3$	42–46, 168–169/2 } (109)			
NCCCl$_2$CCl$_2$N=PCl$_3$	39–40, 76–78/0.03 } (227,230)			IR (230)
	36–38, 77–79/0.01 } (177)			
	81–82/0.05, 37–40 } (133)			

347

(continued overleaf)

TABLE IX (continued)

Compound	Fp or bp/Torr (°C)	^{31}P NMR (85% H$_3$PO$_4$) δ_P (ppm)	Remarks
ClCOCCl$_2$CCl$_2$N=PCl$_3$	88–91/0.015 (226)		IR (226); $n_D^{20} = 1.5611$; $d_{20} = 1.8186$ (226)
Cl$_2$P(O)CCl$_2$CCl$_2$N=PCl$_3$	78/0.05 (37)		$n_D^{20} = 1.5610$; $d_{20} = 1.8470$; mol. refractivity: 75.91 (37)
FClCHCCl$_2$N=PCl$_3$	82–83/3 (109,221); 83–86/3 (126)		$n_D^{20} = 1.5270$ (109,221), 1.5319 (126); $d_{20} = 1.788$ (109,126,221)
CCl$_2$HCCl(COCl)N=PCl$_3$	100–102/1.5 (274,275)		$n_D^{20} = 1.5492$; $d_{20} = 1.7598$ (274,275)
CCl$_3$CHClN=PCl$_3$	106–107/3 (278)		$n_D^{20} = 1.5500$; $d_{20} = 1.7754$ (278)
CCl$_2$=CClN=PCl$_3$	57–57.5/0.07 (221); 63–66/0.1 (221); 92–94/2 (109,149,220); 101–108/3 (219)		$n_D^{20} = 1.5680$ (221); 1.5725 (109,220); $d_{20} = 1.719$ (221); 1.757 (220); 1.821 (109)
ClCH=CClN=PCl$_3$	57–58/0.07 (109)		$n_D^{20} = 1.5680$; $d_{20} = 1.719$ (109)
NCCH=CClN=PCl$_3$	68–71/0.015 (228,230)		$n_D^{20} = 1.5880$; $d_{20} = 1.6322$ (228,230)
PhCCl=CClN=PCl$_3$	159–160/3 (109,126)		$n_D^{20} = 1.6188$; $d_{20} = 1.538$ (109,126)
NCCCl=CClN=PCl$_3$	84–86/0.02 (229,230)		$n_D^{20} = 1.5919$; $d_{20} = 1.6880$ (229,230)
ClCOCCl=CClN=PCl$_3$	92–93/0.015 (226)		$n_D^{20} = 1.5711$; $d_{20} = 1.7553$ (226)
Cl$_2$PClC=CClN=PCl$_3$	125–128/0.05 (219)		$n_D^{20} = 1.6325$; $d_{20} = 1.789$ (219)
Cl$_2$(O)PCCl=CClN=PCl$_3$	110–111/0.03 (219)		
ClCOCH=CClN=PCl$_3$	114–116/0.025 (215)		
Cl$_2$(O)PCH=CClN=PCl$_3$	102–105/0.015 (226); oil (219)		$n_D^{20} = 1.5896$; $d_{20} = 1.7182$ (226)

Compound	b.p./m.p. (°C)	δ_P	Physical constants / notes
ClCH=C(COCl)N=PCl₃	81–82/1 (273,274)		$n_D^{20} = 1.5759$; $d_{20} = 1.6885$ (273,274)
CF₃C(=NPh)N=PCl₃	49–51 (54)		
[CCl₃C(=NH)N=PCl₃]			only the adduct form $CCl_3C(=NH)NH_2 \cdot PCl_5$ (fp = 168–172) is known (52) adduct $CCl_3C(=NMe)NH_2 \cdot PCl_5$ (fp = 139–141) (52)
CCl₃C(=NMe)N=PCl₃	128–131 (52)		
CCl₃C(=NCH₂Ph)N=PCl₃	72–76 (52)		
CCl₃C(=NEt)N=PCl₃	glassy (52)		
CCl₃C(=NBu)N=PCl₃	oil (54)		
CCl₃C(=NPh)N=PCl₃	56–58 (50,55)		
CCl₃C(=NC₆H₄Br-p)N=PCl₃	73–75 (55)		
CCl₃C(=NC₆H₄Me-p)N=PCl₃	38–41 (55)		
CCl₃C(=NC₆H₄OMe-p)N=PCl₃	oil (55)		
CCl₃C(=NCOCCl₃)N=PCl₃	94–96 (54)		
CCl₃C(=NPOCl₂)N=PCl₃	76–78/0.03 (54)		
CCl₃C(=NPMe(O)NMe₂)N=PCl₃	63–66 (54)		
CCl₃C(=NSO₂Ph)N=PCl₃	180–182/0.05 (141)		$d_{20} = 1.6292$ (141)
CCl₃C(=NSO₂C₆H₄Cl-p)N=PCl₃	102–104 (141)		
CCl₃C(=NSO₂C₆H₄Br-p)N=PCl₃	109–111 (141)		
CCl₃C(=NSO₂C₆H₄Me-p)N=PCl₃	104–106 (141)		
CCl₃C(=NSO₂C₆H₄NO₂-m)N=PCl₃	78–80 (141)		
CCl₃C(=NSO₂C₆H₄NO₂-p)N=PCl₃	184–185 (141)		
CCl₃C(=NSO₂C₆H₃NO₂(Cl)-3,4]-N=PCl₃	97–98 (141)		
CCl₃C(=NSO₂C₁₀H₇-2)N=PCl₃	oil (141)		
(PrN=PCl₃)₂	112–115 (271)	+78.7 (79)	IR, ¹H NMR (78,79); $J_{PCH_2} = 28.8$–29 Hz (78,79)
	118–120 (79)	+79.0 (78)	
ClSO₂(CH₂)₃N=PCl₃	160–163 (278)		
Et(Ph)C(CN)N=PCl₃	95–97/0.06 (179)		$n_D^{20} = 1.5480$; $d_{20} = 1.3166$ (179)
MeCCl₂CCl₂N=PCl₃	110–111/0.2 (244)		$n_D^{20} = 1.5509$; $d_{20} = 1.7071$ (244)
	71–73/0.055 (242)		$n_D^{20} = 1.5500$ (242); complex with AlCl₃, FeCl₃, SbCl₅ (140b)

(continued overleaf)

349

TABLE IX (continued)

Compound	Fp or bp/Torr (°C)	^{31}P NMR (85% H_3PO_4) δ_P (ppm)	Remarks
$ClCH_2CCl_2CCl_2N{=}PCl_3$	131–132/0.2 (244) 104–106/0.1 (235) 99–101/0.05 (242)		$n_D^{20} = 1.5668$; $d_{20} = 1.7973$ (244); $n_D^{20} = 1.5660$; $d_{20} = 1.7750$ (235); $n_D^{20} = 1.5658$; $d_{20} = 1.7968$ (242); complex with $SbCl_5$ (140b)
$RCH_2CCl_2CCl_2N{=}PCl_3$	oils (241)		R = Ph, o-, p-ClC_6H_4, p-BrC_6H_4, p-$O_2NC_6H_4$, m-MeC_6H_4, p-$MeOC_6H_4$
p-$MeC_6H_4CH_2CCl_2CCl_2N{=}PCl_3$	72–74 (241)		IR (241)
o-$O_2NC_6H_4CH_2CCl_2CCl_2N{=}PCl_3$	102–105 (241)		IR (241)
$PhN{=}CClCCl_2CCl_2N{=}PCl_3$	158–160/0.03 (177)		$n_D^{20} = 1.6032$; $d_{20} = 1.6217$ (177)
p-$ClC_6H_4N{=}CClCCl_2CCl_2N{=}PCl_3$	159–161/0.03 (177)		$n_D^{20} = 1.6052$; $d_{20} = 1.6547$ (177)
$PrOCH_2CCl_2CCl_2N{=}PCl_3$	oil (235)		
$BuOCH_2CCl_2CCl_2N{=}PCl_3$	oil (235)		
$PhOCH_2CCl_2CCl_2N{=}PCl_3$	oil (235)		
$Cl_2(O)PNRCH_2CCl_2CCl_2N{=}PCl_3$	oils (135)		R = alkyl
$Cl_2(O)PN(Ph)CH_2CCl_2CCl_2N{=}PCl_3$	150–151 (135)		
$Cl_2(O)PN(C_6H_4Me\text{-}p)CH_2\text{-}CCl_2CCl_2N{=}PCl_3$	114–116 (135)		
$MeCCl(CN)CCl_2N{=}PCl_3$	80–82/0.03 (225)		$n_D^{20} = 1.5472$; $d_{20} = 1.6490$ (225)
$MeCCl(COCl)CCl_2N{=}PCl_3$	36–38 (225)		
$PhCH{=}C(CN)CCl_2N{=}PCl_3$	94–97/0.1 (231)		IR, $n_D^{20} = 1.5461$; $d_{20} = 1.6684$ (231)
p-$O_2NC_6H_4CH{=}C(CN)CCl_2N{=}PCl_3$	oil (223)		
$CCl_3CCl(CHCl_2)N{=}PCl_3$	oil (223)		
$MeCCl(CHCl)N{=}PCl_3$	46–47 (273)		
$MeCHClCCl(COCl)N{=}PCl_3$	70–71/2 (278)		$n_D^{20} = 1.5361$; $d_{20} = 1.6281$ (278)
$MeC(CN){=}CClN{=}PCl_3$	96–98/1 (274, 275)		$n_D^{25} = 1.5352$; $d_{25} = 1.6202$ (274, 275)
$MeC(COCl){=}CClN{=}PCl_3$	68–72/0.015 (225)		$n_D^{20} = 1.5848$; $d_{20} = 1.5455$ (225)
$CCl_2{=}C(CCl_3)N{=}PCl_3$	88–90/0.1 (231)		IR, $n_D^{20} = 1.5481$; $d_{20} = 1.5779$ (231)
$CCl_2{=}C(CHCl_2)N{=}PCl_3$	164–165/18 (273) 180–182/30 (273)		$n_D^{20} = 1.5850$; $d_{20} = 1.8084$ (273)
$PhN{=}CClCCl{=}CClN{=}PCl_3$	105–107/2 (270,273) 119–122/0.03 (177)		$n_D^{20} = 1.5752$; $d_{20} = 1.7409$ (270,273); $n_D^{20} = 1.5942$; $d_{20} = 1.5404$ (177)

Compound	b.p./mm (ref)		Physical constants (ref)
$p\text{-ClC}_6\text{H}_4\text{N=CClCCl=CClN=PCl}_3$	133–136/0.03 (177)		$n_D^{20} = 1.6028$ (177); $d_{20} = 1.5907$
$(i\text{-PrN=PCl}_3)_2$	71–73 (79)		IR (79)
$\text{Me}_2\text{C(CN)N=PCl}_3$	62–63/2 (178,181)		$n_D^{20} = 1.4880$; $d_{20} = 1.3300$ (178)
	204–205/760 (178)		
	211–212 (178)		
$(\text{Me}_2\text{C(CN)N=PCl}_3)_2$	77–78/6 (275)		$n_D^{20} = 1.5050$; $d_{20} = 1.4331$ (275)
$\text{Me}_2\text{C(COCl)N=PCl}_3$	93–95/3 (272)		$n_D^{20} = 1.5335$; $d_{20} = 1.5616$ (272)
$(\text{ClCH}_2)_2\text{CHN=PCl}_3$	47–48/10 (138)	+21.2 (99)	IR (99,138); $n_D^{20} = 1.7486$;
$(\text{CF}_3)_2\text{CClN=PCl}_3$			$d_{20} = 1.4125$ (138)
			IR (98,99)
$(\text{CCl}_3)_2\text{CClN=PCl}_3$	52–53	+43.2 (99)	
$(\text{CF}_3)_2\text{CHN=PCl}_3$	140–150/1 (269,273)	+79.3 (79)	IR; $n_D^{20} = 1.3921$; $d_{20} = 1.6760$ (138)
$(\text{BuN=PCl}_3)_2$	136–138/760 (138)		IR, ^{1}H NMR, $J_{\text{PCH}_2} = 28.6$ Hz (79)
	75–78 (271)		
	76–79 (79)		
$(\text{ClSO}_2(\text{CH}_2)_4\text{N=PCl}_3)_2$	145–147 (278)		$n_D^{20} = 1.5434$; $d_{20} = 1.2843$ (179)
$\text{Pr(Ph)C(CN)N=PCl}_3$	95–97/0.05 (179)		$n_D^{20} = 1.5494$; $d_{20} = 1.6519$ (242,244);
$\text{EtCCl}_2\text{CCl}_2\text{N=PCl}_3$	124–125/0.2 (242,244)		complex with AlCl$_3$, FeCl$_3$, SbCl$_6$, (140b)
			$n_D^{20} = 1.5364$; $d_{20} = 1.5551$ (225)
$\text{EtCCl(CN)CCl}_2\text{N=PCl}_3$	87–89/0.02 (225)		IR; $n_D^{20} = 1.5371$; $d_{20} = 1.6161$ (231)
$\text{EtCCl(COCl)CCl}_2\text{N=PCl}_3$	100–102/0.05 (231)		$n_D^{20} = 1.5350$; $d_{20} = 1.5866$ (278)
$\text{EtCCl}_2\text{CHClN=PCl}_3$	83–85/3 (278)		$n_D^{20} = 1.5798$; $d_{20} = 1.5043$ (225)
$\text{EtC(CN)=CClN=PCl}_3$	70–74/0.02 (225)		
$\text{Me(PhCH}_2)\text{C(CN)C(CN)=CClN=PCl}_3$	103–104 (240)		
Et(Me)CHN=PCl_3	58.5–59.5/12 (79)	+38.7 (79)	IR; ^{1}H NMR; $n_D^{20} = 1.4800$ (79)
$\text{Et(Me)C(CN)N=PCl}_3$	60–61/1 (178,181)		$n_D^{20} = 1.4891$; $d_{20} = 1.2960$ (178)
$(\text{Et(Me)C(CN)N=PCl}_3)_2$	181–182 (178)		
$(i\text{-BuN=PCl}_3)_2$	101–104 (63,266a)		
$\text{ClC(Me}_2)\text{CH}_2\text{N=PCl}_3$	63–64/2 (271)		$n_D^{20} = 1.5042$; $d_{20} = 1.3842$ (271)
$i\text{-Pr(Ph)C(CN)N=PCl}_3$	102–104/0.055 (179)		$n_D^{20} = 1.5462$; $d_{20} = 1.3002$ (179)
$\text{Me}_2\text{CClCHClN=PCl}_3$	93–94/5 (278)		$n_D^{20} = 1.5189$; $d_{20} = 1.4749$ (278)
	82–84/2 (271)		$n_D^{20} = 1.5175$; $d_{20} = 1.4685$ (271)
$\text{Me}_2\text{CClCCl}_2\text{N=PCl}_3$	30–31 75–76/0.1 } (232)	−8.8 (149)	
$\text{Me}_2\text{C(CN)CCl}_2\text{N=PCl}_3$	53–58 55–59 86–89/0.03 } (133)		
$\text{ClCH}_2\text{(Me)CClCCl}_2\text{N=PCl}_3$	93–94/0.07 (232)		$n_D^{20} = 1.5541$; $d_{20} = 1.6660$ (232)

351

(continued overleaf)

TABLE IX (*continued*)

Compound	Fp or bp/Torr (°C)	31P NMR (85% H_3PO_4) δ_P (ppm)	Remarks
$BrCH_2(Me)CClCCl_2N{=}PCl_3$	107–108/0.05 (232)		$n_D^{20} = 1.5461$; $d_{20} = 1.7884$ (232)
$Me_2CClCCl(COCl)N{=}PCl_3$	104–106/1.5 (275)		$n_D^{20} = 1.5382$; $d_{20} = 1.5854$ (275)
$Me_2^2C{-}CClN{=}PCl_3$	122–124/0.3 (232)		$n_D^{20} = 1.5309$; $d_{20} = 1.578$ (232)
$Me_3CN{=}PCl_3$	153–154/760 (271)		$d_{20} = 1.2190$ (271)
$(CF_3)_3CN{=}PCl_3$	134–135/760 (138)	+44.8 (99)	IR (99,138); $n_D^{20} = 1.3723$; $d_{20} = 1.7600$ (138)
$(AmN{=}PCl_3)_2$	69–71 (79); 70–72 (271)	+78.4 (79)	IR; ¹H NMR; $J_{P\,CH_3} = 29$ Hz (79)
$Bu(Ph)C(CN)N{=}PCl_3$	104–106/0.03 (179)		$n_D^{20} = 1.5382$; $d_{20} = 1.2480$ (179)
$PrCCl_2CCl_2N{=}PCl_3$	135–136/3 (242,244)		$n_D^{20} = 1.5450$; $d_{20} = 1.6011$ (242,244)
$PrCCl(CN)CCl_2N{=}PCl_3$	98–100/0.025 (225)		$n_D^{20} = 1.5330$; $d_{20} = 1.5264$ (225)
$Et_2C(CN)CCl(CN)CCl_2N{=}PCl_3$	oil (240)		$n_D^{20} = 1.5437$ (240)
$PrCCl(COCl)CCl_2N{=}PCl_3$	90–93/0.04 (231)		IR; $n_D^{20} = 1.5246$; $d_{20} = 1.5635$ (231)
$PrC(CN){=}CClN{=}PCl_3$	88–90/0.015 (225)		$n_D^{20} = 1.5643$; $d_{20} = 1.4179$ (225)
$Me(Et)C(CN)C(CN){=}CClN{=}PCl_3$	135–136/0.018 (240)		$n_D^{20} = 1.5703$; $d_{20} = 1.4875$ (240)
$Et_2C(CN)C(CN){=}CClN{=}PCl_3$	131–132/0.018 (240)		$n_D^{20} = 1.5573$; $d_{20} = 1.3910$ (240)
$Et(i\text{-}Pr)C(CN)C(CN){=}CClN{=}PCl_3$	142–143/0.02 (240)		$n_D^{20} = 1.5541$; $d_{20} = 1.3590$ (240)
$Et_2CHN{=}PCl_3$	94–95/30 (268,272)		$n_D^{20} = 1.4855$; $d_{20} = 1.2280$ (272)
$Et_2C(CN)N{=}PCl_3$	84–85/1 (178)		$n_D^{20} = 1.4909$; $d_{20} = 1.2680$ (178)
$(Et_2C(CN)N{=}PCl_3)_2$	225–226 (178)		
$Me_3CCH_2N{=}PCl_3$	99–100/56 (272)		$n_D^{20} = 1.4766$; $d_{20} = 1.1980$ (272)
$(EtCH(Me)CH_2N{=}PCl_3)_2$	37–39 (79)	+77.9 (79)	IR; ¹H NMR (79)
$EtC(Me)CClCl_2N{=}PCl_3$	82–86/0.1 (232)		
$(Me_2CHCH_2CH_2N{=}PCl_3)_2$	74–77 (79)	+78.7 (79)	IR; ¹H NMR (79)
$i\text{-}PrCHClCCl_2N{=}PCl_3$	73–76/0.02 (243,244)		$n_D^{20} = 1.5445$; $d_{20} = 1.5530$ (243,244)
$i\text{-}PrCCl_2CCl_2N{=}PCl_3$	100–110/0.02(?) (244)		
$i\text{-}PrCCl(CN)CCl_2N{=}PCl_3$	109–110/0.015 (225)		$n_D^{20} = 1.5352$; $d_{20} = 1.5336$ (225)
$i\text{-}PrCCl(COCl)CCl_2N{=}PCl_3$	90–92/0.04 (231)		IR; $n_D^{20} = 1.5271$; $d_{20} = 1.5417$ (231)
$i\text{-}PrCCl_2CCl(COCl)N{=}PCl_3$	136–138/1.5 (275)		$n_D^{20} = 1.5540$; $d_{20} = 1.6200$ (275)
$i\text{-}PrC(CN){=}CClN{=}PCl_3$	80–82/0.02 (225)		$n_D^{20} = 1.5539$; $d_{20} = 1.3846$ (225)
$(AmCH_2N{=}PCl_3)_2$	52–55 (79)	+79.1 (79)	IR; ¹H NMR (79)

Compound	b.p./m.p. (°C)/(mm) (Ref.)	[α]	Physical data
BuCCl₂CCl₂N=PCl₃	115–118/0.05 (242,244)		$n_D^{20} = 1.5003(?)$; $d_{20} = 1.6121$ (242,244)
BuCCl(CN)CCl₂N=PCl₃	106–109/0.015 (225)		$n_D^{20} = 1.5272$; $d_{20} = 1.4736$ (225)
Pr₂C(CN)C(CN)=CClN=PCl₃	40–41 (240)		
BuC(CN)=CClN=PCl₃	81–84/0.015 (225)		$n_D^{20} = 1.5551$; $d_{20} = 1.3711$ (225)
i-BuCCl₂CCl₂N=PCl₃	106–108/0.02 (244)		$n_D^{20} = 1.5310$; $d_{20} = 1.5361$ (244)
Me(i-Bu)C(CN)C(CN)=CClN=PCl₃	141–142/0.018 (240)		$n_D^{20} = 1.5522$; $d_{20} = 1.3613$ (240)
(Me₃CCH₂CH₂N=PCl₃)₂	135–137 (272)		
Me₃CCCl₂CHClN=PCl₃	112–114/1 (272)		$n_D^{20} = 1.5367$; $d_{20} = 1.4901$ (272)
(AmCH₂CH₂N=PCl₃)₂	34–35 (79)	+79.7 (79)	IR; ¹H NMR (79)
Me(Bu)C(CN)C(CN)=CClN=PCl₃	138–139/0.018 (240)		$n_D^{20} = 1.5462$; $d_{20} = 1.3449$ (240)
AmCCl₂CCl₂N=PCl₃	112–114/0.03 (242,244)		$n_D^{20} = 1.5290$; $d_{20} = 1.4816$ (242,244)
(AmCH₂CH₂CH₂N=PCl₃)₂	22–24 (79)	+79.8 (79)	IR (79)
(AmCH₂CH₂CH₂CH₂N=PCl₃)₂	34–36 (79)	+79.8 (79)	IR (79)
Bu₂C(CN)CCl(CN)CCl₂N=PCl₃	oil (240)		$n_D^{20} = 1.5311$ (240)
Bu₂C(CN)C(CN)=CClN=PCl₃	44–45 (240)		
[Am(CH₂)₅N=PCl₃]₂	48–49 (79)	+78.6 (79)	IR (79)
(CH₂)₄C(CN)N=PCl₃	93–94/2 (181)		
	96–97/3 (179)		
	204–205 (178)		
	23–24 (178)		
	97–98/1 (178,181)		$n_D^{20} = 1.5173$; $d_{20} = 1.3700$ (178)

```
Me₂C———CCl
  |       ‖
 MeN      N
    \     /
      P
      Cl₃
```
105–108 (180)

```
  Me
   \
    C(Et)——CCl
    |        ‖
   MeN       N
      \      /
        P
        Cl₃
```
84–86 (180)

(continued overleaf)

TABLE IX (continued)

Compound	Fp or bp/Torr (°C)	31P NMR (85% H$_3$PO$_4$) δ_P (ppm)	Remarks
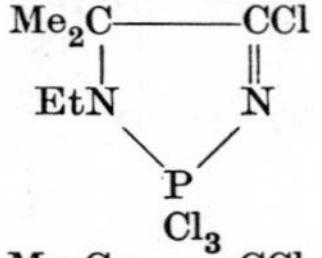 Me$_2$C—CCl EtN P=N Cl$_3$	72 (180)		
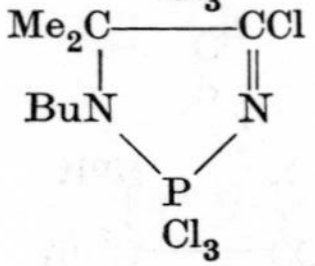 Me$_2$C—CCl BuN P=N Cl$_3$	98–100 (180)		
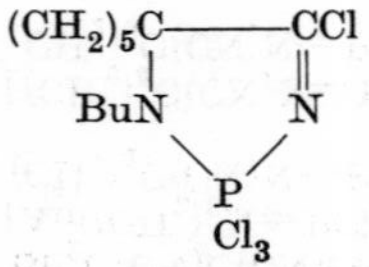 (CH$_2$)$_5$C—CCl BuN P—N Cl$_3$	141–142 (180)		
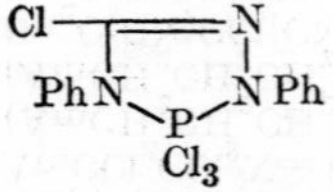 CCl=N Cl—C P—NPh PhN Cl$_3$	156 (dec) (91)	+77.0 (99,148)	
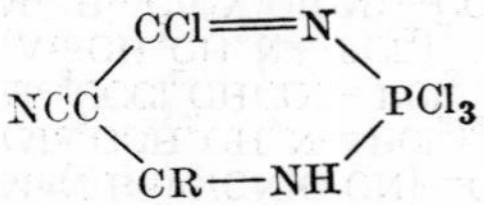 CCl=N PCl$_3$ NCC CR—NH			R = Me, et (132)
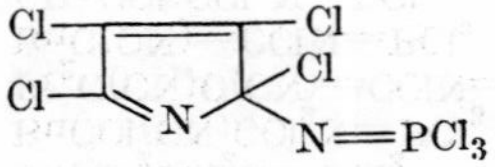 Cl Cl Cl—C N=PCl$_3$ Cl—C=N	96–97 145–147/0.03 (236)		

354

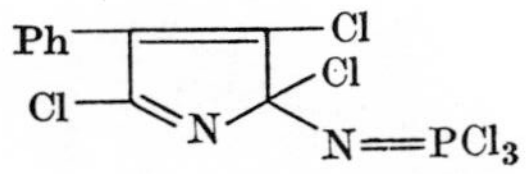

Ph
Cl
Cl N=PCl$_3$ 74–75
128–132/0.05 (233)

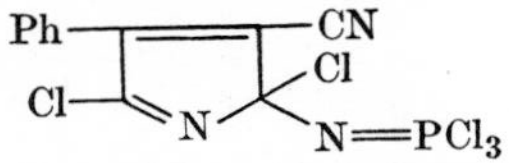

Ph
CN
Cl N=PCl$_3$ 146–147 (239a)

o-ClC$_6$H$_4$
CN
Cl N=PCl$_3$ 123–125 (239a)

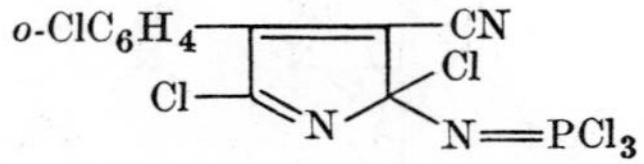

p-ClC$_6$H$_4$
CN
Cl N=PCl$_3$ 129–130 (239a)

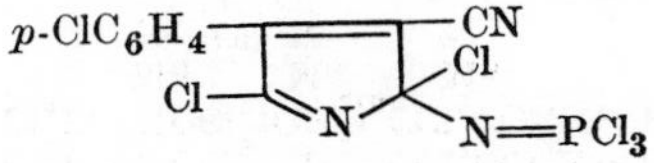

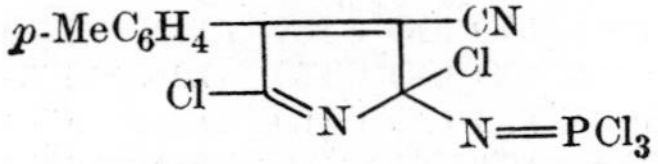

p-MeC$_6$H$_4$
CN
Cl N=PCl$_3$ 126–128 (239a)

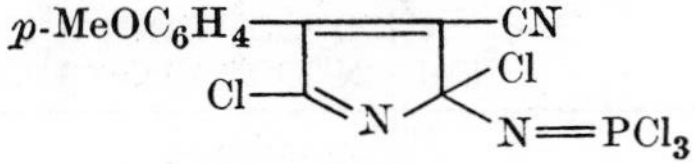

p-MeOC$_6$H$_4$
CN
Cl N=PCl$_3$ 122–123 (239a)

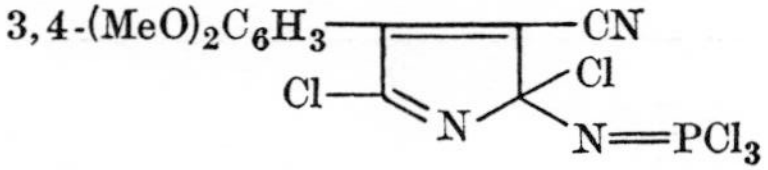

3,4-(MeO)$_2$C$_6$H$_3$
CN
Cl N=PCl$_3$ 155–156 (239a)

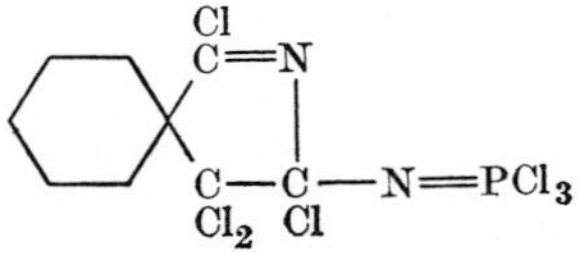

Cl
C=N
C—C—N=PCl$_3$
Cl$_2$ Cl   87–89
163–166/0.04 (236)

355

(continued overleaf)

TABLE IX (continued)

Compound	Fp or bp/Torr (°C)	^{31}P NMR (85% H_3PO_4) δ_P (ppm)	Remarks
$Cl_3P=NCCl_2CCl_2N=PCl_3$	no details (26)	+18 (26)	$n_D^{20} = 1.5827;\ d_{20} = 1.8164$ (177)
$Cl_3P=NCH_2CCl_2CCl_2N=PCl_3$	62–65 (234)		
$Cl_3P=NCCl_2CH_2CCl_2CCl_2N=PCl_3$	131–133 (234)		
$Cl_3P=NCCl_2(CH_2)_2CCl_2CCl_2N=PCl_3$	107–108 (234)		
$Cl_3P=NCCl_2(CH_2)_3CCl_2CCl_2N=PCl_3$	144–145 (234)		
$Cl_3P=NCCl_2(CH_2)_4CCl_2CCl_2N=PCl_3$	83–85 (234)		
$Cl_3P=NCCl_2(CH_2)_5CCl_2CCl_2N=PCl_3$	140–142 (234)		
$Cl_3P=NCCl_2(CH_2)_6CCl_2CCl_2N=PCl_3$			
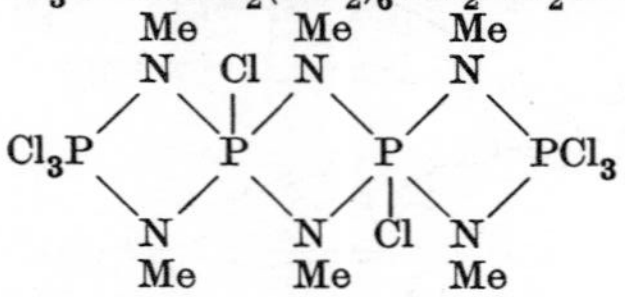 (bicyclic phosphazene structure with P, N, Cl, Me, and PCl₃ groups)	~395 (dec) (18)	+74.5 (17,18)	cryst. struct. (261a)

C. Alkyl- and Arylphosphazo Trifluorides

TABLE X

Compound	Fp or bp/Torr (°C)	31P NMR (85% H$_3$PO$_4$) δ_P (ppm)	Remarks
(MeN=PF$_3$)$_2$	-11 (80) -10.3 (40,41) -8.3 (216) 87.9/760 (40) 88/760 (36,80) 88–90/760 (136) 89–92/760 (63) 91.6/757 (216)	$+69.5$ (41) $+71.5$ (216) $+70.3$ (36) ~73 (92a)	IR (41,265); ^{1}H NMR (41); $J_{PH} = 14.5$ Hz; 14.3 Hz (92a); $J_{PF} = 880$ Hz (216); 894 Hz (23); 890 Hz (92a); ^{19}F NMR (36,92a); $n_D^{20} = 1.3388$ (63); $n_D^{25} = 1.3323$ (216); $d_{20} = 1.5554$ (63); $d_{25} = 1.532$ (216); complete NMR investigation (85); vapor pressure (41,80); heat of fusion etc. (80); electron diffr. (3)
(MeNP)$_2$F$_5$Cl	117–119/760 (257)		IR; $n_D^{20} = 1.3766$ (257); mass sp. (255)
(MeNP)$_2$F$_4$Cl$_2$	155–157/760 (257)		IR; $n_D^{20} = 1.4162$ (257); mass sp. (255)
(MeNP)$_2$F$_3$Cl$_3$	181–183/760 (257)		IR; $n_D^{20} = 1.4526$ (257); mass sp. (255)
(MeNP)$_2$F$_2$Cl$_4$	64–66 (257) 68–70 (32a)	$+59.3$ (32a)	IR (257); mass sp. (255); ^{1}H, ^{19}F NMR (32a)
(MeNP)$_2$FCl$_5$	130–134 (257)		IR (257); mass sp. (255)
(EtN=PF$_3$)$_2$	123–124/760 (63)		$n_D^{20} = 1.3551$; $d_{20} = 1.4030$ (63) IR; ^{19}F NMR;
(PrN=PF$_3$)$_2$	4–7 41/4 } (84) 70–72/21 162–163/760 (63)		$J_{PF} = +897.2$ Hz (84); $n_D^{20} = 1.3750$; $d_{20} = 1.3312$ (63)

(continued overleaf)

TABLE X (continued)

Compound	Fp or bp/Torr (°C)	^{31}P NNR (85% H_3PO_4) δ_P (ppm)	Remarks
$(BuN{=}PF_3)_2$	81–82/12 (63)		$n_D^{20} = 1.3854$; $d_{20} = 1.2580$ (63)
$(i\text{-}BuN{=}PF_3)_2$	71–73/12		$n_D^{20} = 1.3848$; $d_{20} = 1.2573$ (63)
$(Me_3CN{=}PF_3)_2$	186–188/760 (63)		^{19}F NMR (84)
	70–72		
$(PhN{=}PF_3)_2$	subl. 70/4 (84)		$J_{PF} = +918.2$ Hz (84);
	194–204 (83)		
	198–205 (84)		
	subl. 110–130/0.4 (84)		mass sp. (255)
$(PhN{=}PFCl_2)_2$	186 (22)		
$(2,4\text{-}Me_2C_6H_3N{=}PF_3)_2$	164–168 (84)		IR; ^{19}F NMR; $J_{PF} = +914.0$ Hz (84)
	subl. 140–150/0.4 (84)		
$(2,6\text{-}Me_2C_6H_3N{=}PF_3)_2$	255–257 (84)		IR; ^{19}F NMR; $J_{PF} = +900$ Hz (84)
	subl. 130/0.05 (84)		
$(2,6\text{-}Et_2C_6H_3N{=}PF_3)_2$	151–152		IR; ^{19}F NMR (84)
	subl. 125/0.025 (84)		

358

VIII. Miscellaneous Compounds

TABLE XI

Compound	Fp or bp/Torr (°C)	^{31}P NMR (85% H_3PO_4) δ_P (ppm)	Remarks
$\left[\text{MeN}\overset{+}{\underset{-}{\diagdown}}\substack{\text{PCl}_3 \\ \text{AlCl}_3}\right]\cdot\text{HCl}$	195–198 (259a)	-62.7 ± 1 (259a)	
$\text{MeN}\overset{+}{\underset{-}{\diagdown}}\substack{\text{PCl}_3 \\ \text{BCl}_3}$	171 (34) subl. 125/0.1	-55.5 (34)	$\delta_B = -7.5$ (34)
$\text{MeN}\overset{+}{\underset{-}{\diagdown}}\substack{\text{PCl}_3 \\ \text{BCl}_2\text{F}}$	143 (34)	-54.6 (34)	$\delta_B = -5.6$; ^{19}F NMR (34)
$\text{PhN}\overset{+}{\underset{-}{\diagdown}}\substack{\text{PCl}_3 \\ \text{BCl}_3}$	85–88 subl. 75/0.1 (34)	-54.0 (34)	$\delta_B = -6.1$ (34)
$\text{FSO}_2\text{N}{=}\text{S(O)FN}{=}\text{PCl}_3$	106/0.01 (201)	-28.5 (201)	IR; ^{19}F NMR; $J_{PF} = 6.5$ Hz (201)
benzisothiazole–$\text{N}{=}\text{PCl}_3$	165–169 (45)		

(continued overleaf)

TABLE XI (continued)

Compound	Fp or bp/Torr (°C)	^{31}P NMR (85% H_3PO_4) δ_P (ppm)	Remarks
PhCONHN=PCl$_3$	145–148 (161)		
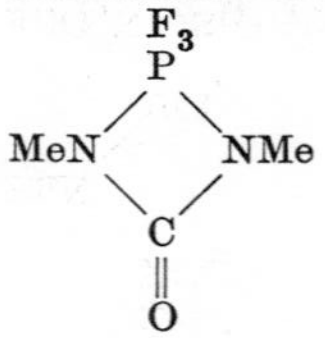 (F$_3$P–NMe ring, C=O)	39/20 (64)	+55.8 (64)	^{19}F NMR (64,167a); J_{PF} = 959 Hz (64); mass sp. (5)
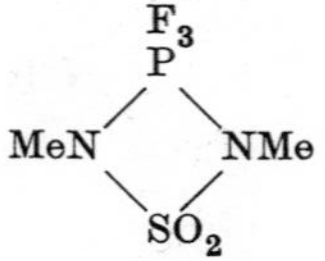 (F$_3$P–NMe ring, SO$_2$)	54 (22,23)	+76.8 (23)	^{1}H NMR; ^{19}F NMR (23); J_{PH} = 15.0 Hz (23), J_{PF} = 968 Hz (22,23)
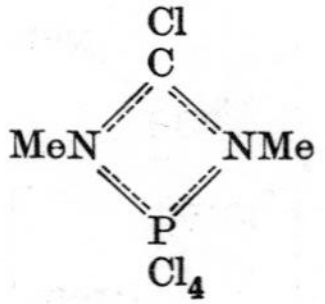 (Cl–NMe ring, C, Cl$_4$)	129 (147)	+202 ± 1 (147,148)	^{1}H NMR; J_{PH} = 20 ± 1 Hz (147); cryst. structure (281)
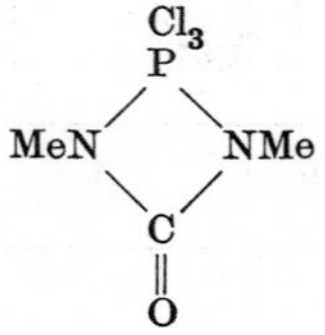 (Cl$_3$P–NMe ring, C=O)	74/0.8 (143) 78–79/1.5 (253,254)	+60.05 ± 0.5 (64,92,145) +59.4 (29)	^{1}H NMR; J_{PH} = 20–21 Hz (29,64,254) mass sp. (5)

Structure		
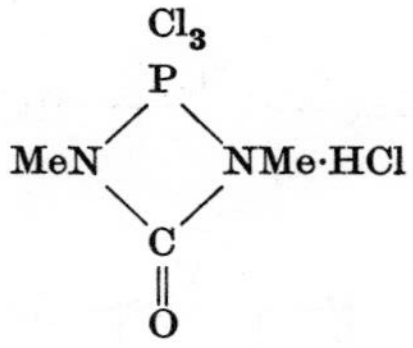 Cl_3P ring with NMe·HCl, MeN, C=O	183 (dec) (144) $\quad$ −50 (144)	
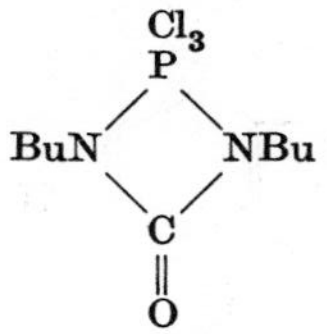 Cl_3P ring with NBu, BuN, C=O	105–108/0.3 (253,254)	^{1}H NMR; J_{PH} = 20 Hz (254)
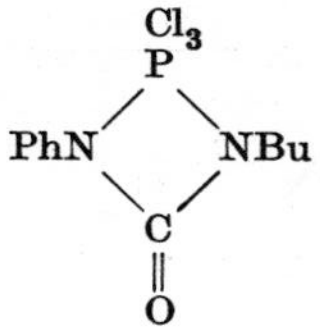 Cl_3P ring with NBu, PhN, C=O	oil (253,254)	^{1}H NMR; J_{PH} = 34 Hz (254)
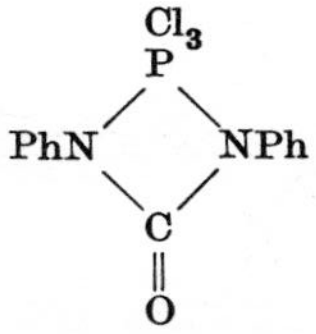 Cl_3P ring with NPh, PhN, C=O	88–90 (49)	

361

(continued overleaf)

TABLE XI (*continued*)

Compound	Fp or bp/Torr (°C)	^{31}P NMR (85% H_3PO_4) δ_P (ppm)	Remarks
$1\text{-}C_{10}H_7N$ ... $\overset{Cl_3}{P}$ ... $NC_{10}H_7\text{-}1$, $C=O$	110–112 (49)		
Ph–C(=N)–NH–P(Cl$_3$)–O (ring)	(162)		not isolated (162)
Ph–C(=N)–N=P(Cl$_3$)–O (ring)	(162)		not isolated (162)
R-substituted pyridine, $N=PCl_3$			R=H, Cl, Br, I, NO$_2$, not isolated (209)
3,6-dichloropyridine, $N=PCl_3$			not isolated (209)
Br, Me-substituted pyridine, $N=PCl_3$			not isolated (209)

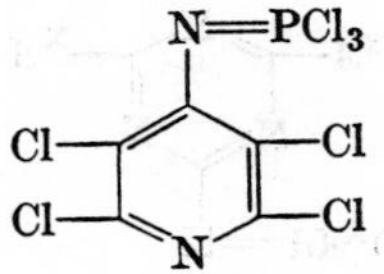

86–87 (176)

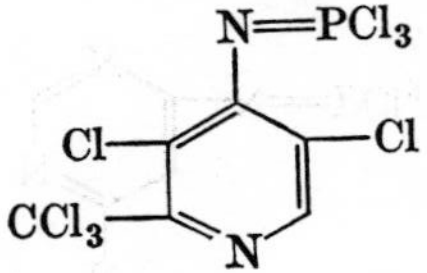

68–69 (176)

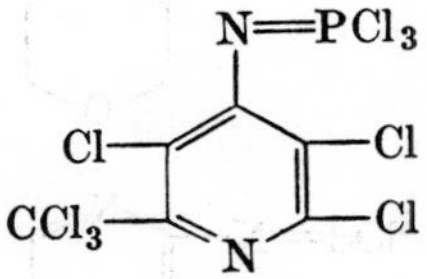

107–108 (176)

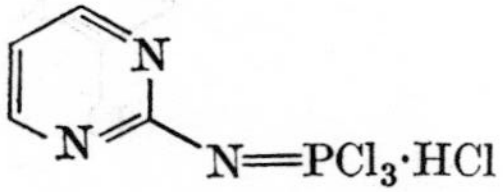

167–169 (140)

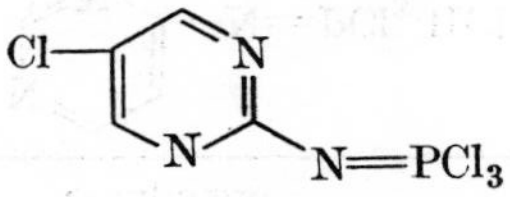

97–100 (140)

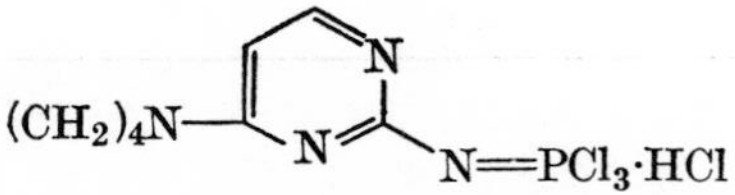

220 (dec) (140)

N=PCl₃ ...

363

(continued overleaf)

TABLE XI (*continued*)

Compound	Fp or bp/Torr (°C)	^{31}P NMR (85% H_3PO_4) δ_P (ppm)	Remarks
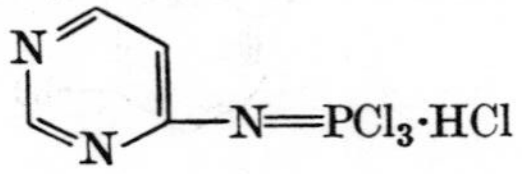 ⬡ N=PCl₃·HCl	140–144 (208)		
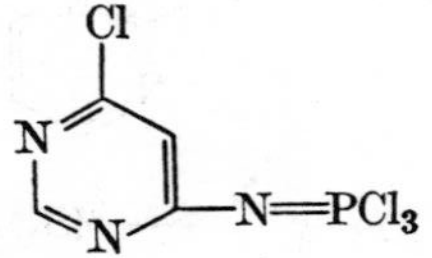 ⬡ Cl N=PCl₃	90–95 (208)		
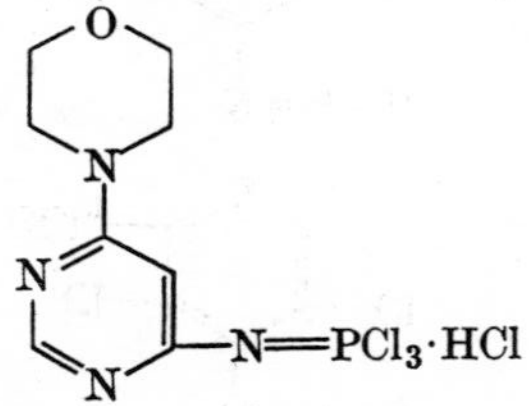 ⬡ O-morpholine N=PCl₃·HCl	239–241 (208)		
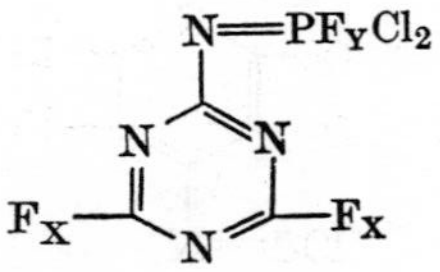 ⬡ triazine N=PF$_Y$Cl$_2$, F$_X$	43/0.01 (199)	+37.5 (199)	IR; ^{19}F NMR; $J_{PF_Y} = 1082$ Hz (199)

Structure		
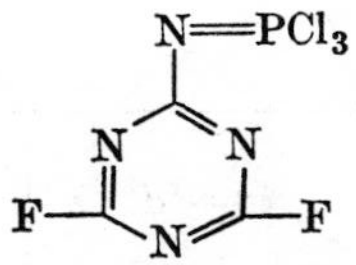 (triazine, 2,4-difluoro-6-($N=PCl_3$)-1,3,5-triazine)	59–60/0.01 (192)	IR (192)
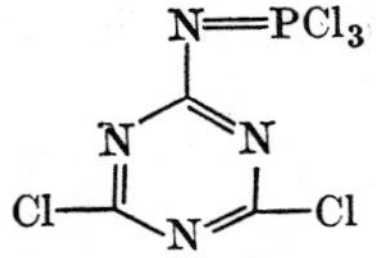 (pyrimidine, 4,6-dichloro-2-($N=PCl_3$)-pyrimidine)	108–109 (43)	
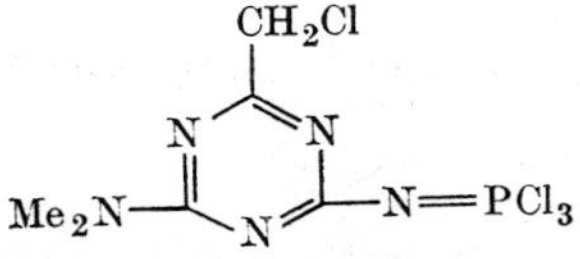 (triazine with CH_2Cl, Me_2N, $N=PCl_3$)	180–181 (149a)	
$Cl_2(O)P_BN=C-N=P_ACl_3$ (with Cl)	oil (13)	$P_A = -23.8$ $P_B = -0.1$ (13)
$Cl_3P=N-N=PCl_3$	134 (dec) (24)	
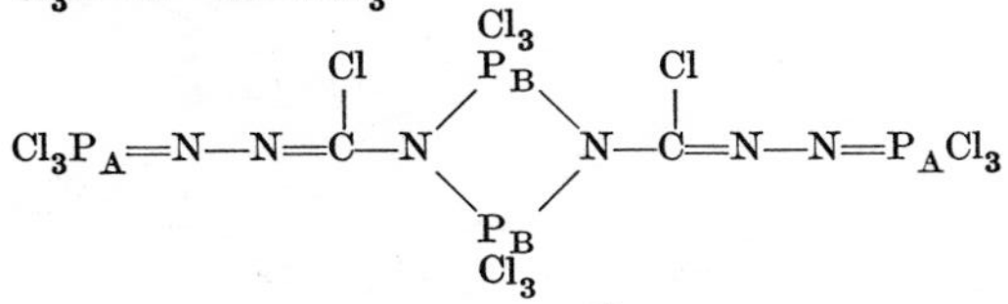 (cyclic structure: $Cl-C=N-N=P_ACl_3$; Cl_3P_B, P_BCl_3)	no details (25)	$P_B = +78$ $P_A = -15$ (25,148)	IR (25)
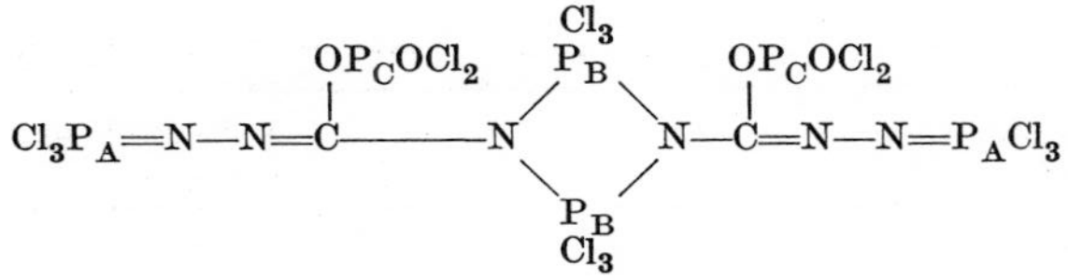 (cyclic structure: OP_COCl_2, Cl_3P_B, $C-N-N=P_ACl_3$; $Cl_3P_A=N-N=C$, OP_COCl_2)	no details (25)	$P_B = +78$ $P_A = -15$ $P_C = -7.6$ (25,148)	IR (25)

(continued overleaf)

TABLE XI (continued)

Compound	Fp or bp/Torr (°C)	^{31}P NMR (85% H_3PO_4) δ_P (ppm)	Remarks
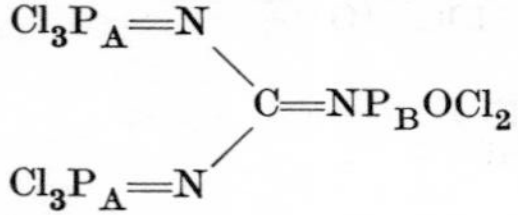 $Cl_3P_A=N$—C=NP_BOCl_2	93–94 (13)	$P_B = -0.6$ (13,148) $P_A = -14.7$	
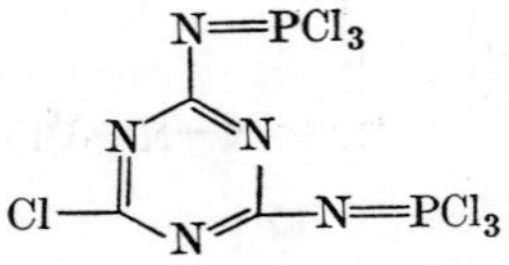 $Cl_3P_A=N$ (triazine, $N=PCl_3$, Cl)	110–114 (42)		
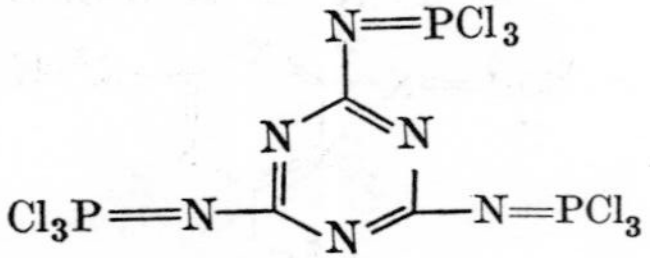 $Cl_3P=N$ (triazine, $N=PCl_3$, $N=PCl_3$)	113–115 (42)		
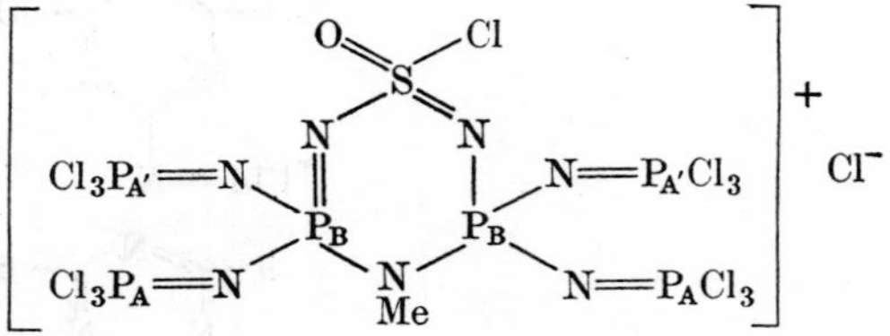 (cyclic structure with $Cl_3P_A=N$, $Cl_3P_{A'}=N$, $N=P_AC l_3$, $N=P_AC l_3$, Me, P_B, S, O, Cl) Cl^-	75 (31)	$P_A = +0.7$ $P_{A'} = -3.4$ (31) $P_B = +22.1$	el. cond. (31)

366

REFERENCES

1. Agahigian, H., G. D. Vickers, J. Roscoe, and J. Bishop, *J. Chem. Phys.*, **39**, 1621 (1963).
2. Aleksankin, M. M., L. I. Samarai, and G. I. Derkach, *Zh. Obshch. Khim.*, **35**, 923 (1965); *Chem. Abstr.* **63**, 6813 (1965).
3. Almenningen, A., B. Andersen, and E. E. Astrup, *Acta Chem. Scand.*, **23**, 2179 (1969).
4. Appel, R. and G. Büchler, *Z. Anorg. Allgem. Chem.*, **320**, 3 (1963).
5. Baldwin, M. A., A. G. Loudon, A. Maccoll, R. E. Dunmur, R. Schmutzler, and I. K. Gregor, *Org. Mass Spectrom.*, **2**, 765 (1969).
6. Baumgärtner, R., W. Sawodny, and J. Goubeau, *Z. Anorg. Allgem. Chem.*, **340**, 246 (1965).
7. Becke-Goehring, M., *Angew. Chem.*, **73**, 246 (1961).
8. Becke-Goehring, M., *Fortschr. Chem. Forsch.*, **10**, 207 (1968); *Chem. Ztg.*, *Chem. App.*, **94**, 179 (1970).
9. Becke-Goehring, M., A. Debo, E. Fluck, and W. Goetze, *Chem. Ber.*, **94**, 1383 (1961).
10. Becke-Goehring, M., E. Fluck, and W. Lehr, *Z. Naturforsch.*, **17B**, 126 (1962).
11. Becke-Goehring, M., W. Gehrmann, and W. Goetze, *Z. Anorg. Allgem. Chem.*, **326**, 127 (1963).
12. Becke-Goehring, M., W. Haubold, and H. P. Latscha, *Z. Anorg. Allgem. Chem.*, **333**, 120 (1964).
13. Becke-Goehring, M. and D. Jung, *Z. Anorg. Allgem. Chem.*, **372**, 233 (1970).
14. Becke-Goehring, M. and W. Lehr, *Chem. Ber.*, **94**, 1591 (1961).
15. Becke-Goehring, M. and W. Lehr, *Z. Anorg. Allgem. Chem.*, **325**, 287 (1963).
16. Becke-Goehring, M. and W. Lehr, *Z. Anorg. Allgem. Chem.*, **327**, 128 (1964).
17. Becke-Goehring, M. and L. Leichner, *Angew. Chem.*, **76**, 686 (1964); *Angew. Chem. Intern. Ed. (Engl.)*, **3**, 590 (1964).
18. Becke-Goehring, M., L. Leichner, and B. Scharf, *Z. Anorg. Allgem. Chem.*, **343**, 154 (1966).
19. Becke-Goehring, M., T. Mann, and H. D. Euler, *Chem. Ber.*, **94**, 193 (1961).
20. Becke-Goehring, M. and H. J. Müller, *Z. Anorg. Allgem. Chem.*, **362**, 51 (1968).
21. Becke-Goehring, M. and A. Slawisch, *Z. Anorg. Allgem. Chem.*, **346**, 295 (1966).
22. Becke-Goehring, M., H. J. Wald, and H. Weber, *Naturwissenschaften*, **55**, 491 (1968).
23. Becke-Goehring, M. and H. Weber, *Z. Anorg. Allgem. Chem.*, **365**, 185 (1969).
24. Becke-Goehring, M. and W. Weber, *Z. Anorg. Allgem. Chem.*, **333**, 128 (1964).
25. Becke-Goehring, M. and W. Weber, *Z. Anorg. Allgem. Chem.*, **339**, 281 (1965).

26. Becke-Goehring, M. and M. R. Wolf, *Naturwissenschaften*, **55**, 543 (1968); *Z. Anorg. Allgem. Chem.*, **373**, 245 (1970).

27. Bermann, M., in *Advances in Inorganic Chemistry and Radiochemistry*, H. J. Emeléus and A. G. Sharpe, Eds., Academic, London, New York, Vol. 14, 1972, p. 1 ff.

28. Bermann, M. and K. Utvary, *Monatsh. Chem.*, **100**, 1041 (1969).

29. Bermann, M. and J. R. Van Wazer, unpublished results.

30. Bieber, T. I. and B. Kane, *J. Org. Chem.*, **21**, 1198 (1956).

31. Bieller, U. and M. Becke-Goehring, *Z. Anorg. Allgem. Chem.*, **380**, 314 (1971).

32. Bieniek, T., G. Burnie, and J. M. Maselli, U.S. Pat. 3,443,913 (1969) (to W. R. Grace & Co.); *Chem. Abstr.*, **71**, 39655 (1969).

32a. Binder, H., *Z. Anorg. Allgem. Chem.*, **384**, 193 (1971).

33. Binder, H. and E. Fluck, *Z. Anorg. Allgem. Chem.*, **381**, 21 (1971).

34. Binder, H. and E. Fluck, *Z. Anorg. Allgem. Chem.*, **381**, 116 (1971).

35. Binder, H. and E. Fluck, *Z. Anorg. Allgem. Chem.*, **381**, 123 (1971).

36. Binder, H. and E. Fluck, *Z. Anorg. Allgem. Chem.*, **382**, 27 (1971).

37. Bodnarchuk, N. D., V. V. Malovik, and G. I. Derkach, *Zh. Obshch. Khim.*, **39**, 168 (1969); *Chem. Abstr.*, **70**, 96857 (1969).

37a. Burness, D. M., R. A. Silverman, and C. J. Wright, Brit. Pat. 1,198,534 (1970) (to Eastman Kodak Co.); *Chem. Abstr.*, **73**, 104312 (1970); U.S. Pat. 3,567,587 (1971).

38. Chapman, A. C., W. S. Holmes, N. L. Paddock, and H. T. Searle, *J. Chem. Soc.*, **1961**, 1825.

38a. Clipsham, R. M., Thesis, McGill University, Montreal, Que. (1970).

38b. Clipsham, R. M. and M. A. Whitehead, *J. Chem. Soc., Faraday Trans. 2*, **1972**, 72.

39. Comte, F., U.S. Pats. 2,775,609; 2,775,610; 2,775,611 (1956) (to Monsanto Chem. Corp.); *Chem. Abstr.*, **51**, 6690 (1957).

40. Demitras, G. C., R. A. Kent, and A. G. MacDiarmid, *Chem. Ind. (London)*, **1964**, 1712.

41. Demitras, G. C. and A. G. MacDiarmid, *Inorg. Chem.*, **6**, 1903 (1967).

42. Derkach, G. I., USSR Pat. 158,880 (1963); *Chem. Abstr.*, **60**, 14395 (1964).

43. Derkach, G. I., M. I. Bukovskii, S. N. Solodushenkov, and A. I. Mosiichuk, *Khim. Org. Soedin. Fosfora, Akad. Nauk SSSR, Otd. Obshch. Tekh. Khim.*, **1967**, 89; *Chem. Abstr.*, **69**, 36082 (1968).

44. Derkach, G. I., G. F. Dregval, and A. V. Kirsanov, *Zh. Obshch. Khim.*, **30**, 3402 (1960); *Chem. Abstr.*, **55**, 21029 (1961).

45. Derkach, G. I., G. F. Dregval, and A. V. Kirsanov, *Zh. Obshch. Khim.*, **32**, 1878 (1962); *Chem. Abstr.*, **58**, 4534 (1963).

46. Derkach, G. I., E. S. Gubnitskaya, V. A. Shokol, and A. A. Kisilenko, *Zh. Obshch. Khim.*, **34**, 82 (1964); *Chem. Abstr.*, **60**, 11502 (1964).

46a. Derkach, G. I., Zh. M. Ivanova, and E. A. Stukalo, USSR Pat. 239,327 (1969); *Chem. Abstr.*, **71**, 50211 (1969).

47. Derkach, G. I. and M. V. Kolotilo, *Zh. Obshch. Khim.*, **36**, 1437 (1966); *Chem. Abstr.*, **66**, 2292 (1967).

48. Derkach, G. I. and A. V. Narbut, USSR Pat. 181,107 (1966); *Chem. Abstr.*, **65**, 8769 (1966).

49. Derkach, G. I. and A. V. Narbut, *Zh. Obshch. Khim.*, **35**, 932 (1965); *Chem. Abstr.*, **63**, 6984 (1965).

50. Derkach, G. I. and V. P. Rudavskii, *Metody Poluch. Khim. Reaktivov Prep.*, **19**, 23 (1969); *Chem. Abstr.*, **74**, 141207 (1971).
51. Derkach, G. I. and V. P. Rudavskii, *Probl. Organ. Sinteza, Akad. Nauk SSSR, Otd. Obshch. Tekh. Khim.*, **1965**, 268; *Chem. Abstr.*, **64**, 12577 (1966).
52. Derkach, G. I. and V. P. Rudavskii, *Zh. Obshch. Khim.*, **35**, 1202 (1965); *Chem. Abstr.*, **63**, 11398 (1965).
53. Derkach, G. I. and V. P. Rudavskii, *Zh. Obshch. Khim.*, **35**, 2200 (1965); *Chem. Abstr.*, **64**, 12580 (1966).
54. Derkach, G. I. and V. P. Rudavskii, *Zh. Obshch. Khim.*, **37**, 1893 (1967); *Chem. Abstr.*, **68**, 21489 (1968).
55. Derkach, G. I., V. P. Rudavskii, and G. F. Dregval, *Zh. Obshch. Khim.*, **34**, 3959 (1964); *Chem. Abstr.*, **62**, 9046 (1965).
56. Derkach, G. I., V. P. Rudavskii, and B. F. Malichenko, *Khim. Organ. Soedin. Fosfora, Akad. Nauk SSSR, Otd. Obshch. Tekh. Khim.* **1967**, 75; *Chem. Abstr.*, **69**, 10169 (1968).
57. Derkach, G. I. and L. I. Samarai, *Zh. Obshch. Khim.*, **34**, 1161 (1964); *Chem. Abstr.*, **61**, 1784 (1964).
58. Derkach, G. I. and E. I. Slyusarenko, *Zh. Obshch. Khim.*, **35**, 532 (1965); *Chem. Abstr.*, **63**, 531 (1965).
59. Derkach, G. I., I. N. Zhmurova, A. V. Kirsanov, V. I. Shevchenko, and A. S. Shtepanek, *Fosfazo Soedineniya*, Izd. Nauk. Dumka, Kiev, 1965.
60. Desai, V. B., R. A. Shaw, and B. C. Smith, *Angew. Chem.*, **80**, 910 (1968); *Angew. Chem. Intern. Ed. (Engl.)*, **7**, 887 (1968).
61. Douglas, W. M., M. Cooke, M. Lustig, and J. K. Ruff, *Inorg. Nucl. Chem. Lett.*, **6**, 409 (1970).
62. Drach, B. S. and I. N. Zhmurova, *Metody Poluch. Khim. Reaktivov Prep.*, **19**, 63 (1969); *Chem. Abstr.*, **75**, 5144 (1971).
63. Drach, B. S. and I. N. Zhmurova, *Zh. Obshch. Khim.*, **37**, 892 (1967); *Chem. Abstr.*, **68**, 2533 (1968); *Metody Poluch. Khim. Reaktivov Prep.*, **19**, 65 (1969); *Chem. Abstr.*, **75**, 140164 (1971).
64. Dunmur, R. E. and R. Schmutzler, *J. Chem. Soc. (A)*, **1971**, 1289.
64a. Dyadyusha, G. G., D. P. Khomenko, and E. S. Kozlov, *Zh. Strukt. Khim.*, **13**, 155 (1972).
64b. Emsley, J., J. Moore, and P. B. Udy, *J. Chem. Soc. (A)*, **1971**, 2863; J. Emsley and P. B. Udy, Ger. Pat. Appl. (DOS) 2,117,055 (1972) (to Castrol Ltd.); *Chem. Abstr.*, **76**, 101801 (1972).
64c. Emsley, J. and P. B. Udy, *J. Chem. Soc. (A)*, **1970**, 3025.
65. England, D. C., R. V. Lindsay, Jr., and L. R. Melby, *J. Amer. Chem. Soc.*, **80**, 6442 (1958).
66. Feistel, G. R. and T. Moeller, *J. Inorg. Nucl. Chem.*, **29**, 2731 (1967).
67. Fleig, H. and M. Becke-Goehring, *Z. Anorg. Allgem. Chem.*, **376**, 215 (1970).
68. Fluck, E., *Chem. Ber.*, **94**, 1388 (1961).
69. Fluck, E., *Z. Anorg. Allgem. Chem.*, **315**, 181 (1962).
70. Fluck, E., *Z. Anorg. Allgem. Chem.*, **320**, 64 (1963).
71. Fluck, E., *Z. Naturforsch.*, **20B**, 505 (1965).
72. Fluck, E. and F. L. Goldmann, *Z. Anorg. Allgem. Chem.*, **356**, 307 (1968).
73. Fluck, E. and G. Heckmann, *Z. Naturforsch.*, **24B**, 953 (1969).
74. Gadekar, S. M. and E. Ross, *J. Org. Chem.*, **26**, 606 (1961).

75. Glemser, O., U. Biermann, and S. P. v. Halasz, *Inorg. Nucl. Chem. Lett.*, **5**, 501 (1969).
76. Glemser, O., U. Biermann, and S. P. v. Halasz, *Inorg. Nucl. Chem. Lett.*, **5**, 643 (1969).
77. Glemser, O., H. W. Roesky, and P. R. Heinze, *Inorg. Nucl. Chem. Lett.*, **4**, 179 (1968).
77a. Graves, G. E., D. W. McKennon, and M. Lustig, *Inorg. Chem.*, **10**, 2083 (1971).
78. Green, M., R. N. Haszeldine, and G. S. A. Hopkins, *J. Chem. Soc. (A)*, **1966**, 1766; Brit. Pat. 1,185,462 (1970) (to National Research Development Corp.); *Chem. Abstr.*, **72**, 122182 (1970); U.S. Pat. 3,531,422 (1970).
79. Gutmann, V., K. Utvary, and M. Bermann, *Monatsh. Chem.*, **97**, 1745 (1966).
80. Haasemann, P., Thesis, Technische Hochschule Stuttgart (1963).
80a. Hagemann, H., Ger. Pat. Appl. (DOS) 1,803,784 (1970) (to Farbenfabriken Bayer A. G.); *Chem. Abstr.*, **73**, 34838 (1971).
81. Haiduc, I., *The Chemistry of Inorganic Ring Systems*, Interscience, London, New York, Sydney, Toronto, Vol. 2, 1970, p. 624 ff.
82. v. Halasz, S. P., Thesis, Göttingen (1969); quoted in E. Niecke, O. Glemser, and H. Thamm, *Chem. Ber.* **103**, 2864 (1970).
83. Harris, J. J., U.S. Pat. 3,304,160 (1967) (to Koppers Co. Inc.); *Chem. Abstr.*, **66**, 85852 (1967).
84. Harris, J. J. and B. Rudner, *Abstr. Papers, 147th ACS Meeting*, Philadelphia, Pa., 25 L (1964); *J. Org. Chem.*, **33**, 1392 (1968).
85. Harris, R. K. and C. M. Woodman, *Mol. Phys.*, **10**, 437 (1966).
86. Hart, R. M. and M. A. Whitehead, *J. Chem. Soc. (A)*, **1971**, 1738.
87. Hart, R. M. and M. A. Whitehead, *Mol. Phys.*, **19**, 383 (1970).
88. Haubold, W. and M. Becke-Goehring, *Z. Anorg. Allgem. Chem.*, **352**, 113 (1967).
89. Hess, H. and D. Forst, *Z. Anorg. Allgem. Chem.*, **342**, 240 (1966).
90. Hoard, L. G. and R. A. Jacobson, *J. Chem. Soc. (A)*, **1966**, 1203.
91. Hormuth, P. B. and M. Becke-Goehring, *Z. Anorg. Allgem. Chem.*, **372**, 280 (1970).
92. Hormuth, P. B. and H. P. Latscha, *Z. Anorg. Allgem. Chem.*, **365**, 26 (1969).
92a. Horn, H. G., *Chem. Ztg., Chem. App.*, **95**, 849 (1971).
93. Horn, H. G. and M. Becke-Goehring, *Z. Anorg. Allgem. Chem.*, **367**, 165 (1969).
94. John, K. and T. Moeller, *J. Inorg. Nucl. Chem.*, **22**, 199 (1961).
95. Kabachnik, M. I., V. A. Gilyarov, Cheng-Te Chang, and E. I. Matrosov, *Izv. Akad. Nauk SSSR, Otd. Khim. Nauk*, **1962**, 1589; *Chem. Abstr.*, **58**, 6673 (1963).
96. Kahler, E. J., U.S. Pat. 2,925,320 (1960) (to Pennsalt Chem. Corp.); *Chem. Abstr.*, **54**, 17820 (1960).
97. Kaplansky, M., R. Clipsham, and M. A. Whitehead, *J. Chem. Soc. (A)*, **1969**, 584.
98. Khomenko, D. P., G. G. Dyadyusha, and E. S. Kozlov, *Spectroscop. Lett.*, **1**, 245 (1968); *Zh. Strukt. Khim.*, **11**, 660 (1970); *Chem. Abstr.*, **74**, 8002 (1971).
99. Khomenko, D. P., E. S. Kozlov, and G. G. Dyadyusha, *Spectroscop. Lett.*, **3**, 129 (1970).

100. Kirsanov, A. V., *Izv. Akad. Nauk SSSR, Otd. Khim. Nauk*, **1950**, 426; *Chem. Abstr.*, **45**, 1503 (1951).
101. Kirsanov, A. V., *Izv. Akad. Nauk SSSR, Otd. Khim. Nauk*, **1954**, 646; *Chem. Abstr.*, **49**, 13161 (1955).
102. Kirsanov, A. V., *Metody Poluch. Khim. Reaktivov Prep.*, **19**, 5 (1969); *Chem. Abstr.*, **74**, 106704 (1971).
103. Kirsanov, A. V., *Metody Poluch. Khim. Reaktivov Prep.*, **19**, 7 (1969); *Chem. Abstr.*, **74**, 141165 (1971).
104. Kirsanov, A. V., *Sborn. Stat. Obshch. Khim.*, **2**, 1046 (1953); *Chem. Abstr.* **49**, 3051 (1955).
105. Kirsanov, A. V., *Zh. Obshch. Khim.*, **22**, 88 (1952); *Chem. Abstr.*, **46**, 6984 (1952).
106. Kirsanov, A. V., *Zh. Obshch. Khim.*, **22**, 269 (1952); *Chem. Abstr.*, **46**, 11135 (1952).
107. Kirsanov, A. V., *Zh. Obshch. Khim.*, **22**, 1346 (1952); *Chem. Abstr.*, **47**, 5836 (1953); *Metody Poluch. Khim. Reaktivov Prep.*, **19**, 9 (1969); *Chem. Abstr.*, **74**, 106703 (1971).
108. Kirsanov, A. V. and E. A. Abrazhanova, *Sborn. Stat. Obshch. Khim.*, **2**, 865 (1953); *Chem. Abstr.*, **49**, 6820 (1955).
109. Kirsanov, A. V., N. D. Bodnarchuk, and V. I. Shevchenko, *Dopov. Akad. Nauk Ukr. RSR*, **1963**, 221; *Chem. Abstr.*, **59**, 12666 (1963).
110. Kirsanov, A. V. and G. I. Derkach, *Zh. Obshch. Khim.*, **26**, 2009 (1956); *Chem. Abstr.*, **51**, 1821 (1957).
111. Kirsanov, A. V. and G. I. Derkach, *Zh. Obshch. Khim.*, **27**, 3248 (1957); *Chem. Abstr.*, **52**, 8997 (1958).
112. Kirsanov, A. V. and G. I. Derkach, *Zh. Obshch. Khim.*, **28**, 1887 (1958); *Chem. Abstr.*, **53**, 1208 (1959).
113. Kirsanov, A. V. and N. L. Egorova, *Zh. Obshch. Khim.*, **25**, 187 (1955); *Chem. Abstr.*, **50**, 1647 (1956).
114. Kirsanov, A. V. and N. G. Feshchenko, *Zh. Obshch. Khim.*, **27**, 2817 (1957); *Chem. Abstr.*, **52**, 8069 (1958).
115. Kirsanov, A. V. and N. G. Feshchenko, *Zh. Obshch. Khim.*, **30**, 3389 (1960); *Chem. Abstr.*, **55**, 18112 (1961).
116. Kirsanov, A. V. and N. A. Kirsanova, *Zh. Obshch. Khim.*, **29**, 1802 (1959); *Chem. Abstr.*, **54**, 8693 (1960).
117. Kirsanov, A. V. and R. G. Makitra, *Zh. Obshch. Khim.*, **26**, 907 (1956); *Chem. Abstr.*, **50**, 14633 (1956).
118. Kirsanov, A. V. and M. S. Marenets, *Zh. Obshch. Khim.*, **29**, 2256 (1959); *Chem. Abstr.*, **54**, 10855 (1960).
119. Kirsanov, A. V. and M. S. Marenets, *Zh. Obshch. Khim.*, **31**, 1605 (1961); *Chem. Abstr.*, **55**, 23339 (1961).
120. Kirsanov, A. V. and V. P. Molosnova, *Zh. Obshch. Khim.*, **25**, 772 (1955); *Chem. Abstr.*, **50**, 2416 (1956).
121. Kirsanov, A. V. and V. P. Molosnova, *Zh. Obshch. Khim.*, **28**, 30 (1958); *Chem. Abstr.*, **52**, 12760 (1958).
122. Kirsanov, A. V. and V. P. Molosnova, *Zh. Obshch. Khim.*, **28**, 347 (1958); *Chem. Abstr.*, **52**, 13663 (1958).
123. Kirsanov, A. V. and Z. D. Nekrasova, *Zh. Obshch. Khim.*, **27**, 1253 (1957); *Chem. Abstr.*, **52**, 3666 (1958).
124. Kirsanov, A. V. and Z. D. Nekrasova, *Zh. Obshch. Khim.*, **28**, 1595 (1958); *Chem. Abstr.*, **53**, 1206 (1959).

125. Kirsanov, A. V. and V. I. Shevchenko, *Zh. Obshch. Khim.*, **24**, 474 (1954); *Chem. Abstr.*, **49**, 6164 (1955).
126. Kirsanov, A. V., V. I. Shevchenko, and N. D. Bodnarchuk, USSR Pat. 162,843 (1964); *Chem. Abstr.*, **61**, 10592 (1964).
127. Kirsanov, A. V. and I. N. Zhmurova, *Zh. Obshch. Khim.*, **28**, 2478 (1958); *Chem. Abstr.*, **53**, 3118 (1959).
128. Kirsanov, A. V. and Yu. M. Zolotov, *Zh. Obshch. Khim.*, **24**, 122 (1954); *Chem. Abstr.*, **49**, 3052 (1955).
129. Klement, R. and E. Rother, *Naturwissenschaften*, **45**, 489 (1958); M. Baudler, R. Klement, and E. Rother, *Chem. Ber.*, **93**, 149 (1960).
130. Kongpricha, S. and W. C. Preusse, *Abstr. Papers, 148th ACS Meeting*, Chicago, Ill., 3 K (1964); S. Kongpricha, U.S. Pat. 3,445,513 (1969) (to Olin Mathieson Chem. Corp.); *Chem. Abstr.*, **71**, 50218 (1969).
131. Kongpricha, S. and W. C. Preusse, *Inorg. Chem.*, **6**, 1915 (1967).
132. Kornuta, P. P., A. I. Kalenskaya, and V. I. Shevchenko, *Zh. Obshch. Khim.*, **41**, 2390 (1971).
133. Kornuta, P. P. and V. I. Shevchenko, *Zh. Obshch. Khim.*, **40**, 788 (1970); *Chem. Abstr.*, **73**, 34707 (1970).
134. Kornuta, P. P., V. I. Shevchenko, and A. V. Kirsanov, *Zh. Obshch. Khim.*, **37**, 2788 (1967); *Chem. Abstr.*, **68**, 113971 (1968).
135. Kosinskaya, I. M., A. M. Pinchuk, and V. I. Shevchenko, *Zh. Obshch. Khim.*, **41**, 105 (1971); *Chem. Abstr.*, **75**, 35088 (1971).
135a. Kosinskaya, I. M., A. M. Pinchuk, and V. I. Shevchenko, *Zh. Obshch. Khim.*, **41**, 2396 (1971); *Chem. Abstr.*, **76**, 112592 (1972).
136. Kozlov, E. S. and B. S. Drach, *Zh. Obshch. Khim.*, **36**, 760 (1966); *Chem. Abstr.*, **65**, 8742 (1966); *Metody Poluch. Khim. Reaktivov Prep.*, **19**, 70 (1969); *Chem. Abstr.*, **75**, 88033 (1971).
137. Kozlov, E. S., S. N. Gaidamaka, and A. V. Kirsanov, *Zh. Obshch. Khim.*, **40**, 991 (1970); *Chem. Abstr.*, **73**, 77332 (1970).
138. Kozlov, E. S., A. A. Kisilenko, A. I. Sedlov, and A. V. Kirsanov, *Zh. Obshch. Khim.*, **37**, 1611 (1967); *Chem. Abstr.*, **68**, 48835 (1968).
139. Kropacheva, A. A., G. I. Derkach, and A. V. Kirsanov, *Zh. Obshch. Khim.*, **31**, 1601 (1961); *Chem. Abstr.*, **55**, 22274 (1961).
140. Kropacheva, A. A. and N. V. Sazonov, *Puti Sin. Izyskaniya Protivoopukholevykh Prep.*, *2nd Tr. Symp.*, *Moscow*, **1965**, 94 (1967); *Chem. Abstr.*, **70**, 11660 (1969); *Khim. Geterosikl. Soedin.*, *Sb. 1:* "Azotsoderzhashchie Geterosikly", S. Hillers, Ed., Izd. Zinatne, Riga, 1967, p. 372; *Chem. Abstr.*, **70**, 106453 (1969).
140a. Kukhar, V. P., V. Ya. Semenii, and A. V. Kirsanov, *Zh. Obshch. Khim.*, **41**, 1459 (1971); *Chem. Abstr.*, **75**, 140418 (1971).
140b. Kukhar, V. P., V. Ya. Semenii, and A. V. Kirsanov, *Zh. Obshch. Khim.*, **42**, 98 (1972).
141. Kukhar, V. P., V. Ya. Semenii, and N. P. Pisanenko, *Zh. Obshch. Khim.*, **40**, 557 (1970); *Chem. Abstr.*, **73**, 14368 (1970).
141a. Kulibaba, N. K., V. I. Shevchenko, and A. V. Kirsanov, *Zh. Obshch. Khim.*, **41**, 2105 (1971); *Chem. Abstr.*, **76**, 24605 (1972).
142. Lapidot, A. and D. Samuel, *J. Chem. Soc.*, **1962**, 2110.
143. Latscha, H. P., *Z. Anorg. Allgem. Chem.*, **346**, 166 (1966).
144. Latscha, H. P., *Z. Anorg. Allgem. Chem.*, **355**, 73 (1967).
145. Latscha, H. P., *Z. Naturforsch.*, **23B**, 139 (1968).

146. Latscha, H. P., W. Haubold, and M. Becke-Goehring, Z. Anorg. Allgem. Chem., **339**, 82 (1965).
147. Latscha, H. P. and P. B. Hormuth, Angew. Chem., **80**, 281 (1968); Angew. Chem. Intern. Ed. (Engl.), **7**, 299 (1968).
148. Latscha, H. P., P. B. Hormuth, and H. Vollmer, Z. Naturforsch., **24B**, 1237 (1969).
149. Latscha, H. P., W. Weber, and M. Becke-Goehring, Z. Anorg. Allgem. Chem., **367**, 40 (1969); but see E. Fluck and W. Steck, Z. Anorg. Allgem. Chem., **387**, 349 (1972).
149a. Lazukina, L. A., N. G. Kotlyar, V. P. Kukhar, and S. N. Solodushenkov, Zh. Obshch. Khim., **41**, 2386 (1971).
150. Lehr, W., Z. Anorg. Allgem. Chem., **350**, 18 (1967).
151. Lehr, W. and J. Pietschmann, Chem. Ztg., Chem. App., **94**, 362 (1970).
152. Levchenko, E. S. and L. V. Budnik, Zh. Org. Khim., **6**, 2239 (1970); Chem. Abstr., **74**, 42065 (1971).
153. Levchenko, E. S. and A. V. Kirsanov, Zh. Obshch. Khim., **27**, 3078 (1957); Chem. Abstr., **52**, 8070 (1958).
154. Levchenko, E. S. and A. V. Kirsanov, Zh. Obshch. Khim., **29**, 1813 (1959); Chem. Abstr., **54**, 8694 (1960).
155. Levchenko, E. S., I. E. Sheinkman, and A. V. Kirsanov, Zh. Obshch. Khim., **30**, 1941 (1960); Chem. Abstr., **55**, 6426 (1961).
156. Levchenko, E. S., I. E. Sheinkman, and A. V. Kirsanov, Zh. Obshch. Khim., **33**, 3315 (1963); Chem. Abstr., **60**, 4043 (1964).
157. Levchenko, E. S., I. N. Zhmurova, and A. V. Kirsanov, Zh. Obshch. Khim., **29**, 2262 (1959); Chem. Abstr., **54**, 10926 (1960).
158. Lustig, M., Inorg. Chem., **8**, 443 (1969).
159. Mao, T. J., R. D. Dresdner, and J. A. Young, J. Inorg. Nucl. Chem., **24**, 53 (1962).
160. Markovskii, L. N., G. S. Fedyuk, and E. S. Levchenko, Zh. Org. Khim., in press 1972; quoted in E. S. Levchenko and L. N. Markovskii, Usp. Khim. Fosfororgan. Seraorgan. Soedin., Izd. Nauk. Dumka, Kiev, No 2, p. 181 (1970), loc. cit. p. 199.
161. Matyushecheva, G. I., A. V. Narbut, G. I. Derkach, and L. M. Yagupol'skii, Zh. Org. Khim., **3**, 2254 (1967); Chem. Abstr., **68**, 59236 (1968).
162. Mikhailov, G. I., G. I. Matyushecheva, G. I. Derkach, and L. M. Yagupol'skii, Zh. Org. Khim., **6**, 149 (1970); Chem. Abstr., **72**, 90373 (1970).
163. Mochalina, E. P., B. L. Dyatkin, and I. L. Knunyants, Izv. Akad. Nauk SSSR, Ser. Khim., **1966**, 2247; Chem. Abstr., **66**, 76089 (1967).
164. Moeller, T. and A. Vandi, J. Org. Chem., **27**, 3511 (1962).
165. Moeller, T. and A. H. Westlake, J. Inorg. Nucl. Chem., **29**, 957 (1967).
166. Moran, E. F., J. Inorg. Nucl. Chem., **30**, 1405 (1968).
167. Murch, R. M. and D. C. De Vore, Canad. Pat. 884,204 (1971) (to W. R. Grace & Co.); Ger. Pat. Appl. (DOS) 1,802,016 (1969); Chem. Abstr., **71**, 50677 (1969).
167a. Murray, M. and R. Schmutzler, Z. Chem., **8**, 241 (1968).
168. Nannelli, P., A. Failli, and T. Moeller, Inorg. Chem., **4**, 558 (1965).
169. Nichols, G. M., ASD Tech. Rept. 61–2 Contract No. AF 33(616)-7158 Project No. 8128 (1961); ASD Tech. Rept. 61–2 Part II Contract No. AF 33(616)-7158 Project No. 1(8-7340) (1962).

170. Nichols, G. M., U.S. Pat. 3,449,091 (1969) (to E. I. du Pont de Nemours & Co.); *Chem. Abstr.*, **71**, 31905 (1969); see also U.S. Pat. 3,249,397 (1966); *Chem. Abstr.*, **65**, 1831 (1966).

170a. Niedenzu, K., I. A. Boenig, and E. B. Bradley, *Z. Anorg. Allgem. Chem.*, in press (1972).

171. Niedenzu, K. and G. Magin, *Z. Naturforsch.*, **20B**, 604 (1965).

172. Nielsen, M. L. and J. V. Pustinger, Jr., *J. Chem. Phys.*, **68**, 152 (1964).

173. Nielsen, M. L., J. V. Pustinger, Jr., and J. Strobel, *J. Chem. Eng. Data*, **9**, 169 (1964).

174. Norris, W. P. and H. B. Jonassen, *J. Org. Chem.*, **27**, 1449 (1962).

175. Oyamada, K. and S. Morimura, *Takamine Kenkyusho Nempo*, **12**, 41 (1960); *Chem. Abstr.*, **55**, 6459 (1961).

176. Pavlenko, A. F., V. P. Akkerman, G. A. Zalesskii, and Ya. N. Ivashchenko, *Zh. Obshch. Khim.*, **39**, 1516 (1969); *Chem. Abstr.*, **71**, 112762 (1969).

177. Pinchuk, A. M. and I. M. Kosinskaya, *Zh. Obshch. Khim.*, **40**, 546 (1970); *Chem. Abstr.*, **73**, 35295 (1970).

178. Pinchuk, A. M., I. M. Kosinskaya, and V. I. Shevchenko, *Zh. Obshch. Khim.*, **37**, 856 (1967); *Chem. Abstr.*, **67**, 108169 (1967).

179. Pinchuk, A. M., I. M. Kosinskaya, and V. I. Shevchenko, *Zh. Obshch. Khim.*, **37**, 2693 (1967); *Chem. Abstr.*, **69**, 67069 (1968).

180. Pinchuk, A. M., I. M. Kosinskaya, and V. I. Shevchenko, *Zh. Obshch. Khim.*, **39**, 583 (1969); *Chem. Abstr.*, **71**, 61300 (1969).

181. Pinchuk, A. M., L. N. Markovskii, E. S. Levchenko, and V. I. Shevchenko, *Zh. Obshch. Khim.*, **37**, 852 (1967); *Chem. Abstr.*, **67**, 108407 (1967).

182. Prons, V. N., M. P. Grinblat, and A. L. Klebanskii, *Zh. Obshch. Khim.*, **40**, 589 (1970); *Chem. Abstr.*, **73**, 25595 (1970).

183. Protsenko, L. D., G. I. Derkach, and A. V. Kirsanov, *Zh. Obshch. Khim.*, **31**, 3433 (1961); *Chem. Abstr.*, **57**, 3355 (1962).

184. Protsenko, L. D. and K. A. Kornev, *Ukr. Khim. Zh.*, **24**, 636 (1958); *Chem. Abstr.*, **53**, 12266 (1959).

185. Raetz, R. F. W. and E. H. Kober, U.S. Pat. 2,981,734 (1961) (to Olin Mathieson Chem. Corp.); *Chem. Abstr.*, **56**, 10170 (1962).

186. Rice, R. G., R. M. Murch, and D. C. De Vore, U.S. Pat. 3,545,942 (1970) (to W. R. Grace & Co.); *Chem. Abstr.*, **74**, 54398 (1971); Ger. Pat. Appl. (DOS) 1,801,945 (1969); *Chem. Abstr.*, **71**, 39654 (1969).

187. Rodionov, V. M. and E. V. Yavorskaya, *Zh. Obshch. Khim.*, **18**, 110 (1948); *Chem. Abstr.*, **42**, 4976 (1948).

188. Roesky, H. W., personal communication.

189. Roesky, H. W., *Chem. Ber.*, **101**, 3679 (1968).

190. Roesky, H. W., *Inorg. Nucl. Chem. Lett.*, **6**, 807 (1970).

191. Roesky, H. W., *Z. Anorg. Allgem. Chem.*, **367**, 151 (1969).

192. Roesky, H. W. and H. H. Giere, *Inorg. Nucl. Chem. Lett.*, **4**, 639 (1968).

193. Roesky, H. W. and L. F. Grimm, *Angew. Chem.*, **82**, 255 (1970); *Angew. Chem. Intern. Ed. (Engl.)*, **9**, 244 (1970).

194. Roesky, H. W. and L. F. Grimm, *Chem. Ber.*, **102**, 2319 (1969).

195. Roesky, H. W. and L. F. Grimm, *Chem. Ber.*, **103**, 1664 (1970).

196. Roesky, H. W. and L. F. Grimm, *Inorg. Nucl. Chem. Lett.*, **5**, 13 (1969).

196a. Roesky, H. W., L. F. Grimm, and E. Niecke, *Z. Anorg. Allgem. Chem.*, **385**, 102 (1971).

197. Roesky, H. W. and W. Grosse Böwing, *Inorg. Nucl. Chem. Lett.*, **5**, 597 (1969).

198. Roesky, H. W. and W. Grosse Böwing, *Inorg. Nucl. Chem. Lett.*, **6**, 781 (1970).
199. Roesky, H. W. and W. Grosse Böwing, *Z. Naturforsch.*, **24B**, 1250 (1969).
200. Roesky, H. W., W. Grosse Böwing, and E. Niecke, *Chem. Ber.*, **104**, 653 (1971).
201. Roesky, H. W. and G. Holtschneider, *Z. Anorg. Allgem. Chem.*, **378**, 168 (1970).
202. Roesky, H. W., G. Holtschneider, and H. H. Giere, *Z. Naturforsch.*, **25B**, 252 (1970).
203. Roesky, H. W. and E. Niecke, *Inorg. Nucl. Chem. Lett.*, **4**, 463 (1968).
204. Roesky, H. W. and E. Niecke, *Z. Naturforsch.*, **24B**, 1101 (1969).
205. Roesky, H. W. and G. Remmers, *Z. Naturforsch.*, **26B**, 75 (1971).
205a. Rudavskii, V. P. and V. I. Kondratenko, *Zh. Obshch. Khim.*, **41**, 2151 (1971); *Chem. Abstr.*, **76**, 72149 (1972).
205b. Rudavskii, V. P., N. A. Litoshenko, and E. P. Babin, *Khim. Prom. Ukr.*, **1970**, 46; *Chem. Abstr.*, **73**, 14054 (1970).
206. Rudavskii, V. P., N. A. Litoshenko, and V. P. Kukhar, *Zh. Obshch. Khim.*, **40**, 1002 (1970); *Chem. Abstr.*, **73**, 76813 (1970).
206a. Rudavskii, V. P., D. M. Zagnibeda, and V. I. Kondratenko, *Farm. Zh. (Kiev)*, **26**, 14 (1971); *Chem. Abstr.*, **76**, 3496 (1972).
207. Ruff, J. K., *Inorg. Chem.*, **6**, 2108 (1967).
208. Sazonov, N. V. and A. A. Kropacheva, *Khim. Geterosikl. Soed.*, *Sb. 1: "Azotsoderzhashchie Geterosikly"*, S. Hillers, Ed., Izd. Zinatne, Riga, 1967, p. 377; *Chem. Abstr.*, **70**, 87725 (1969).
209. Sazonov, N. V. and A. A. Kropacheva, *Khim. Geterosikl. Soed.*, **1970**, 55; *Chem. Abstr.*, **72**, 111393 (1970).
210. Schmidpeter, A. and K. Düll, *Chem. Ber.*, **100**, 1116 (1967).
211. Schmidpeter, A., K. Düll, and R. Böhm, *Z. Anorg. Allgem. Chem.*, **362**, 58 (1968).
212. Schmidpeter, A., J. Ebeling, and H. Groeger, *Chem. Ber.*, in preparation.
213. Schmidpeter, A. and N. Schindler, *Chem. Ber.*, **102**, 2201 (1969).
214. Schmidpeter, A. and C. Weingand, personal communication; A. Schmidpeter and K. Schumann, *Z. Naturforsch.*, **25B**, 1364 (1970).
215. Schmidpeter, A. and C. Weingand, *Angew. Chem.*, **81**, 573 (1969); *Angew. Chem. Intern. Ed. (Engl.)*, **8**, 615 (1969).
216. Schmutzler, R., *Chem. Commun.*, **1965**, 19.
217. Seglin, L., M. R. Lutz, and H. Stange, U.S. Pat. 3,231,327 (1966) (to FMC Corp.); *Chem. Abstr.*, **64**, 10810 (1966).
218. Sherif, F. G., *J. Inorg. Nucl. Chem.*, **30**, 1707 (1968).
219. Shevchenko, V. I. and N. D. Bodnarchuk, *Zh. Obshch. Khim.*, **36**, 1645 (1966); *Chem. Abstr.*, **66**, 54962 (1967).
220. Shevchenko, V. I., N. D. Bodnarchuk, and A. V. Kirsanov, *Zh. Obshch. Khim.*, **33**, 1342 (1963); *Chem. Abstr.*, **59**, 11239 (1963).
221. Shevchenko, V. I., N. D. Bodnarchuk, and A. V. Kirsanov, *Zh. Obshch. Khim.*, **33**, 1591 (1963); *Chem. Abstr.*, **59**, 12839 (1963).
222. Shevchenko, V. I., M. El'Dik, and A. M. Pinchuk, *Zh. Obshch. Khim.*, **38**, 1527 (1968); *Chem. Abstr.*, **69**, 95848 (1968).
223. Shevchenko, V. I., M. El'Dik, and A. M. Pinchuk, *Zh. Obshch. Khim.*, **40**, 1949 (1970); *Chem. Abstr.*, **74**, 141201 (1971).
224. Shevchenko, V. I. and A. V. Kirsanov, *Usp. Khim. Fosfororg. Seraorg.*

Soedin., Izd. Nauk. Dumka, Kiev, No. 1, 44 (1969); *Chem. Abstr.* **72,** 110394 (1970).

225. Shevchenko, V. I. and P. P. Kornuta, *Zh. Obshch. Khim.*, **36,** 1254 (1966); *Chem. Abstr.*, **65,** 15381 (1966).

226. Shevchenko, V. I. and P. P. Kornuta, *Zh. Obshch. Khim.*, **36,** 1642 (1966); *Chem. Abstr.*, **66,** 65032 (1967).

227. Shevchenko, V. I., P. P. Kornuta, and N. D. Bodnarchuk, *Metody Poluch. Khim. Reaktivov Prep.*, **19,** 89 (1969); *Chem. Abstr.*, **74,** 140802 (1971).

228. Shevchenko, V. I., P. P. Kornuta, and N. D. Bodnarchuk, *Metody Poluch. Khim. Reaktivov Prep.*, **19,** 96 (1969); *Chem. Abstr.*, **74,** 141989 (1971).

229. Shevchenko, V. I., P. P. Kornuta, and N. D. Bodnarchuk, *Metody Poluch. Khim. Reaktivov Prep.*, **19,** 98 (1969); *Chem. Abstr.*, **74,** 140793 (1971).

230. Shevchenko, V. I., P. P. Kornuta, N. D. Bodnarchuk, and A. V. Kirsanov, *Zh. Obshch. Khim.*, **36,** 730 (1966); *Chem. Abstr.*, **65,** 8912 (1966).

231. Shevchenko, V. I., P. P. Kornuta, and A. V. Kirsanov, *Zh. Obshch. Khim.*, **35,** 1598 (1965); *Chem. Abstr.*, **63,** 17891 (1965).

232. Shevchenko, V. I., P. P. Kornuta, and A. V. Kirsanov, *Zh. Obshch. Khim.*, **35,** 1970 (1965); *Chem. Abstr.*, **64,** 8027 (1966).

233. Shevchenko, V. I. and V. P. Kukhar, *Zh. Obshch. Khim.*, **36,** 735 (1966); *Chem. Abstr.*, **65,** 8858 (1966).

234. Shevchenko, V. I. and V. P. Kukhar, *Zh. Obshch. Khim.*, **36,** 1260 (1966); *Chem. Abstr.*, **65,** 16845 (1966); *Metody Poluch. Khim. Reaktivov Prep.*, **19,** 102 (1969); *Chem. Abstr.*, **75,** 76046 (1971).

235. Shevchenko, V. I., V. P. Kukhar, and A. V. Kirsanov, *Zh. Obshch. Khim.*, **36,** 467 (1966); *Chem. Abstr.*, **65,** 615 (1966).

236. Shevchenko, V. I., V. P. Kukhar, A. A. Koval, and A. V. Kirsanov, *Zh. Obshch. Khim.*, **38,** 1270 (1968); *Chem. Abstr.*, **69,** 106391 (1968).

237. Shevchenko, V. I., N. K. Kulibaba, and A. V. Kirsanov, *Zh. Obshch. Khim.*, **38,** 326 (1968); *Chem. Abstr.*, **69,** 18498 (1968).

238. Shevchenko, V. I., N. K. Kulibaba, and A. V. Kirsanov, *Zh. Obshch. Khim.*, **38,** 850 (1968); *Chem. Abstr.*, **69,** 58906 (1968).

239. Shevchenko, V. I., N. K. Kulibaba, and A. V. Kirsanov, *Zh. Obshch. Khim.*, **39,** 1689 (1969); *Chem. Abstr.*, **72,** 3132 (1970).

239a. Shevchenko, V. I. and N. R. Litovchenko, *Zh. Obshch. Khim.*, **41,** 1243 (1971); *Chem. Abstr.*, **75,** 88422 (1971).

240. Shevchenko, V. I., N. R. Litovchenko, and V. P. Kukhar, *Zh. Obshch. Khim.*, **40,** 1229 (1970); *Chem. Abstr.*, **74,** 76249 (1971).

241. Shevchenko, V. I. and E. E. Nizhnikova, *Zh. Obshch. Khim.*, **40,** 1219 (1970); *Chem. Abstr.*, **74,** 31632 (1971).

242. Shevchenko, V. I., E. E. Nizhnikova, N. D. Bodnarchuk, and P. P. Kornuta, *Metody Poluch. Khim. Reaktivov Prep.*, **19,** 82 (1969); *Chem. Abstr.*, **74,** 140801 (1971).

243. Shevchenko, V. I., E. E. Nizhnikova, N. D. Bodnarchuk, and P. P. Kornuta, *Metody Poluch. Khim. Reaktivov Prep.*, **19,** 100 (1969); *Chem. Abstr.*, **74,** 140794 (1971).

244. Shevchenko, V. I., E. E. Nizhnikova, N. D. Bodnarchuk, and P. P. Kornuta, *Zh. Obshch. Khim.*, **37,** 1358 (1967); *Chem. Abstr.*, **68,** 39055 (1968).

245. Shevchenko, V. I. and V. I. Stadnik, *Probl. Organ. Sinteza, Akad. Nauk SSSR, Otd. Obshch. Tekh. Khim.*, **1965**, 281; *Chem. Abstr.*, **64**, 4933 (1966).
246. Shevchenko, V. I., V. I. Stadnik, and A. V. Kirsanov, *Probl. Organ. Sinteza, Akad. Nauk SSSR, Otd. Obshch. Tekh. Khim.*, **1965**, 272; *Chem. Abstr.*, **64**, 5071 (1966).
247. Shokol, V. A., G. A. Golik, and G. I. Derkach, *Zh. Obshch. Khim.*, **38**, 871 (1968); *Chem. Abstr.*, **69**, 52211 (1968).
248. Shokol, V. A., G. A. Golik, V. T. Tsyba, and G. I. Derkach, *Zh. Obshch. Khim.*, **40**, 931 (1970); *Chem. Abstr.*, **73**, 35465 (1970).
249. Shokol, V. A., A. A. Kisilenko, and G. I. Derkach, *Zh. Obshch. Khim.*, **39**, 874 (1969); *Chem. Abstr.*, **71**, 43880 (1969).
250. Smolina, A. I., G. N. Tsybul'skaya, V. P. Rudavskii, and G. I. Derkach, *Fiziol. Aktiv. Veshchestva, Akad. Nauk Ukr. SSR, Respub. Mezhvedom. Sb.*, **1966**, 96; *Chem. Abstr.*, **67**, 53861 (1967).
251. Sunagawa, G. and K. Onoyama, Japan. Pat. 9,779 (1960) (Sankyo Co.); *Chem. Abstr.*, **55**, 9348 (1961).
252. Trippett, S., *J. Chem. Soc.*, **1962**, 4731.
253. Ulrich, H. and A. A. R. Sayigh, *Angew. Chem.*, **76**, 647 (1964); *Angew. Chem. Intern. Ed. (Engl.)*, **3**, 585 (1964).
254. Ulrich, H. and A. A. R. Sayigh, *J. Org. Chem.*, **30**, 2779 (1965).
255. Utvary, K., personal communication.
256. Utvary, K. and M. Bermann, *Monatsh. Chem.*, **99**, 2369 (1968).
257. Utvary, K. and W. Czysch, *Monatsh. Chem.*, **100**, 681 (1969).
258. Vandi, A. and T. Moeller, *Inorg. Synth.*, **8**, 116 (1966).
259. Vandi, A. and T. Moeller, *Inorg. Synth.*, **8**, 119 (1966).
259a. Vollmer, H. and M. Becke-Goehring, *Z. Anorg. Allgem. Chem.*, **382**, 281 (1971).
260. Volodin, A. A., V. V. Kireev, G. S. Kolesnikov, and S. S. Titov, *Zh. Obshch. Khim.*, **40**, 2202 (1970); *Chem. Abstr.*, **74**, 141955 (1971).
261. Wald, H. J., Thesis, University of Heidelberg (1968).
261a. Weiss, J. and G. Hartmann, *Z. Anorg. Allgem. Chem.*, **351**, 152 (1967).
262. Whitehead, M. A., *Canad. J. Chem.*, **42**, 1212 (1964).
263. Wiegräbe, W. and H. Bock, *Chem. Ber.*, **101**, 1414 (1968).
264. Yagupol'skii, L. M. and V. I. Troitskaya, *Zh. Obshch. Khim.*, **29**, 552 (1959); *Chem. Abstr.*, **54**, 356 (1960).
265. Yagupsky, M. P., *Inorg. Chem.*, **6**, 1770 (1967).
266. Yakubovich, A. Ya., I. M. Filatova, E. L. Zaitseva, and A. P. Simonov, *Zh. Obshch. Khim.*, **39**, 2213 (1969); *Chem. Abstr.*, **72**, 55579 (1970).
266a. Zhmurova, I. N., *Metody Poluch. Khim. Reaktivov Prep.*, **19**, 57 (1969); *Chem. Abstr.*, **75**, 140401 (1971).
266b. Zhmurova, I. N. and B. S. Drach, *Metody Poluch. Khim. Reaktivov Prep.*, **19**,60 (1969); *Chem. Abstr.*, **75**, 151296 (1971).
267. Zhmurova, I. N. and B. S. Drach, *Metody Poluch. Khim. Reaktivov Prep.*, **19**, 68 (1969); *Chem. Abstr.*, **75**, 19668 (1971).
268. Zhmurova, I. N. and B. S. Drach, *Metody Poluch. Khim. Reaktivov Prep.*, **19**, 72 (1969); *Chem. Abstr.*, **74**, 140792 (1971).
269. Zhmurova, I. N. and B. S. Drach, *Metody Poluch. Khim. Reaktivov Prep.*, **19**, 74 (1969); *Chem. Abstr.*, **74**, 140791 (1971).
270. Zhmurova, I. N. and B. S. Drach, *Metody Poluch. Khim. Reaktivov Prep.*, **19**, 76 (1969); *Chem. Abstr.*, **74**, 140796 (1971).

271. Zhmurova, I. N. and B. S. Drach, *Zh. Obshch. Khim.*, **34**, 1441 (1964); *Chem. Abstr.*, **61**, 5499 (1964).

272. Zhmurova, I. N. and B. S. Drach, *Zh. Obshch. Khim.*, **34**, 3055 (1964); *Chem. Abstr.*, **62**, 14480 (1965).

273. Zhmurova, I. N. and B. S. Drach, *Zh. Obshch. Khim.*, **35**, 718 (1965); *Chem. Abstr.*, **63**, 4149 (1965).

274. Zhmurova, I. N., B. S. Drach, and A. V. Kirsanov, *Ukr. Khim. Zh.*, **31**, 223 (1965); *Chem. Abstr.*, **63**, 1689 (1965).

275. Zhmurova, I. N., B. S. Drach, and A. V. Kirsanov, *Zh. Obshch. Khim.*, **35**, 344 (1965); *Chem. Abstr.*, **62**, 13030 (1965).

276. Zhmurova, I. N. and A. V. Kirsanov, *Zh. Obshch. Khim.*, **30**, 3044 (1960); *Chem. Abstr.*, **55**, 17551 (1961).

277. Zhmurova, I. N. and A. V. Kirsanov, *Zh. Obshch. Khim.*, **32**, 2576 (1962); *Chem. Abstr.*, **58**, 7848 (1963); see also V. A. Shokol, N. K. Mikhailyuchenko, G. I. Derkach, and A. V. Kirsanov, USSR Pat. 271,520 (1970); *Chem. Abstr.*, **73**, 119810 (1970).

278. Zhmurova, I. N. and A. P. Martynyuk, *Khim. Organ. Soedin. Fosfora, Akad. Nauk SSSR, Otd. Obshch. Tekh. Khim.*, **1967**, 191; *Chem. Abstr.*, **69**, 2442 (1968).

279. Zhmurova, I. N. and A. P. Martynyuk, *Zh. Obshch. Khim.*, **37**, 896 (1967); *Chem. Abstr.*, **68**, 29350 (1968).

280. Zhuravleva, L. P. and A. V. Kirsanov, *Zh. Obshch. Khim.*, **32**, 3752 (1962); *Chem. Abstr.*, **58**, 11253 (1963).

281. Ziegler, M. L. and J. Weiss, *Angew. Chem.*, **81**, 430 (1969); *Angew. Chem. Intern. Ed. (Engl.)*, **8**, 455 (1969).

Peroxophosphates

I. I. CREASER and JOHN O. EDWARDS

Metcalf Chemical Laboratories, Brown University, Providence, Rhode Island

CONTENTS

I. Introduction

The peroxophosphates are compounds of considerable potential importance. They are active oxidants and could be useful in oxidation processes for synthesis, in bleaching, and in decontamination. Their preparation can be inexpensive because phosphates are common and the peroxide can be formed inexpensively; for example, the peroxodiphosphates can be made by electrolysis.

The peroxophosphates are stable with respect to intramolecular decomposition. The phosphate group can be reduced only with difficulty, therefore the process does not occur (as it would for peroxo-

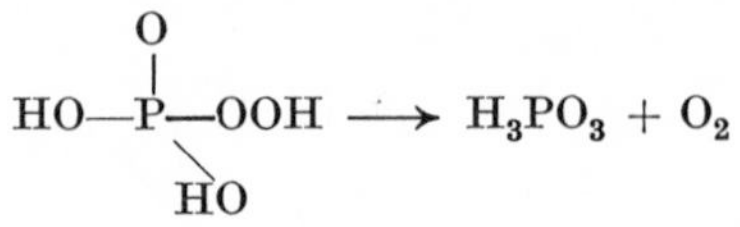

$$\text{HO}-\overset{\overset{\displaystyle O}{\|}}{\underset{\displaystyle \text{HO}}{P}}-\text{OOH} \longrightarrow H_3PO_3 + O_2$$

nitrate or peroxoperchlorate). Also, the phosphorus atom is in its highest oxidation state so that any reaction analogous to the behavior of peroxonitrous acid cannot take place.

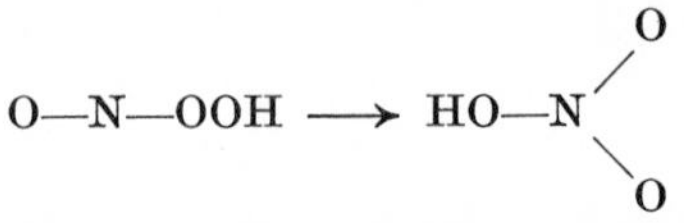

$$O-N-OOH \longrightarrow HO-N\overset{\displaystyle O}{\underset{\displaystyle O}{\big<}}$$

The hydrolysis of a bound peroxophosphate (a derivative of orthophosphoric acid where one oxygen is substituted by a peroxo group)

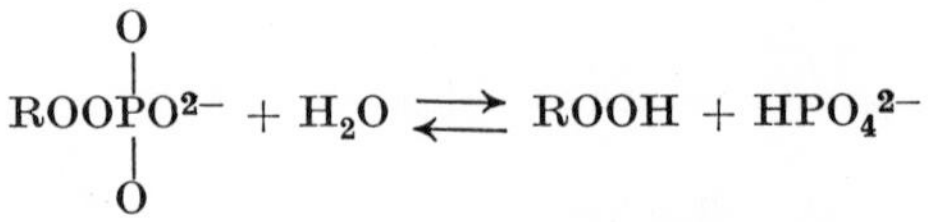

$$\text{ROO}\overset{\overset{\displaystyle O}{\|}}{\underset{\displaystyle O}{P}}O^{2-} + H_2O \rightleftharpoons ROOH + HPO_4^{2-}$$

is a slow process as is also the case for the true peroxosulfates and substituted phosphates. The hydrolysis of many types of phosphates has been studied extensively. For example, the pyrophosphates (1)

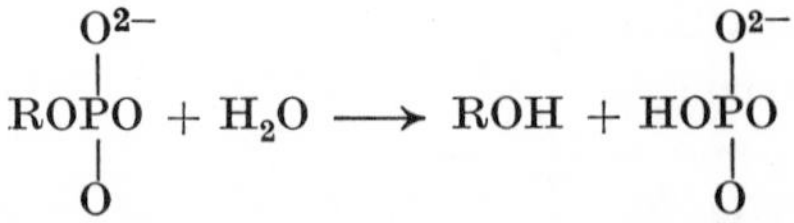

$$\text{RO}\overset{\overset{\displaystyle O^{2-}}{\|}}{\underset{\displaystyle O}{P}}O + H_2O \longrightarrow ROH + HO\overset{\overset{\displaystyle O^{2-}}{\|}}{\underset{\displaystyle O}{P}}O$$

and the higher polyphosphates (2) hydrolyze slowly even in acidic solution. The exchange of oxygen between phosphate ions and water is a type of replacement reaction, and this also takes place slowly (3). Monosubstituted organic phosphates under mild conditions are inert to substitution (4). The unique place of phosphate derivatives in biological systems (such as nucleic acids and adenosine triphosphate) is only possible because of their hydrolytic stability. The fact that replacements in phosphates are slow is of no small importance to the chemistry of the peroxophosphates, also.

Both the cleavage and the formation of the P—O—O link are slow. The bound peroxides, peroxodiphosphate ion,

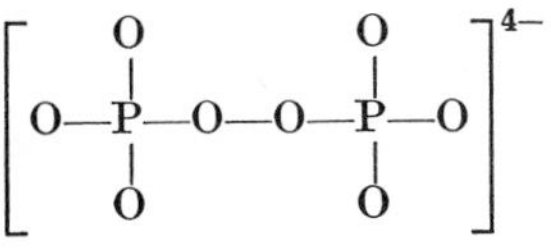

$$\left[\begin{array}{ccc} & O & O \\ & \| & \| \\ O—&P—O—O—P&—O \\ & \| & \| \\ & O & O \end{array} \right]^{4-}$$

and peroxomonophosphoric acid

$$\begin{array}{c} O \\ \| \\ HO—P—OOH \\ \| \\ OH \end{array}$$

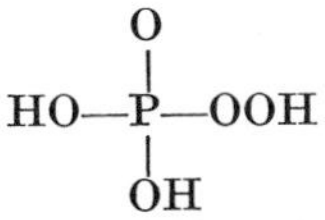

are stable towards hydrolysis under mild conditions and persist in aqueous solution. By way of comparison, the solid hydroperoxidates (i.e., phosphates with hydrogen peroxide of crystallization) such as $Na_4P_2O_7 \cdot 2\ H_2O_2$ dissolve in water to form solutions containing hydrogen peroxide and phosphate ions.

In this chapter the bound peroxides, the hydroperoxidates, and the organoperoxophosphates are discussed. Aspects covered include preparations, structures, and chemical properties including reaction mechanisms.

II. Peroxodiphosphates

A. Discovery and Preparations*

The first true peroxophosphates were prepared in 1910 (5). (All phosphates containing peroxide prepared before that time appear to have been phosphate hydroperoxidates.) By adding a small amount of 30% hydrogen peroxide to sirupy pyrophosphoric acid, Schmidlin and Massini obtained a mixture of hydrogen peroxide, peroxomonophosphoric acid, and peroxodiphosphoric acid in solution. While an excess of pyrophosphoric acid increased the yield of peroxodiphosphoric acid, the main product was always peroxomonophosphoric acid.

* Note added in proof. Three recent patents dealing with the preparation of potassium peroxodiphosphate are noted [P. R. Mucenieks, U.S.P. #3,607,142 (U.S.P. Gazette, 1115, Sept. 21, 1971), U.S.P. #3,616,325 (U.S.P. Gazette, Oct. 26, 1971), and German patent #1,809,799 (corresponding to U.S.P. #3,616,325)]. Proton nuclear resonance spectra of some peroxo salts in the solid state at 90°K have been investigated [T. M. Connor and R. E. Richards, J. Chem. Soc., 289 (1958)]. The conclusion that sodium pyrophosphate peroxides contain the peroxide as hydrogen peroxide is in agreement with other evidence that these compounds are hydroperoxidates.

Yields of peroxodiphosphoric acid up to 40% were claimed. The peroxodiphosphoric acid was characterized by the darkening of an aqueous solution of aniline and by the slow liberation of iodine from a hydrogen iodide solution. These reactions are analogous to those of peroxodisulfuric acid and the compound was therefore assigned the formula $H_4P_2O_8$. Attempts to prepare peroxodiphosphoric acid by electrolysis of meta- or pyrophosphoric acid or their salts failed. Also attempts to make peroxodiphosphoric acid from ozone and phosphorus pentoxide analogous to the formation of peroxodisulfuric acid from sulfur trioxide and ozone were in vain. These results were confirmed by D'Ans and Friederich (6).

Fichter and Müller (7) succeeded in preparing mixtures of an unstable peroxomonophosphate and a stable peroxodiphosphate in solution by electrolysis of a dipotassium phosphate solution at low temperature and low anodic current density using platinum electrodes. In order to avoid reduction of the products at the cathode a trace of potassium chromate was added to the electrolyte while a large quantity ($4N$) of potassium fluoride was added in order to raise the potential of oxygen molecule formation at the anode. A yield of about 30% of peroxodiphosphate was reported. Similar results were obtained with ammonium and rubidium salts. The sodium salt gave poor yields presumably because of the low solubility of both sodium phosphate and sodium fluoride at low temperature. Free phosphoric acid gave no measurable peroxodiphosphoric acid by electrolysis. It is important to note that in some electrolysis studies none of the products were isolated as compounds; identification is based solely on solution properties. Fichter and Rius y Miro (8) studied the factors influencing the yields of peroxophosphates prepared by electrolysis such as current density, time of electrolysis, acidity, anode potential, and the presence of fluoride. They were able to isolate solid potassium peroxodiphosphate by evaporation of the electrolyte, and by repeated recrystallizations obtained a product which contained a small amount of fluoride. The analysis confirmed the formula $K_4P_2O_8$ suggested by Schmidlin and Massini (5). None of the typical reactions of hydrogen peroxide with chromic acid, permanganate ion, or titanic acid occurred, whereas an acidified solution of manganous sulfate developed an intense purple color in a few hours and ferrous salt solutions were oxidized by the peroxodiphosphate. It was therefore concluded that the isolated salt was not an addition compound of hydrogen peroxide but a true peroxo compound. It was proposed that the electrolysis proceeds by the dehydration of K_2HPO_4 to $K_4P_2O_7$ since peroxoyphosphates were also produced by electrolysis of pyrophosphates. However more recent studies by Franchuk and Brodskii (9) do not support this hypothesis.

The interaction between elemental fluorine and phosphoric acid and phosphate solution was studied by Fichter and Bladergroen (10), who claimed to have obtained a very small amount of peroxodiphosphoric acid together with some peroxomonophosphoric acid by treating solutions of phosphoric acid with fluorine for 45–90 min. Peroxodiphosphoric acid was supposed to be formed first according to

$$2\ H_3PO_4 + F_2 \rightarrow H_4P_2O_8 + 2\ HF$$

This primary product then underwent a fast hydrolysis in the acid solution to give a new oxidizing agent presumed to be the peroxomonophosphate:

$$H_4P_2O_8 + H_2O \rightarrow H_3PO_5 + H_3PO_4$$

In basic solution the main product was thought to be the peroxomonophosphoric acid which however decomposed

$$2\ H_3PO_5 \rightarrow O_2 + 2\ H_3PO_4$$

while the more stable peroxodiphosphate remained in solution. By utilizing this process, a product was isolated which was assumed to be a mixture of potassium peroxodiphosphate and potassium fluoride. It is possible that instead of peroxophosphates one (or more) oxygen fluoride(s) was formed according to the equations

$$PO_4{}^{3-} + F_2 \longrightarrow O-\overset{\displaystyle O}{\underset{\displaystyle O}{\overset{|}{\underset{|}{P}}}}-O-F^{2-} + F^-$$

and

$$PO_4F^{2-} + H_2O \rightarrow HOF + HPO_4{}^{2-}$$

This would be expected to give the same kind of reactions (oxidizing species in solution and O_2 release).

Several phosphates containing active oxygen have been prepared by the action of hydrogen peroxide. They are described as peroxophosphates in patent literature but are really compounds containing hydrogen peroxide of crystallization. Husain and Partington (11) undertook a thorough investigation of the different methods and pointed out the distinction between true peroxophosphates and hydroperoxidates. They found that true peroxophosphoric acids are obtained by the action of hydrogen peroxide on phosphoric acid while the action of the same reagent on phosphates gives rise to hydroperoxidates.

A claim by Siebold (12) that the action of 30% hydrogen peroxide on phosphorus oxychloride at 0° gives rise to the formation of peroxodiphosphoric acid according to the equation

$$H_2O_2 + 4 H_2O + 2 POCl_3 \rightarrow H_4P_2O_8 + 6 HCl$$

could not be substantiated by Husain and Partington. The later workers obtained a mixture of phosphoric acid and free hydrogen peroxide.

An improved technique for preparing potassium peroxodiphosphate was published by Fichter and Gutzwiller (13). A solution of potassium phosphates, $2 K_2HPO_4 + 3 K_3PO_4$, potassium fluoride, and potassium chromate was electrolyzed for 3 hr at low temperature (<14°). The electrolyte was left overnight at room temperature. The peroxomonophosphate formed in the electrolysis then decomposed with the formation of oxygen and orthophosphate leaving the more stable peroxodiphosphate in solution. By repeating this procedure three times on successive days, a yield of 80% potassium peroxodiphosphate was obtained on evaporation of the electrolyte to crystallization. Upon recrystallization purities of 96.4–99.8% were obtained. A similar procedure produced the very soluble $Rb_4P_2O_8$. The ammonium salt $(NH_4)_4P_2O_8$ was produced in a yield of 97% in solution but all attempts at isolating the compound failed owing to decomposition.

The action of an electric discharge on phosphorus pentoxide vapor, especially when mixed with oxygen, resulted in an oxygen-rich phosphorus compound in 6% yield which exhibited strong oxidizing properties. When this substance was dissolved in water, peroxodiphosphoric acid was formed; for this reason the substance formed by gas discharge was supposed to be P_2O_6 (14). Later investigations (15) indicated the existence of two peroxides, a purple one which is stable only at low temperature and a colorless one with the formula P_4O_{11} which is obtained by heating the purple compound. On hydrolysis the colorless peroxide gives a mixture of pyrophosphoric and peroxodiphosphoric acid.

$$P_4O_{11} + 4 H_2O \rightarrow H_4P_2O_7 + H_4P_2O_8$$

Probably the best laboratory method for preparing potassium and lithium peroxodiphosphates today has been developed by Chulski (16) and by Carlin and Walker (17). It involves electrolysis of the electrolyte described by Fichter and Gutzwiller (13) employing a platinum dish as the anode and a rapidly rotating platinum wire as the cathode. A current density of 0.015 A/cm^2 at the anode and a temperature of 8–12° of the electrolyte were used. The electrolysis was carried out for 3 hr and the electrolyte was left to stand quietly overnight at room temperature. The next day electrolysis was carried out for 2 hr and the electrolyte was again left overnight. It was finally electrolyzed for one hour on the third day. The potassium peroxodiphosphate was isolated after evaporation of the electrolyte to a small volume. On recrystallization a yield of 27% potassium peroxodiphosphate of 98.9% purity was obtained. The salt contains some chromate which cannot be removed by recrystallization. More peroxodiphosphate can be isolated from the electrolyte by adding solid potassium hydroxide. A yield of 42% (having 88% purity) has been claimed by this procedure. The peroxodiphosphate can be purified by converting it to the less soluble lithium salt. The air-dried lithium salt exists as $Li_4P_2O_8 \cdot 4 H_2O$. By adding more potassium hydroxide to the electrolyte after a day, the yield can be improved to 69% (18). Details of a slightly different method useful for small scale preparations are given in Appendix I.

There have been several reinvestigations, published as patents (19), of the electrolytic preparation of peroxodiphosphates from phosphate salts.

B. Known Compounds

The most common peroxodiphosphate is the potassium salt but several others have been prepared. Electrolysis of orthophosphoric acid or of phosphates of lithium, sodium, and thallium does not give rise to peroxophosphates while secondary phosphates of potassium, rubidium, cesium, and ammonium on electrolysis produce two peroxophosphates in solution, namely peroxomonophosphate and peroxodiphosphate (11). In the case of rubidium and cesium, peroxophosphates may be obtained without the addition of fluoride, but potassium and ammonium phosphates do not give good yields except with such addition. Rubidium and cesium peroxodiphosphates have been isolated as the solids but in an impure state due to their high solubility. Both salts are stable in solution.

Barium peroxodiphosphate can be prepared by treating a concentrated solution of the potassium salt with barium chloride whereby $Ba_2P_2O_8 \cdot 4 H_2O$ precipitates. Zinc sulfate solutions give a white microcrystalline precipitate with a potassium peroxodiphosphate solution. The product analyzes as $Zn_2P_2O_8$. A slightly soluble, white, microcrystalline $Pb_2P_2O_8$ is formed similarly with lead nitrate. Magnesium sulfate solutions do not precipitate potassium peroxodiphosphate solutions and this has been suggested as a method of separating peroxodiphosphates from phosphates. When silver nitrate is added to a peroxodiphosphate solution, a brown to black precipitate is formed which may be silver peroxodiphosphate. It decomposes giving off oxygen and turning yellow forming silver phosphate (13). The sodium salt precipitates with varying amounts of lattice water, and the acid salts $K_2H_2P_2O_8$, $Na_2H_2P_2O_8 \cdot 2\ H_2O$, and $(NH_4)_2H_2P_2O_8$ have been prepared. Attempts to prepare secondary salts of the types $MH_3P_2O_8$ and $M_3HP_2O_8$ where M represents a monovalent metal cation have failed. The ammonium salt $(NH_4)_4P_2O_8 \cdot 2\ H_2O$ can be prepared by treating a potassium peroxodiphosphate solution with ammonium perchlorate, filtering off the insoluble $KClO_4$, and then treating the filtrate with methanol (20).

Cohen (21) has prepared a number of peroxodiphosphate salts containing mixed cations. This class of solids has the general formula $M_mN_{4-(m+n)}H_nP_2O_8$ where M and N may be Na^+, K^+, Li^+, or NH_4^+, and where n may be either 0 or 2.

The potassium peroxodiphosphate is the most thoroughly investigated of the peroxodiphosphate salts. It is a white, crystalline solid. The crystal is orthorhombic and the unit cell contains two $P_2O_8^{4-}$ groups. The lattice constants are $a = 11.119$, $b = 7.199$, and $c = 5.968$ Å. The salt is thermally stable at normal temperatures. It does not melt but undergoes an exothermic decomposition starting at 387° with the evolution of oxygen and the formation of pyrophosphate.

$$K_4P_2O_8\ (\text{solid}) \rightarrow K_4P_2O_7 + \tfrac{1}{2}O_2$$

The energy for this reaction is 7.06 kcal/mole. The solubility of the compound in water is 47% over the range 0 to 40° and 51% at 61°. It is insoluble in organic solvents. The heats of solution and neutralization have been determined to be

	ΔH, kcal/mol (22)
$K_4P_2O_8\ (\text{solid}) + H_2O \rightarrow K_4P_2O_8\ (\text{solution})$	4.4
$K_4P_2O_8\ (\text{solution}) + HCl(aq) \rightarrow H_4P_2O_8(aq) + 4\ KCl(aq)$	-1.8
$K_4P_2O_8\ (\text{solid}) + HCl(aq) \rightarrow H_4P_2O_8\ (aq) + 4\ KCl(aq)$	2.6

No X-ray studies have been undertaken to determine the structure of the peroxodiphosphate ion. However from the similarity in reactions its structure has been considered to be similar to that of peroxodisulfate with two PO_4 tetrahedra bonded through an —O—O— bond

$$\overset{\overset{\displaystyle O}{\|}}{\underset{\underset{\displaystyle O^-}{|}}{{}^-O-P}}-O-O-\overset{\overset{\displaystyle O}{\|}}{\underset{\underset{\displaystyle O^-}{|}}{P-O^-}}$$

There is chemical as well as physical evidence for this structure. Upon hydrolysis orthophosphate and peroxomonophosphate are formed.

Assuming cleavage at the P—OO site this would be expected from the proposed structure. The alternative structure

$$-O-\underset{\underset{O_-}{\overset{\overset{O}{\|}}{P}}}{}-O-\underset{\underset{O_-}{\overset{\overset{O}{\|}}{P}}}{}-O-O^-$$

could give pyrophosphate and hydrogen peroxide; pyrophosphate has not been reported as a hydrolysis product. Infrared and Raman studies (23–25) indicate the structure to be angular with C2h or C1 symmetry. The appearance of only one peak in the ^{31}P NMR spectrum also indicates that the two phosphorus atoms are electronically equivalent (18,26).

The dipotassium dihydrogen peroxodiphosphate, $K_2H_2P_2O_8$, a white crystalline solid, is soluble in water to the extent of 37.9% at 0° and 43.2 wt % at 25°. At higher temperatures decomposition occurs. The salt is not soluble in common organic solvents. It is not as thermally stable as the tetrapotassium salt. Decomposition starts at 160°. The mode of decomposition appears to be the loss of oxygen to give pyrophosphate

$$K_2H_2P_2O_8 \rightarrow K_2H_2P_2O_7 + \tfrac{1}{2}O_2$$

C. Acid Ionization and Ion-Pairing Constants

The acid ionization constants of peroxodiphosphoric acid have been measured by titrating the tetramethylammonium salt with acid; starting at low pH is not favored because the acid undergoes a fast hydrolysis in acid solution (27,28). A further complication is that aqueous peroxodiphosphate ions form complex ions with alkali metal ions. The third and fourth acid dissociation constants have been measured at 25° by pH titrations with hydrochloric acid at a series of ionic strengths (Fig. 1). Extrapolation to infinite dilution yields the values $K_3 = 6.6 \pm 0.3 \times 10^{-6}$ and $K_4 = 2.1 \pm 0.1 \times 10^{-8}$. The first and second dissociations are so strong that inflection points were not observed. Also hydrolysis to peroxomonophosphate and oxidation of the chloride ion by the peroxodiphosphate interfered at low pH and only rough estimates for pK_1 and pK_2 could be made. The approximate values of $K_1 \sim 2$ and $K_2 \sim 3 \times 10^{-1}$ were obtained by extrapolation from the corresponding values for hypophosphoric acid, $H_4P_2O_6$, and pyrophosphoric acid, $H_4P_2O_7$.

Venturini, Indelli, and Raspi (29) studied the voltammetric behavior of H_3PO_5 and $H_4P_2O_8$. The dependence of electroreduction rate of

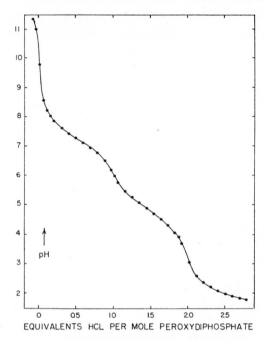

Fig. 1. Titration curve (addition of acid to tetramethylammonium peroxodiphosphate) in order to evaluate ionization constants for the acid in aqueous
solution. Taken from M. M. Crutchfield and J. O. Edwards, *J. Amer Chem.
Soc.*, **82**, 3533 (1960); reprinted by permission.

peroxodiphosphate upon hydrogen ion concentration suggests that for
$0.01M < [H^+] < 0.7M$ only one ionic species is predominant through
all this range of acidity and that the corresponding conjugate acid is the
species undergoing reduction.   Using their data plus results from the
earlier studies (28), it was estimated that the dissociation constant for
$H_3P_2O_8^-$ is at least 3.0 at an ionic strength of 1.0.   The value of K_1
would then be still larger.   Uncertainty remains as to the correct values
for K_1 and K_2.

Formation constants for peroxodiphosphate complexes with Li^+,
Na^+, K^+, and Mg^{2+} were determined (27,28) from the pH shift in the
titration curves caused by the presence of the added metal ions.   The
measured ion-pairing constants (Table I) are similar to those of
pyrophosphate (30) with differences being primarily attributed to the
greater freedom of movement in the P—O—O—P structure than in the
P—O—P structure of $P_2O_7^{4-}$.

It is to be expected that the peroxodiphosphate ion should form
complex ions with transition metal cations.   Unpublished work in our

TABLE I

Formation Constants for Peroxodiphosphate and Pyrophosphate Complex Ions[a]

Complex ion[b]	$K_f(\text{Pr})$[c]	$\log K_f(\text{Pr})$[c]	$\log K_f(\text{Py})$[d]
KP	10.3	1.01	0.80
NaP	10.5	1.02	1.00
NaHP	1.8	0.25	
LiP	22	1.34	2.39
LiHP	5	0.70	1.03
MgP	2160	3.33	5.41
Mg$_2$P	21	1.32	2.34
MgHP	58	1.76	3.06

[a] Data at 25°.

[b] The complex ion symbol represents metal ion plus P (where P can be either $P_2O_8^{4-}$ or $P_2O_7^{4-}$). For example, NaHP stands for $NaHP_2O_8^{2-}$.

[c] Pr stands for $P_2O_8^{4-}$. All measurements at ionic strength adjusted to 1.0 with $(CH_3)_4NCl$. Data from Ref. 28.

[d] Py stands for $P_2O_7^{4-}$. Data of S. M. Lambert and J. I. Watters, *J. Amer. Chem. Soc.*, **79**, 4262, 5606 (1957).

laboratory on Ni^{2+} and VO^{2+} indicates that such is the case. Quantitative and detailed studies are needed.

D. Spectra

The infrared spectra of the lithium, sodium, potassium, strontium, barium, and ammonium salts of peroxodiphosphate have been recorded (23,24). They are distinguished by the absence of any absorption in the P—O—P stretching region in accordance with the structural formula usually assumed for the peroxodiphosphate. The Raman spectra (23,25) together with the IR spectra of the alkali salts exclude all stretched (i.e., linear P—O—O—P) structures and suggest that the peroxodiphosphate ion has an angular structure and occurs in transconfiguration with symmetry C2h or Ci.

In the NMR spectrum of lithium peroxodiphosphate only one peak was observed; this is shifted -7.4 ppm relative to 85% orthophosphoric acid. The single peak indicates that the two phosphorus atoms are in identical electronic environments (18,26).

E. Hydrolysis

The hydrolyses of peroxodiphosphoric acid to peroxomonophosphoric acid and orthophosphoric acid and of peroxomonophosphoric acid to hydrogen peroxide and orthophosphoric acid were observed by early

$$H_4P_2O_8 + H_2O \rightarrow H_3PO_5 + H_3PO_4$$
$$H_3PO_5 + H_2O \rightarrow H_2O_2 + H_3PO_4$$

workers but not studied in detail at that time. Franchuk and Brodskii (9) carried out hydrolysis reactions in ^{18}O-labeled water and showed that oxygen in the hydrogen peroxide comes entirely from the peroxodiphosphoric acid. They proposed the following mechanism for the hydrolysis:

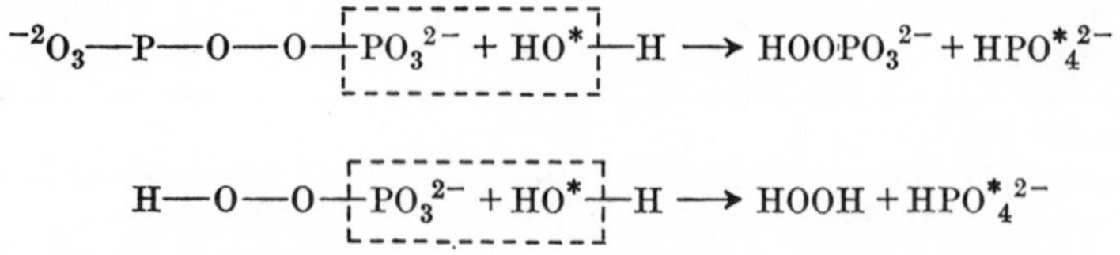

$$^{-2}O_3\!-\!P\!-\!O\!-\!O\!-\!\overline{|PO_3^{2-} + HO^*|}\!-\!H \longrightarrow HOOPO_3^{2-} + HPO^{*\,2-}_4$$

$$H\!-\!O\!-\!O\!-\!\overline{|PO_3^{2-} + HO^*|}\!-\!H \longrightarrow HOOH + HPO^{*\,2-}_4$$

This mechanism, however, does not explain the rate-enhancing effect of hydrogen ions as observed in the kinetics of the reaction. The hydrolysis of peroxodiphosphate to peroxomonophosphate was found by Crutchfield (26,27) to be pseudo first order in peroxodiphosphate concentration at each hydrogen ion concentration studied (Table II).

$$\frac{-d[P_2O_8^{4-}]_t}{dt} = k_{obs}[P_2O_8^{4-}]_t$$

In this equation, $[P_2O_8^{4-}]_t$ indicates total concentration of peroxodiphosphate regardless of ionic form. If each species hydrolyzes at a different specific rate, the rate law may be rewritten as

$$\frac{-d[P_2O_8^{4-}]_t}{dt} = \sum_n k_n[H_nP_2O_8^{(n-4)}]$$

where k_n is the rate constant associated with the nth ionic species. When the concentration of each of the ionic species of peroxodiphosphate is expressed in terms of the hydrogen ion concentration, the total concentration of peroxodiphosphate and the acidity constants for peroxodiphosphoric acid, k_{obs} can be expressed as

$$k_{obs} = \frac{K_1K_2k_2 + K_1k_3[H^+] + k_4[H^+]^2}{K_1K_2 + K_1[H^+] + [H^+]^2}$$

for the pH range where $H_2P_2O_8^{2-}$ is the predominant species (Fig. 2). The species $H_3P_2O_8^-$ and $H_4P_2O_8$ are present in smaller concentration and undergo hydrolysis as well; i.e., the rate of the hydrolysis of each

TABLE II
Pseudo First Order Rate Constants for Hydrolysis of Peroxodiphosphate (26)

Temp. (°C)	Initial pH	k_{obs} (min^{-1})	μ	$P_2O_8{}^{4-}$ salt	"Buffer"
0.0	0.00	9.40×10^{-2}	1.1 ± 0.2	Li$^+$	HClO$_4$
25.0	0.00	1.47×10^{-2}	1.1	Li$^+$	HClO$_4$
25.0	0.00	1.47×10^{-2}	1.5[a]	Li$^+$	HClO$_4$
25.0	0.00	1.42×10^{-2}	2.0[a]	Li$^+$	HClO$_4$
25.0	0.00	1.21×10^{-3}	3.0[a]	Li$^+$	HClO$_4$
25.0	0.28	6.76×10^{-3}	0.8	Li$^+$	H$_2$SO$_4$
25.0	0.56	2.58×10^{-3}	0.5	Li$^+$	H$_2$SO$_4$
25.0	0.74	2.80×10^{-4}	0.5	Li$^+$	H$_2$SO$_4$
25.0	1.28	3.41×10^{-4}	0.5	Li$^+$	sulfate
25.0	1.56	1.02×10^{-5}	0.5	Li$^+$	sulfate
25.0	1.81	6.66×10^{-5}	0.5	Li$^+$	phosphate
25.0	2.05	2.00×10^{-1}	0.5	Li$^+$	phosphate
50.0	0.18	1.47×10^{-2}	0.9	K$^+$	H$_2$SO$_4$
50.0	0.41	7.58×10^{-2}	0.5	K$^+$	HClO$_4$
50.0	0.65	3.22×10^{-2}	0.5	K$^+$	H$_2$SO$_4$
50.0	0.71	3.24×10^{-3}	0.5	K$^+$	H$_2$SO$_4$
50.0	1.00	9.94×10^{-3}	0.5	K$^+$	H$_2$SO$_4$
50.0	1.26	3.35×10^{-3}	0.5	K$^+$	sulfate
50.0	1.40	3.14×10^{-3}	0.5	K$^+$	sulfate
50.0	1.65	1.15×10^{-3}	0.5	K$^+$	phosphate
50.0	2.14	2.60×10^{-4}	0.5	K$^+$	phosphate
50.0	2.60	1.05×10^{-4}	0.5	K$^+$	phosphate
50.0	3.67	4.47×10^{-5}	0.5	K$^+$	phosphate
50.0	4.91	1.60×10^{-5}	0.5	K$^+$	acetate
50.0	4.94	2.24×10^{-5}	0.5	K$^+$	acetate
63.0	0.00	3.83×10^{-1}	1.1	Li$^+$	HClO$_4$
63.0	1.10	2.83×10^{-2}	0.5	Li$^+$	H$_2$SO$_4$
63.0	2.22	8.66×10^{-4}	0.5	Li$^+$	phosphate
63.0	3.90	2.99×10^{-4}	0.5	Li$^+$	phosphate

NOTE: $[P_2O_8{}^{4-}]_0 = 0.025$ M.
[a] NaClO$_4$ added to increase ionic strength.

form has its own specific rate constant. The individual rate constants at 25° are

$$k_2 = 2.0 \times 10^{-6} \text{ min}^{-1}$$
$$k_3 = 6.2 \times 10^{-3} \text{ min}^{-1}$$
$$k_4 = 5.3 \times 10^{-1} \text{ min}^{-1}$$

These are calculated for $K_1 = 2$ and $K_2 = 0.3$ by Crutchfield (27,28); employment of different K values would, of course, lead to different k values.

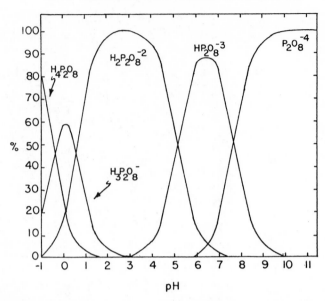

Fig. 2. Population of the various species $H_nP_2O_8^{n-4}$ as a function of pH. The data employed were taken from the Ph.D. thesis of M. M. Crutchfield at Brown University (1960); reprinted by permission.

The activation energy varies with pH and increases from 18 to 28 kcal/mole over the pH range 0 to 3. A similar range of activation energies is observed for the acid-catalyzed hydrolysis of P—O—P linkages in chain polyphosphates (31) and indicates that the reactions are similar.

Goh, Heslop, and Lethbridge (32) studied the hydrolysis of peroxodiphosphoric acid in the pH range 0–7.6 and the dependence of rate constant on ionic strength over a pH range of 1.6–3.8 (Table III). The results are substantially in agreement with those of Crutchfield (26) at the same ionic strengths. However they found their results to be correlated better by the addition of a term $k_5[H_4P_2O_8][H^+]$ for the reaction

$$H_4P_2O_8 + H_3O+ \rightarrow products$$

giving

$$k_{obs} = \frac{K_1K_2k_2 + K_1k_3[H^+] + k_4[H^+]^2 + k_5[H^+]^3}{K_1K_2 + K_1[H^+] + [H^+]^2}$$

The hydrolysis reaction was studied polarographically by Venturini, Indelli, and Raspi (29) at $[H^+] = 0.1M$ and $0.2M$ and ionic strength $1.0M$. They found rate constants which were in good agreement both with the ones of Crutchfield and with a value for k_2 found by Indelli and

TABLE III
Velocity Constants for the Hydrolysis of Peroxodiphosphoric
Acid: Dependence on Ionic Strength and pH[a]

Temp. (°C)	pH	μ	$F(\mu)$	$k_{obs} \times 10^4$ (sec^{-1})
70	1.62	0.0280	0.138	4.18 ± 0.03
70	1.62	0.0422	0.162	3.71 ± 0.10
70	1.62	0.0558	0.180	3.28 ± 0.02
70	1.62	0.0731	0.198	3.05 ± 0.03
70	1.62	0.0888	0.212	2.78 ± 0.07
70	1.62	0.1282	0.238	2.42 ± 0.05
70	2.00	0.0105	0.090	1.12
70	2.00	0.0155	0.106	1.03
70	2.00	0.0205	0.121	0.94
70	2.00	0.0255	0.133	0.95
70	2.00	0.0305	0.141	0.87
70	2.00	0.0405	0.159	0.73
70	2.40	0.0045	0.062	0.59
70	2.40	0.0065	0.074	0.60
70	2.40	0.0085	0.083	0.56
70	2.40	0.0125	0.098	0.50
70	2.40	0.0165	0.110	0.49
70	2.40	0.0285	0.139	0.43
80	2.03	0.0102	0.090	3.30
80	2.03	0.0272	0.136	2.60
80	2.03	0.0569	0.181	2.05
80	2.03	0.1038	0.223	1.78
80	2.25	0.0068	0.074	1.55
80	2.25	0.0068	0.074	1.58
80	2.25	0.0227	0.126	1.36
80	2.25	0.0429	0.163	1.15
80	2.25	0.0703	0.196	1.02
80	2.25	0.0944	0.216	1.02
81.4	2.54	0.0033	0.053	0.97
81.4	2.54	0.0359	0.152	0.79
81.4	2.54	0.0533	0.177	0.73
81.4	2.54	0.0973	0.218	0.71
80	2.86	0.0018	0.040	0.64
80	2.86	0.0185	0.116	0.60
80	2.86	0.0365	0.153	0.56
80	2.86	0.0616	0.187	0.55
80	2.86	0.1036	0.223	0.55
80	3.81	0.0019	0.041	0.47
80	3.81	0.0136	0.102	0.47
80	3.81	0.0327	0.147	0.47
80	3.81	0.0648	0.190	0.48

[a] Data of Ref. 32.

Bonora (33) in a kinetic study of the oxidation of iodide ion. Activation energies of 20.8 ± 0.3 kcal/mole and 21.0 ± 0.1 kcal/mole at $[H^+] = 0.2$ and 0.1 respectively also agree with those found earlier if the combined experimental errors are considered.

The kinetics of the hydrolysis of peroxodiphosphate are similar to those of pyrophosphate (1b) and a similar mechanism is suggested, i.e., a nucleophilic attack on phosphorus by the oxygen of water with the over-all rate of the reaction being governed by the factors which affect the accessibility of the phosphorus. The activated complex for undissociated peroxophosphoric acid might be

$$
\begin{array}{ccc}
\text{H} & \text{O} & \text{O} \\
\diagdown & \| & \| \\
\text{O}\text{---}\text{P}\text{---}\text{O}\text{---}\text{O}\text{---}\text{P}\text{---}\text{O}\text{---}\text{H} \\
\diagup \quad \diagdown \quad \diagdown & & | \\
\text{H} \quad \text{O} \quad \text{O} & & \text{O} \\
\diagup \quad \diagdown & & | \\
\text{H} \qquad\qquad \text{H} & \text{H} &
\end{array}
$$

The approach of a properly oriented water molecule would be facilitated by the neutralization of negative charge on the anion due to the increasing association of protons in acid solution. Also the availability of the d orbitals of phosphorus for sp^3d bonding in the activated complex is enhanced by the presence of the bound hydrogens.

The rate of formation of peroxomonophosphate from peroxodiphosphate near pH 4.6 at 62° is enhanced by at least a factor of ten when fluoride ion is added to the hydrolysis solution. The presence of FPO_3^{2-} in the spent reaction solution was shown by precipitation of Ag_2FPO_3 and by NMR spectra (34). It is presumed that the increase in rate is due to the reaction as fluoride ion is known to be a powerful

$$
F^- + H_2P_2O_8^{2-} \rightarrow H_2PO_5^- + FPO_3^{2-}
$$

nucleophile toward tetrahedral phosphorus. Further study of this interesting reaction should be carried out, for the data suggest that peroxomonophosphates may be derived from peroxodiphosphates by nucleophilic attack (of both F^- and H_2O).

Several enzymes known as phosphatases are effective in catalyzing the hydrolysis of peroxodiphosphate through cleavage of the O—P bond. Acid phosphatase is most effective over the pH range 4–7 whereas alkaline phosphatase is especially effective between pH 8–10 (35).

F. Decomposition

The decomposition of peroxodiphosphate in basic solution is very slow. By measuring the volume of oxygen gas evolved as a function of time from a boiling solution of peroxodiphosphate, the over-all decomposition was found to be approximately first-order in total peroxodiphosphate concentration (26). The rate is dependent on pH, concentration and specific nature of the cations present, and the condition of the glass surface. Oxygen evolution in the presence of the strongly complexing Li^+ ion is much faster than with the weakly complexing K^+ ion. Taking the observed rate of decomposition of $0.025M$ $K_4P_2O_8$ in $0.1M$ KOH as the upper limit for the rate of homolytic cleavage, the estimated first-order rate constant at $100°$ is 3×10^{-5} min^{-1}. The corresponding rate for peroxodisulfate is 6.0×10^{-3} min^{-1} (36), and therefore any uncatalyzed decomposition by a homolytic cleavage of the peroxide bond in $P_2O_8^{4-}$ to form free radicals must be at least 200 times slower than the corresponding reaction of peroxodisulfate ion.

G. Redox Reactions

Peroxodiphosphate is a strong oxidizing agent but kinetically inhibited. For example it will oxidize —SH bonds to —S—S— and —NH$_2$ to —NO slowly, but in general organic materials are not attacked. From the preparative method, a measurement of the cell potential using a salt bridge calomel electrode showed the minimum value of the oxidation potential to be -2.07 V (22). Also the fact that peroxodiphosphate oxidizes silver ions, copper(II) ions, and vanadyl ions (37,38) indicates that the potential is not much different from that of the peroxodisulfate couple which is -2.01 V (39a). A recent attempt to measure the redox potential directly did not give conclusive results (39b).

The peroxodiphosphates react with iodide ion in acid solution to form phosphoric acid and iodine. The rate is approximately given by the equation (Figs. 3 and 4) (33):

$$\text{Rate} = k_1[H^+][H_2P_2O_8^{2-}][I^-] + k_2[H^+][H_2P_2O_8^{2-}] + k_3[H_2P_2O_8^{2-}][I^-]$$

The kinetics of the reaction suggest that the reaction takes place through at least three main rate-determining paths which have been represented as

$$H_3P_2O_8^- + I^- \xrightarrow{k_1} HPO_4^{2-} + H_2PO_4^- + I^+$$
$$H_3P_2O_8^- + H_2O \xrightarrow{k_2} H_2PO_4^- + H_2PO_5^- + H^+$$
$$H_2P_2O_8^{2-} + I^- \xrightarrow{k_3} 2\,HPO_4^{2-} + I^+$$

For k_1, k_2, and k_3, the following values were obtained: $k_1 = 0.021$ liter^2/(mole)^2(sec); $k_2 = 0.06 \times 10^{-3}$ liter/(mole)(sec); and $k_3 = 0.01 \times 10^{-3}$ liter/(mole)(sec) at $25°$ and ionic strength 0.1.

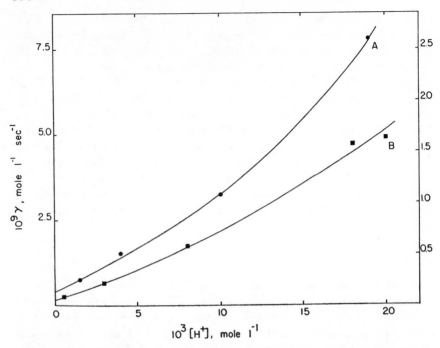

Fig. 3.   Rates of iodide ion oxidation as a function of the hydrogen ion concentration for two different peroxodiphosphate concentrations. Taken from A. Indelli and P. L. Bonora, *J. Amer. Chem. Soc.*, **88**, 924 (1966); reprinted by permission.

The mechanisms given for the iodide oxidation do not express all of the complications which must obtain. For example, it is doubtful if I^+ is ever formed as a discrete entity; it would probably remain complexed to $H_2PO_4^-$ until transferred directly to a second I^- and thus becomes I_2. Also the second path is the hydrolysis of $H_3P_2O_8^-$ to peroxomonophosphoric acid which then oxidizes iodide to iodine in a rapid, post-rate-determining step.

$$H_2PO_5^- + 2\,I^- + 2\,H^+ \rightarrow I_2 + H_2PO_4^- + H_2O$$

In sufficient excess of oxidant, peroxodiphosphate reacts with iron(II) complexes such as $Fe(phen)_3^{2+}$, $Fe(bipy)_3^{2+}$, and $Fe(terpy)_2^{2+}$ in essentially neutral solutions (40,41). The kinetics are first order with respect to complex ion and zero order with respect to peroxodiphosphate. At low concentrations of peroxodiphosphate, the initial first-order rate constants show dependence on the peroxodiphosphate concentration. The reactions of $Fe(phen)_3^{2+}$ with copper(II) sulfate, nickel(II) sulfate, sodium pyrophosphate, and disodium–EDTA produce the same kind of

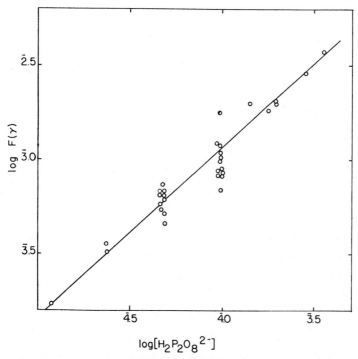

Fig. 4. The kinetic order with respect to $H_2P_2O_8^-$ concentration in the oxidation of iodide ion. Taken from A. Indelli and P. L. Bonora, *J. Amer. Chem. Soc.*, **88**, 924 (1966); reprinted by permission.

rate process and almost the same rate constants as for the oxidation reaction. For $Fe(phen)_3^{2+}$ and $Fe(bipy)_3^{2+}$ the rates of oxidation are very close to those for dissociation of the ligands in acid solution (42–44). In the case of $Fe(terpy)_2^{2+}$ the rate of oxidation at 35° is approximately one power of ten greater than the ligand exchange rate in neutral solution but about 10^{-3} times the limiting rate for dissociation; i.e., the breaking of one Fe—N bond (45). Also the activation parameters for the oxidations and the dissociations of the same complexes are similar (Table IV), and added free ligand depresses the oxidation rates. These results strongly indicate that the peroxodiphosphate oxidations of the iron(II) complexes proceed by way of a dissociative mechanism

$$Fe(N)_6^{2+} \underset{k_{-1}}{\overset{k_1}{\rightleftharpoons}} Fe(N)_4^{2+} + N\text{—}N$$

$$Fe(N)_4^{2+} + P_2O_8^{4-} \overset{k_2}{\longrightarrow} products$$

TABLE IV

Rates and Activation Parameters for Reactions

Complex	k (sec^{-1})	E_a(kcal/mole)	ΔS^*[cal/(mole)(deg)]	Ref.
	A. Dissociation of Metal Complexes			
Fe(phen)$_3{}^{2+}$	3.7×10^{-4} (34.6°)	32.1	$+28$	42
Fe(bipy)$_3{}^{2+}$	1.32×10^{-4} (25°)	28.4	$+17$	43,44
Fe(terpy)$_2{}^{2+}$	1.1×10^{-6} (35°, pH 6)	28.7		45
	7.3×10^{-3} (35°, $3M$ H$^+$)	28.7		45
Os(bipy)$_3{}^{2+}$	extremely slow			
Os(phen)$_3{}^{2+}$	extremely slow			
	B. Oxidation of Complexes by Peroxodiphosphate			
Fe(phen)$_3{}^{2+}$	3.8×10^{-4} (35°)	30.9	$+26$	38,40
Fe(bipy)$_3{}^{2+}$	1.0×10^{-4} (25°)	30.6	$+24$	38,41a
Fe(terpy)$_2{}^{2+}$	6.1×10^{-5} (35°, pH 5.9)	29.1	$+13.4$	38,41a
Os(bipy)$_3{}^{2+}$	a			37b,38
Os(phen)$_3{}^{2+}$	b			37b,38

a No reaction for 14 days at 60°.
b No reaction for 14 days at 60° in the presence of EDTA.

Applying the steady state approximation to the dissociated intermediate, the following rate law is obtained

$$\text{Rate} = \frac{k_1 k_2 [\text{Fe(N)}_6{}^{2+}][\text{P}_2\text{O}_8{}^{4-}]}{k_{-1}[\text{N—N}] + k_2[\text{P}_2\text{O}_8{}^{4-}]}$$

At high concentrations of peroxodiphosphate and low concentrations of free ligand the rate law takes on the simple form

$$\text{Rate} = k_1[\text{Fe(N)}_6{}^{2+}]$$

wherein the observed rate constant approaches k_1, that of the dissociation of the complex. In the case of the Fe(terpy)$_2{}^{2+}$ complex which has the more flexible ligand, the oxidation may take place before a ligand is totally removed

$$\text{Fe(N)}_6{}^{2+} \underset{k_{-1}}{\overset{k_1}{\rightleftharpoons}} \text{Fe(N)}_5\text{—N}^{2+} \underset{k_{-3}}{\overset{k_3}{\rightleftharpoons}} \text{Fe(N)}_4\text{—N—N}^{2+} \underset{k_{-5}}{\overset{k_5}{\rightleftharpoons}} \text{Fe(N)}_3{}^{2+} + \text{N—N—N}$$

$$k_2'' \downarrow \text{P}_2\text{O}_8{}^{4-} \qquad k_2' \downarrow \text{P}_2\text{O}_8{}^{4-} \qquad k_2 \downarrow \text{P}_2\text{O}_8{}^{4-}$$

$$\text{products} \qquad\qquad \text{products} \qquad\qquad \text{products}$$

Assuming that the oxidation occurs only through the k_2 path, the rate law is

$$\text{Rate} = \frac{k_1 k_3 k_5 k_2 [\text{Fe(N)}_6{}^{2+}][\text{P}_2\text{O}_8{}^{4-}]}{k_{-1}k_{-3}k_{-5}[\text{N—N—N}] + k_2(k_{-1}k_{-3} + k_{-1}k_5 + k_3 k_5)[\text{P}_2\text{O}_8{}^{4-}]}$$

A change in the observed first-order rate constant as the peroxodiphosphate concentration increases suggests that the reaction may be shifted from taking place primarily through oxidation of $Fe(N)_3^{2+}$ to primarily oxidation of $Fe(N)_5$—N as the oxidant concentration increases.

A less than 2:1 stoichiometry and depression of the observed rates in the presence of 2-propanol and acrylonitrile indicate that the reaction mechanism involves the formation of a free radical. A phosphate radical anion $PO_4^{\cdot 2-}$ may be produced in the oxidation step

$$Fe(N)_4^{2+} + P_2O_8^{4-} \rightarrow Fe(III)\ species + PO_4^{\cdot 2-} + PO_4^{3-}$$

A similar mechanism has been proposed for the peroxodisulfate oxidations of iron(II) complexes and $Fe^{2+}(aq)$ and for the hydrogen peroxide oxidation of $Fe^{2+}(aq)$ (46–51). In the presence of 2-propanol or acrylonitrile, the reaction may involve the interaction of the Fe(III) products with the organic radical to partially reform the original complex and to reduce the observed reaction rate. Some possible steps are

$$Fe(N)_4^{2+} + P_2O_8^{4-} \xrightarrow{k_2} PO_4^{\cdot 2-} + Fe(N)_4^{3+} + PO_4^{3-}$$
$$PO_4^{\cdot 2-} + (CH_3)_2CHOH \longrightarrow (CH_3)_2\dot{C}OH + HPO_4^{2-}$$
$$(CH_3)_2\dot{C}OH + Fe(N)_4^{3+} \longrightarrow (CH_3)_2CO + Fe(N)_4^{2+} + H^+$$

Although the formation of iron(III) species could not be definitely detected spectrophotometrically in the reaction mixture owing to the low concentration of complex employed, several tests confirm that oxidations rather than mere displacements of a ligand for $P_2O_8^{4-}$ do take place. For example, SCN^- tests for Fe^{3+} were positive, and the color of the original Fe(II) complexes returned upon addition of $S_2O_4^{2-}$ to the product solution. Also it is known that peroxodiphosphate oxidizes aqueous ferrous ion very rapidly. Rapid electron transfer must transpire once the peroxodiphosphate enters the coordination sphere.

Reactions of peroxodiphosphate with Os(II) complexes (38,41b) indicate that a dissociative mechanism operates for these complexes as well. The very inert $Os(bipy)_3^{2+}$ and $Os(phen)_3^{2+}$ complexes react extremely slowly with peroxodiphosphate in the absence of trace metal ions. Silver(I) and copper(II) ions catalyze the reactions strongly. In addition there is indication of autocatalysis by the Os(III) product. The $[Os(terpy)(bipy)Cl]^+$ complex both hydrolyzes and reacts with peroxodiphosphate at the same rate, $k = 4.0 \times 10^{-6}\ sec^{-1}$ at pH 7 and 60°. The rate of oxidation is independent of $[P_2O_8^{4-}]$ for concentrations at least larger than $2 \times 10^{-3} M$. This result is especially interesting as the dissociating ligand here is the monodentate chloride ion leaving a single coordination site for attack by peroxodiphosphate. Electron transfer to peroxodiphosphate acting as a monodentate ligand seems to be the appropriate conclusion.

It is of interest to compare these results with those from the oxidation of the same complexes with peroxodisulfate. Peroxodisulfate usually reacts with these complexes and with the methyl substituted complexes by second-order kinetics (48,52,53), first order in each reactant, with rate constants which generally are faster than replacement and are little dependent on the nature of the ligand. In addition the activation parameters are very different from those from the peroxodiphosphate reactions. While peroxodisulfate apparently prefers an outer-sphere mechanism the results from the peroxodiphosphate reactions are best explained by an inner-sphere mechanism where the rate-determining step is the partial (or full) dissociation of a ligand with $P_2O_8^{4-}$ or the PO_4^{2-} radical approaching the vacant position followed by a fast oxidation reaction. Roughly correcting for the temperature difference, it was found that peroxodisulfate oxidizes $Os(phen)_3^{2+}$ at least 10^8 times more rapidly than does peroxodiphosphate. This is equivalent to a difference of 11 kcal/mole in E_a and seems too large to attribute to a difference in bond dissociation energy in the peroxide bond. One possible explanation is that the peroxodiphosphate ion is more solvated than the perodisulfate ion and that desolvation prior to insertion between the ligand planes would be required as appears to be the case with peroxodisulfate. The energy difference could be rationalized as due to about three hydrogen bonds.

For tris(1,10-phenanthroline)iron(II) complexes containing nitro or sulfato substituents the reactions with peroxodisulfate proceed by parallel second-order oxidation and aquation (54). The electron-withdrawing substituents may decrease the iron–ligand bond strength and facilitate aquation while the electron density is reduced at the iron making oxidation slower. The two rates in this way become comparable. Results for the two oxidations and the aquation may be summarized as follows. The aquation (54a) of tris(5-nitrophen-anthroline)iron(II) and the oxidation of this complex by peroxodiphosphate (41) have rates in good agreement; the rate constant for the first-order path for peroxodisulfate oxidation (54b) is less by a factor of four. This is probably larger than the combined experimental errors but is not considered as evidence for a different mechanism.

The reactions of Ru(II) complexes are complicated by hydrolysis of reactant and product as well as by photosensitivity of the solutions. However it is established that the complex $[Ru(phen)_2pyCl]^+$ reacts with peroxodiphosphate at a faster rate than the rate of dissociation but the complications are too severe to justify any firm mechanistic conclusions.

Chong (18) noticed that the vanadyl ion, VO^{2+}, reacts with peroxodiphosphate and studied the possibility of analytical determination of the vanadyl ion with peroxodiphosphate. The kinetics of the reaction were studied spectrophotometrically by Edwards et al. (37). The reaction is complicated by the interaction of the vanadyl ion with other species in solution such as vanadium(V) and pyrophosphate to form complexes, but it is established that two vanadyl ions are oxidized by one peroxodiphosphate ion and that the kinetics are first order each in oxidant and reductant with a rate constant of $5 \times 10^{-2} \pm 1 \times 10^{-2}$ liter/(mole)-(sec) at 25° and ionic strength 0.6 in phosphate buffer. The rate is not markedly dependent on pH in the range from 1.1 to 3.5. Despite the relatively large errors in the rate constants, the results lead to reasonable conclusions concerning a mechanism which involves a two-step reaction where a free radical is formed in a first, slow step

$$VO^{2+} + P_2O_8^{4-} + H_2O \rightarrow VO_2^+ + PO_4^{3-} + PO_4^{2-} + 2\,H^+$$

This step is followed by a fast step in which the radical attacks another VO^{2+} ion

$$VO^{2+} + PO_4^{2-} + H_2O \rightarrow VO_2^+ + PO_4^{3-} + 2\,H^+$$

In view of the ability of peroxodiphosphate to form complexes, a transition state is postulated in which the peroxoanion acts as a chelate ligand and is bound to the vanadyl ion. This postulation is supported by the fact that the peroxodisulfate ion, $S_2O_8^{2-}$, which is isoelectronic with $P_2O_8^{4-}$ but does not form complexes, reacts very slowly with vanadyl ion (55).

H. Photochemistry

Peroxodiphosphate does not undergo homolytic scission as readily as

$$P_2O_8^{4-} \rightarrow 2\,PO_4^{2-}$$

peroxodisulfate. For example, peroxodiphosphate does not readily oxidize water at 100° or below nor are alcohols oxidized at convenient temperatures as is the case with peroxodisulfate. However upon irradiation at 253.7 nm, peroxodiphosphate is activated to radical reactions with water and alcohols (56). The photolysis of peroxodiphosphate in water is first order in peroxodiphosphate throughout the pH range 3 to 11. Molecular oxygen is produced according to

$$2\,P_2O_8^{4-} + 2\,H_2O \rightarrow O_2 + 4\,HPO_4^{2-}$$

Isotope studies in $H_2^{18}O$ show that the evolved oxygen is derived from the solvent water. The quantum yield is essentially independent of the peroxodiphosphate concentration and of pH in the absence of added phosphate or metal ions other than alkali ions. It is depressed by added PO_4^{3-} in basic solution. A study of the extent of exchange between PO_4^{3-} and $P_2O_8^{4-}$ in photolyzed and nonphotolyzed solutions using ^{32}P-labeled phosphate at pH values from 9.0 to 9.38 shows that the apparent exchange is only about 4% thus excluding any appreciable interference in the photochemical studies. In analogy to the oxidation of water by peroxodisulfate (57), a number of steps involving free radicals have been suggested. The steps are detailed elsewhere (56).

The primary products of the photochemical oxidations of 2-propanol and ethanol are acetone and acetaldehyde respectively. The loss of peroxodiphosphate is again first order. The quantum yield is dependent on both the initial pH and the alcohol concentration. The proposed mechanisms in both weakly basic solution and acid solution involve free radical intermediates.

I. Potential Uses

Peroxodiphosphate has found practical use as initiator in emulsion polymerization, as oxidative starch modifier, as metal cleaning compound, in water treatment, and in several bleaching, germicidal, and biocidal compounds. The phosphatase-catalyzed hydrolysis of peroxodiphosphate allows the generation of peroxomonophosphate to be controlled by adjusting the $K_4P_2O_8$/phosphatase ratio to coincide with the rate of consumption over long periods (35). This opens possibilities for applications requiring a continuous long term generation of peroxomonophosphate which is a powerful bleaching, oxidizing, and germicidal agent.

III. Peroxomonophosphates

A. Discovery and Preparations

The discovery and early preparations of peroxomonophosphoric acid are closely connected with those of peroxodiphosphoric acid. The first preparation of peroxomonophosphoric acid in solution was published together with the first preparation of peroxodiphosphoric acid by Schmidlin and Massini (5). By treating phosphorus pentoxide with 30% hydrogen peroxide they obtained a solution which oxidized iodide to iodine, aniline to nitroso- and nitrobenzene, and manganous salts to permanganate at room temperature. The content of active oxygen corresponded to the formula H_3PO_5. They presented the reaction as

$$P_2O_5 + 2 H_2O_2 + H_2O \rightarrow 2 H_3PO_5$$

That the new acid was a derivative of orthophosphoric acid and not of meta- or pyrophosphoric acid was established by treating the neutralized solution with silver nitrate which caused the precipitation of a white salt which quickly turned yellow during the formation of silver phosphate and evolution of oxygen:

$$2 Ag_3PO_5 \rightarrow 2 Ag_3PO_4 + O_2$$

In acid solution no precipitation of the peroxomonophosphoric acid by silver, ferrous, nickelous, manganous, or other heavy metal ions was obtained. In neutral solution any precipitation was followed by immediate decomposition preventing the isolation of the acid itself or its salts. Schmidlin and Massini found that the peroxomonophosphoric acid was unstable except in very dilute solution.

In some later preparations peroxomonophosphoric acid was prepared in solution together with peroxodiphosphoric acid by electrolysis of phosphate solutions. By changing (7,8,11,13) the variables (such as current density, anode potential, time of electrolysis, concentration, acidity, etc.) which affected the yield, it was possible to maximize the production of either of the peroxoacids. However the solutions always contained some chromate and fluoride and these preparations are therefore of limited synthetic usefulness for the preparation of peroxomonophosphoric acid. A group of patents and publications (58–63) have dealt with the preparations of peroxomonophosphoric acid and its salts utilizing the reaction between phosphorus pentoxide and hydrogen peroxide. As the direct reaction between the two reactants is very violent, different methods to modify the reaction have been attempted. One consists of reacting aqueous hydrogen peroxide with phosphorus pentoxide which has been heated at an elevated temperature to reduce substantially its reactivity (60). In other preparations (58,59,61,62) the phosphorus pentoxide is suspended in an inert solvent to which the hydrogen peroxide is added. Toennies (58) studied the effect of ether, isoamyl alcohol, and acetonitrile as solvents. Only the last one gave satisfactory results. He claims yields of 56–58% of peroxomonophosphoric acid. The solutions were relatively stable even at room temperature. After three days at −11° about 90% of the original peroxo acid was still present. No isolation of the acid or its salts was attempted.

Solid potassium dihydrogen peroxomonophosphate, KH_2PO_5, was prepared (59) from the reaction between phosphorus pentoxide and hydrogen pentoxide in Freon-113 (1,1,2-trichloro-1,2,2-trifluorethane) followed by neutralization of the H_3PO_5–H_2O_2 layer with KOH. The isolated compound was found to oxidize manganese(II) to MnO_2 in neutral solution and to permanganate in $0.1N$ H_2SO_4, and chloride to chlorine. No yield is reported.

For the use in the studies of some reactions with peroxomonophosphoric acid, the peroxoacid was produced (64–67) in solution by acid hydrolysis of peroxodiphosphate

$$H_4P_2O_8 + H_2O \rightarrow H_3PO_5 + H_3PO_4$$

As the stable salt $Li_4P_2O_8 \cdot 4\ H_2O$ can be prepared in high purity, it is a good source for peroxomonophosphoric acid but produces an equimolar amount of orthophosphoric acid.

Schenk and Dommain (68) studied the reaction between phosphorus pentoxide and hydrogen peroxide under various conditions and found that it was more complicated than suggested by Schmidlin and Massini. Instead of following the simple equation

$$P_2O_5 + 2\ H_2O_2 + H_2O \rightarrow 2\ H_3PO_5$$

they found that the reaction produced a mixture of phosphates besides peroxomonophosphoric acid. With 100% H_2O_2, mono-, di-, and triphosphate were formed. In 90% H_2O_2 some tetrametaphosphate was also formed. With 70% and 50% H_2O_2 considerable amounts of tetrametaphosphate and monophosphate were formed in addition to di-, tri-, and tetraphosphate. In aqueous solution metaphosphates were rapidly decomposed by water in the presence of high concentrations of hydrogen peroxide forming lower phosphates and peroxomonophosphoric acid, for example

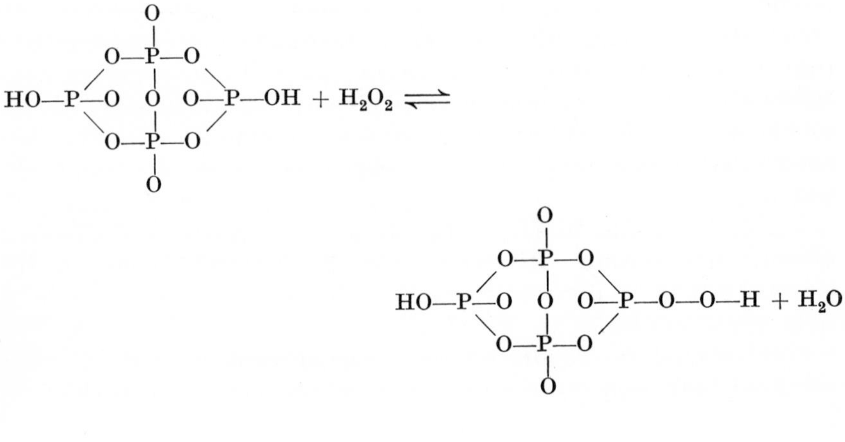

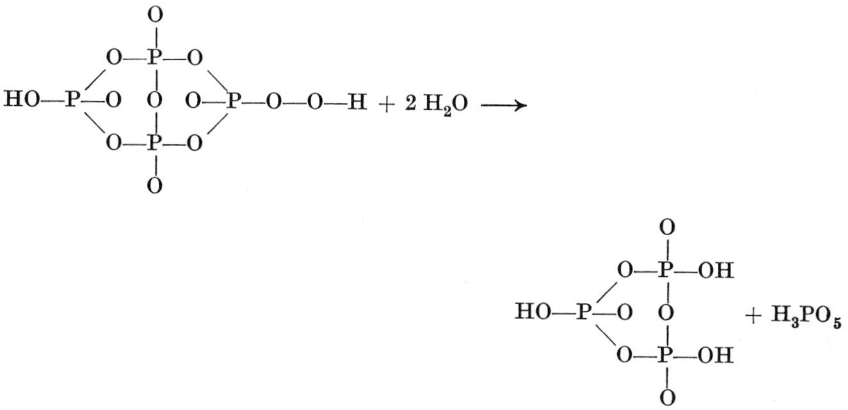

$$
\begin{array}{c}
\quad\quad\quad\text{O} \\
\quad\quad\quad| \\
\quad\quad\text{O—P—O} \\
\quad\quad\diagup\quad\quad| \\
\text{HO—P—O}\quad\text{O}\quad\text{O—P—O—O—H} + 2\,\text{H}_2\text{O} \longrightarrow \\
\quad\quad\diagdown\quad\quad| \\
\quad\quad\text{O—P—O} \\
\quad\quad\quad| \\
\quad\quad\quad\text{O}
\end{array}
$$

$$
\begin{array}{c}
\quad\quad\quad\text{O} \\
\quad\quad\quad| \\
\quad\quad\text{O—P—OH} \\
\quad\quad\diagup\quad\quad| \\
\text{HO—P—O}\quad\text{O}\quad\quad\quad + \text{H}_3\text{PO}_5 \\
\quad\quad\diagdown\quad\quad| \\
\quad\quad\text{O—P—OH} \\
\quad\quad\quad| \\
\quad\quad\quad\text{O}
\end{array}
$$

The trimetaphosphate is then decomposed similarly to diphosphate and peroxomonophosphoric acid. At lower concentrations of hydrogen peroxide, the usual mechanism of hydrolysis of the phosphates takes place (69). The introduction of the peroxide O_2 group is believed to introduce a polarization of the neighboring P—O—P bond which then becomes easier to hydrolyze. This mechanism explains why only peroxomonophosphoric acid and peroxodiphosphoric acid but not other peroxophosphoric acids including the isoperoxodiphosphoric acid

$$
\begin{array}{c}
\text{HO}\quad\quad\text{OH} \\
|\quad\quad\quad| \\
\text{OH—P—O—P—O—O—H} \\
|\quad\quad\quad| \\
\text{O}\quad\quad\quad\text{O}
\end{array}
$$

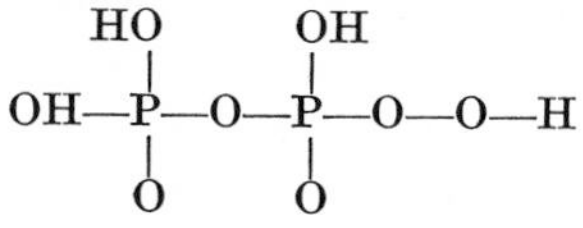

seem to exist.

Neutral or acid ammonium, alkali, or alkaline earth salts of peroxomonophosphoric acid were produced mixed with salts of other phosphoric acids by adding the oxides, hydroxides, carbonates, or bicarbonates of the cation to the reaction mixture of phosphorus pentoxide, water, and hydrogen peroxide (70). The salts were isolated either by evaporation to dryness or by adding an organic solvent soluble in water to the concentrated reaction mixture. The neutral ammonium peroxomonophosphate was purified by recrystallization.

Heslop and Lethbridge (71) separated phosphoric acid and peroxomonophosphoric acid by anion-exchange chromatography of the reaction mixture in acetonitrile described by Toennies (58). The acids were separated using an acetate buffer as eluant. The eluate was concentrated by vacuum distillation and run through a cation-exchange column to give a solution of peroxomonophosphoric acid in aqueous acetic acid. The acetic acid could be removed by further vacuum distillation and freeze drying, leaving a viscous nonvolatile residue; however more than 25% of peroxo oxygen was lost during the final concentration. Peroxomonophosphates were obtained by the addition of one equivalent of alkali hydroxide to the cold concentrated aqueous acid followed by concentration in vacuum.

Analytical methods for the determination of peroxomonophosphoric acid, peroxodiphosphoric acid, and hydrogen peroxide in mixtures of these compounds have been developed and include an electrometric method by Rius (72) and a cerimetric determination by Csányi and Solymosi (73).

B. Known Compounds

The neutral peroxomonophosphoric acid, H_3PO_5, is known only in solution. Very few peroxomonophosphates have been isolated in the solid state, and those are generally in an impure condition. The best characterized salt is the potassium dihydrogen salt, KH_2PO_5, which has been reported to be produced in 96–98% purity (59). The salt was characterized by elemental analysis and by active oxygen analysis. It is a white, crystalline material which is stable, although hygroscopic. A single crystal X-ray diffraction study showed the crystals to be orthorhombic with $a = 4.67$, $b = 5.49$, and $c = 8.43$ Å. The space group was deduced to be $P222_1$. A powder pattern of the salt showed it to contain a small amount of KH_2PO_4 and possibly a hydrate of KH_2PO_5 or potassium peroxodiphosphate.

Dihydrogen peroxomonophosphates of lithium, sodium, and potassium were isolated as deliquescent, amorphous, white solids which contained up to 70% of the active oxygen required for a salt of the formula MH_2PO_5. These salts lost 40% of their active oxygen in eight weeks when stored over phosphorus pentoxide at 0°.

The formation of peroxomonophosphoric acid from the hydrolysis of peroxodiphosphoric acid together with other chemical and physical evidence suggest the structure of peroxomonophosphoric acid to be

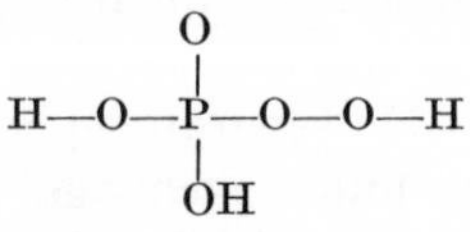

$$
\begin{array}{c}
\text{O} \\
\| \\
\text{H—O—P—O—O—H} \\
| \\
\text{OH}
\end{array}
$$

Four oxygen atoms are considered to be tetrahedrally bound to the phosphorus atom as in other phosphates.

C. Acid-Ionization Constants

Peroxomonophosphoric acid is a tribasic acid:

$$H_3PO_5 \underset{}{\overset{K_1}{\rightleftharpoons}} H^+ + H_2PO_5^-$$

$$H_2PO_5^- \underset{}{\overset{K_2}{\rightleftharpoons}} H^+ + HPO_5^{2-}$$

$$HPO_5^{2-} \underset{}{\overset{K_3}{\rightleftharpoons}} H^+ + PO_5^{3-}$$

K_1 and K_2 represent the ionization constants for the P—O—H hydrogen ions and K_3 that for the peroxidic P—O—O—H hydrogen ion. The acid ionization constants have been determined (64,66) spectrophotometrically on an equimolar mixture of H_3PO_5 and H_3PO_4 containing about 5% of hydrogen peroxide formed by the hydrolysis of $Li_4P_2O_8$.

TABLE V

Summary of Dissociation Constants of Peroxomonophosphoric Acid at 25°

Acid	λ (nm)	Dissociation Constant[a]	Remarks
H_3PO_5	230	$10^2K_1 = 7.5 \pm 1.3$	$I \simeq 0.2$
H_3PO_5	240	$10^2K_1 = 8.9 \pm 1.9$	$I \simeq 0.2$
$H_2PO_5^-$	230	$10^6K_2 = 3.16 \pm 0.42$	$I = 0.14 \pm 0.01$
$H_2PO_5^-$	240	$10^6K_2 = 3.01 \pm 0.65$	$I = 0.14 \pm 0.01$
$H_2PO_5^-$	250	$10^6K_2 = 3.57 \pm 1.15$	$I = 0.14 \pm 0.01$
HPO_5^{2-}	240	$10^{13}K_3 = 1.6 \pm 0.5$	I varies[b]
HPO_5^{2-}	250	$10^{13}K_3 = 1.6 \pm 0.5$	I varies[b]

[a] Uncertainties given for K_1 and K_2 are standard deviations and for K_3 are estimated uncertainties.   Data from Ref. 64.

[b] Average $I \simeq 0.15$.

When base is added to the peroxo acid solution, the UV absorption spectrum of hydrogen peroxide and other peroxo acids shifts toward longer wavelengths due to the formation of the conjugate base, i.e.,

$$ROOH + OH^- \rightleftharpoons ROO^- + H_2O$$

This shift makes it possible to evaluate the dissociation constants of peroxomonophosphoric acid in the presence of phosphoric acid which does not absorb at the same wavelengths.   The only absorbing species in the test solutions are peroxomonophosphate and hydrogen peroxide. The absorbance of the latter was calculated and subtracted from the total.

A summary of the dissociation constants at 25° is shown in Table V. Only concentration ionization constants were evaluated.   Because of the limited precision of the spectrophotometric method used, no attempt was made to measure the acid ionization constants as a function of ionic strength.

Comparing the acid strength of peroxomonophosphoric acid ($pK_1 =$ 1.1, $pK_2 = 5.5$, and $pK_3 = 12.8$) to phosphoric acid ($pK_1 = 2.1$, $pK_2 = 7.1$, and $pK_3 = 12.3$) (74) it is seen that K_1 and K_2 for peroxomonophosphoric acid are larger than the corresponding constants for phosphoric acid. The same trend is observed for telluric acid and boric acid and their peroxo acids (75–77). This acid-strengthening effect can be explained by the higher electronegativity of the OOH group as compared to an OH group. Also the acid strength of hydrogen peroxide is much greater than that of water.

Estimates of the ionization constants for peroxomonophosphoric acid have also been obtained from kinetic data (65,66). At ionic strength 1.5, $K_1 \simeq 5 \times 10^{-2}$ and $K_2 \simeq 1.4 \times 10^{-5}$. K_2 from the rate data is larger than from the spectrophotometric data. As the ionic strength was higher in the kinetic studies such a difference is expected. In a study of the rate of decomposition (67), pK_3 was found to be 12.8 ± 0.2.

D. Hydrolysis and Decomposition

While peroxomonophosphate is rapidly decomposed in basic solution, it is quite stable in acid solution. The change that occurs in strongly acid solution is primarily hydrolysis to hydrogen peroxide and phosphoric acid

$$H_3PO_5 + H_2O \rightarrow H_2O_2 + H_3PO_4$$

while a decomposition

$$H_3PO_5 \rightarrow H_3PO_4 + \tfrac{1}{2}O_2$$

becomes the most important reaction at lower acid strength.

The observed rate of the hydrolysis obeys the kinetic law (64,66)

$$\frac{-d[H_3PO_5]}{dt} = k_1[H_3PO_5]$$

where k_1 is the pseudo-first-order rate constant dependent upon the acidity of the solution over the concentration range of 1.0 to $9.42M$ $HClO_4$ (Table VI). The constant k_1 was found to increase markedly with increase in hydrogen ion concentration. However the dependence of $[H^+]$ fits neither of the two limiting forms of the Hammett-Zucker hypothesis (78), i.e., a linear dependence of rate on $[H^+]$ or on Hammett's acidity function H_0. When $\log k_1$ was plotted against $\log [H^+]$ at each temperature, a reasonably straight line of slope 1.3 was obtained.

TABLE VI

Pseudo-First-Order Rate Constants for Hydrolysis of
Peroxomonophosphoric Acid

Temp. (°C)	[H$^+$]^a	$10^5 k_1$ (sec^{-1})
25.0	9.42	4.80
25.0	9.42	4.91
25.0	7.62	3.91
25.0	7.62	3.95
25.0	5.06	2.26
25.0	5.06	2.10
25.0	4.05	1.54
35.4	9.42	12.9
35.4	7.62	10.7
35.4	5.07	6.03
35.4	4.05	4.22
35.4	2.90	2.55
35.4	2.0	1.55
35.4	1.4	1.00
35.4	1.0	0.622
48.5	9.42	42.2
48.5	7.59	33.4
48.5	5.07	19.8
48.5	4.05	15.2
48.5	2.90	9.44
48.5	2.0	5.45
48.5	1.4	3.63
48.5	1.0	2.27
61.0	2.90	29.4
61.0	2.0	17.7
61.0	1.4	12.9
61.0	1.0	7.83

^{a} Perchloric acid concentration in moles/liter. Data
from Ref. 64.

When the data were treated in the manner suggested by Bunnett (79), i.e., a plot of (log $k_1 + H_0$) against the activity of water, a straight line with a slope of 5.2 was obtained in the acid range from 1 to 5M. At higher acidities the plot shows significant curvature in the direction of decreasing slope. The value of the slope was similar to those found for other phosphate hydrolyses in this acidity range suggesting a similar mechanism. The addition of sodium perchlorate resulted in both a positive salt effect on the hydrolysis rate and an increased peroxide decomposition; a thorough study of salt effects was not possible.

When the ionic strength is kept constant ($I = 7.5$) the rate is approximately linear with [H$^+$] with an apparent constant of 4.5×10^{-5}

liter/(mole)(sec) at 48.5°.   Thus the slope of 1.3 obtained when log k_1 is plotted against log $[H^+]$ reflects a salt effect due to the increasing $HClO_4$ concentration in addition to the first-order dependence of the rate on $[H^+]$.   Applying the equation (80)

$$\ln k = \ln k° + \beta I$$

(where k and k_0 are rate constants at ionic strengths I and zero, respectively, and β is an empirical constant) characteristic for hydrolysis reactions in concentrated salt solutions of a neutral molecule and an ion, k_1 may be expressed as

$$k_1 = k_2[H^+]e^{\beta I}$$

where k_2 is a second-order constant.   The values of k_2 and β were found to be 2.62×10^{-5} liter/(mole)(sec) and 0.079 respectively at 48.5°.

The activation energy E_a values decrease from 20.2 to 17.5 kcal/mole with increase in the acid concentration from $1.0M$ to $9.42M$ and are similar to values found for the acid-catalyzed hydrolysis of peroxodiphosphoric acid (26–28) and of simple P—O—P linkages in chain polyphosphates (81).   For the rate of hydrolysis in $1M$ $HClO_4$, the thermodynamic activation parameters are $\Delta H^* = 19.6$ kcal/mole and $\Delta S^* = -19$ cal/(mole)(deg) at a standard state of $1M$ and 48.5°.

The rate of peroxomonophosphate decomposition increases as the acidity decreases and hydrolysis is not experimentally observable at acidities much less than $1M$.   The rate of hydrolysis of the neutral H_3PO_5 molecule therefore cannot be measured directly.   In the range from $1.0M$ to $9.4M$ $HClO_4$ peroxomonophosphoric acid is present predominantly as the neutral species with a small contribution from the monoanion $H_2PO_5^-$.   From comparison with the rates of hydrolysis of other phosphate monoanions (82) where hydrolyses are found to be little dependent on the nature of the leaving anion, it is estimated that the contribution of the $H_2PO_5^-$ ion to the observed rate is negligible in the range of 1 to $9M$ $HClO_4$.   In the range of acidities studied, the major part of the reaction is carried by the conjugate acid species and the data suggest that it is the only species that contributes significantly to the rate of hydrolysis.   The rate is approximately proportional to $[H^+]$ at constant ionic strength and the hydrolysis is formulated as a nucleophilic attack of a water molecule on the phosphorus atom in the conjugated acid in the rate-determining step, i.e.,

$$H_3PO_5 + H^+ \rightleftharpoons H_4PO_5^+$$
$$H_4PO_5^+ + H_2O \rightarrow H_3PO_4 + H_2O_2 + H^+$$

The study of the decomposition of peroxomonophosphoric acid was complicated by trace metal catalysis over the pH interval 2.5 to 8.9 (66,67). Above pH 8.9 the rate of decomposition appeared second order in peroxomonophosphate concentration. A maximum rate of decomposition of 7.1×10^{-4} liter/(mole)(sec) was observed at pH 12.5 ± 0.2, $35.8°$. From the spectrophotometric estimate of pK_3 it seems that the maximum rate is observed under conditions where $[HPO_5^{2-}] = [PO_5^{3-}]$ (Fig. 5).

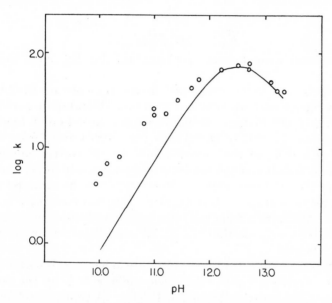

Fig. 5. The rate of decomposition of peroxomonophosphoric acid as a function of pH in the basic range at $35.8°$. Data are taken from the Ph.D. thesis of C. J. Battaglia Brown University (1962); reprinted by permission.

These results are similar to those found for the spontaneous decomposition of peroxocarboxylic acids and peroxomonosulfuric acid and a similar mechanism has been suggested. The rate law seems to be

$$R = k[^{-2}O_3POOH][^{-2}O_3POO^-] \tag{1}$$

and the postulated mechanism is

$$HPO_5^{2-} \overset{K_3}{\underset{\rightleftharpoons}{}} H^+ + PO_5^{3-} \tag{2}$$

$$HPO_5^{2-} + PO_5^{3-} \rightleftharpoons \ddagger \rightarrow HPO_6^{2-} + PO_4^{3-} \tag{3}$$

$$[HPO_6^{2-}] + OH^- \rightarrow H_2O + O_2 + PO_4^{3-} \tag{4}$$

Using

$$[P] = [HPO_5^{2-}] + [PO_5^{3-}] \tag{5}$$

and

$$K_3 = \frac{[H^+][PO_5^{3-}]}{[HPO_5^{2-}]} \tag{6}$$

the derived rate law becomes

$$\frac{-d[P]}{dt} = \left\{ \frac{k[H^+]}{K_3(1 + [H^+]/K_3)^2} \right\} [P]^2$$

where

$$\frac{k[H^+]}{K_3(1 + [H^+]/K_3)^2}$$

is the theoretical second-order rate constant for the spontaneous reaction.

The reactions had to be performed in the presence of EDTA to suppress the trace metal catalysis and even then the catalytic pathway was not entirely eliminated. Deviations from second-order kinetics became serious as the pH of the system was shifted away from that corresponding to pK_3 of the peroxo acid. At pH corresponding to pK_3, the rate is fastest and the catalytic rate presumably forms a small percentage of the total rate while at lower total rates the catalytic rate will be a larger percentage of the total rate. The fact that there is still considerable contribution from the trace metal catalysis even with the addition of EDTA suggests that the EDTA complexes are not entirely inactive as catalysts.

The decomposition of peroxomonosulfuric acid has been reported (83) to take place during a bimolecular nucleophilic attack by the anion SO_5^{2-} on the outer peroxide oxygen of HSO_5^-. A similar mechanism for the phosphate analog is supported by ^{18}O isotope studies of the decomposition of peroxomonophosphoric acid at 20°. Studies of the decomposition of normal peroxide in isotopically labeled water (84) in alkaline solution indicated that the oxygen from the peroxomono-phosphate is mixed with oxygen from the water. In the phosphate product no oxygen from the water was found. The decomposition of peroxomonophosphate containing about 1% $H^{18}O^{18}OPO_3H_2$ gave $^{18}O_2/^{16}O_2$ and $^{18}O_2/^{16}O^{18}O$ ratios which agree satisfactorily with values calculated for a mechanism which involves a nucleophilic attack of the PO_5^{3-} ion on the external peroxidic oxygen of HPO_5^{2-}:

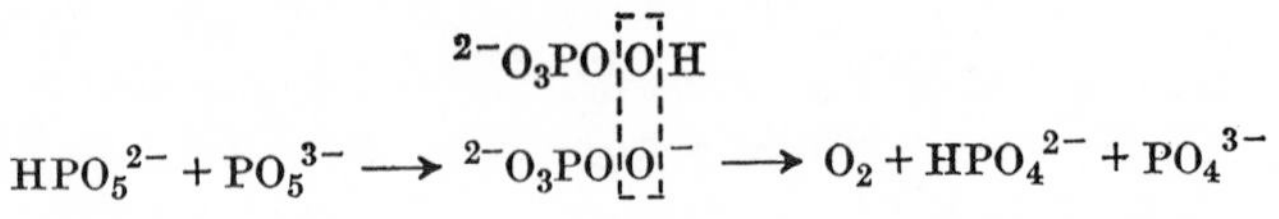

$$HPO_5^{2-} + PO_5^{3-} \longrightarrow {}^{2-}O_3PO\overset{\ulcorner}{\underset{\llcorner}{|O|}}H\overset{|}{\underset{|}{O}}{}^{-} \longrightarrow O_2 + HPO_4^{2-} + PO_4^{3-}$$

The results for peroxomonophosphoric acid and peroxomonosulfuric acid are comparable not only in rate law but also in transition state configuration (oxygen attack at oxygen) (85). The alternative transition state configuration (peroxide oxygen attack at tetrahedral phosphorus) was a significant alternative, for peroxoanion attack at phosphorus is a well established process (see Section IV below) and peroxoacetic acid does not decompose with the above type of transition state.

E. Redox Reactions

One of the first methods used to characterize peroxomonophosphoric acid was its ability to oxidize manganous ion in acid solution to permanganate. In another analytical oxidation reaction, iodide ion is

$$5\ H_3PO_5 + 2\ Mn^{2+} + 3\ H_2O \rightarrow 2\ MnO_4^- + 5\ H_3PO_5 + 6\ H^+$$

instantaneously oxidized to iodine. Aniline is oxidized to nitrosobenzene and then to nitrobenzene. Very few reactions have been examined in detail. Boyland and Manson (86) studied the oxidation of aromatic amines by peroxomonophosphoric acid in different solvents. It appeared that carbonyl compounds facilitate the oxidation of aniline to p-aminophenol. They suggested a mechanism similar to that proposed for the reaction between benzoyl peroxide and aniline, i.e., a primary step which is a transfer of one electron from the unshared pair of the nitrogen atom to one of the peroxidic oxygen atoms resulting in fission of the —O—O— bond and formation of a radical and a negative ion. In the case of peroxomonophosphoric acid the transfer of an electron from the amino nitrogen might give hydroxyl or phosphate radicals. The fact that the oxidation seems to require the presence of a carbonyl compound suggests that the reaction proceeds by way of an addition compound between the carbonyl compound and peroxomonophosphoric acid.

The oxidation of bromide ion by peroxomonophosphoric acid was studied by Battaglia (66). The reaction was found to be first order in peroxomonophosphate and first order in bromide ion, the rate law being

$$\frac{d[Br_2]}{dt} = k_{obs}[P][Br^-]$$

where [P] represents peroxomonophosphate in its various ionized forms and k_{obs} is a second-order rate constant dependent on the pH of the solution. Figure 6 shows a logarithmic plot of k_{obs} against pH. The

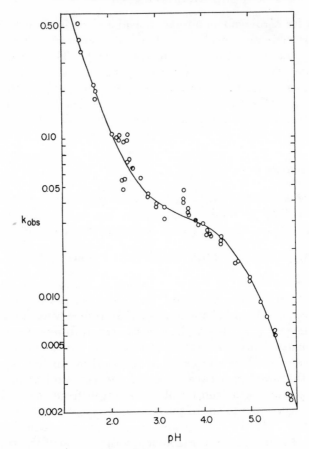

Fig. 6.   Change in second-order rate constant with pH for the oxidation of bromide ion by peroxomonophosphate at 25.0° and $\mu = 1.5$.   Data are taken from the Ph.D. thesis of C. J. Battaglia Brown University (1962); reprinted by permission.

rate law over the pH range from 1.37 to 5.83 was found to be

$$R = k_1[\text{H}_2\text{PO}_5^-][\text{Br}^-] + k_2[\text{H}_2\text{PO}_5^-][\text{Br}^-][\text{H}^+]$$

so that

$$k_{\text{obs}} = \frac{k_1[\text{H}^+] + k_2[\text{H}^+]^2}{[\text{H}^+] + K_2}$$

where K_2 is the second dissociation constant of peroxomonophosphoric acid at ionic strength 1.5.   Using the values 1.4×10^{-5} for K_2, 3.2×10^{-2} liter/(mole)(sec) for k_1, and 9 liter²/(mole²)(sec) for k_2 at 25° in this equation, a fair agreement with the experimental results was obtained.

A similar rate law is found for other halide ion and peroxo acid reactions and the mechanism is postulated to be similar, i.e.,

$$\text{ROOH} + \text{Br}^- \rightarrow \text{RO}^- + \text{HOBr} \text{ (slow)}$$
$$\text{HOBr} + \text{Br}^- + \text{H}^+ \rightarrow \text{Br}_2 + \text{H}_2\text{O} \text{ (fast)}$$

where $R = PO_3H^-$ or PO_3H_2.

The rate step appears to be a nucleophilic displacement on the peroxide oxygen. The activated complex may be represented as

$$\text{R—O} \ldots \text{O} \ldots \text{Br}^-$$
$$\setminus$$
$$\text{H}$$

An attempt to investigate the oxidation of nitrite ion by peroxo-monophosphoric acid was frustrated by side reactions which indicated peroxide decomposition (87). However at 25° and pH = 6.0, a second-order rate constant was estimated to be 7×10^{-4} liter/(mole)(sec) for the reaction assumed to have the stoichiometry

$$\text{HPO}_5{}^{2-} + \text{NO}_2{}^- \rightarrow \text{HPO}_4{}^{2-} + \text{NO}_3{}^-$$

Comparison of the estimated second-order rate constant with that of the bromide ion oxidation under similar conditions (20×10^{-4}) shows that the nitrite ion is less reactive than bromide ion as a nucleophile with peroxide oxygen.

In an ^{18}O isotope study of the oxidation of nitrite, selenite, and sulfite (88) by peroxomonophosphate and peroxomonosulfate at 18–20° and approximately neutral solution it was found that one oxygen atom from the oxidizing agent entered the oxo anion product. Oxygen from water did not enter the oxidation products. The kinetics of the oxidation of nitrite ion by peroxomonophosphoric acid were second order in the pH range 4.5 to 5.8, first order in oxidant, and first order in reductant. The dependence on $[\text{H}^+]$ was explained by a mechanism in which the nitrite ion reacts with the undissociated and dissociated forms of peroxomonophosphoric acid at different rates in three parallel reactions:

$$\text{NO}_2{}^- + \text{H}_3\text{PO}_5 \xrightarrow{k_1} \text{NO}_3{}^- + \text{H}_3\text{PO}_4$$
$$\text{NO}_2{}^- + \text{H}_2\text{PO}_5{}^- \xrightarrow{k_2} \text{NO}_3{}^- + \text{H}_2\text{PO}_4{}^-$$
$$\text{NO}_2{}^- + \text{HPO}_5{}^{2-} \xrightarrow{k_3} \text{NO}_3{}^- + \text{HPO}_4{}^{2-}$$

Thus

$$R = \frac{-d[\text{NO}_2{}^-]}{dt} = k_1[\text{NO}_2{}^-][\text{H}_3\text{PO}_5] + k_2[\text{NO}_2{}^-][\text{H}_2\text{PO}_5{}^-]$$
$$+ k_3[\text{NO}_2{}^-][\text{HPO}_5{}^{2-}]$$

Introducing $K_1 = 8 \times 10^{-2}$ and $K_2 = 3 \times 10^{-6}$ (64):

$$\frac{R}{[NO_2^-][H_2PO_5^-]} = k_1\left(\frac{[H^+]}{8 \times 10^{-2}}\right) + k_2 + k_3\left(\frac{3 \times 10^{-6}}{[H^+]}\right)$$

The mean rate constants at 25° were found to be $k_1 = 2.5 \times 10^3$ liter/(mole)(min); $k_2 = 2.9 \times 10^{-3}$ liter/(mole)(min); and $k_3 = 1.9 \times 10^{-3}$ liter/(mole)(min). Since neither investigation (87,88) gave precise results, the data for rate of oxidation of nitrite ion by HPO_4^{2-} may be considered in satisfactory agreement.

The isotopic data together with the kinetic data (i.e., the marked difference in the reaction rates for the oxidation by the undissociated and the dissociated form of peroxomonophosphoric acid) agree with the mechanism

$$^{-m}O_nA: + HOOR \longrightarrow {}^{-m}O_nA:\overset{\overset{\displaystyle H}{|}}{O}OR \longrightarrow {}^{-m}O_nA{=}O + HOR$$

where A = N, S, P, or Se and R = PO_3^{2-} or SO_3^-. The outer peroxide oxygen of the peroxomonoacid attacks the atom A of the oxo anion at the position with greatest electron density. The most favorable conditions for these reactions should then be those in which the oxidizing agent is present in the form of the undissociated acid and the reducing agent in the anionic, completely dissociated form. This is in agreement with the rate constants that were found.

The oxidative properties of peroxomonophosphoric acid have found practical use in the detoxification of poisons which affect platinum hydrogenation catalysts (89). The acid oxidizes the poison to nontoxic products.

IV. Other True Peroxophosphates

A. Inorganic Peroxophosphates

Although peroxomonophosphates and peroxodiphosphates are the only well-established (to these authors' knowledge) bound inorganic phosphates, other possibilities can be hypothesized. For example, the ions

$$O{-}\underset{\underset{\displaystyle O}{|}}{\overset{\overset{\displaystyle O}{|}}{P}}{-}O{-}\underset{\underset{\displaystyle O}{|}}{\overset{\overset{\displaystyle O}{|}}{P}}{-}OOH^{3-} \qquad O{-}\underset{\underset{\displaystyle O}{|}}{\overset{\overset{\displaystyle O}{|}}{P}}{-}O{-}\underset{\underset{\displaystyle O}{|}}{\overset{\overset{\displaystyle O}{|}}{P}}{-}O{-}\underset{\underset{\displaystyle O}{|}}{\overset{\overset{\displaystyle O}{|}}{P}}{-}OOH^{4-}$$

can, in principle, be made.

Two fluoroperoxophosphoric acids were reported (90) from anodic oxidation of monofluorophosphoric acid or the sodium, potassium, or

ammonium monofluorophosphates.   They are monofluoroperoxomono-
phosphoric acid

$$
\begin{array}{c}
\text{F} \\
| \\
\text{O—P—O—O—H} \\
| \\
\text{OH}
\end{array}
$$

and difluoroperoxodiphosphoric acid

$$
\begin{array}{ccc}
\text{F} & & \text{F} \\
| & & | \\
\text{HO—P—O—O—P—OH} \\
| & & | \\
\text{O} & & \text{O}
\end{array}
$$

The peroxo acids decompose in alkaline solution, the monoperoxo acid
losing oxygen and forming monofluorophosphate, the diperoxo acid
hydrolyzing to peroxophosphate.   Confirmation of this result would be
welcome.

B. Organic Peroxophosphates

When an organic phosphorylating compound is treated with alkaline
peroxide, rapid "hydrolysis" occurs.   For example, the compound
isopropoxymethylphosphoryl fluoride (Sarin)

$$
\begin{array}{cc}
\text{CH}_3 & \text{O} \\
| & || \\
\text{H—C—O—P—F} \\
| & | \\
\text{CH}_3 & \text{CH}_3
\end{array}
$$

forms the acid

$$
\begin{array}{cc}
\text{CH}_3 & \text{O} \\
| & || \\
\text{H—C—O—P—OH} \\
| & | \\
\text{CH}_3 & \text{CH}_3
\end{array}
$$

The peroxide enhancement of hydrolysis rate has been ascribed to the
formation of a peroxidic intermediate

$$
\begin{array}{cc}
\text{CH}_3 & \text{O} \\
| & || \\
\text{H—C—O—P—OOH} \\
| & | \\
\text{CH}_3 & \text{CH}_3
\end{array}
$$

by nucleophilic attack of OOH^- on Sarin.   The intermediate is presumed to react with another peroxide according to the equation

$$
\begin{array}{cc}
\text{CH}_3 & \text{O} \\
| & || \\
\text{H—C—O—P—OOH} & + \text{OOH}^- \longrightarrow \\
| & | \\
\text{CH}_3 & \text{CH}_3
\end{array}
\qquad
\begin{array}{cc}
\text{CH}_3 & \text{O} \\
| & || \\
\text{H—C—O—P—O}^- + \text{O}_2 + \text{H}_2\text{O} \\
| & | \\
\text{CH}_3 & \text{CH}_3
\end{array}
$$

because peroxide is used up in the hydrolysis and because the intermediate has not been isolated.

The kinetics of this reaction have been studied (91) and it has been
found to follow the rate law

$$v = k[\text{Sarin}][\text{OOH}^-]$$

The rate constant has a value of 1.34×10^3 liter/(mole)(sec) at 25°.
The activation parameters are 7.7 kcal/mole and -20 cal/(mole)(deg)
for S_a and ΔS^*, respectively. These data are fully consistent with a
mechanism involving nucleophilic attack by OOH^- on the tetrahedral
phosphorus (92). The low value of E_a suggests that an empty d orbital
on phosphorus aids information of the transition state. The negative
value of ΔS^* indicates that the nucleophile OOH^- is bound quite
closely to the phosphorus substrate in the transition state.

The peroxy intermediate which is formed in the rate step oxidizes
another mole of hydrogen peroxide fairly readily, for the over-all
reaction stoichiometry has been found to be

$$\text{Sarin} + 2\,\text{OOH}^- \longrightarrow \underset{\underset{\text{CH}_3}{|}}{\overset{\overset{\text{CH}_3}{|}}{\text{H—C}}}—\text{O}—\underset{\underset{\text{O}}{|}}{\overset{\overset{\text{CH}_3}{|}}{\text{P}}}—\text{O}^{\ominus} + \text{O}_2 + \text{H}_2\text{O} + \text{F}^-$$

At higher concentrations of Sarin, the intermediate reacts with a second
molecule of Sarin to give a peroxydiphosphate compound

$$\underset{\underset{\text{CH}_3}{|}}{\overset{\overset{\text{CH}_3}{|}}{\text{H—C}}}—\text{O}—\underset{\underset{\text{O}}{|}}{\overset{\overset{\text{CH}_3}{|}}{\text{P}}}—\text{O}—\text{O}^{\ominus} + \text{Sarin} \longrightarrow$$

$$\underset{\underset{\text{CH}_3}{|}}{\overset{\overset{\text{CH}_3}{|}}{\text{H—C}}}—\text{O}—\underset{\underset{\text{O}}{|}}{\overset{\overset{\text{CH}_3}{|}}{\text{P}}}—\text{O}—\text{O}—\underset{\underset{\text{O}}{|}}{\overset{\overset{\text{CH}_3}{|}}{\text{P}}}—\text{O}—\underset{\underset{\text{CH}_3}{|}}{\overset{\overset{\text{CH}_3}{|}}{\text{C—H}}} + \text{F}^-$$

Although Sarin is a notorious example, all phosphorylating agents
react with peroxide ion at a rapid rate. A large number of investigations (93) dealing with peroxoanion attack on phosphorylating agents
have been carried out. The rate constant for attack by OOH^- is in
each case about two powers of ten larger than the rate constant for
hydroxide ion attack on the phosphorylating compound. Other
peroxoanions ROO^- are reactive but less so than OOH^- (93d).

The intermediate phosphorus peroxo anion is decomposed fairly
rapidly in aqueous solution, however being reactive it can undergo
characteristic reactions. Very sensitive tests for anticholinesterase

compounds depend on this reactivity (94). For instance, in the Schoenemann reaction, the peroxoacid is made with peroxoborate in the presence of indole, which is rapidly oxidized to indigo via indoxyl and diindoxyl:

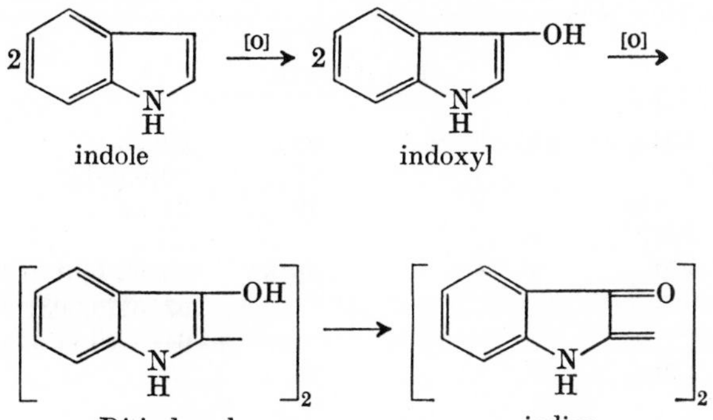

indole   indoxyl

Di-indoxyl   indigo

The two indoxyls are fluorescent, and their formation provides a test sensitive to 1–10 μg of phosphorus compound. The oxidation of benzidine is also used in some variations of the test. As the phosphorus peroxo acids are about equally unstable irrespective of the parent compounds from which they are formed, and the fluorescent compounds are unstable, the test is only applicable to compounds which react rapidly with OOH⁻. Since OOH⁻ reacts about 100 times more rapidly than an equal concentration of OH⁻, the test is useful only for compounds which are unstable to alkalis. The limit appears to be reached when k_{OH} is less than 0.1. Compounds hydrolyzing more slowly than this are not powerful inhibitors of cholinesterase, so the test is applicable to most poisons which act directly, but not to those which have to be metabolized first (95).

Organoperoxophosphates have been prepared in order to learn their chemistry. The two types of phosphorus peroxides that can be isolated are esters and diphosphate compounds. Some of these are listed in Table VII, which is taken from the review of Sosnovsky and Brown (96).

Rieche and his co-workers (97) have prepared monoperoxo triesters of phosphoric acid by reactions

$$(RO)_2POCl + R'OOH \xrightarrow{Py} (RO)_2PO(OOR') + PyH^+Cl^-$$

where R $= CH_3$ or C_2H_5, and R′ is $C(CH_3)_3$, carried out at -10 to $-20°$ in the presence of excess pyridine. Mageli and Harrison (98) prepared peroxoesters from chlorophosphates and alkyl hydroperoxides in aqueous solutions containing alkali metal hydroxides.

TABLE VII

Organophosphorus Peroxides

R	R′	Method of preparation	Yield (%)	Bp (°C) (mm) or mp (°C)	Ref.
		$R_2P(O)OOR'$			
MeO	CMe_3	a	65	23–25	97b
			65	70(0.001)	97b
			43	24–25	97c,d
EtO	CMe_3	a	~65–85	60–65(0.01–0.001)	97b
				53–54(0.003)	97a,c,d
				64–67(0.1)	99a
EtO	CMe_3	b	76		98
i-PrO	CMe_3	b			99a
n-BuO	CMe_3	b	88.6		98
n-BuO	Hexyl	b	89.3		98
n-BuO	CMe_2Ph	b	90		98
n-BuO	Pinanyl	b	98		98
n-OctylO	CMe_3	b	78.7		98
n-OctylO	CMe_2Ph	b	76		98
n-OctylO	Pinanyl	b	99		98
PhO	CMe_3	b	80		98
PhO	CMe_2Ph	b	78		98
		$R_2P(O)OOP(O)R_2$			
Ph		c	57	88–89	100

$$(H_3C)_2C(CH_2)_2C(CH_3)_2$$
$$| \qquad\qquad |$$
$$R_2P(O)OO \qquad OOP(O)R_2$$

R		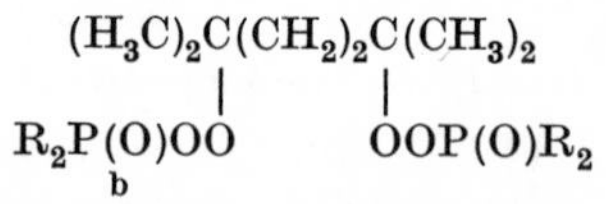 Method of preparation	Yield (%)		Ref.
n-Bu		b			98
Ph		b	81		98

---

a $R_2P(O)Cl + R'OOH \xrightarrow[\text{pyridine}]{\text{anhydrous}} R_2P(O)OOR' + HCl.$

b $R_2P(O)Cl + R'OOH \xrightarrow{\text{MOH,H}_2\text{O}} R_2P(O)OOR' + MCl$ (M = Li, Na, K, Cs).

c $2 R_2P(O)Cl + H_2O_2 \rightarrow R_2P(O)OOP(O)R_2 + 2 HCl.$

The peroxoester can be cleaved without decomposition by strong base (97b), whereas

$$2 \text{ OH}^- + (\text{RO})_2\text{PO}(\text{OOR}') \longrightarrow (\text{RO})_2\text{PO}_2^- + \text{R}'\text{OO}^- + \text{H}_2\text{O}$$

In a water solution without base decomposition one obtains

$$(\text{RO})_2\text{RO}(\text{OOR}') + \text{H}_2\text{O} \longrightarrow (\text{RO})_2\text{PO}_2\text{H} + \overset{\text{CH}_3}{\underset{\text{CH}_3}{\diagdown}} \text{C}{=}\text{O} + \text{CH}_3\text{OH}$$

where $\text{R}' = \text{C}(\text{CH}_3)_3$. These peroxophosphates decompose rapidly on heating or in the presence of amines (99).

Dannley and Kabre (100) have prepared the compound

$$\phi_2\overset{|}{\underset{\text{O}}{\text{P}}}{-}\text{OO}{-}\overset{|}{\underset{\text{O}}{\text{P}}}\phi_2$$

by reaction of diphenylphosphoric chloride with sodium peroxide in water. The disubstituted peroxide is stable at $-80°$ but rearranges at higher temperatures to an unsymmetrical anhydride

$$\phi_2\overset{|}{\underset{\text{O}}{\text{P}}}{-}\text{OO}{-}\overset{|}{\underset{\text{O}}{\text{P}}}\phi_2 \longrightarrow \phi{-}\overset{\phi}{\underset{\text{O}}{\overset{|}{\text{P}}}}{-}\text{O}{-}\overset{\phi}{\underset{\text{O}}{\overset{|}{\text{P}}}}{-}\text{O}\phi$$

Kinetic studies of this rearrangement in a number of solvents and in the presence of amine catalyst have been carried out. Decomposition is also accelerated by acid and by light. Mechanisms for the several reaction paths have been proposed by the authors (100), but details remain to be resolved.

The formation of peroxohydrates of organic phosphonates and phosphates has been reported (101).

Peroxoesters of phosphoric acid are reported to initiate the polymerization of vinyl monomers.

V. Phosphate Hydroperoxidates

A. Types of Compounds

The phosphate hydroperoxidates represent a quite different kind of phosphate peroxide compound. While the "true" (i.e., bound) peroxophosphates are substitution products of hydrogen peroxide having the kinetically inert P—O—O linkage, the hydroperoxidates are addition products with hydrogen peroxide playing the same role as water in some hydrates. The bonding of peroxide to phosphate in the lattice is believed to involve the peroxidic hydrogens as in the linkage

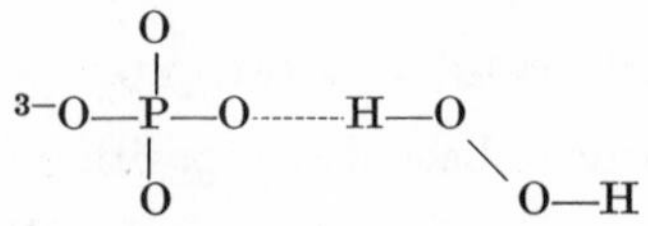

$$^{3-}O-\underset{\underset{O}{|}}{\overset{\overset{O}{\|}}{P}}-O\text{------}H-O\diagdown_{O-H}$$

and they are comparable in type to hydrates such as $Na_2HPO_4 \cdot 12\ H_2O$ wherein hydrogen bonding to the anion is important. However there does not seem to be any simple relationship between water of crystallization and hydrogen peroxide of crystallization; some phosphates which do not form hydrates give compounds containing hydrogen peroxide of crystallization, and vice versa.

Some of the early literature does not distinguish between true peroxophosphates and phosphates with hydrogen peroxide of crystallization. Especially in patent literature, one finds the term "perphosphate" applied to both bound peroxophosphates and hydroperoxidates. Nevertheless the fundamental distinction between these two types of peroxophosphates has been pointed out (11,102). Further, the distinction is clearly consistent with the known slow rates of replacements in phosphate coordination spheres as discussed in the introduction to this chapter.

B. Synthesis

The experimental distinction between the two types of perphosphate can usually be made on the basis of the synthetic method. Because of the inert nature of the phosphate ion, synthesis of bound peroxides can be accomplished only by electrolysis of phosphate or reaction of hydrogen peroxide with an anhydride. By way of comparison, the formation of a hydroperoxidate is accomplished simply by mixing of hydrogen peroxide and a phosphate followed by removal of water under mild conditions.

A great number of patents describe preparations of phosphate hydroperoxidates. These preparations are in most instances more or less the same: a phosphate is treated with a concentrated hydrogen peroxide solution and the product isolated by precipitation with alcohol or by slow evaporation to dryness of the reaction mixture either on a water bath or over a drying agent. Appendix II contains a list of patents and publications dealing with specific preparative methods.

C. Known Compounds

Most alkali and alkaline earth phosphates react with hydrogen peroxide to give products containing hydrogen peroxide of crystallization.

TABLE VIII

Some Phosphate Hydroperoxidates

Probable formula of product	Remarks	Ref.
$Li_3PO_4 \cdot 0.25\ H_2O_2$	Completely decomposes in a month.	11
$Li_4P_2O_7 \cdot 0.25\ H_2O_2$	Completely decomposes in a month.	11
$KH_2PO_4 \cdot 1.25\ H_2O_2$	Loses 4.4% of active oxygen in 3 days.	11
$K_2HPO_4 \cdot 2.5\ H_2O_2$	Very stable, no loss of active oxygen in a week.	11
$K_4P_2O_7 \cdot 3\ H_2O_2$	Loses 0.6% of active oxygen in a fortnight.	11
$Rb_2HPO_4 \cdot 2.5\ H_2O_2$	Loses 0.47% of active oxygen in a week.	11
$Cs_2HPO_4 \cdot 1.5\ H_2O_2$	Loses 0.81% of active oxygen in 3.5 weeks.	11
$(NH_4)_2HPO_4 \cdot 0.75\ H_2O_2$	Completely decomposes in a month.	11
$NaNH_4HPO_4 \cdot 0.75\ H_2O_2$	Loses 1.77% of active oxygen in 2 months.	11
$(NH_4)_2H_2P_2O_7 \cdot 2\ H_2O_2$	Loses 1.84% of active oxygen in 11 days.	11
$CaHPO_4 \cdot 0.5\ H_2O_2$	Loses 3.14% of active oxygen in a month.	11
$SrHPO_4 \cdot 1.5\ H_2O_2$	Loses 6.26% of active oxygen in a fortnight.	11
$BaHPO_4 \cdot 1.5\ H_2O_2$	Loses 3.18% of active oxygen in a fortnight.	11
$Na_2HPO_4 \cdot 2\ H_2O_2$	Very stable, loses 1% H_2O_2 in 6 months.	102
$Na_4P_2O_7 \cdot 2\ H_2O_2$	Very stable, loses less than 1% of active oxygen per month.	103
$Na_3P_3O_9 \cdot H_2O_2$	Stable for several weeks in the absence of air.	104

An investigation of the action of 30% hydrogen peroxide on different phosphates has shown that primary acidic phosphates do not react with hydrogen peroxide or give very unstable compounds (11). Secondary, less acidic phosphates of alkaline earth metals also give unstable compounds while secondary alkali metal phosphates or pyrophosphates give quite stable compounds. Strongly alkaline phosphates like tertiary alkali phosphates decompose hydrogen peroxide. Table VIII lists some hydroperoxidates with remarks concerning their stability (11,102–104). The patent literature usually gives only the preparation and the content of active oxygen and the postulated formulas for the products may be somewhat accidental (i.e., artifacts) in some cases.

A more thorough study of the systems $Na_3PO_4-H_2O_2-H_2O$, $Na_2HPO_4-H_2O_2-H_2O$, and $Na_4P_2O_7-H_2O_2-H_2O$ has been carried out by Ukraintseva (105). Phase diagram studies showed the presence of solid peroxidic compounds of composition $Na_3PO_4 \cdot H_2O_2$, $Na_3PO_4 \cdot 4.5$ H_2O_2, $Na_2HPO_4 \cdot 1.5$ H_2O_2, $Na_2HPO_4 \cdot 2.5$ H_2O_2, and $Na_4P_2O_7 \cdot 6$ H_2O_2. The thermal stability of these compounds was investigated. Early data on the three systems were referenced.

Agreement between different research groups on the compositions of matter from the same system is lacking. For example, the relatively stable solid obtained from mixtures of tetrasodium pyrophosphate and hydrogen peroxide has been reported as $Na_4P_2O_7 \cdot 2$ H_2O_2, $Na_4P_2O_7 \cdot 2.5$ H_2O_2, $Na_4P_2O_7 \cdot 3$ H_2O_2, and $Na_4P_2O_7 \cdot 6$ H_2O_2. The composition of the solid phase must be significantly influenced by the conditions employed for its preparation.

D. Physical Properties

One of the better characterized hydroperoxidates is the material $Na_4P_2O_7 \cdot 2$ H_2O_2 prepared by the FMC Corporation. The solubility of $Na_4P_2O_7 \cdot 2$ H_2O_2 is 8.3% by weight in water at 20°. At room temperature the loss of active oxygen content averages less than 1% per month. At higher temperature decomposition sets in. For example about 10% of active oxygen was lost at 100° and 67% at 130° after 2 hr. Decomposition is caused by the same substances which cause decomposition of hydrogen peroxide. Among these are heavy metals and their ions such as Fe, Cu, Pb, Mn, and Co and easily oxidizable organic compounds. Another group includes ferments and enzymes. Solutions of $Na_4P_2O_7 \cdot 2$ H_2O_2 are quite stable at room temperature. A 5% aqueous solution lost 4% of active oxygen in five days. At 80°, 92% was lost in one day.

The hydrogen peroxide is not held strongly in these compounds. On treatment with ether, hydrogen peroxide is given off. On heating in vacuum no measurable change is noticed at 70° for 3–4 hr, while at 130° all hydrogen peroxide disappears leaving the anhydrous Na_2HPO_4 and $Na_4P_2O_7$ behind (102). These methods for removal of hydrogen peroxide are often used to establish whether a given compound is a true peroxo compound or a hydroperoxidate; such tests must be carefully considered for bound peroxides and hydroperoxidates can react similarly in certain analytical procedures (such as titrimetric analysis of H_2O_2 by ceric ion).

The stability of the solid hydroperoxidates increases in the sequence primary phosphate < secondary phosphate < pyrophosphate.

Lithium salts form less stable solids than sodium salts, and these in turn are less stable than potassium salts (102).

It appears that hydrogen peroxide of crystallization is bound more firmly than water of crystallization. When Na_3PO_4 containing 17.8% H_2O_2 and 29.2% H_2O of crystallization was kept in vacuum over KOH, about 50% of the hydrogen peroxide was still present after 42 hr while all the water was lost. In the dehydration of $Na_4P_2O_7$ having 15.5% H_2O_2 and 32.5% water, a still greater tendency to hold hydrogen peroxide was noted. Even after 15 days, over $\frac{2}{3}$ of the original hydrogen peroxide was retained whereas 90% of the water had disappeared (106).

E. Chemical Properties

On dissolution, the hydroperoxidates immediately yield solutions showing reactions typical of those in which phosphate and hydrogen peroxide are inserted separately. The peroxide reacts with permanganate ion and ceric ion in the manner characteristic of hydrogen peroxide rather than that of the true peroxophosphates.

Solubility and cryoscopic studies along with studies of hydrogen peroxide distribution between the aqueous phase and amyl alcohol suggest that there is some bonding between hydrogen peroxide and alkali phosphates in aqueous solution although much less than in the solid state (107).

In summary, the hydroperoxidates are different from the bound peroxides in chemical characteristics and they present no solution properties different from simple mixtures of hydrogen peroxide and the analogous phosphate.

Acknowledgments

We are indebted to the U.S. Air Force Office of Scientific Research (GRANT #71–1961) for continuing support, and to Dr. Eleanor Chaffee, Dr. Marvin M. Crutchfield, and Dr. Leonard R. Darbee for suggestions and comments.

VI. Appendix I. Laboratory Preparation of Lithium Peroxidiphosphate (1)

Electrolyze (in a 400-ml beaker) an aqueous solution containing 45.3 g KH_2PO_4, 29.7 g KOH, 18 g KF, and 0.053 g K_2CrO_4 plus enough

H_2O to dissolve the salts. A Pt gauze (or wire of 2-cm length and about 0.3-mm diameter) electrode serves as anode and a piece of Pt foil (about 5 cm^2) as cathode (2,3). Maintain a current of 0.7 to 0.8 A across the electrodes (this usually requires a potential of 6 to 7 V). During the periods of electrolysis, the electrolyte should be cooled in an ice bath and should be stirred mechanically.

Electrolyze the solution for 6 hr on each of three successive days. Between electrolyses, allow it to stand overnight at room temperature so that the unstable peroxomonophosphate, which is also formed, will decompose. The day after the last electrolysis, transfer the solution to a 500-ml Erlenmeyer flask. Heat to 50°, add 70 g of KOH pellets, stopper with a rubber stopper, and shake until the KOH has dissolved. Heat is evolved. Allow the solution to stand for at least 12 hr, decant the supernatant liquor, and collect the crystals on a medium porosity glass frit. Dry (in an oven at about 70° overnight) the thick wet precipitate as much as possible.

To purify the peroxodiphosphate, dissolve the solid in 150 ml water. Stir mechanically for 15 min. (If solution is not complete, add up to 50 ml more water and stir until complete solution takes place.) Cool in an ice bath and add 100 ml of a $LiClO_4$ solution which contains 54 g salt/100 ml water. The solution should be mechanically stirred and cooled in an ice bath during this addition. Filter and discard the solid $KClO_4$. Test for completeness of precipitation of $KClO_4$ by adding more of the above $LiClO_4$ solution drop-wise to the cold filtrate. After the last trace of solid $KClO_4$ is removed, warm the solution to 45°, and add 50 ml methanol rapidly. Stir the solution mechanically and add dropwise at a rapid rate another 100 ml methanol. Stir for 30 min more. Filter and wash free of chromate with 1:1 water–methanol until wash solution is colorless. Air-dry on the filter and then on a watch glass. The product is the tetrahydrate $Li_4P_2O_8 \cdot 4 H_2O$. Yields on the basis of the original KH_2PO_4 are around 45%.

NOTES.

1. This preparation is adapted from the thesis of T. Chulski, Michigan State University (1953). As written here, it has been used as a laboratory experiment for inorganic students at Brown University. Among those who have worked on the details are Professor R. Carlin, Dr. F. A. Walker, and Mr. M. Hoffmann.

2. For somewhat small quantities, a Pt dish used both as container and as cathode simplifies the setup (M. M. Crutchfield and R. J. Lussier, Ph.D. theses at Brown University in 1960 and 1969, respectively).

3. Copper fittings and wire are to be avoided wherever possible as trace amounts of copper can influence the chemistry of peroxodiphosphates in undesirable manners (e.g., stoichiometry change and/or nonreproducible rates).

4. The potassium salt contains potassium fluoride as an impurity difficult to remove by recrystallization. Conversion to $Li_4P_2O_8 \cdot 4 H_2O$ gives a product relatively free of fluoride as LiF precipitates along with the $KClO_4$.

VII. Appendix II. Some Hydroperoxidate Preparations

Orthophosphate Hydroperoxidates

S. Aschkenasi, Ger. Pat. 296,796, see Ref. 11; Ger. Pat. 296,888 (1917), *Chem. Abstr.*, **17**, 3577 (1923); Brit. Pat. 156,713 (1921), *Chem. Abstr.*, **15**, 1786 (1921); Ger. Pat. 299,300 (1914), *Chem. Abstr.*, **13**, 365 (1919); Ger. Pat. 316,997 (1915), see Ref. 11; Ger. Pat. 318,219 (1918), see Ref. 11.

M. Sareson, Ger. Pat. 325,155.

Chemische Werke vorm. H. Byk, Ger. Pat. 287,588, *Chem. Abstr.*, **10**, 2128 (1916); Austrian Pat. 63,288.

Petrenko, *J. Russ. Phys. Chem. Soc.*, **34**, 204 (1902).

Rudenko, *J. Russ. Phys. Chem. Soc.*, **44**, 1215 (1912).

G. Schönberg, Fr. Pat. 700,132 (1930), C., **1931I**, 3593.

S. Husain and R. J. Partington, Brit. Pat. 300,946, C., **1929I**, 1599.

Österreichische Chemische Werke, Fr. Pat. 750,125 (1932), C., **1933II**, 2720; Austrian Pat. 140,836 (1932), C., **1935II**, 260; Austrian Pat. 143,281 (1932); Austrian Pat. 140,553 (1932), C., **1936II**, 146.

F. Munzberg, *Lotus*, **76**, 351 (1928); *Chem. Abstr.*, **26**, 2132.

Henkel & Cie, G.M.B.H., V. Habernickel, Ger. Pat. 1,066,190 (1956); C., **1960**, 5610.

Chemische Fabrik Weissenstein, G.M.B.H., Austrian Pat. 140,185 (1928), C., **1935I**, 2422.

H. Menzel and G. Gäbler, *Z. Anorg. Allgem. Chem.*, **177**, 187 (1929).

S. Husain and R. J. Partington, *Trans. Faraday Soc.*, **24**, 235 (1928).

Diphosphate Hydroperoxidates

Chemische Werke vorm. H. Byk, Ger. Pat. 287,588 (1911); *Chem. Abstr.*, **10**, 2128 (1916).

Ges. für Chemische Industrie, Ger. Pat. 293,786; *Chem. Abstr.*, **11**, 2947 (1917); Fr. Pat. 482,785; Brit. Pat. 15,749.

Deutsche Gold- und Silber-Scheideanstalt vorm. Roessler, K. Viewig, Ger. Pat. 555,055 (1928); C., **1933I,** 376; Brit. Pat. 355,016; U.S. Pat. 1,914,312.

S. Husain and R. J. Partington, Brit. Pat. 300,946 (1927); C., **1929I,** 1599.

Chemische Fabrik Budenheim Rudolf A. Oetker K.-G, T. Roessel, Ger. Pat. 1,219,452; *Chem. Abstr.,* P4532 (1967).

Kali-Chemie A.-G., L. Pellens, E. Kranel, Brit. Pat. 679,908 (1951); *Chem. Abstr.,* 4567 (1953); Ger. Pat. 877,446 (1950); C., **1959,** 5376.

Hagan Chemicals & Controls, Inc., K. Lindner, Ger. Pat. 1,118,167 (1957); *Chem. Abstr.,* 15630 (1962); U.S. Pat. 3,037,838 (1958); *Chem. Abstr.,* 9459 (1962).

Chemische Fabrik Budenheim A.-G., O. Pfrengle, R. Hubold, Brit. Pat. 767,636 (1954); C. **1958,** 5481; Swiss Pat. 343,379 (1954); C. **1960,** 15872.

Staüffer Chemical Co., H. W. Majewski, Fr. Pat. 2,016,050.

Triphosphate Hydroperoxidates

Henkel & Cie, G.M.B.H., V. Habernickel, Ger. Pat. 1,048,265 (1956); *Chem. Abstr.,* 2038 (1961).

Metaphosphate Hydroperoxidates

Chemische Werke Albert, H. Huber, R. L. v. Reppert, Ger. Pat. 743,832 (1940); C. **1944I,** 1118; Swedish Pat. 107,378 (1941).

G. Bonneman-Bémia, *Ann. Chim. (Paris)* **16** (11) 395 (1941).

REFERENCES

1. (a) Bunton, C. A. and H. Chaimovich, *Inorg. Chem.,* **4,** 1763 (1965); (b) Osterheld, R. K., *J. Chem. Phys.,* **62,** 1133 (1958).
2. Griffith, E. J. and R. L. Buxton, *A. Amer. Chem. Soc.,* **89,** 2884 (1967); this article gives an extensive list of references on polyphosphate hydrolyses.
3. (a) Keisch, B., J. W. Kennedy, and A. C. Wahl, *J. Amer. Chem. Soc.,* **80,** 4778 (1958); (b) Bunton, C. A., D. R. Llewellyn, C. A. Vernon, and V. A. Welch, *J. Chem. Soc.,* **1961,** 1636.
4. (a) Jencks, W. P., in *Brookhaven Symposia in Biology,* No. 15, 134 (1962); (b) Kirby, A. J. and A. G. Varvoglis, *J. Chem. Soc. (B),* **1968,** 135; (c) Kirby, A. J. and A. G. Varvoglis, *J. Amer. Chem. Soc.,* **89,** 415 (1967); (d) Bunton, C. A., D. R. Llewellyn, K. G. Oldham, and C. A. Vernon, *J. Chem. Soc.,* **1958,** 3574, 3588; (e) Degani, Ch. and M. Halmann, *J. Amer. Chem. Soc.,* **88,** 4075 (1966); (f) Bunton, C. A. and H. Chaimovich, *J. Amer. Chem. Soc.,* **88,** 4082 (1966); (g) Cox, J. R. and O. B. Ramsey, *Chem. Rev.,* **64,** 317 (1964); and (h) "Phosphoric Esters and Related Compounds," *Chem. Soc. (London),* Spec. Pub. #8 (1957).

5. Schmidlin, J. and P. Massini, *Ber.*, **43**, 1162 (1910).
6. D'Ans, J. and W. Friederich, *Ber.*, **43**, 1880 (1910).
7. Fichter, F. and J. Müller, *Helv. Chim. Acta*, **1**, 297 (1918).
8. Fichter, F. and A. Rius y Miro, *Helv. Chim. Acta*, **2**, 3 (1919).
9. Franchuk, I. F. and A. I. Brodskii, *Doklady Akad. Nauk SSSR*, **118**, 128 (1958).
10. Fichter, F. and W. Bladergroen, *Helv. Chim. Acta*, **10**, 559 (1927).
11. Husain, S. and J. R. Partington, *Trans. Faraday Soc.*, **24**, 235 (1928); see also Yu. P. Rudenko, *Zh. Russk. Fiz-Khim. Obshch.*, **44**, 1209 (1912).
12. Siebold, H., Ger. Pat. 279,306; *Chem. Abstr.*, **9**, 1402 (1915).
13. Fichter, F. and E. Gutzwiller, *Helv. Chim. Acta*, **11**, 323 (1928).
14. Schenk, P. and H. Rehaag, *Z. Anorg. Allgem. Chem.*, **233**, 403 (1937).
15. Schenk, P. W. and H. Vietzke, *Z. Anorg. Allgem. Chem.*, **326**, 152 (1963).
16. Chulski, T., Ph.D. Thesis, Michigan State University, 1953.
17. Carlin, R. L. and F. A. Walker, unpublished preparation; see Appendix I.
18. Chong, C. H. H., Ph.D. Thesis, Michigan State University, 1958.
19. (a) Voelkel, W., U.S. Pat. 2,135,545 (to DEGUSSA) (Nov. 8, 1938); (b) Voelkel, W., U.S. Pat. 2,162,655 (to DEGUSSA) (June 13, 1939); (c) Mucenicks, P. R., Ger. Pat. 1,809,799 (to FMC Corp.) (Feb. 25, 1971).
20. Simon, A. and H. Richter, *Z. Anorg. Allgem. Chem.*, **302**, 165 (1959).
21. Cohen, B., U.S. Pat. 3,547,580 (to FMC Corp.) (Dec. 15, 1970).
22. FMC Corporation Technical Data. Communication from Dr. L. R. Darbee.
23. Simon, A. and H. Richter, *Z. Anorg. Allgem. Chem.*, **304**, 1 (1960).
24. Corbridge, D. E. C. and E. J. Lowe, *J. Chem. Soc. (London)*, **1954**, 4555.
25. Simon, A. and H. Richter, *Naturwissenschaften*, **44**, 178 (1957).
26. Crutchfield, M. M., *Peroxide Reaction Mechanisms*, J. O. Edwards, Ed., Interscience, New York, 1962, p. 41.
27. Crutchfield, M. M., Ph.D. Thesis, Brown University, 1960.
28. Crutchfield, M. M. and J. O. Edwards, *J. Amer. Chem. Soc.*, **82**, 3533 (1960).
29. Venturini, N., A. Indelli, and G. Raspi, private communication, 1970.
30. Lambert, S. M. and J. I. Watters, *J. Amer. Chem. Soc.*, **79**, 4262, 5606 (1957).
31. Van Wazer, J. R., *Phosphorus and Its Compounds*, Vol. I, Interscience, New York, 1958, p. 452–454.
32. Goh, S. H., R. B. Heslop, and J. W. Lethbridge, *J. Chem. Soc. (A)*, **1966**, 1302.
33. Indelli, A., and P. L. Bonora, *J. Am. Chem. Soc.*, **88**, 924 (1966).
34. Unpublished data of C. J. Battaglia, B. R. Fitch, and R. Paju; for some details, see Ref. 64 below.
35. FMC Corporation Technical Data—"Potassium Peroxodiphosphate Activation by the Enzyme Phosphates" (7/70).
36. Kolthoff, I. M. and I. K. Miller, *J. Amer. Chem. Soc.*, **73**, 3055 (1951).
37. Andersen, M., J. O. Edwards, Sr. A. A. Green, and Sr. M. D. Wiswell, *Inorg. Chem. Acta*, **3**, 655 (1969).
38. Chaffee, E., I. I. Creaser, and J. O. Edwards, *Inorg. Nucl. Chem. Lett.*, **7**, 1 (1971).
39. (a) Latimer, W. M., *Oxidation Potentials*, Prentice-Hall, Englewood Cliffs, New Jersey, 2nd ed., 1952, p. 78; (b) Vaskelis, A., *Lietuvos TSR Mokslu Akad. Darbai, Ser. B.*, **4**, 41 (1964).

40. Green, Sr. A. A., J. O. Edwards, and P. Jones, *Inorg. Chem.*, **5**, 1858 (1966).
41. (a) Chaffee, E., Ph.D. Thesis, Brown University, 1971; (b) Creaser, I. I., unpublished data, Brown University.
42. Basolo, F., J. C. Hayes, and H. M. Neumann, *J. Amer. Chem. Soc.*, **76**, 3807 (1954).
43. Burgess, J. and R. H. Prince, *J. Chem. Soc.*, **1965**, 6061.
44. Hogg, R. and R. G. Wilkins, *J. Chem. Soc.*, **1962**, 341.
45. Farina, R., R. Hogg, and R. G. Wilkins, *Inorg. Chem.*, **7**, 170 (1968).
46. See Refs. p. 186 cited in Haim, A. and W. K. Wilmarth, *Peroxide Reaction Mechanisms*, J. O. Edwards, Ed., Interscience, New York, 1962, p. 176.
47. Irvine, D. H., *J. Chem. Soc.*, **1958**, 2166.
48. Raman, S. and C. H. Brubaker, Jr., *J. Inorg. Nucl. Chem.*, **31**, 1091 (1969).
49. Kolthoff, I. M., A. I. Medalia, and H. P. Raaen, *J. Amer. Chem. Soc.*, **73**, 1733 (1951).
50. Barb, W. G., J. H. Baxendale, P. George, and K. R. Hargrove, *Trans. Faraday Soc.*, **47**, 462 (1951).
51. Conocchioli, T. J., E. J. Hamilton, Jr., and N. Sutin, *J. Amer. Chem. Soc.*, **87**, 926 (1965).
52. Burgess, J. and R. H. Prince, *J. Chem. Soc. (A)*, **1966**, 1772.
53. Irvine, D. H., *J. Chem. Soc.*, **1959**, 2977.
54. (a) Burgess, J. and R. H. Prince, *J. Chem. Soc.*, **1963**, 5752; (b) Burgess, J. and R. H. Prince, *J. Chem. Soc. (A)*, **1970**, 2111.
55. Yost, D. M. and W. H. Claussen, *J. Amer. Chem. Soc.*, **53**, 3349 (1931).
56. Lussier, R. J., W. M. Risen, Jr., and J. O. Edwards, *J. Phys. Chem.* **74**, 4039 (1970).
57. See references cited in reference 5 of Ref. 56.
58. Toennies, G., *J. Amer. Chem. Soc.*, **59**, 555 (1937).
59. Mamantov, G., J. H. Burns, J. R. Hall, and D. B. Lake, *Inorg. Chem.*, **3**, 1043 (1964).
60. Heiderich, E. W., U.S. Pat. 2,765,216 (October 9, 1953).
61. Lake, D. B. and G. Mamantov, U.S. Pat. 3,085,856 (March 11, 1958).
62. Beer, F. and J. Müller, U.S. Pat. 3,036,887 (July 23, 1959).
63. Cremer, J. and G. Müller-Schiedmayer, Ger. Pat. 1,148,215 (August 9, 1961).
64. Battaglia, C. J. and J. O. Edwards, *Inorg. Chem.*, **4**, 552 (1965).
65. Fortnum, D. H., C. J. Battaglia, S. R. Cohen, and J. O. Edwards, *J. Amer. Chem. Soc.*, **82**, 778 (1960).
66. Battaglia, C. J., Ph.D. Thesis, Brown University, 1962.
67. Koubek, E., M. L. Haggett, C. J. Battaglia, K. M. Ibne-Rasa, H.-Y. Pyun, and J. O. Edwards, *J. Amer. Chem. Soc.*, **85**, 2263 (1963).
68. Schenk, P. W. and K. Dommain, *Z. Anorg. Allgem. Chem.*, **326**, 139 (1963).
69. Thilo, E. and W. Wieker, *Z. Anorg. Allgem. Chem.*, **277**, 27 (1959).
70. Beer, F. and J. Müller, Ger. Pat. 1,143,491 (June 24, 1960).
71. Heslop, R. B. and J. W. Lethbridge, *J. Chromatog.*, **13**, 199 (1964).
72. Rius, A., *Trans. Amer. Electrochem. Soc.*, **54**, 347 (1928).
73. Csányi, L. J. and F. Solymosi, *Acta Chim. Acad. Sci. Hung.*, **13**, 275 (1957).
74. Van Wazer, J. R., *Phosphorus and Its Compounds*, Vol. 1, Interscience, New York, 1958, p. 481.
75. Earley, J. E., D. Fortnum, A. Wojcicki, and J. O. Edwards, *J. Amer. Chem. Soc.*, **81**, 1295 (1959).
76. Menzel, H., *Z. Phys. Chem.*, **105**, 402 (1923).

77. Edwards, J. O., *J. Amer. Chem. Soc.*, **75**, 6154 (1953).

78. Long, F. A. and M. A. Paul, *Chem. Rev.*, **57**, 1 (1957).

79. Bunnett, J. F., *J. Amer. Chem. Soc.*, **82**, 499 (1960); *ibid.*, **83**, 4956, 4968, 4973, 4978 (1961).

80. Frost, A. A. and R. G. Pearson, *Kinetics and Mechanism*, Wiley, New York, 1953, p. 139.

81. Ref. 74, pp. 452–458.

82. Vernon, C. A., *Special Publication No. 8*, The Chemical Society, London, 1957, p. 17.

83. (a) Ball, D. L. and J. O. Edwards, *J. Amer. Chem. Soc.*, **78**, 1125 (1956); and (b) Koubek, E., G. Levey, and J. O. Edwards, *Inorg. Chem.*, **3**, 1331 (1964).

84. Lunenok-Burmakina, V. A., A. P. Potemskaya, G. P. Aleeva, *Teor. Eksp. Khim.*, **2**, 549 (1966).

85. Two reviews which discuss the isotope method and results are as follows: (a) Edwards, J. O. and P. D. Fleischauer, *Inorg. Chim. Acta Rev.*, **2**, 53 (1968); and (b) Brown, S. B., P. Jones, and A. Suggett, *Progr. Inorg. Chem.*, **13**, 159 (1970).

86. Boyland, E. and D. Manson, *J. Chem. Soc.*, **1957**, 4689.

87. See footnote 7 in Edwards, J. O. and J. J. Mueller, *Inorg. Chem.*, **1**, 696 (1962).

88. Lunenok-Burmakina, V. A., G. P. Aleeva, and T. M. Franchuk, *Rus. J. Inorg. Chem.*, **13** (4), 509 (1968).

89. (a) Maxted, E. B., *J. Chem. Soc.*, **1945**, 204; (b) Maxted, E. B., *J. Chem. Soc.*, **1945**, 763; (c) Maxted, E. B. and A. Marsden, *J. Chem. Soc.*, **1946**, 23; and (d) Maxted, E. B. and G. T. Ball, *J. Chem. Soc.*, **1953**, 1509.

90. Garcia Marquina, J. M., *Rev. Acad. Cienc. Madrid*, **30**, 382 (1933); *Chem. Abstr.*, **28**, 712 (1934).

91. (a) Larsson, L., *Acta Chem. Scand.*, **12**, 723 (1958); (b) Aksnes, G., *Acta Chem. Scand.*, **14**, 2075 (1960).

92. For a brief review of characteristics of nucleophilic displacements, see Edwards, J. O., *J. Chem. Educ.*, **45**, 386 (1968).

93. (a) Green, A. L., G. L. Sainsbury, B. Saville, and M. Stansfield, *J. Chem. Soc.*, **1958**, 1583; (b) Epstein, J., M. M. Demek, and D. H. Rosenblatt, *J. Org. Chem.*, **21**, 796 (1956); (c) Ginjaar, L., Ph.D. thesis at the University of Leiden, 1960; and (d) Behrman, E. J., M. J. Biallas, H. J. Brass, J. O. Edwards, and M. Isaks, *J. Org. Chem.*, **35**, 3069 (1970).

94. (a) Gehauf, B., J. Epstein, G. B. Wilson, B. Witten, S. Sass, V. E. Bauer, and W. H. C. Rueggeberg, *Anal. Chem.*, **29**, 278 (1957); (b) Cherry, R. H., G. M. Foley, C. O. Badgett, R. D. Eames, and H. R. Smith, *Anal. Chem.*, **30**, 1239 (1958); (c) Marsh, D. J. and E. Neale, *Chem. Ind.*, **1956**, 494; and (d) Gehauf, B. and J. Goldenson, *Anal. Chem.*, **29**, 276 (1957).

95. Heath, D. F., *Organophosphorus Poisons*, Pergamon, London, 1961, pp. 62–63.

96. Sosnovsky, G. and J. H. Brown, *Chem. Rev.*, **66**, 529–566 (1966).

97. (a) Rieche, A., G. Hilgetag, and G. Schramm, *Angew. Chem.*, **71**, 285 (1959); (b) Rieche, A., G. Hilgetag, and G. Schramm, *Ber.*, **95**, 381 (1962); (c) Rieche, A., G. Hilgetag, and G. Schramm, Ger. Pat. (East) 21,489 (April 21, 1959); *Chem. Abstr.*, **56**, 9967a (1962); (d) Rieche, A., G. Hilgetag, and G. Schramm, Ger. Pat. 1,082,895 (June 9, 1960); *Chem. Abstr.*, **55**, 16422b (1961).

98. Mageli, O. L. and J. B. Harrison, U.S. Pat. 2,960,526 (November 15, 1960).

99. (a) Sosnovsky, G. and E. Zaret, *Chem. Ind.* (*London*), **1966**, 628; (b) Sosnovsky, G. and E. H. Zaret, *Chem. Ind.* **1967**, 1297.
100. Dannley, R. L. and K. R. Kabre, *J. Amer. Chem. Soc.*, **87**, 4805 (1965).
101. Bliznyuk, N. K., A. F. Kolomiets, and S. G. Zhemchuzhin, USSR Pat. 159,526 (Dec. 28, 1963); *Chem. Abstr.*, **60**, 15913c (1964).
102. Menzel, H. and C. Gäbler, *Z. Anorg. Allgem. Chem.*, **177**, 187 (1928).
103. *Bulletin No. 8*, Becco Chemical Division of FMC Corporation, Research Laboratories, Princeton, New Jersey 1952.
104. Bonneman-Bémia, P., *Ann. Chim.* (*Paris*), **16** (11), 395 (1941).
105. Ukraintseva, E. A., *Akad. Nauk. SSR, Sibirskoe Otd. Izvestia Ser. Khim Nauk.*, No. 1, 14 (1943).
106. Levi, G. R. and F. Battaglino, *Gazz. Chim. Ital.*, **67**, 659 (1937).
107. Husain, S., *Z. Anorg. Allgem. Chem.*, **177**, 215 (1928).

Subject Index

A

Acetylated 1-hydroxy-1, 1-diphosphonate, formation of, 57

Acetylated hydroxydiphosphonates, preparation of, 55

Acetyl phosphorous acid, detection by ^{31}P NMR, 61

Acid-Base Titrations, 273

Activation parameters for phosphate hydrolysis, *see* compound class headings

Acylphosphonates, reported dimerization of, 56

Addition reactions, of primary and secondary phosphorus compounds, 32

of tertiary phosphorus compounds, with benzyne, 31

with electrophilic acetylenes, 17

with electrophilic olefins, 4

Adenosine-5'-methylenediphosphonate, 47

Adenosine triphosphate hydrolysis, 136, 138

Alkane-1-amino-1, 1-diphosphonates, preparation of, 66

Alkane bis(phosphazo trichlorides), physical data, 356

Alkane-1-hydroxy-1, 1-diphosphonic acids, 54

Alkyl 1-amino-1, 1-diphosphonates pK_2 values, 68-70

Alkyl-(aryl) Phosphazo Trihalides, monomericdimeric forms, 316

Alkylenediphosphonates, formation constants of complexes, 71-73

Alkylenediphosphonic acids, x-ray powder patterns, 74

Alkyl-1-hydroxy-1, 1-diphosphonate thermal stability, 66-67

Alkylphosphazo trichlorides, correlation of physical data with structure, 317

physical data, 346

Alkylphosphazo trifluorides, physical data, 357

Alkynylphosphines, reaction of, 25

Amines, oxidation by peroxomonophosphate, 413

Aminodiphosphonates, preparation from nitriles, 79

1-Amino-1, 1-diphosphonates, 39

Anhydrizing ratio, 59

Aqueous cyclotetraphosphate, apparent molar refraction, 286

Aqueous cyclotriphosphate, apparent molar refraction, 286

Arabinofuranosylcytosine, phosphonate derivative of, 47

Arbuzov reaction, variations, 39

Arsenophosphate hydrolysis, 194

Arylphosphazo trichlorides, physical data, 344

Arylphosphazo trifluorides, physical data, 358

B

Benzyne, reactions with tertiary phosphorus compounds, 31

'Biphilicity' of tervalent phosphous compounds, 3

cis-1, 2-Bisdiphenylphosphinoethane, reaction with dimethyl acetylenedicarboxylate, 21

Bisdiphenylphosphinomethane, reaction with dimethyl acetylene-dicarboxylate, 19

Bromide ion, oxidation by peroxomonophosphate, 413-415

Butane-1,2,3,4-tetraphosphonate, preparation of, 53

t-Butyl methanehydroxydiphosphonate,

Iodide ion, peroxodiphosphate oxidation of, 395-396, 397

Ionic phosphazo trihalides with $(N=P)_2 C(R)$ units, correlation of chemical shift with substituent, 314

  physical data, 322

  with symmetrical cation, correlation of chemical shift with chain length, 314

  physical data, 319

Ionic unsymmetrical phosphazo trihalides, physical data, 323

Iron(II) complexes, oxidation by peroxodiphosphate, 396-399, 400

Isopropyl esters, pyrolysis of, 40

K

$K_4 P_2 O_8$, preparation, 384-385, 425-427

L

$Li_4 P_2 O_8 \cdot 4 H_2 O$, preparation, 384-385, 425-427

M

Metal complexes, oxidation, by peroxodiphosphates, 396-401

  by peroxodisulfate, 400

Metaphosphate abstraction, 145-146, 148-149, 151, 154

Metaphosphates, 104

Methylbenzylbutylphosphine, resolution by reaction with acrylonitrile, 4

Methylenediphosphonates, as extractants, 71

  reduction of, 46

Methylenediphosphonic acid, enzyme inhibition by, 47

  I.R. spectrum of derivatives, 74

  reaction with phosphorus chloride, 46-47

  thermal stability, 66

Methylenediphosphonic tetrachloride, preparation of, 46-47

  reaction with phosphorus sulfide, 47

Methyl ethylene phosphate hydrolysis, 234

Michaelis-Arbuzov Reaction, see Arbuzov reaction

Michaelis-Becker-Nylen Reaction, 39

Middle group phosphate, acid-base titrations, 273

molten salts, titration in, 274

Molten Salts, Titration in, 274

Monometaphosphate mechanism, see Hydrolysis of phosphates, monometaphosphate mechanism

N

$Na_4 P_2 O_7 \cdot xH_2 O_2$, 424

Nitrite ion, oxidation by peroxomonophosphate, 415-416

bis(4-Nitrophenyl) phosphate hydrolysis, 214

p-Nitrophenyl phosphate hydrolysis, 225

Nucleophilic addition, 1

  by tervalent phosphorus compounds, 2

Nucleophilicity of tervalent, phosphorus compounds, 2

O

Octaphosphate hydrolysis, 147

Oligophosphates, 104

  hydrolysis, 146

Organic compounds, oxidation by peroxodiphosphates, 395, 399, 400-401

Organoperoxophosphates, 417-421

Organophosphorus peroxides, 417-421

Orthophosphate, Acid-base titrations, 273

  molten salts, titration in, 274

$Os(bipy)_3$, $^{2+}$, dissociation of, 398

  oxidation by peroxodiphosphate, 398-399

$Os(phen)_3$ $^{2+}$, dissociation of, 398

  oxidation by peroxodiphosphate, 398-399

$Os(terpy)(bipy) Cl^+$, dissociation of, 399-400

  hydrolysis of, 399-400

  oxidation by peroxodiphosphate, 399-400

Osmium complexes, dissociation of, 398-400

  hydrolyses of, 398-400

  oxidation, by peroxodiphosphate, 398-400

  by peroxodisulfate, 400

Oxoanions, oxidation by peroxomonophosphate, 415-416

Oxygen exchange between phosphates and water, 234, 237

  activation energy, 239, 241

  effect of pH, 237

  mechanism, 238

Topics in Phosphorus Chemistry

Cumulative Index, Volumes 1—7